Earth's Natural Hazards and Disasters

Advanced Textbook Series

1. **Unconventional Hydrocarbon Resources: Techniques for Reservoir Engineering Analysis**
   Reza Barati and Mustafa M. Alhubail
2. **Geomorphology and Natural Hazards: Understanding Landscape Change for Disaster Mitigation**
   Tim R. Davies, Oliver Korup, and John J. Clague
3. **Remote Sensing Physics: An Introduction to Observing Earth from Space**
   Rick Chapman and Richard Gasparovic
4. **Geology and Mineralogy of Gemstones**
   David P. Turner and Lee A. Groat
5. **Data Analysis for the Geosciences: Essentials of Uncertainty, Comparison, and Visualization**
   Michael W. Liemohn
6. **Earth System Geophysics**
   Steven R. Dickman
7. **Earth's Natural Hazards and Disasters**
   Bethany D. Hinga

Advanced Textbook 7

Earth's Natural Hazards and Disasters

Bethany D. Hinga
University of Nebraska Kearney, USA

This Work is a co-publication of the American Geophysical Union and John Wiley and Sons, Inc.

AGU
ADVANCING EARTH AND SPACE SCIENCES

WILEY

This edition first published 2024

Published under the aegis of the AGU Publications Committee

Matthew Giampoala, Vice President, Publications
Steven A. Hauck, II, Chair, Publications Committee
For details about the American Geophysical Union visit us at www.agu.org.

Registered Office
John Wiley & Sons, Inc., 111 River Street, Hoboken, NJ 07030, USA

Editorial Office
111 River Street, Hoboken, NJ 07030, USA

For details of our global editorial offices, customer services, and more information about Wiley products visit us at www.wiley.com.

Wiley also publishes its books in a variety of electronic formats and by print-on-demand. Some content that appears in standard print versions of this book may not be available in other formats.

Library of Congress Cataloging-in-Publication Data
Names: Rinard Hinga, Bethany D., author.
Title: Earth's natural hazards and disasters / Bethany D. Hinga.
Description: Hoboken, NJ : Wiley-American Geophysical Union, 2024. | Series: Advanced textbooks series | Includes bibliographical references and index.
Identifiers: LCCN 2024003429 (print) | LCCN 2024003430 (ebook) | ISBN 9781119217718 (paperback) | ISBN 9781119217787 (adobe pdf) | ISBN 9781119217725 (epub)
Subjects: LCSH: Natural disasters. | Hazard mitigation.
Classification: LCC GB5014 .R56 2024 (print) | LCC GB5014 (ebook) | DDC 363.34–dc23/eng/20240214
LC record available at https://lccn.loc.gov/2024003429
LC ebook record available at https://lccn.loc.gov/2024003430

Cover Design: Wiley
Cover Image: © Courtesy of Michael Coniglio/NOAA

Set in 9.5/12.5pt STIXTwoText by Straive, Pondicherry, India

Printed and bound by CPI Group (UK) Ltd, Croydon, CR0 4YY
C9781119217718_280324

Contents

Contributors

Joel Berrier
University of Nebraska at Kearney
Kearney, Nebraska, USA

Jillian Gregg
Oregon State University
Corvalis, Oregon, USA

John Helms
Sterling College
Craftsbury Common, Vermont, USA

Bethany D. Hinga
Academic Affairs
University of Nebraska at Kearney
Kearney, Nebraska, USA

Anni Moore
Natural and Mathematical Sciences
Morningside University
Sioux City, Iowa, USA

Austin Nuxoll
University of Nebraska at Kearney
Kearney, Nebraska, USA

Kelli Wakefield
Mesa Community College
Mesa, Arizona, USA

Preface

There is no place on the planet that is free from natural hazards. They are found on every continent and in every country. Rapid population growth means that more people are moving into areas that were previously unoccupied. Many of those places, such as steep slopes, floodplains, and coastal plains, which are vulnerable to storm surges in hurricanes or typhoons, thus have inherent hazards associated with them. Economic losses due to natural disasters are growing higher, routinely in the tens to hundreds of billions of dollars even in relatively quiescent years. There have been numerous very large natural disasters within our lifetime. These include the 2004 Indonesian earthquake and tsunami, which killed roughly 230,000 people, a 2010 earthquake in Haiti, which killed a quarter of a million people, and the COVID-19 pandemic that began in 2019, spreading worldwide in 2020. The COVID-19 pandemic alone has killed millions of people, and the general public has struggled to understand the science behind the disease and how best to protect themselves. A devastating earthquake in Turkey in 2023 killed more than 54,000 people, and left more than 1.5 million people homeless. Building collapse was the main reason for the high death toll. This begs the question: What can we do to prevent such loss of life in earthquakes? These are all good reasons to study natural hazards and disasters.

Why does a specific place or population experience a particular hazard? What caused a natural event to turn into a disaster and what were the contributing factors? Why didn't a similar natural event in another location turn into a disaster? What were the effects of this disaster on the population and on the environment? What were the economic losses? What is involved in the recovery process? What have we learned from this event as scientists, as a general public, as public officials? How can scientists use data to determine better mitigation strategies for the future? How can we be better prepared the next time an event like this occurs so there is less loss of life, property, and infrastructure?

In this book, I seek to answer these questions and to challenge readers to think critically about disasters. Using the benefit of hindsight, readers are guided through a discussion of what went wrong so they can pinpoint contributing factors to each type of disaster. Of course, we cannot prevent natural events from occurring. There will always be extreme weather, volcanic eruptions, earthquakes, and disease outbreaks. Rather, people need to know the risks in their location, how they should prepare to deal with these risks, and how they can keep their families safe. This book provides some insight and food for thought about what increases the risk of natural disaster and how the reader can prepare for the next natural event in their area that has the potential to turn deadly.

The Earth is ruled by awesome forces in all of the interrelated components of the Earth System. For that reason, I have taken an Earth System approach in writing this book, beginning with a discussion of this concept in Chapter 1. How does each natural event manifest itself in the four realms of the Earth System and how does each part of the system contribute to a population's risk? The book is divided into three parts, each tied to an aspect of the Earth System. Hazards that have their roots in the solid Earth (geosphere) are covered in Part I. Hazards primarily related to the hydrosphere and atmosphere are explored in Part II. Biological hazards (which affect the biosphere) are addressed in Part III.

The format of each chapter follows a similar formula. Learning outcomes are identified for the chapter so the reader can keep those objectives in mind. The text of each chapter begins with a case study of a disaster. The reader is drawn into a narrative that includes an introduction to the location, site characteristics, and overriding hazards. As the case study unfolds, the reader follows a cascade of events that turned this natural event into a disaster. In these case studies, the reader is introduced to the science behind the disaster that concludes with a summary of the event, lessons learned, and future directions. The remainder of the chapter presents a much broader view of each natural event. It includes a discussion of the science of hazards in accessible terms, acknowledgement of other factors that contribute to hazards, further case studies and examples of natural events, and mitigation strategies. Jargon with which the reader may be unfamiliar is given in italics in each chapter and defined in the glossary.

This book is suitable for undergraduates in a general education class in natural hazards, but may also be appropriate for higher level students or disaster responders who are seeking a general foundation in natural hazards and a more applied approach to the natural sciences in which geology, geography, weather, and microbes create hazardous conditions for us as well as for animals, plants, and the environment.

Glossary

'A'a	A lava surface that is jagged and sharp.
Accidental host	A host that gets sick with a disease from a pathogen, but cannot transmit that disease.
Accretion	The process of creating larger bodies from several smaller ones.
Achondrite	A stony meteorite that does not contain spherical chondrules, or collections of minerals.
Aerial fuels	Fuel for fires found higher than 6 ft (1.8 m) above the surface of the ground.
Aerobiology	A branch of biology in which scientists study the microbes in the air.
Aerosol	A small droplet or gas molecule in the atmosphere.
Aftershock	A smaller earthquake that occurs in the same area, but after a major earthquake. Aftershocks may be nearly as large as the main shock.
Air mass	A body of air with a set of characteristics that is distinct from the air around it.
Albedo	Reflectiveness of a surface. Surfaces with high albedo are strongly reflective.
Algal bloom	Rapid reproduction of algae in a body of water.
Alluvial fan	A fan-shaped deposit of debris left behind when a steep stream changes slope and starts flowing on flat land. Debris carried in the fast-moving steep stream can no longer be carried by the slower-moving stream on flat land, and thus deposits most of its sediment as an alluvial fan.
Alpine glacier	A glacier found in a mountainous region.
Alternative fuels	Fuels other than fossil fuels. This includes nuclear, wind, tidal, hydroelectric, and geothermal energy.
Amplitude	The height of a wave, measured from the midpoint to the top (crest) or bottom (trough) of the wave.

Andesite	An extrusive or volcanic igneous rock. It is typically gray or brown in color and crystals may be visible within the rock. Andesite has between 55% and 69% silica.
Anthropogenic	Created by humans.
Antibiotic resistance	The ability of bacteria to resist antibiotic treatment. Antibiotic resistance is a growing problem because many strains of bacteria are now resistant to most antibiotics, so people infected with these bacteria are very difficult or impossible to treat.
Anticyclone	Rotation around a high pressure system. The rotation is clockwise in the Northern Hemisphere and counterclockwise in the Southern Hemisphere as a result of the Coriolis effect.
Aphanitic	An igneous rock texture that describes a rock composed primarily of microscopic crystals.
Aphelion	The Earth's farthest distance from the Sun in its elliptical orbit.
Archaea	A group of single-celled organisms whose cells lack a defined nucleus that have distinct molecular characteristics separating them from bacteria. They are among the most ancient life forms on the planet, and can inhabit a number of environments that are inhospitable to other life forms.
Aspect	The direction a slope faces.
Asteroid	A rocky or stony body in space that is smaller than a dwarf planet.
Asthenosphere	The mechanical layer of the Earth found directly beneath the lithosphere. The asthenosphere is part of the upper mantle compositional layer. It is a plastic solid, which means it is flexible and able to circulate. Circulation in this layer keeps the tectonic plates in motion.
Astronomical high tide	High tide during new Moon and full Moon phases, when the Sun and the Moon have the greatest effect on tidal levels.
Astronomical unit (AU)	The distance from the Earth to the Sun, approximately 93 million mi (149.6 million km).
Atmosphere	A relatively thin layer of air that surrounds a planet. The Earth's atmosphere is composed of about 78% nitrogen, 21% oxygen, 0.9% argon, and 0.04% carbon dioxide.
Atmospheric pressure	The weight of a column of air from the Earth's surface to the edge of the atmosphere. It is measured in millibars or inches of mercury.
Background noise	Small movements of the Earth caused by things such as the wind, thunder, waves crashing on the shoreline, people walking, traffic, animals, or other everyday activity.
Bacteria	The smallest life forms on the planet, they are composed of cells without a defined nucleus.
Basal cavity	A place where an ice shelf is undercut by seawater.

Basalt An igneous rock composed of microscopic crystals, which has between 45% and 54% silica (SiO_2).

Basaltic An igneous rock composition that has between 45% and 54% silica.

Base flow The level of sustained flow in a river during dry or fair weather conditions. It usually represents the amount of water supplied to a stream by groundwater, or water that is stored and migrates below the surface of the Earth.

Base isolation A construction practice in which the weight of a building is carried on flexible pilings or large casters with roller bearings. This allows the building to remain more or less stationary as the ground beneath it moves during an earthquake. Base isolation is a method used to reduce or eliminate damage to a building in an earthquake.

Biogeochemical cycle A description of a cycle that describes the interaction of organisms, the Earth, and chemical compounds and how they move through the Earth system.

Biosphere The component of the Earth system that includes all life on Earth.

Black smoker A hydrothermal vent on the ocean floor that emits hot, mineral laden water. Black smokers are found at midocean ridges in oceans around the world. They were discovered in the 1970s.

Blast deflection A strategy to change the trajectory of an asteroid so it does not impact the Earth. Blast deflection uses a nuclear explosion, detonated very near the asteroid, causing the asteroid to remain intact but change its trajectory.

Block Solid piece of rock that is hurled out of a volcano's crater.

Body wave Seismic waves that travel through the body of the interior of the Earth.

Bolide An extraterrestrial body that strikes a larger body, such as a meteor striking the Earth.

Bomb A blob of molten lava or a chunk of rock hurled through the air by an erupting volcano.

Braided stream A stream that consists of multiple interwoven channels separated by sand or gravel bars. It is common at the base of a mountain range or in any other setting with high sediment supply and nearly flat terrain.

Bread crust bomb A bomb in which the surface of the rock is cracked like a crusty loaf of bread.

Building codes A set of laws that govern how a building must be constructed. Some areas of the world require enhanced building codes to ensure that buildings can survive severe shaking from an earthquake or high winds from hurricanes.

Calculated risk The decision to live with the hazards in a region, even though you know that a natural disaster could happen in that area.

Caldera	A depression on a volcano, caused by the collapse of the ground surface over an area that magma has evacuated.
Calving	A piece of the front of a glacier or ice sheet breaks off of the main body of ice. If the ice falls into the water it becomes an iceberg, or floating block of ice.
Carbon capture	A collection of techniques that allow the removal of carbon dioxide from flue gas and other industrial sources.
Carbon footprint	An individual or country's contribution to excess carbon dioxide in the atmosphere.
Carbon sequestration	A natural or artificial process by which carbon dioxide is removed from the atmosphere and held in solid or liquid form.
Carbon sink	A reservoir that takes up and stores carbon dioxide. One such reservoir is forest biomass, which is the plant material within a forest.
Cellulose	An insoluble substance, which is the main constituent of plant cell walls and of vegetable fibers such as cotton.
Char	Solid material left after vegetation burns.
Chemolithotrophs	Archaea that derive their nutrients directly from rocks. The rock provides them with electrons for energy production.
Chemosynthesis	The process in which bacteria or other organisms produce food from energy derived from oxidation of inorganic chemicals. Carbon dioxide, hydrogen sulfide, and oxygen combine to create sugar, sulfur, and water.
Chondrite	A type of stony meteorite that contains spherical structures called chondrules.
Chondrule	A spherical structure within a stony meteorite, usually composed of mineral crystals.
Cinder	A small block of hardened lava filled with circular air pockets. Cinder is usually black or red in color. It is the main component of a volcanic feature called a cinder cone.
Climate	The prevailing or typical weather patterns for a given region based on long-term (30 year) averages of weather.
Climate feedback	Phenomena of slightly warming or cooling of the planet, leading to more significant warming or cooling by another mechanism.
Coastal desert	A desert generally found on the western edges of continents near about 23° north and south of the equator. They are affected by cold ocean currents that parallel the coast. These cold water zones offshore do not produce a great deal of evaporation, so precipitation on the adjacent coast is low.
Coastal flood	A type of flood that occurs in coastal areas when extremely high tide carries water farther inland than usual, causing water to cover low-lying areas with elevations just above the average high tide. This

	tends to happen most often at new and full Moon phases, when the Moon and Sun are pulling together on the Earth's waters, causing more extreme tides than usual.
Collision zone	A type of convergent boundary in which two pieces of continental lithosphere move toward each other. They collide, crumpling the fusing the two pieces of lithosphere. The result is frequent earthquake activity and a tall mountain range.
Combustion	A self-sustaining process in which there is rapid oxidation of fuel. Combustion produces both light and heat, and often involves flames.
Comet	A ball of ice, dust, and other material that orbits the Sun on a highly elliptical orbit.
Complex crater	A crater formed by the impact of a medium sized meteorite. It contains a central peak within a circular depression.
Conduction	Transfer of heat from one object to another through direct contact.
Conflict disaster	A disaster caused by war, civil unrest (such as riots), mass shootings, terrorist attacks, and genocide
Conservation agriculture	A farming system that can prevent losses of arable land while regenerating degraded lands.
Continental rift	A plate boundary within the interior of a continent. At a continental rift, the two sides of a continent are moving away from each other. Eventually, the continent will split into two or more parts and narrow oceans will form between the portions of the former continent.
Convection	Vertical circulation of a material. When material is heated, it rises. As it cools, it sinks. Convection occurs in air, liquids, and the solid Earth.
Convergent boundary	A type of plate boundary in which two plates move toward each other. Common features at convergent boundaries include volcanoes, earthquakes, deep ocean trenches, and mountain ranges.
Coral bleaching	A process by which coral expel symbiotic phytoplankton from their polyps. Since corals depend on these phytoplankton to survive, this often kills the coral. Coral bleaching happens most commonly when water temperatures become too warm.
Core	The innermost compositional layer of the Earth. The core is composed primarily of iron, with a small amount of nickel.
Coriolis effect	A phenomenon in which the Earth's air is diverted to the right in the Northern Hemisphere and to the left in the Southern Hemisphere due to the rotation of the Earth.
Coronal mass ejection (CME)	An eruption of plasma, currents, and magnetic fields from the Sun's corona. CMEs can reach a quarter of the distance from the Sun to the Earth and cause serious disruptions in satellites, radio, and electrical grids.

Coronavirus	A class of viruses, a few of which cause respiratory diseases such as COVID-19 and SARS.
Cosmic dust	Dust from space that falls to the Earth's surface.
COVID-19	A new coronavirus disease that appeared in 2019. It grew to a global pandemic, infecting nearly 300 million people and killing 5.4 million, as of January 2022.
Crater	A depression on a volcano, caused by explosion of material from the volcano; a depression in the surface of a planet, moon, or asteroid caused by a meteor or asteroid impact.
Craton	The oldest, most stable part of a continent. It is often composed of granite and metamorphic rocks.
Creep	The slow downhill movement of soil due to the effect of gravity.
Crest	The high point of a wave.
Crown	The top of a tree, above about 6 ft (1.8 m).
Crown fire	Fires that burn aerial fuels, such as the crowns of trees.
Crust	The Earth's outermost compositional layer. The crust is composed primarily of lightweight silicate minerals.
Cut bank	The steep outer bank of a meander in a stream. Water velocity is fastest on the outside of the meander, and erosion occurs here.
Cyanobacteria	Previously known as blue-green algae. This aquatic bacteria produces oxygen through the process of photosynthesis.
Dead-end host	A host that gets sick with a disease from a pathogen, but cannot transmit that disease.
Debris flow	A mass of mud and rock that moves downhill under the influence of gravity.
Delta	A fan-shaped deposit that forms when sediment is dropped out of the river as it enters another body of water, often the ocean.
Dendrochronology	The study of tree rings, usually undertaken to determine past climate.
Derecho	Straight-line winds produced by squall lines or mesoscale convective complexes. They are at least 240 mi (400 km) long, have gusts of at least 58 mi/hr (93 km/hr), and typically last between 10 and 20 min. They can, however, last several hours.
Desertification	The process of fertile land transforming into desert, typically as a result of deforestation, drought, or improper agricultural practices.
Dew point	The temperature to which the air must be cooled to reach 100% relative humidity.
Differentiation	In the solar system, denser elements are concentrated in the inner solar system, while lighter, less dense elements are dominant in the outer solar system.

Dike	A human-made flood control structure that runs parallel to a shoreline or a stream channel.
Disaster subculture	A heightened awareness of a hazard in a given area. People in the area acknowledge the hazard and are prepared to face it.
Discharge	Volume of water moving past a specific point in the stream in a given period of time (typically each second).
Disease	An adverse reaction to a pathogen.
Divergent boundary	A type of plate boundary in which two plates move away from each other. Common features at divergent boundaries include volcanoes, earthquakes, midocean ridges, rift zones, and ocean basins.
Divide	The high points in elevation that separate one drainage basin from another.
Doppler weather radar	A type of weather radar in which instruments are able to record not only the presence of clouds but also the direction those clouds are moving. It can be used to detect rotation in clouds, which is present when tornadoes are forming.
Downburst	A powerful downdraft of air, found in or around the edges of a thunderstorm.
Drainage basin	An area that contributes water to a specific stream.
Drought	A period of below average rainfall in a region.
Dry line	A situation in which warm, dry air from Mexico meets moist air from the Gulf of Mexico. The resulting weather includes high winds and low humidity.
Duff	Decaying plant material such as leaves and twigs in the upper soil horizons.
Dust Bowl	A period of profound drought in the High Plains region of the United States, during the 1930s.
Dwarf planet	A body in a solar system that is too small to be considered a planet because it is too small to clear debris from its orbit.
Earth System	A concept of the Earth that recognizes four components of the Earth. These are the geosphere (solid Earth), hydrosphere (water on the Earth), atmosphere (air and water vapor above the Earth), and biosphere (all life on the planet).
Easterly wave	A disruption in the area's trade winds featuring a weak low pressure system that moves from east to west.
Earthquake	Sudden movement along a break in the Earth's crust.
Eccentricity	The distance from the Earth to the Sun changes over time because the Earth's orbit is elliptical. At some times the Earth is farther from the Sun. At other times the Earth is closer to the Sun. While this is always true, the distances change on a period of a few hundred thousand years.

Elastic rebound	The process of increasing the stress and strain on either side of a fault, until it ruptures and recovers to an undeformed state.
Electromagnetic spectrum	The range of wavelengths of energy that includes gamma rays, X-rays, ultraviolet radiation, visible light, infrared energy, microwaves, and radio waves.
El Niño Southern Oscillation	A phenomenon in which sea surface temperatures in the equatorial Pacific Ocean are warmer than usual.
Endemic	A disease always present in the population at a low level.
Epicenter	The location of an earthquake projected onto the surface of the Earth, above the earthquake's focus.
Epidemic	A disease outbreak that persists over a longer period of time and spreads to a wider area.
Eukaryote	An organism that has cells with defined nuclei and organelles. Eukaryotes can be unicellular or multicellular.
Eutrophic dead zone	An area in a body of water in which low oxygen conditions exist, making the area uninhabitable by most forms of life.
Extraterrestrial hazards	Hazards to the Earth that come from space. Examples are meteors and asteroids.
Extratropical cyclone	Counterclockwise rotation around a low pressure system (in the Northern Hemisphere) in the midlatitudes. Rotation is clockwise in the Southern Hemisphere.
Extrusive	Used to describe an igneous rock formed at or above the Earth's surface.
Eye	A relatively calm and clear zone at the center of rotation in a hurricane.
Eye wall	The innermost band of clouds that surrounds the eye of the hurricane. A hurricane's maximum wind speeds are measured in the eye wall.
Fall	A type of mass wasting that occurs when a rock becomes dislodged from a slope and falls downhill with very little to no contact with the ground.
Fault	A break in the Earth's crust along which movement occurs.
Fault scarp	A slope created when a fault breaks to the surface during an earthquake or series of earthquakes.
Feedback	Something that speeds up or slows down a warming trend. A positive feedback accelerates an increase. A negative feedback slows or dampens an increase.
Ferrel cell	A large scale circulation in the Earth's atmosphere, found between about 30° and 60° north and south latitude. Air descends around 30° north and south of the equator, creating a high pressure zone. This high pressure can keep clouds from forming in some areas of the

	world, creating the right conditions for deserts. Air rises again near 60° north and south latitude. There it meets cooler air carried away from the polar regions.
Filament	A trace of a prominence on the Sun. It appears as a dark line on the surface of the Sun because it is cooler than the Sun's surface.
Firebrand	An ember that escapes a fire. It can become a source of a new fire.
Fire line	An area where firefighters cut and remove vegetation all the way down to "mineral soil," which is the lower soil horizons that do not contain appreciable organic material. This is done in an effort to remove fuel from the fire's path.
Fire whirl	A spinning column of rising hot air and gases creates what looks like a small tornado of fire.
First Law of Thermodynamics	Energy cannot be created or destroyed but may change form.
Fissure	A crack in the ground. On volcanoes, fissures can be eruption sites.
Flaming combustion	A material is burning with a flame.
Flash flood	A flood that occurs quickly, within minutes or hours after a rainfall event, release of water from an ice jam, or dam or levee failure.
Flood basalt	A long-lived catastrophic fissure eruption of basalt that can cover a vast landscape with a great thickness of lava.
Flood frequency curves	A diagram containing a specific stream's data in which recurrence interval in years is plotted along the X- axis, usually on a logarithmic scale. (Equal distances along the X -axis are between powers of 10.) Discharge is plotted along the Y-axis. Peak discharges are plotted as points on the graph. A best fit line is drawn through the peak discharges. This allows a user to estimate the peak discharge for a flood with any specific recurrence interval on that particular stream.
Floodplain	A low-lying area next to a river that is covered with water during a flood.
Flow	The chaotic movement of fluid and solid material downhill under the influence of gravity.
Fluvial	Pertaining to a river.
Focus	The location on a fault at which an earthquake begins. An earthquake's focus is always below the surface of the Earth. Another term for focus is hypocenter.
Foehn winds	Sometimes called gravity winds. These can occur over relatively small areas, but are perhaps best known as regional wind patterns. They form when there is a high pressure gradient that forces warm air up and over mountains. The air cools as it rises in elevation, and the cooled air rushes down the mountain slopes creating strong winds. Examples in the United States include Chinook winds, Santa Ana winds, and Diablo winds.

Foot wall	The fault block that exists below a fault.
Forecast	A warning to the public that a threat is imminent. It should contain the location, timing, and estimated magnitude of the threat.
Foreshock	An earthquake that occurs before the main shock in an earthquake sequence.
Fossil fuels	Organic compounds created in the Earth that we use as fuel, including coal, oil, and natural gas.
Front	A boundary between two air masses.
Fuel	Vegetation, leaf litter, roots, and other flammable material available to burn in a fire.
Fuel continuity	How well spaced or connected fuel (vegetation) is in a wildland area.
Fuel load	Description of how much fuel (both alive and dead), exists in a wildland area, and is available to burn.
Fungi	Spore-producing organisms feeding on organic matter, including molds, yeast, mushrooms, and toadstools.
Fusion	A nuclear reaction in which lighter elements are combined to make heavier elements. Fusion releases a great deal of energy.
Gabbro	An intrusive igneous rock with a basaltic composition, or between 45% and 54% silica.
Geomagnetic storm	Severe disturbance in the Earth's magnetic field as a result of a coronal mass ejection (CME). Geomagnetic storms cause disruptions in radio and satellite communications, disruptions to power grids, and can endanger astronauts in space.
Geosphere	The solid Earth component of the Earth System.
Geothermal	Pertaining to hot water and other hot fluids within rocks. Geothermal features include geysers, hot springs, and mud pots. Geothermal fluids within the Earth can also alter rocks, causing them to change colors. This happens because the hot fluids change the chemistry of the rocks.
Glacial maxima	Intense phases of cold, also referred to as Ice Ages.
Global carbon cycle	The movement of carbon through different environmental regimes, and transformation of carbon from one chemical form to another.
Global climate change	Combined impacts of global warming and the concomitant impacts on droughts, floods, extreme weather, fires, ocean acidification, and sea level rise.
Global warming	Warming of the planet using temperature as the only variable.
Gradient	A slope. Used in studies of streams to denote the slope of a stream channel segment. After defining a segment of the stream to be studied, the elevation at the top and bottom of this stream segment is measured. Then the length of the stream segment, following each

bend in the stream, is measured. The gradient is the change in elevation divided by the length of the stream segment. It is usually reported in feet per mile (or meters per kilometer).

Granite — A plutonic igneous rock that contains more than 69% silica.

Gravity tractor — A strategy to change the trajectory of an asteroid or meteor so it does not impact the Earth. It involves sending a dense spacecraft to the asteroid. The gravitational attraction between the dense spacecraft and the asteroid will change the trajectory of the asteroid slightly, causing it to avoid the Earth.

Greenhouse effect — Warming of the Earth due to absorption of infrared radiation by greenhouse gases.

Greenhouse gases — Gases in the atmosphere that are capable of absorbing infrared radiation from the Sun and warming the Earth.

Ground fuel — Fuel for a fire found beneath the surface of the Earth.

Hadley cell — A large scale circulation that occurs in the Earth's atmosphere between the equator and about 30° north and south latitude. It pulls warm, moist air from the equatorial region away from the equator. Air becomes drier and somewhat cooler, descending around 30° north and south, creating a high pressure zone there.

Hanging wall — The fault block that exists above a fault.

Headwaters — The location where a stream begins. It is found at the highest elevation along the stream channel.

Heat index — An apparent temperature calculated using actual temperature and relative humidity. Heat index is traditionally reported on hot days.

Heat transfer — The transfer of heat from one substance to another.

Heat wave — A period of time in which temperatures rise far above the average for a particular location.

Heavy fuels — In a fire, trees more than 6 ft (1.8 m) in height, continuous brush, and slash.

Helminth — A group of parasitic worms that includes tapeworms, hookworms, and other worms, often infecting the digestive system.

Host — An organism that will harbor a pathogen; the host can, but does not necessarily, get sick from the pathogen.

Hot drought — A set of conditions in which higher soil and air temperatures mean that a greater amount of rainfall is needed to recover from drought conditions due to the higher rate of evaporation and transpiration.

Hot spot — A phenomenon in which a part of the mantle is extra hot. Volcanoes and earthquakes are frequent occurrences above a hot spot. Hot spots often create chains of volcanoes as the hot spot stays more or less stationary and a tectonic plate moves over it.

Hurricane	A tropical cyclonic storm that forms over warm ocean waters and has a minimum sustained wind speed of at least 74 mi/hr (119 km/hr).
Hydrograph	A record of a stream's discharge at a specific point on that stream, plotted over time.
Hydrosphere	The component of the Earth System that includes rivers, lakes, glaciers, and groundwater.
Hydrothermal vent	A location on the ocean floor at which hot, mineral-rich water is erupted. Hydrothermal vents in ocean settings are found at midocean ridges.
Hygiene	A set of practices that help to maintain health and prevent the spreading of diseases. Hygiene is directly linked to sanitation, the availability of clean water and working waste disposal, but also to the actual habit of keeping the body and surroundings clean.
Hypocenter	The location on a fault at which an earthquake begins. Also goes by the name focus. The hypocenter of an earthquake is always beneath the Earth's surface.
Ice age	A period in Earth history during which glaciers were present for a prolonged period of time. These periods are also known as glacial maxima.
Ice jam	A situation in which ice blocks the flow of water in a river. This is a common phenomenon during the beginning of the spring melt season.
Ice wedging	A process in which water invades cracks in rocks, then freezes. The water expands as it turns into ice, wedging pieces of rock away from the bedrock.
Igneous rock	Rock formed when magma, or molten rock, cools and solidifies.
Immunocompromised	A person that has a weakened immune system and is vulnerable to disease.
Impactor	An extraterrestrial body that strikes a larger body, such as a meteor striking the Earth.
Infections molecules	Nucleic acids or polymers of amino acid molecules that cause infectious diseases.
Infrared	Electromagnetic energy with wavelengths longer than visible light. We often experience infrared energy as heat.
Inland flooding	Flooding that takes place when a tropical or extratropical storm drops large volumes of precipitation as it moves ashore.
Inner core	The innermost mechanical layer of the Earth. It is composed of solid iron and nickel.
Insolation	The amount of solar radiation reaching a given area on the Earth.

Instrument record	The time period since about 1880 when we have had sufficient weather station across the globe to calculate global average temperatures.
Intensity	A subjective measure of earthquake size. Intensity is determined by the amount of damage done by an earthquake, so it is dependent not only on the size of an earthquake but also by population density, building practices, and the composition of underlying ground.
Interglacial period	A warmer period between ice ages.
Intermittent stream	A stream channel that only has water flowing within it when rain falls.
Intertropical Convergence Zone (ITCZ)	The location where Hadley cells on either side of the equator meet. It is not always located exactly at the equator. The ITCZ migrates northward in the Northern Hemisphere's summer, and southward during the Southern Hemisphere's summer.
Intraplate earthquake	An earthquake that occurs in the interior of a tectonic plate rather than along its boundary.
Intrusive	Used to describe an igneous rock formed below the Earth's surface. Intrusive igneous rocks are composed of mineral crystals that are large enough to see with the naked eye.
Inundation zone	The area of land that is covered by the waters of a tsunami.
Iron meteorite	A meteorite composed primarily of iron with small amounts of nickel. When cut open and etched with acid, it will display a Widmanstatten pattern, which is a crisscrossing pattern of iron crystals.
Isotope	Isotopes are different varieties of an element, determined by the number of neutrons in the element's nucleus. The isotope is named for the number of protons + neutrons in the nucleus of an atom.
Isotropic	Equal in all directions.
Jet stream	A zone of fast-moving air in the upper troposphere that moves from west to east across the Earth.
Karst	A landscape created by the erosion of limestone by surface water and groundwater. Sinkholes and caves are karst features.
Kinetic energy	Energy of motion.
Kinetic impactor	A strategy for changing the trajectory of a near-Earth object so it does not impact the Earth. It involves the impact of a spacecraft and explosives on the surface of the asteroid.
Ladder fuels	Vegetation in a wildland fire that leads from the surface environment progressively upward to the aerial environment. They can cause a fire to spread from the surface of the ground to the canopies of trees in very short order.
Lahar	A flow of mud and other debris formed from a mixture of volcanic ash and water.

La Niña	A phenomenon in the equatorial Pacific Ocean in which sea surface temperatures are cooler than usual.
Latent heat	A quantity of heat energy that is involved in water changing from one phase to another (evaporating or condensing).
Latitudinal desert	A desert created by the prevailing air circulation patterns on Earth. Most latitudinal deserts lie between 30° and 60° N and S of the equator.
Lava	Molten rock above the surface of the Earth.
Lava dome	A thick lava flow that sits over a vent.
Lava shield	A broad, shallowly sloping lava dome.
Levee	A ridge of material that runs parallel to a river or stream. It may be natural or artificial. Levees are often used as flood control structures.
Light fuels	Grass, light brush, and other vegetation that is less than 2 ft (0.6 m) tall. It also includes pine needles, twigs, and other small-diameter vegetation or vegetative debris near the surface of the ground.
Lignin	A complex organic material in the cell walls of many plants, making them rigid and woody.
Liquefaction	A phenomenon in which ground shaking causes soil and sediment to either expel a great deal of water or to act like more like a liquid than a solid.
Lithosphere	The outermost mechanical layer of the Earth. The lithosphere is composed of the crust and uppermost part of the mantle. The Earth's tectonic plates are made of lithosphere.
Love wave	A type of seismic surface wave. Love waves travel in a horizontal S-shaped motion through the lithosphere. Surface waves travel slower than body waves.
Macroburst	A downburst greater than about 2.5 mi (4 km) across. They can last up to 30 minutes and have wind speeds of 130 mi/hr (210 km/hr).
Made land	Land created by filling in a portion of a body of water with soil, sand, and other debris.
Magma	Molten rock below the surface of the Earth.
Magnitude	The size of an earthquake calculated using parameters of an earthquake as recorded by a modern seismometer.
Magnitude scale	A set of modern equations that allows scientists to calculate the size of an earthquake and how much energy was released by that earthquake.
Mantle	The middle compositional layer of the Earth. The mantle is composed primarily of medium density silicate minerals.
Mass wasting	The downslope movement of soil, regolith, rocks, and other debris under the influence of gravity.

Meander	A bend in a stream.
Meandering stream	A stream that has many bends and turns (meanders).
Medium fuels	Shrubs and saplings up to 6 ft (1.8 m) tall, as well as the grassy understory.
Mesopause	Boundary in the Earth's atmosphere between the mesosphere below and thermosphere above.
Mesoscale convective complex	A multicell cluster of thunderstorms, often found in the midlatitudes during the summer months.
Mesosphere (in the atmosphere)	The third layer of the Earth's atmosphere, located between 31 mi (50 km) above the Earth's surface to an altitude of 50 mi (80 km). Temperatures drop in the mesosphere with increasing altitude. The top of the mesosphere is the coldest part of the Earth's atmosphere.
Mesosphere (in the solid Earth)	The third mechanical layer in the Earth. It corresponds to the lower part of the mantle in the compositional Earth.
Meteor	A small to medium sized rocky or iron body in the solar system.
Meteorite	A meteor that has landed on a planet.
Meteoroid	A very small meteor that has landed on a planet.
Methanogenesis	Production of methane by bacteria from organic acids or carbon dioxide, in an anoxic environment.
Methanotrophy	Microbial use of methane as food.
Microburst	A small downdraft, typically less than 2.5 mi (4 km) across. It can carry winds of 170 mi/hr (270 km/hr) but usually only lasts up to 10 min.
Midlatitude desert	A desert typically found between about 30° and 50° latitude (north or south of the equator) in areas inland of or otherwise distant from sources of moisture that could bring precipitation.
Midocean ridge	A volcanic mountain range found where two tectonic plates are moving away from each other. As the name implies, these are found most often in ocean settings, though there are a few notable instances where midocean ridges appear above sea level. Iceland is one of these locations.
Milankovitch cycles	Periodic changes in the Earth's distance from the Sun, the "wobble" in the Earth's rotation, and the tilt of the Earth's axis that have an impact on climate.
Mitigation	Any action or item that is meant to reduce the impact of a future natural disaster.
Modified Mercalli Intensity Scale	A numerical scale from I to X in which each Roman numeral represents a level of damage or shaking during an earthquake.
Monogenetic	A term used in volcanology that refers to a volcano formed from one eruption. It is not expected to erupt again after the volcano-forming eruption.

Monsoon desert	A desert formed inland of areas that experience monsoonal rains.
Mouth	The end of a stream. The location where a stream empties into a larger body of water, found at the lowest elevation along the stream channel.
mRNA vaccine	Messenger RNA (mRNA) vaccines teach our cells how to make a protein that will trigger an immune response inside our bodies.
Multi-ring crater or multi-ring basin	A very large impact crater that has multiple concentric slopes formed due to faulting and slumping along the edges of the structure.
Mutation	A change in a virus or organism that occurs because there a mistake is made during the replication of that virus or cell within an organism. Mutations in viruses lead to new variants, which have different properties of transmission and pathology than the original virus.
Mycorrhizae	A symbiotic relationship between a fungus and plant root systems.
Natural disaster	Any natural event that has a profound negative effect on a population.
Natural event	Any event that occurs in nature, such as a flood, a volcanic eruption, an earthquake, a landslide, a disease outbreak, or a meteor impact.
Near Earth object (NEO)	An object in space orbiting within 1.3 AU of the Sun.
Negative feedback	A situation in which a set of interacting processes serve to reduce the effects of a set of conditions.
Nomogram	A diagram that allows you to determine one parameter if you know two more. A nomogram can help determine the approximate magnitude of an earthquake if the earthquake's maximum amplitude and distance from the seismometer are known.
Normal fault	A type of fault in which the hanging wall (area above the fault) moves down relative to the foot wall (area below the fault).
Novel	In the case of pathogens, this means a particular pathogen is new and has not been seen before.
Nucleus	The center of an object such as a comet, cell, or atom.
Obliquity	Changes in the tilt of the Earth's axis, from 21.5° to 24.5°.
Ocean acidification	The process of making the ocean's waters more acidic. This happens at higher temperatures when excess carbon dioxide in the atmosphere leads to more carbon dioxide dissolved in seawater. Ocean acidification can cause shells of marine organisms to dissolve, and make it more difficult for crustaceans to form robust exoskeletons.
Oscillation	A term used in climate science to describe a cycle, particularly one that varies between one extreme and the other, such as hot and cold, high atmospheric pressure and low atmospheric pressure.
Outbreak	A disease that suddenly arises in a population.

Outer core	A layer in the Earth that corresponds to the top part of the compositional core. The outer core is composed of liquid iron and nickel.
Oxidation	The process of a substance combining chemically with oxygen.
Ozone	O_3, a molecule present in the Earth's stratosphere that helps absorb ultraviolet radiation. Closer to the Earth's surface, it is a pollutant that is a major component of smog.
Ozone layer	A protective layer within the Earth's stratosphere in which ozone absorbs much of the Sun's incoming ultraviolet radiation. This dramatically reduces the amount of ultraviolet radiation that reaches the Earth's surface, making it more hospitable for life.
Pahoehoe	A lava surface that is smooth or ropy.
Paleoclimatology	The study of the Earth's past climates.
Paleodesert	A landscape that was once a desert under prior climatic conditions, but today receives greater annual rainfall and supports vegetation.
Pandemic	A large scale epidemic that spreads between continents.
Pangaea	Name given to a previous configuration of the Earth's continents into a single supercontinent. Pangaea existed from about 150 million to 300 million years ago. Pangaea received its name from Alfred Wegener, the first scientist to provide a wide range of evidence of its existence.
Parasite	An organism that lives in or on another organism, deriving its nutrients from the host.
Pathogen	A bacterium, virus, or other microorganism that can cause disease.
Pathology	The study of the causes and effects of disease.
Pele's hair	Thin, hair-like filaments of glass created when blobs of lava are stretched away from each other. Named for the Hawaiian goddess of the volcano, Pele.
Perihelion	The Earth's closest location to the Sun in its elliptical orbit.
Period	The time that elapses between the arrival of the crest of one wave and the crest of the next wave.
Permafrost	Frozen soil in arctic regions.
Phenology	The study of the timing of biological responses to the seasons.
Photoperiod	The length of the day, which varies with latitude and the seasons.
Photosynthesis	The process in which some organisms use energy from sunlight to create food (simple sugars) from carbon dioxide and water. Byproducts are oxygen and water.
Pillow lava	A lava flow made of pillow shapes. Pillow lavas are only produced when lava flows occur underwater.
Planetesimal	A protoplanet, a mass of material that will become a planet.

Plate tectonics	A revolutionary geologic theory that states the Earth's lithosphere is broken into several pieces called plates. These plates are in constant motion, and they interact at their boundaries. These interactions at plate boundaries are responsible for creating volcanoes, earthquakes, mountain ranges, and deep ocean trenches.
Plinian column	A vertical column of volcanic ash and gases that rises from an erupting volcano. The top of the column eventually spreads out much like a mushroom.
Plutonic	Used to describe an igneous rock formed beneath the Earth's surface. Plutonic rocks are composed of mineral crystals that are large enough to see with the naked eye.
Pluvial	Related to or characterized by rainfall.
Point bar	A sand or gravel bar found on the inside of a bend in a stream. A stream's velocity is low on the inside of a meander, so sediment is deposited here.
Polar amplification	Cooler high latitude regions are warming more rapidly than areas at lower latitudes.
Polar cell	Large scale atmospheric circulation between 60° north and south latitude and the poles. Polar cells pull cold air from the poles toward the equator. At about 60° north and south, they meet warmer air from Ferrel cells.
Polar desert	A desert that occurs near the poles. They often receive precipitation, but only in the form of ice or snow.
Polar jet stream	A fast-moving stream of air that moves from west to east across the globe at around 60° north and south of the equator, though its exact location varies with time. It separates cold air from the poles from more temperate air of the midlatitudes. It is found at the boundaries between Ferrel cells and Polar cells.
Polar vortex	Southward excursions of the polar jet stream to lower latitudes. A polar vortex brings extreme cold to the midlatitudes.
Positive feedback	A situation in which a set of interacting processes enhance the effects of a set of conditions.
Potential energy	Stored energy.
Precautionary principle	States that a new product or process whose ultimate effects are disputed or unknown should be resisted.
Precession	A wobble in the Earth's rotation that causes the Northern Hemisphere to occur alternately when the Earth is nearest the Sun and farthest away from the Sun.
Predictable hazard	A hazard that is known to exist in an area. One example is the earthquake hazard in western California as a result of the many faults along the plate boundary there, including the San Andreas, Hayward, and Garlock faults.

Preignition	Phase of combustion in which a pilot heat source begins to act on organic material such as vegetation.
Primary decomposer	An organism that helps recycle nutrients back into the environment. Many species of bacteria are primary decomposers.
Primary host	The first host needed in the life cycle of a pathogen.
Prion	An infectious protein molecule. Prions can cause fatal neurodegenerative diseases in people and animals.
Prominence	A large arch of plasma that is aligned with the Sun's magnetic field.
Protostellar disk	The rotating disk-shaped mass of gas, dust, and solid materials that will become a star and its solar system.
Protozoans	Single-celled animals that have a defined nucleus and organelles.
Psychological First Aid	A set of actions relief workers can use to engage survivors of a disaster. This includes making sure the survivors know they are safe, attending to their immediate needs, and guiding them toward resources to assist them in recovery.
P wave	A type of seismic body wave. P stands for primary. This is a compressional wave and it can travel through solids, liquids, and gases. P waves are the fastest of all seismic waves.
Pyroclastic flow	A deadly, fast-moving avalanche of hot gases, volcanic ash, and rocks.
Pyrolysis	A chemical change brought about by heat. The process of pyrolysis begins with chemical reactions within the fuel that break the fuel down into its various components. This usually includes volatile materials, waxes, resins, lignins, and tar.
Radiance	Light or heat as emitted or reflected by something, such as the Sun.
Radiant energy	Energy of electromagnetic waves. It includes heat and light energy as well as microwaves and x-rays.
Radiant heat	Heat energy that is emitted by a hot body, such as the Sun.
Radiation	Transfers heat from one place to another through radiant energy.
Radiative forcing	A direct measure of the amount that the Earth's energy budget is out of balance. The Earth is absorbing more energy from the Sun than it emits, causing warming of the Earth.
Rain shadow desert	A desert that forms in the lee of a tall mountain range.
Rayleigh waves	A type of seismic surface wave. They travel in a retrograde elliptical motion, meaning that they travel in a nearly circular motion with particles rotating opposite the direction of wave motion. Surface waves travel slower than body waves.
Recurrence interval	A theoretical length of time between events of the same size in the same location.
Regolith	Weathered rock.

Relative humidity	The percentage of moisture the air is carrying compared to how much it can carry at a specific temperature. Warmer air can carry more moisture than colder air. Relative humidity is measured by an instrument called a psychrometer, and is reported as a percentage.
Renewable energy	Energy sources that may be produced in the absence of fossil fuels. This includes wind energy, hydroelectricity, geothermal energy, hydrogen fuel, and biofuels.
Reservoir host	An organism that harbors a pathogen.
Retreat	Melting of a glacier that occurs seasonally or more permanently. The glacier advances, or grows, during the winter, but retreats, or melts at progressively higher elevations, in the summer or as the region warms.
Return period	A theoretical length of time between events of the same size in the same location.
Reverse fault	A type of fault in which the hanging wall (area above the fault) moves up relative to the foot wall (area below the fault).
Rhyolite	A volcanic igneous rock that contains more than 69% silica. It is usually light gray, white, pink, or orange in color and may have visible crystals.
Rhyolitic	An igneous rock composition that is used to describe rocks with more than 69% silica.
Ribonucleic acid (RNA)	A long, single strand of cells that process protein.
Richter Scale	A set of equations determined by scientists Charles Richter and Beno Gutenberg to describe the size of an earthquake based on the amplitude of seismic waves corrected for distance from the epicenter. The Richter Scale is no longer used to describe the size of an earthquake. Instead, earthquake size is described based on the Magnitude scale.
Rift zone	A zone of weakness on a volcano. These are most common on shield volcanoes. Many shield volcanoes have two or three rift zones emanating from the summit area.
Risk	The possibility or probability that a person or population will be negatively affected by a hazard or natural event.
River flood	A flood that occurs over a period of weeks. It is usually the result of an unusually high amount of rainfall, but may be exacerbated by melting of heavy snowfall or other weather events.
Rotational slide	A type of mass movement in which material slides downhill along a curved surface.
Run-up	The maximum height reached by a tsunami wave as it rushes onshore.

Rupture	In the context of a fault, the surface area of the fault that breaks and moves during an earthquake. The term *rupture* can also refer to the process of a fault breaking.
Saffir-Simpson Hurricane Wind Scale (also known as the Saffir-Simpson Scale)	A scale for classification of hurricanes into five categories. Category 1 hurricanes are weakest, and Category 5 storms are strongest.
Salinization	The process of making salty.
Saltwater incursion	Invasion of saltwater into freshwater lenses.
Sanitation	A broad term that includes drinking water purification as well as proper treatment and disposal of sewage.
SARS	Severe Acute Respiratory Syndrome caused by a coronavirus.
SARS-CoV-2	The coronavirus that causes Covid-19.
Scoria	A red or brown piece of basalt that has been erupted into the air, cooled, and landed back on the ground as a solid piece of rock. Scoria has numerous holes throughout the rock.
Seafloor spreading	The first part of plate tectonic theory to be proposed. In some areas of the world, two adjacent plates are moving away from each other. At these plate boundaries, new oceanic lithosphere is being created between the two plates, causing the oceans to become wider. This is the mechanism for moving continents away from each other, at the same time widening oceans.
Seamount	An underwater volcano. The volcano may be active or extinct.
Secondary host	The second host needed in the lifecycle of a pathogen. A secondary host is infected with the pathogen from the primary host.
Segments	Used in the context of plate tectonics, a segment is a portion of a much larger plate boundary. Because the Earth is a sphere, long linear features are usually broken into smaller pieces called segments. A midocean ridge, for example, may have a number of segments that are connected to each other by transform boundaries.
Seiche	Sloshing motion of water that occurs in closed basins during an earthquake. Seiches may occur even thousands of miles from the location of an earthquake.
Seismic waves	Energy released by an earthquake. It travels through the interior of the Earth as P waves, S waves, Love waves, and Rayleigh waves.
Seismogram	A paper or electronic record produced by a seismometer.
Seismologist	A scientist who studies earthquakes and seismic waves.
Seismometer	An instrument used to detect and record earthquakes.
Shear wall	A solid wall that connects vertical studs and ties a wall together. A shear wall is a construction element that is meant to reduce damage during an earthquake.

Shield volcano	A volcano whose edifice appears broad with gentle slopes. This type of volcano is frequently composed of the rock basalt.
Shocked quartz	Grains of the mineral quartz that have undergone a transformation in structure due to the enormous force of a meteorite impact.
Silica	A component of igneous rock composition. Its chemical formula is SiO_2.
Simple crater	A relatively small circular crater with a single depression, created by the impact of a small meteor.
Sink	A reservoir, such as a reservoir for carbon dioxide in the Earth.
Sinkhole	A type of subsidence that occurs when material in the subsurface is eroded, leaving the surface of the Earth unsupported. The surface of the Earth collapses, creating a hole.
Slash	The material left over after trees have been harvested. Includes branches, treetops, stumps, and logs.
Slide	A type of mass wasting event in which material moves downhill along an inclined plane.
Slump	A type of mass wasting event in which material slides downhill along a curved surface.
Slushball Earth	A time in the Earth's history in which ice and snow reached nearly to the equator, postulated by scientists.
Smoldering fire	A fire in which material burns without a flame.
Snowball Earth	A postulated time period in the Earth's history when the planet was nearly entirely covered with ice and snow.
Soil	Unconsolidated mineral and organic material on the immediate surface of the Earth.
Solar flare	A massive release of energy caused by the rotation of equatorial and polar regions of the Sun rotating at different speeds. This causes magnetic lines to tangle. Occasionally, sudden changes in the magnetic field in these tangled regions cause a massive release of energy, known as a solar flare.
Solar radiation	Electromagnetic energy that travels outward in all directions from the Sun.
Solar wind	A constant stream of charged particles that is emitted in all directions from the Sun.
Solifluction	A type of creep that occurs in arctic climates, where permafrost partially melts in the summer months and starts moving downhill under the influence of gravity. It gives the landscape a "molten" look.
Spatter	Semimolten lava. Spatter often starts as molten lava that is tossed out of a vent. It sails through the air and becomes more viscous. When it lands on the ground, it flattens.

Spillover	A pathogen infects a new host. The COVID-19 pandemic was the result of spillover of a novel coronavirus into the human population, probably from the bat population.
Spindle bomb	A type of bomb in which molten lava spins through the air, creating a shape that is thick in the middle and tapered on the ends.
Spore-forming bacteria	A group of bacteria that live in the soil and can cause disease. Examples of spore-forming bacteria include those that cause anthrax, botulism, and tetanus.
Spotting	The process of starting a new fire from firebrands, which are embers that escape a fire.
Squall line	An elongated line of interacting thunderstorm cells.
Stage	The height of a stream's surface above a specific datum, such as the base of the stream channel.
Storm surge	A dome of water created when the high winds of a hurricane pushes water ashore as the storm approaches shore.
Strain	Deformation within an object.
Stratigraphy	The study of the Earth's rock and soil record, usually for the purpose of determining past climates and environments.
Stratopause	Boundary in the Earth's atmosphere between the stratosphere below and mesosphere above.
Stratosphere	Second layer of the Earth's atmosphere, which extends from about 10 mi (16 km) above the Earth's surface to an altitude of 31 mi (50 km). Located within the stratosphere is the Earth's ozone layer. Temperature in the stratosphere increases with altitude.
Stream	A body of flowing water confined within a channel, regardless of the size of the channel or volume of flow.
Stress	A force applied to an object.
Stony-iron meteorite	A meteorite that contains iron and nickel as well as rocky minerals.
Stony meteorite	A meteorite that contains rocky materials such as minerals or angular particles of rock.
Strike-slip fault	A type of fault in which movement is lateral, or side-to-side. The fault is usually vertical or nearly so. Strike-slip faults are described by noting which side move toward the observer if the observer straddles the fault. Strike-slip faults are either right-lateral or left-lateral.
Subduction zone	A type of convergent boundary in which one plate dives beneath the other. This happens anytime oceanic lithosphere exists on at least one side of the boundary. Common features at subduction zones include earthquakes, volcanoes, and deep ocean trenches. The largest earthquakes in history have occurred at subduction zones.
Sublimate	A material goes directly from solid to gas form.

Subsidence	Lowering of ground level, usually due to the erosive action of groundwater.
Subtherapeutic	A level of medicine within the body that is insufficient to cure or prevent disease.
Subtropical jet stream	A fast moving stream of air that moves from west to east across the globe. It is found around 30° north and south latitude, at the boundary between warmer air in the equatorial region and cooler air of the temperate regions. It is found at the boundary between Hadley cells and Ferrel cells.
Sunspot	A dark area of the Sun's photosphere that is thousands of degrees cooler than the surrounding areas. Sunspots tend to appear in 11 year cycles from minimum to maximum, then back to minimum.
Superbolide	A meteor that is at least 100 times brighter than the Moon as it descends through the Earth's atmosphere.
Supercell thunderstorm	The most energetic and dangerous of thunderstorm types. Hazards of supercell thunderstorms include lightning, high winds, torrential rainfall, hail, and tornadoes.
Surface fuel	Fuel for a fire located at or near the surface of the ground.
Surface wave	A type of seismic wave that travels through the lithosphere, close to the surface of the Earth. Surface waves are slower than body waves. They come in two varieties: Love waves and Rayleigh waves.
S wave	A type of seismic body wave. S stands for secondary or shear. S waves travel in an S shape perpendicular to the direction of wave motion. They are the second fastest type of seismic wave. S waves can only travel through solids.
Symbiotic	A mutually beneficial relationship between two organisms.
Talus	Pieces of rock found at the base of a rocky slope.
Talus slope	A steep slope at the base of a rocky outcrop, composed of talus.
Technological disaster	A disaster that occurs due to a failure of technology. One example is the melt-down of the Chernobyl nuclear plant in Pripyat, Ukraine, in 1986.
Texture	In an igneous rock, texture is usually used to describe the size of the crystals that make up the rock. However, other features can also be used. For example, if a rock is full of round holes, it is vesicular in texture. If it is full of tiny holes that make the rock light enough to float on water, it is pumicious in texture.
Thermosphere	The fourth layer of the Earth's atmosphere. The base of the thermosphere is found at an altitude of about 50 mi (80 km) and it extends to an altitude of 311 to 621 mi (500 to 1,000 km).
Thin section	A thin slice of rock mounted on a glass microscope slide. It is viewed through a microscope with polarized light shining through the slice of rock. It reveals minerals and other structures within the rock.

Thrust fault	A type of fault in which the hanging wall (area above the fault) moves up relative to the foot wall (area below the fault). The term *thrust fault* is usually reserved for a fault of this type in which the angle on the fault is low, typically less than 40° from horizontal. Otherwise the fault is called a reverse fault.
Tipping point	The point at which a series of small changes or incidents becomes significant enough to cause a larger, more important change. It is effectively a point of no return.
Topography	The shape of the Earth's surface.
Torino Scale	A scale that communicates the likelihood of an impact from an asteroid or meteor.
Trade wind desert	A desert caused when warm trade winds blow over land toward the equator, evaporating and dissipating clouds. The Sahara and Kalahari deserts are trade wind deserts.
Transform boundary	A plate boundary in which two plates move laterally past each other. Earthquakes are a common occurrence at transform boundaries.
Transpiration	The passage of water from a plant to the atmosphere.
Trench	A linear feature in the ocean where the water is very deep. Trenches are created most commonly at subduction zones, and they mark the location where one plate dives beneath the other.
Tributary	A smaller stream that empties into a larger stream.
Trophic level	A level on the food web.
Tropical cyclone	A rotating storm within the tropics. In the Northern Hemisphere, it rotates counterclockwise. In the Southern Hemisphere, it rotates clockwise.
Tropical depression	An organized group of storms over a warm ocean in which maximum sustained winds are between 23 mi/hr (37 km/hr) and 39 mi/hr (63 km/hr). When a storm strengthens into a tropical depression, it receives a number.
Tropical disturbance	An organized cluster of thunderstorms in a warm or tropical ocean.
Tropical storm	An organized, rotating storm system over a warm ocean in which maximum sustained winds are between 39 mi/hr (63 km/hr) and 74 mi/hr (119 km/hr). When a tropical depression strengthens into a tropical storm, it is given a name.
Tropopause	Boundary in the Earth's atmosphere between the troposphere below and stratosphere above.
Troposphere	The lowermost level of the Earth's atmosphere. It extends from the Earth's surface to about 10 mi (16 km). It contains about 99% of all of the atmosphere's air molecules. Temperature decreases in the troposphere with increased altitude.

Trough	The low point of a wave.
Tsunami	A water wave created by a disturbance of the ocean floor.
Tsunami bore	A steep, turbulent water flow typically found in an estuary or in the mouth of a river when a tsunami comes ashore.
Typhoon	A name used for a hurricane off the east coast of Asia.
Ultraviolet	Electromagnetic energy with wavelengths shorter than that of visible light.
Updraft	Heated air rises and flows upward and outward away from a fire.
Urban heat island effect	Urban areas experience warmer temperatures than the surrounding landscape.
Urbanization	Covering a landscape with nonporous surfaces, leaving nowhere for water to soak into the ground. Urbanization tends to enhance the speed and severity of flooding in urban areas.
Vaccine	A prepared biological agent that causes acquired immunity to a disease. It usually contains either a disabled pathogen, or something similar to the pathogen that is able to stimulate the body into making antibodies to fight the pathogen.
Variants	Mutated forms of viruses or other pathogens that may have different properties than the original, causing them to be more or less transmissible and virulent.
Vector	An organism that transports a pathogen from one host to another. Often the vector does not get sick.
Vent	An eruption site on a volcano. A single volcano can have hundreds of individual vents or a single one.
Vesicular	An igneous rock texture in which the rock is full of holes. The holes are caused by gas bubbles in the lava when the rock was molten.
Viral antigens	The surface molecules of a virus.
Viral vector vaccine	Viral vector vaccines use a modified version of a virus that is different from the virus being targeted to deliver important instructions to our cells. The modified version of the virus is called a vector virus.
Viroid	An infectious molecule that resembles a virus in a sense that the infectious part is the nucleic acid RNA. Viroids lack the protein coat that protects the nucleic acid in viruses.
Virus	DNA or RNA surrounded by a coat of protein. They are not living organisms, but contain some of the same components of life such as DNA or RNA. They are unable to reproduce on their own, or to produce their own food. They instead invade other cells and use those cells to replicate.
Viscosity	A measure of a material's resistance to flow.

Volatilize	Bring to a vapor form.
Volcanic ash	Pulverized rock and magma. Volcanic ash is formed in violent, explosive volcanoes. It is gray to black in appearance. Individual particles are the size of sand or smaller.
Volcanic field	An area that has many individual volcanoes.
Volcanic rock	An igneous rock formed at or above the Earth's surface. Also used to describe rocks that were erupted from a volcano.
Volcanologist	A scientist who studies volcanoes.
Watershed	An area that contributes water to a specific stream.
Wavelength	The length of a wave, measured from one crest to the next crest.
Weather	Short-term or instantaneous set of atmospheric conditions in a given location.
Widmanstatten pattern	A crisscrossing pattern of iron crystals observed when an iron meteorite is sawed open and etched with acid. Widmanstatten pattern can be used to differentiate between iron formed on Earth and iron meteorites because a Widmanstatten only appears in iron formed in space.
Wildland fire	A fire involving natural vegetation and fuels.
Wildland/urban interface	An area in which human-made structures, towns, and cities are located in or next to wildland environments.
Wind chill	An apparent temperature calculated using actual temperature and wind speed. Wind chill is typically reported on cold days.

Acknowledgments

Thank you to my coauthors (in alphabetical order) Drs. Joel Berrier, Jillian Gregg, John Helms, Anni Moore, Austin Nuxoll, and Kelli Wakefield. They had the good grace to accompany me on this project and their contributions have vastly improved the book.

Christine Bodnar-Anderson at NISEE-PEER, Berkeley's Earthquake Engineering Online Archive, was instrumental in retrieving images for use in the Plate Tectonics and Earthquakes chapters. SGT Tawny Schmit from the Iowa National Guard was kind enough to provide some of her photographs for the Floods chapter. Kevin Eckstrom provided access to dramatic photos of the Washington National Cathedral in Washington, D.C., documenting damage that occurred as a result of an earthquake in 2011. The photo was used in the Earthquakes chapter. Dr. Mike Coniglio of the National Severe Storms Laboratory provided the breathtaking image used for the front cover. Nathan Stillwell was very kind to chat with me about his experience as a volunteer firefighter in the Texas Panhandle. Two of his images are featured in the Wildland Fires chapter. Thanks also to archive staff at Yale University, the National Archives, the Smithsonian Institution, and the Nebraska History Association. These organizations provided other images used in the text. Thank you also to Dr. Marli Miller who provided a number of photographs.

Members of the editorial and production team have been wonderful to work with, and more patient than I could have asked them to be. Many thanks to Dr. Kate Lajtha, Neena Ganjoo, Noel McGlinchey, Mandy Collison, Ritu Bose, Carol Kromminga, and Naveen Kumaran Shanmugam.

I would like to thank Dr. Gilbert Hinga, Semira Hinga, and Ben Hinga for their unwavering encouragement. Erica Ferrell and Andrea Darr, your kind support helped me get to the finish line. Dr. John Mosley, our conversations about geologic events and the circumstances that turn them into disasters have inspired me to continue writing, hoping in some indirect way I might help someone make a decision that will save a few lives.

Author Joel Berrier would like to thank Heather Berrier for his support.

1

Introduction to Natural Disasters

Bethany D. Hinga

University of Nebraska at Kearney, USA

1.1 Introduction

Tonga is an island nation in the southern Pacific Ocean. It is composed of about 100 islands roughly 500 mi (800 km) east of Fiji, and 1,500 mi (2,400 km) north of New Zealand. The volcano Hunga Tonga-Hunga-Ha'apai was well known to residents and researchers. The volcano is located between the two small islands of Hunga-Tonga and Hunga-Ha'apai. It emerged from the sea in 2015, and had been active many times in the recent past. It rose about 2,000 m (6,500 ft) from the sea floor, with about 328 ft (100 m) visible above sea level, joining the two islands on either side.

In December 2021, earthquake activity beneath the volcano increased. Small eruptions carried volcanic ash and gases 7.5 mi (12 km) into the atmosphere. This activity put scientists and residents on alert. Activity waned in the early days of 2022, causing scientists to believe the volcano's eruptive activity may have started to slow down. However, on 15 January 2022, the volcano erupted with tremendous force. An ash cloud rose more than 12.4 mi (20 km) into the atmosphere and spread to a diameter of approximately 425 mi (680 km) within 1 hr (Fig. 1.1). The volcano's eruption also created an unexpectedly large explosion, sending strong pressure waves outward through the atmosphere at tremendous speed. The explosion was heard as far away as New Zealand, more than 1,250 mi (2,000 km) from the volcano. Over the Mediterranean Sea, the air pressure wave from this explosion was measured at 697 mi/hr (312 m/s). It was measured all around the globe by both barometric pressure readings and infrasound instruments. (Infrasound arrays are used by scientists to track changes in the atmosphere due to volcanic eruptions and explosions.) Atmospheric pressure gauges measured the wave not once, but twice. This means the air pressure wave circled the globe at least twice.

The eruption occurred at sea level, which led to the transfer of tremendous amounts of energy into the ocean, which created a tsunami. The enormous energy of the explosion coupled with the atmospheric pressure wave to enhance the energy of the tsunami so it crossed the entire Pacific Ocean. Waves near the eruption site reached nearly 50 ft (15 m) in height, overwhelming coastal communities on Tongataup Islands, 'Eua, and Ha'apai Islands. Throughout the rest of Tonga, coastal flooding from 1 to 4 ft (0.3 to 1.2 m) tsunamis destroyed homes, docks, and infrastructure, leaving people homeless and without some vital services. The tsunami spread outward across the Pacific, causing measurable tsunami waves as far away as Peru and Alaska. Two swimmers on the coast of Peru were drowned by high waves and strong currents that accompanied the tsunami's arrival.

Earth's Natural Hazards and Disasters, Advanced Textbook 7, First Edition. Bethany D. Hinga.

Figure 1.1 Eruption cloud of Hunga Tonga-Hunga-Ha'apai volcano in Tonga, as captured by satellite. Image courtesy of the National Oceanographic and Atmospheric Administration.

Nearer to the volcano, the island nation of Tonga experienced a number of effects from the eruption. The islands adjacent to the eruption site experienced ash fall that coated everything, including vegetation and open-air water supplies (Fig. 1.2). Some vegetation was stripped of leaves, and people were unable to use surface water sources for drinking or other household uses. They had to rely on bottled water and groundwater sources.

The eruption also severed a submarine cable that connected the islands of Tonga to each other, to Fiji, and to the rest of the world via telephone and internet connectivity. Communications were completely knocked out at the time of the eruption. A few days later, some communications were restored with high frequency radio and satellite phones. Telephone and internet were available only through satellite communications for weeks. The cable near Tonga was repaired on 22 February 2022. The company employed to repair the submarine cable was unable to locate it for several weeks. The lack of communications hampered relief efforts. People living on outlying islands within the Kingdom of Tonga were unable to communicate with officials in the government to report damage and ask for help. Even when the main cable to Tonga was repaired in late February, the domestic cables connecting the outer islands to Tonga took two more months to reconnect to the cable network.

This eruption occurred during the Covid-19 pandemic. Prior to relief efforts, only one case of Covid-19 had been reported in the islands, and the person was isolated immediately. Therefore, the islands had remained Covid-19 free for the first 2 years of the pandemic. In an effort to prevent spreading Covid to the islands, relief agencies were careful about testing relief workers. Nobody wanted to spread the virus to those already in dire need of help. However, even with enhanced precautions, community spread of Covid-19 was reported in early February following initial relief efforts, and the affected islands went into temporary lockdown to try to prevent further spread. This delayed the delivery of goods and services to the locked down islands.

Many effects of the eruption of 15 January 2022 were quite unexpected. Though the explosion that accompanied the eruption was extremely powerful, the actual eruption was over in about 10 min. That is an extremely short period of time for an eruption that produced such a tremendous explosion. The tsunami also was much more widespread than scientists expected. They now believe that the tremendous air pressure wave from the eruption helped propel the tsunami to greater distances than it might otherwise have reached. Scientists use computer models to predict the extent and size of waves resulting from a tsunami-forming event, and their models did not predict this. After review, they reasoned that their models were inadequate because they did not have enough data incorporated into their models from volcano-generated tsunamis. They now have data from this eruption that they can use to quantify how tsunamis generated by volcanoes behave differently than those created by earthquakes. (Earthquakes cause about 90% of tsunamis.) They will be updating their models with this new information.

Volcanic eruptions can be characterized by a number from 0 to 8 on a scale called the

(a)

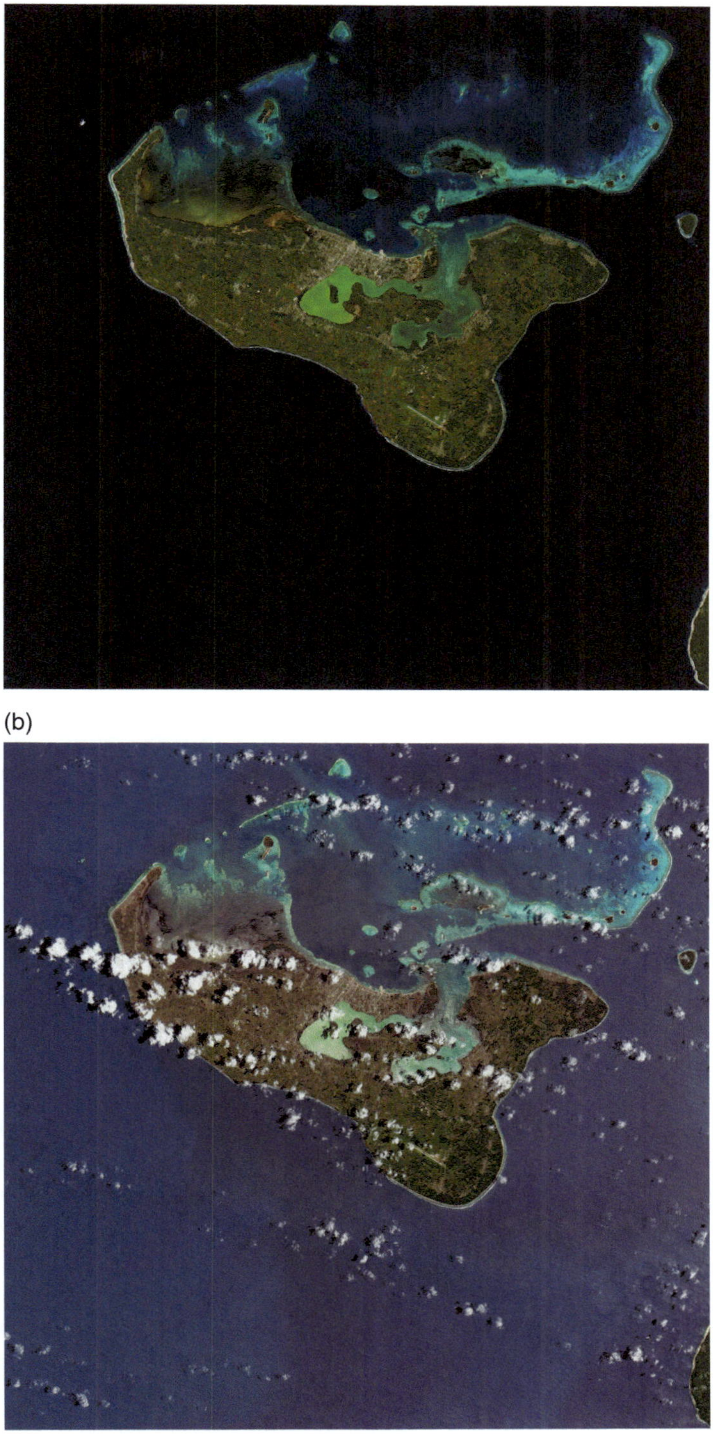

(b)

Figure 1.2 (a) Satellite image of Hunga Tonga, one of the islands adjacent to the Hunga Tonga-Hunga-Ha'apai volcano, in May 2020. (b) Satellite image of the same area, after the eruption of the volcano on 15 January 2022. The eruption of Hunga Tonga-Hunga-Ha'apai volcano blanketed much of the surrounding islands in volcanic ash. In this image, the green foliage of the island has been coated by ash, leaving the landscape a dull gray/brown. White dots in the image are clouds. Images courtesy of NASA's Earth Observatory. https://earthobservatory.nasa.gov/images/149391/island-nation-hit-hard-by-eruption

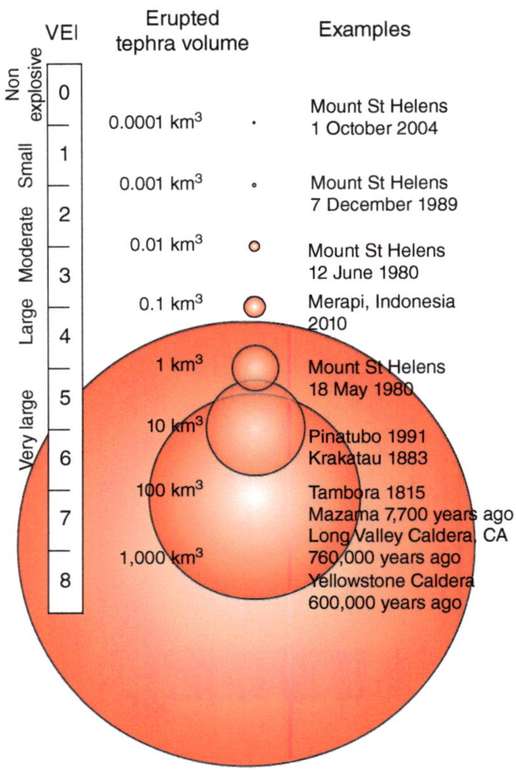

Figure 1.3 The Volcano Explosivity Index. Courtesy of the U.S. Geological Survey; https://www.usgs.gov/media/images/volcanic-explosivity-index-vei-a-numeric-scale-measures-t

Volcanic Explosivity Index (Fig. 1.3; Table 1.1). It is a logarithmic scale in which eruptions of a VEI 0 are very gentle, and eruptions with a VEI of 8 are regionally devastating with global consequences. Scientists have calculated the VEI of this eruption as a 6. Eruptions of this size only occur once or perhaps twice per century. The eruption of Mount Pinatubo in the Philippines in 1991 was a VEI 6. It released enough aerosols and volcanic ash into the atmosphere to cool atmospheric temperatures for a few years afterward. Scientists do not foresee this eruption having the same effect on the atmosphere because fewer particulates and much less sulfur dioxide were erupted during the event in Tonga than at Pinatubo.

A VEI of 4 means between 0.1 km^3 and 1 km^3 of material were erupted from the volcano. The 2010 eruption of Eyjafjallajokull in Iceland, which disrupted air traffic in Europe for weeks, and the 1902 eruption of Pelèe on the Caribbean island of Martinique, which killed over 20,000 people in the city of St. Pierre, were VEI 4 eruptions. A VEI of 5 means between 1 and 10 km^3 of material were erupted. The eruption of Mt. St. Helens in 1980 had a VEI of a very weak 5.

1.2 The Earth System

The Earth is an active planet. It is geologically active, meaning that earthquakes and volcanic eruptions are commonplace. Tectonic plates are constantly in motion, generating the forces needed to produce such events. The Earth's atmosphere is active. Air is always in motion, creating conditions that are ripe for drought, flood, thunderstorms, hurricanes, and extremes in temperature. The Earth's waters are active. Tsunamis, which are water waves created by geologic disturbances, have decimated shorelines around the world time and again. The Earth's waters evaporate to form clouds, which organize into storms, and provide both rainfall and strong storms. Life on Earth is active as well. The planet's ecosystems are teeming with microbes that assist plants and animals with metabolism. Occasionally, microbes also cause disease and death.

The *Earth System* approach to viewing our planet recognizes that all parts of the Earth, and events that occur on the Earth, are related. The solid Earth, which includes the rocky surface and soil, is called the lithosphere, or *geosphere*. Earthquakes, volcanic eruptions, and landslides originate within the geosphere. The *atmosphere* is another component of the Earth System, and it is composed of the air that surrounds the planet. Weather systems originate in the atmosphere. The *hydrosphere* is composed of the waters of the Earth. The Earth's water includes rivers, lakes, oceans, and glaciers, as well as groundwater stored beneath the Earth's surface. Floods, droughts, hurricanes, and ground subsidence originate

Table 1.1 Volcanic Explosivity Index.

VEI	Ejecta volume	Description	Plume height	Frequency of eruption	Examples	Occurrences in last 10,000 yr*
0	< 10,000 m^3	Effusive An outpouring of lava on the ground (as compared with eruptions of ash into the air)	< 100 m	Persistent	Kilauea, Piton de la Fournaise	Many
1	> 10,000 m^3	Gentle Low-level, small to medium volume	100–1,000 m	Daily	Nyiragongo (2002)	Many
2	> 1,000,000 m^3	Explosive Dense cloud of ash and gases with volcanic bombs (2–3 m in diameter)	1–5 km	Weekly	Ruapehu, New Zealand (1971), Mount Sinabung (2010)	3,477
3	> 10,000,000 m^3	Severe Glowing avalanche of hot ash and pyroclastic flows	3–15 km	Few months	Soufriere Hills (1995), Nevado del Ruiz, Colombia (1985)	868
4	> 0.1 km^3	Cataclysmic Columns of gas and ash, extends to stratosphere	10–25 km	≥ 1 yr	Mount Pelee, West Indies (1902), Eyjafjallajokull (2010)	421
5	> 1 km^3	Paroxysmal	20–35 km	≥ 10 yr	Mount Vesuvius, Mount St. Helens (1980)	166
6	> 10 km^3	Colossal	> 30 km	≥ 100 yr	Krakatoa, Indonesia (1883), Mount Pinatubo, Philippines (1991)	51
7	> 100 km^3	Super-colossal	> 40 km	≥ 1,000 yr	Tambora (1815)	5 (+2 suspected)
8	> 1,000 km^3	Megacolossal	> 50 km	≥ 10,000 yr	Yellowstone (Pleistocene)	0

Source: http://ete.cet.edu/gcc/?/volcanoes_explosivity/
Note: Volcanic Explosivity Index, as devised by Chris Newhall of the U.S. Geological Survey and Stephen Self at the University of Hawaii. The index places volcanic eruptions on a scale from 0–8, with a VEI of 0 assigned to gentle, low-volume eruptions, and a VEI of 8 assigned only to those volcanic eruptions that produce more than 1,000 km^3 of eruptive products and have devastating effects on a regional and perhaps a global scale.

within the hydrosphere. The fourth component of the Earth System is the *biosphere*. The biosphere is all of the life on Earth, including everything from the smallest bacteria to large, complex organisms such as human beings. Disease outbreaks and wildland fires originate in the biosphere.

From examples given above, it follows that any part of the Earth System can and does impact all other parts of the Earth System (Fig. 1.4). For example, evaporation of water (from the hydrosphere) in the Caribbean Sea leads to formation of clouds in the atmosphere. Given the right temperature, atmospheric pressure, and air circulation, the clouds can organize into a storm. If the storm further organizes, produces counterclockwise rotation, and grows in size and strength, it becomes a hurricane. The hurricane can move ashore. High winds uproot trees, strip leaves from plants, and

Figure 1.4 The Earth System has four components: The geosphere (lithosphere), hydrosphere, atmosphere, and biosphere. Image from Strahler, 2013.

disturb animals on land (in the biosphere). Heavy surf can stir up sand in the water (the hydrosphere) and bury parts of coral reefs (the biosphere). Heavy rainfall saturates soil and creates mudslides (in the geosphere). Heavy rainfall also floods low-lying areas, which are part of the hydrosphere. Wind (in the atmosphere) creates a storm surge, which floods coastal areas, again part of the hydrosphere. A single event, in this case a hurricane originating in the atmosphere, affects all parts of the Earth System. Hurricanes are not unique in this way. Almost all natural events touch on all parts of the Earth System.

On occasion, a threat to the Earth will arise from outside the Earth. Space is filled with meteors and dust that are on a collision course with the Earth. We refer to these as *extraterrestrial hazards*. Dust from space routinely falls to the Earth's surface. When it does, it is known as *cosmic dust*. Meteors are larger pieces of rock, metal, or some combination of those two substances. Many meteors burn up in the Earth's atmosphere due to friction between the meteor and the molecules in the atmosphere. However, meteors do occasionally survive to impact the Earth's surface. When they do, they are termed meteorites. Every so often, a meteor enters the Earth's surface and explodes within the atmosphere. If the meteor is large enough, it can cause extensive damage on the ground in the area where it explodes. In February 2013, a 33 to 66 ft (10 to 20 m) meteor exploded above the city of Chelyabinsk, Russia. About 1,500 people were injured and more than 7,000 buildings were damaged. The injuries and damage were not caused by falling pieces of the meteor hitting the ground, but by the shock wave from the meteor breaking apart explosively in the Earth's atmosphere. Meteors travel at tremendous speeds and therefore carry a great deal of kinetic energy. This is why such small objects can cause surprising amounts of damage.

1.3 Natural Disasters

Natural disasters affect millions of people worldwide each year, and occur in every country on the planet. A natural disaster is any *natural event* that has a profound negative effect on a population. Most commonly, we refer to natural disasters as natural events that affect human populations. This includes the usual suspects: volcanoes, earthquakes, tsunamis, landslides, floods, droughts, and so on. However, natural disasters can also affect animal and plant populations.

Imagine, for example, a disease outbreak among poultry on farms in Iowa. The disease spreads rapidly and there is no way to safely and economically treat the birds. The birds are unsafe for consumption so they must be culled to stop the spread of the disease. While this is an economic problem for the farmers who cannot sell chickens or the eggs they normally produce, it is undoubtedly much worse for the millions of birds who lose their lives in this disease outbreak. Such an outbreak occurred on poultry farms and in wild bird populations throughout the Midwest in 2014–2015.

The focus of this book will be disasters caused by natural events, and their effects on human populations. We cover the broad variety of natural events that can bring negative consequences to both human populations and populations of other organisms throughout the world. In Chapter 11, we venture into the realm of climate change, which is a profound change in the natural world caused by humans, which in turn causes more natural disasters to occur. Climate change is altering weather patterns, causing heat waves, polar vortexes, droughts, floods, and both more and stronger hurricanes. Nearly every year, we are setting new records for hottest day, month, and year. Climate change is also fueling a rise in global sea level, melting glaciers around the world, and is redefining animal and plant habitats. Clearly these changes are having a profound negative effect on humans around the world, so we would be remiss if we did not address climate change in this text.

1.3.1 Types of Disasters

Disasters can come in many forms. The first type of disaster, and the focus of this book, is a natural disaster. Natural disasters are those caused by an event in nature, such as a volcanic eruption, an earthquake, a tsunami, a wildfire, a hurricane, a tornado, a disease outbreak, or a meteorite impact (Fig. 1.5).

A second type of disaster is technological. *Technological disasters* are those that occur due to a failure of technology. The 1986 nuclear plant meltdown at Chernobyl in Pripyat, Ukraine (then part of the Soviet Union) is an example of a technological disaster. Derailment of a train carrying toxic gas, such as chorine, collapse of bridges, dam failures, plane crashes, and chemical plant explosions are other examples of technological disasters.

A third category of disaster is *conflict disaster*. This includes war, civil unrest, mass shootings, terrorist attacks, and genocide. Technological and conflict disasters can be every bit as terrifying, deadly, environmentally damaging, and tragic as natural disasters. Responses to these disasters often mirror responses to natural disasters as well. First responders, community volunteers, and in many cases local and international aid organizations are called upon to respond to all types of disasters. They shelter, feed, and comfort victims, and help people move forward in their lives after great loss.

Almost any type of disaster can have an anthropological component. *Anthropological disasters* are those that are caused by humans. Of the three types of disasters mentioned above, conflict disasters and many technological disasters are caused directly by humans. Natural disasters can also be made worse by

Figure 1.5 A wildland fire near Sylmar in the San Fernando Valley, California, in October 2008. From Strahler, 2013.

humans. In this book, we will address multiple examples of natural disasters that were made worse by poor planning, poor engineering, poor management of natural resources, and incredibly poor decision making by officials in charge of situations.

No place in the world is free from all hazards. Every location on the planet experiences weather, for example, and extremes in weather can occur anywhere. Mountainous areas are prone to landslides. Valleys near rivers are in danger of flooding. Coastal areas experience storms, flooding, and in some cases tsunamis. Severe thunderstorms and tornadoes occur quite often in North America. The Pacific Rim is known for its volcanic activity and earthquakes. As we learned in 2020 with the Covid-19 pandemic, only the most isolated human populations are able to escape outbreaks of highly contagious diseases. Each location on the planet has hazards associated with it. The key then is to know which hazards exist in your environment, and be prepared to deal with a natural event when it occurs, to prevent it from becoming a disaster.

1.4 Definitions

In order to appropriately address the topic of natural disasters, it is helpful to have some familiarity with terminology that is commonly used to discuss natural events that can turn into disasters. A *natural event* is any type of occurrence in nature. Earthquakes, volcanic eruptions, tsunamis, landslides, extremes in weather, disease outbreaks, and meteorite impacts are examples of natural events. A *natural disaster* is what occurs when a natural event negatively impacts a population.

A *hazard* is an event or set of natural events that occurs and can cause a disaster. For example, a tornado is a hazard. When it comes in contact with people, livestock, and infrastructure, it becomes a disaster. We may discuss a wildland fire hazard in a certain area. The National Weather Service routinely declares Red Flag Days, which are days with above average temperature with low humidity and high winds. These conditions enhance wildland fire hazards. However, we also use the term *hazard* to describe the set of events that may occur around or contribute to a natural disaster, such as the dry conditions, high winds, and high temperatures that often contribute to the formation and spread of wildland fires.

Risk refers to the possibility or probability that a person or population will be negatively affected by a hazard or natural event. In general, the more people in harm's way, or the less prepared a population is to meet the challenges faced by a hazard, the greater the risk. There are always ways that a population can reduce the effects of a natural event and therefore diminish the severity of a disaster. In some cases, it's as simple as posting signs telling people the location of a tornado shelter. In other cases, it's a major engineering project that reduces the magnitude of flooding. Perhaps it is a yearly reminder of best practices in the event of a severe thunderstorm. All of these actions are referred to as mitigation of the natural hazard. *Mitigation* is any action or item that is meant to reduce the impact of a future natural disaster.

1.5 Disasters Are Predictable

Scientists have been studying the Earth for millennia, and have learned a great deal about how the Earth and atmosphere work. We know which areas are prone to earthquakes, and therefore which areas of the world carry an earthquake hazard. The same is true of nearly every natural hazard possible, including severe weather, volcanic eruptions, landslides, floods, wildland fires, and tsunamis. The fact that we know what hazards are faced by a population in a specific geographic area means that these hazards are *predictable*. The ability to know the approximate or exact timing and location of an event based on scientific observations is called *forecasting*. A *forecast* is a warning to the public that a threat is imminent. It should contain the location, timing, and estimated magnitude of the threat. Some events can be forecast days in advance, such as severe thunderstorms, tornadoes, heat waves, blizzards, and hurricanes (Fig. 1.6). Those that are unable

Figure 1.6 The aftermath of a tornado that struck Joplin, Missouri on May 24, 2011. The outbreak of storms that produced this tornado was forecast by the National Weather Service. Photo by Kansas City District U.S. Army Corps of Engineers, Public domain, via Wikimedia Commons.

Figure 1.7 Fault systems within California. Each of the yellow lines is a fault. The San Andreas, Hayward, and Garlock faults are marked. From Arbogast, 2017.

to be forecast at this particular time include most disease outbreaks, many volcanic eruptions, most landslides, and earthquakes.

We know that western California has several active faults such as the San Andreas, Hayward, and Garlock (Fig. 1.7). Therefore, the fact that earthquakes will occur in western California is predictable. The precise day or time during which the next earthquake will occur on one of these faults is unknown, therefore we are unable to forecast the event. We do know that on average, major earthquakes occur on the San Andreas Fault about once every 100 years, but that is only a statistical average gathered from very few data points. In reality, major earthquakes can occur at any time, at any time interval.

The last major earthquake on the northern section of the San Andreas Fault occurred in San Francisco in 1906. No major earthquake occurred in 2006, and has not occurred as of January 2024. Another major earthquake, however, will occur. Every small earthquake is a reminder to members of the population that they should do everything they can to prepare themselves and their families to be self-sufficient for a few days after a large earthquake in which power and water are interrupted, stores are unable to operate, and roads may be blocked with debris. A home earthquake kit can easily be prepared with stored water, nonperishable food, blankets, and a light source.

In the meantime, we go about our lives and do everything we can to mitigate the hazard. California has some of the strictest building codes in the world, such that new homes, businesses, and government buildings must be built to withstand severe shaking. Owners of older buildings work to retrofit those buildings to bring them up to date with respect to current codes.

In October each year, an event called the Great Shakeout takes place, in which entire communities across the earthquake-prone regions of the United States rehearse a response to a large earthquake. (See https://www.shakeout.org.) This exercise allows first responders, city managers, and hospitals to explore their various roles and relationships in a disaster, and to update their response plans.

Some natural hazards are easier to forecast. We know that the midsection of the United States has occasional severe thunderstorms and periodic tornadoes. They can occur at any time of the year, but are most prevalent in the spring and fall as the weather changes from one season to the next and bodies of air move through the region. With modern forecasting tools and advanced weather radar, we have hours to days of warning before thunderstorms develop, and often several minutes of warning between when a storm shows signs

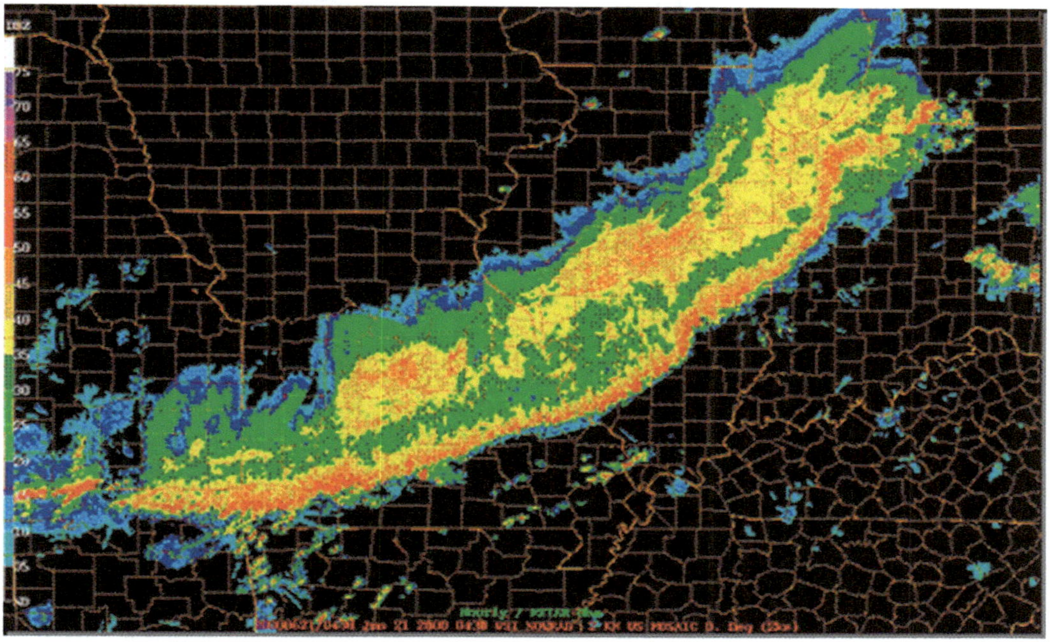

Figure 1.8 A radar image of thunderstorms in the midsection of the United States. A line of thunderstorms is located from southwest Missouri west-northwestward through Illinois and into Indiana. Scientists use images such as these to issue forecasts and alerts as severe weather threatens. From Arbogast, 2017.

of tornado formation and when tornadoes touch down. In this way, severe thunderstorms and tornadoes are not only predictable, in that we generally know what areas of the United States they occur, but they are also forecastable. Atmospheric scientists can tell us where and when to expect thunderstorms, and where tornadoes may occur once a severe storm develops (Fig. 1.8).

Another routinely forecast natural event is the hurricane (Fig. 1.9). Each year in the Northern Hemisphere, oceans become warm enough to support the growth and development of hurricanes in June, and cool sufficiently to prevent

Figure 1.9 Satellite image of Hurricane Katrina in the central Gulf of Mexico. Florida is in the upper center of the image, and Texas is to the left. This satellite image was captured one day before landfall in August 2005. From Arbogast, 2017.

formation of hurricanes in November. Thus hurricane season is June through November. Residents of coastal areas of tropical and temperate oceans and seas in the Northern Hemisphere know to be alert for these storms during that time period. Satellite imagery is used to monitor cloud formation over the eastern Atlantic Ocean, as air currents move offshore from the coast of North Africa. Those clouds are tracked as they move westward across the Atlantic. If the cloudy areas turn into storms, scientists can forecast a storm's strength a day or two in advance based on atmospheric conditions in the region. They can also forecast the path of the storm's movement based on atmospheric pressure surrounding the storm. Coastal communities often have several days' warning before a tropical storm or hurricane hits their shores.

Likelihood of a natural event occurring at some point in the future, and the ability to forecast a natural event in the short term are only two components of disaster predictability. Remember that a disaster requires some negative impact to a population. The human impact is somewhat predictable as well based on the size of the population at risk, the economic status of that population, the quality of communications in the region, how long it has been since the last similar incident, and the overall preparedness of the population. In the best of circumstances, a population knows that an event, such as an earthquake, is likely to strike at some point in the future. Building codes are strict and are strictly enforced. As a result, most buildings survive the earthquake and the death toll is minimal. People have information about what to do in the event of an earthquake to keep themselves and their families safe, and emergency crews are prepared to restore power, natural gas, water, and sewer service quickly.

In the worst of circumstances, building codes are nonexistent, and where they do exist they are not enforced. When even a moderate-sized earthquake occurs, many buildings collapse, burying and crushing tens of thousands of people both inside and outside the buildings. Fires start from overturned cooking stoves and ruptured gas lines. Ambulances and fire trucks are unable to get to those needing help because of large amounts of debris from collapsed buildings clogging streets. Emergency crews are overwhelmed by damage to the infrastructure, and it takes months or even years for basic services to be restored.

The city of Istanbul, Turkey, is a modern city with an ancient past. Its metropolitan area has a population of about 15 million. Though new buildings are engineered to withstand earthquakes, the city has thousands of buildings built hundreds of years before earthquake engineering was invented. Many people and government agencies are working to retrofit these buildings, but not all can afford the expense. Istanbul is located on a major east-west fault that runs through Turkey, called the North Anatolian Fault (Fig. 1.10). The fault is very similar to the San Andreas in California. The area is expected to experience a large and damaging earthquake at some point in the future. Based on the age of the city, the distribution of old buildings within the city, the narrow streets in much of the city, and the age of its infrastructure, it is likely that the next major earthquake in Istanbul will be a true disaster. Thousands of people will be killed when those old buildings that have not gone through the retrofitting process collapse. First responders will find it difficult to navigate their way through the narrow streets of the city to get to those in trouble. Scientists estimate that a large earthquake of magnitude 7.5 that occurred directly beneath the city could kill between 70,000 and 90,000 people and cause $40 billion in damages. It would also make 500,000 to 600,000 people homeless. A certain percentage of those will not have earthquake insurance and will have no means to purchase a new home.

A 1999 earthquake in the city of Izmit, Turkey, may give us a preview of what to expect in an Istanbul earthquake. A magnitude 7.6 earthquake struck the city of Izmit, Turkey, around 3 a.m. local time on 17 August 1999. Shaking lasted less than a minute. In that time,

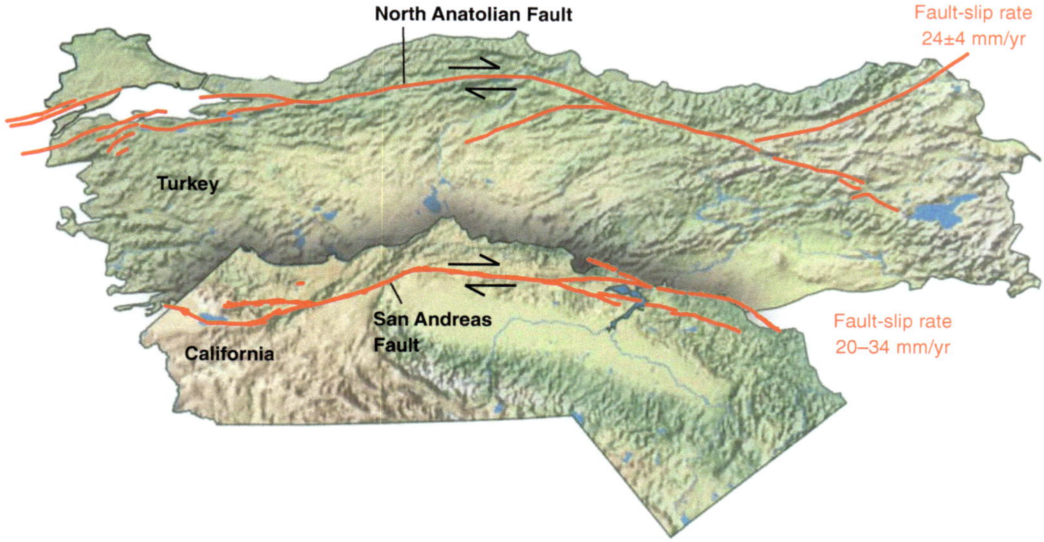

Figure 1.10 A comparison of the North Anatolian Fault in Turkey and the San Andreas Fault in California. The two faults are remarkably similar in their length and rate of slip. From Fletcher, 2014.

thousands of buildings in the city and its suburbs collapsed, and thousands of others were damaged beyond repair. Authorities in the area blamed poor building practices for the loss of life and destruction of homes. Approximately 300,000 people were left homeless. A fire started in an oil refinery and spread to nearby chemical storage tanks. It took 5 days to control the fire. The official death toll was 17,127, and 43,959 people were injured. The city of Izmit and surrounding area had a population of approximately 1.5 million people. If we scale the disaster up to the size of metropolitan Istanbul, it is a frightening picture indeed.

Both of these earthquakes are predictable. They will happen, but we have no way of knowing when. We can predict the outcome of each event, including how many deaths to expect within an order of magnitude, and how many will be left homeless with the same degree of accuracy. It is distressing to know the outlook for Istanbul during the next major earthquake. It is even more distressing to know that there are many other cities throughout the world like Istanbul that face a predictable disaster with a predictably dire outcome. What can we do? We can work to eliminate poverty in a myriad of different ways. We can help cities with old buildings retrofit those buildings so they are safer in earthquakes. We can encourage city governments to be thoughtful in design of infrastructure, and to enforce appropriate building codes even though that makes buildings more expensive to construct. In the end, unfortunately, many communities in ancient cities around the world will not have the money necessary to prepare their buildings for major earthquakes. It is a tragic reality of the world we live in.

1.6 The Human Factor

"I think it's important to recognize that there's no such thing as a natural disaster. What you have is a natural hazard that overlaps with a vulnerable system." This quote is from Wendy Bohon, a geologist with IRIS, the Incorporated Research Institutions for Seismology, a research and outreach organization located in the United States (Associated Press, 2021. https://www.voanews.com/a/americas_why-are-earthquakes-so-devastating-haiti/6209611.html)

Those who study the physical sciences can tell you a great deal about natural events. Scientists can describe the physical mechanisms at work within the Earth and its atmosphere that create a volcanic eruption, a tsunami, a thunderstorm, or a blizzard. They can tell you how much energy is released during a magnitude 5 earthquake, or how fast the winds blow in a Category 4 hurricane. They have a huge array of instruments they use to monitor the Earth system so they can forecast events. They can even detect near-Earth objects in outer space and determine if those objects are on a trajectory to impact the Earth decades from now. However, the physical sciences do not provide a full picture of why natural events turn into disasters. Yes, large earthquakes can cause horrific damage. But why does one magnitude 7 earthquake leave relatively little damage and incur relatively minor loss of life, and another completely decimates a major city leaving tens or even hundreds of thousands of people dead? The two earthquakes occur via the same mechanisms, and release the same amount of energy. What is different about one location versus another? Why do we see death tolls from Category 4 hurricanes vary so widely from one country to the next? Category 4 wind speeds are the same no matter which country the storm touches. The answers to these and many other questions are in the human factor.

Natural events such as earthquakes, volcanic eruptions, landslides, and severe weather have been happening as long as the Earth has existed. What makes a natural event turn into a disaster? For the purposes of this book, a natural event turns into a natural disaster if it adversely affects a vulnerable human population. Thus, in the study of natural disasters, humans and the environments they create for themselves cannot be ignored. The study of natural disasters is as much a study in human vulnerability as it is a study of natural forces on our planet.

1.7 Increasing Risk

Generally speaking, the human population is at greater risk today than at any point in its past. This is due to a mixture of physical, social, and technological trends. Why are people at greater risk now? First, the population of the world is increasing at an alarming rate. This means people must find new places to live. In some cases, this means more people are moving into areas that have been left unsettled in the past. For example, floodplains, steep slopes, or other undesirable locations may now be seen as acceptable places to live, while in the past this would not have been the case because of the obvious hazards associated with these environments. In 2017, 113 people, including many children, were killed in Ethiopia's capital of Addis Ababa when a landfill they had been living on collapsed. Extreme poverty drove them to live in such an environment in substandard housing. Lack of government oversight allowed people to live on and near the heaps of trash in the landfill. Both of these were contributing factors to the tragedy.

In other cases, populations are streaming from the country into the city where jobs are more plentiful. Thus, populations of cities are rising much faster than populations in rural areas. Rapid construction of substandard housing puts people at risk in earthquakes and storms. The enormous death toll of the 2010 earthquake near Port-au-Prince, Haiti (over 220,000 people died) can be largely attributed to lack of building codes and substandard housing within the city. Many homes and business were built from unreinforced cinder block walls and tin roofs. When the magnitude 7 earthquake struck in January of 2010, the walls literally came tumbling down. Many people were crushed to death by falling debris, and several thousand who survived lost limbs due to traumatic crushing injuries.

The second reason more people are at risk today is because of the effects of climate change. It is now commonplace for locations all over the planet to set new record high temperatures each year. Sweltering heat has replaced milder summer temperatures, and people, plants, and animals are having a hard time adapting. A heat wave in Europe in August 2003 led to at least 35,000 deaths, nearly half of which occurred in France. Though heat is rarely thought of as a natural disaster, heat waves kill more people in most years than floods, tornadoes, and hurricanes combined. Weather patterns are also changing as a result of global warming, leading to more torrential rainfall in some areas and more extreme droughts in others. Increased rainfall translates to a heightened danger of flash flooding, and extreme droughts put people at risk because of fire. Most species of plants and animals have specific water needs, and prolonged severe droughts can cause mass death of plants, and the animals that eat those plants, within the region.

Hurricane season is starting slightly earlier and ending slightly later than in the past. It is no longer uncommon to see hurricanes forming in late May, when in the past they rarely started before the beginning of June. Warmer sea surface and air temperatures lead to an increased level of evaporation, which feeds hurricanes. Atmospheric scientists are warning that a likely result of climate change will be more hurricanes each year, as well as more severe hurricanes overall. Even if the percentage of strong, damaging hurricanes stays the same as it is now, the fact that we are seeing more hurricanes each year means there are more damaging hurricanes in total. Higher sea level also means that more coastal communities are in danger of damaging floods from storm surges that come ashore during hurricanes and other storms.

As counterintuitive as it may seem, a third reason people are more at risk today is technology. Technology can, in some cases, make people more vulnerable to disaster by making them feel safer than they actually are. Floodplains are known to be dangerous environments. Annual floods during rainy seasons or periods of high snow melt are seasonal events and have been well documented for thousands of years. Modern engineering allows us to design and build dams, canals, levees, and other flood control structures that may contain floodwaters that are typical for the average year. As a result, people may feel comfortable building in floodplains because they feel that the flood hazard has been eliminated. Unfortunately, in years where floodwaters exceed the design limits of the levees, dams, or canals, flooding is still possible in the old floodplains. People who may have been reassured that their homesites were safe from annual floods, or even a 100 yr flood, may be unwittingly in harm's way when a 500 or 1,000 yr flood takes place.

Failure of flood control technology occurred in the New Orleans area during Hurricane Katrina in 2005. Much of New Orleans and the surrounding area is below sea level. The city is protected through a series of canals and levees, as well as several large-volume pumps that work continuously to keep water out of the city. If the levees on either side of the Mississippi River had held, much of New Orleans would have stayed relatively dry. However, with the extra influx of rainfall and storm surge from the hurricane, several levees failed and flooded populated areas that had been protected during seasonal floods and lesser storms (Fig. 1.11). The result was a devastating loss of entire neighborhoods, thousands of people trapped on rooftops trying to escape floodwaters, displacement of tens of thousands of residents, and the tragic loss of over 1,500 lives within the New Orleans area alone. The total death toll for Hurricane Katrina is officially 1,833.

Figure 1.11 Extreme flooding in a New Orleans neighborhood following landfall of Hurricane Katrina nearby. This flooding was in large part caused by a technological failure of the levee system within the city. The levees have since been repaired and withstood a 2021 hurricane remarkably well. Photo credit: https://commons.wikimedia.org/wiki/File:FEMA_-_18178_-_Photograph_by_Jocelyn_Augustino_taken_on_08-30-2005_in_Louisiana.jpg

1.8 Calculated Risks

Why do people choose to live in hazardous areas? An easy answer might be that some people do not realize the risk to their well-being, and do not recognize the hazard at hand. This may be the case in a small handful of cases. In most cases, however, living in an area in which there is some threat of a natural disaster is a *calculated risk*. People live on the slopes of volcanoes like Vesuvius near Naples, Italy, because it is exceedingly beautiful. Volcanic soil is rich in nutrients, which means it is fertile and grows bountiful crops. However, a devastating eruption of Vesuvius occurred in the year 79 A.D. When it erupted in August of that year, it destroyed the cities of Herculaneum and Pompeii. An estimated 16,000 people were killed (Fig. 1.12). Some died from suffocation by inhaling too much volcanic ash. Others died as they were poisoned by toxic gases. Still others were killed by a phenomenon known as a pyroclastic flow. Pyroclastic flows are furious, fast-moving, superheated clouds of ash, rocks, and toxic gases. Nothing in their path typically survives.

Those who live in the area around Vesuvius know without a doubt the history of the volcano. Pompeii is one of the most famous and well-preserved of the world's archaeological sites. The modern city of Naples, located just to the west of Vesuvius, has a population of about 4 million people. Part of the city of Naples also sits on top of the dangerous volcano Campi Flegrei (Flegreian Fields) (Fig. 1.13). However, millions of people make the area their home. The likelihood of an eruption during a single person's lifetime is small, but it is not zero. Those in the vicinity of Vesuvius and Campi Flegrei are taking a calculated risk by living in this area. They rely on intensive monitoring of the volcano by scientists to give them ample warning of an impending eruption so they can evacuate. And perhaps there is a bit of denial that the risk is real.

People live near coasts in tropical areas that can be affected by hurricanes because coasts

Figure 1.12 Plaster casts of people who perished in the 79 A.D. eruption of Vesuvius in the city of Pompeii near Naples, Italy. These individuals died of asphyxiation due to heavy ash fall. Their bodies were covered in volcanic ash and pumice and eventually decayed. A hollow area in the ash was pumped full of plaster to make these casts. From Murck & Skinner, 2011.

are places of commerce, travel, and beauty. Confident that they will have enough warning to flee if necessary, people feel that it is worth the risk to live in areas prone to hurricanes. Why do people live on the prairies of Kansas and Oklahoma, where tornadoes are such a common occurrence? The soil is fertile, natural resources are abundant, jobs are available, cost of living is reasonable, and the odds of a tornado destroying their homes are small. Thus, people take a risk and build their lives on the prairie.

In truth, all people live in an area where there is a risk of natural disaster. Natural disasters not only include volcanic eruptions, landslides, and earthquakes, but also extreme weather events, disease outbreaks, and meteor impact. Nowhere is without risk. Therefore people have to determine which risks they feel they can manage.

Occasionally, a disaster makes us ask about an unimaginable situation that led to loss of life. Why did those people in Addis Ababa, Ethiopia, choose to live on a landfill, which is not only unpleasant, but inherently very risky? The answer is both simple and complex. The simple answer is likely that they could not afford to live anywhere else in the city. They found this unregulated area where they could build a home from scraps of lumber and metal they found, and where they did not have to pay anyone for the land on which they built. The more complex answer for many families may be a story years in the making. Perhaps they used to live on a farm in the country. They had crops, a garden for their own household use, goats, cows, and chickens. A severe multi-year drought led to failure of their crops and death of their livestock. This meant that not only did they have a lack of income from sale of their crops, or of meat, milk, and eggs from their livestock, but also that they had nothing to eat. Ethiopia has no social safety net such as crop insurance, food stamps, social security, or Medicaid. In order to survive, this family needed to move to the city so at least one adult could work for wages and buy food for the family. The first natural disaster that impacted this particular family was the drought. The fact that there is no social safety net is partially due to priorities set by the Ethiopian government, but it is also an economic issue. There must be sufficient tax income to pay for social services. Taxes come from natural resources, income taxes from the country's citizens, and taxes on businesses. If the population does not have sufficient income to create a sufficient tax base, government services are limited. Economics therefore has a major part to play in this disaster. If aid is unavailable, and there is no one to help subsistence farmers maintain their way of life in their community in the face of the drought, then they have to move. People in poverty have to resort to living in areas they can afford. In this example, it was in a landfill.

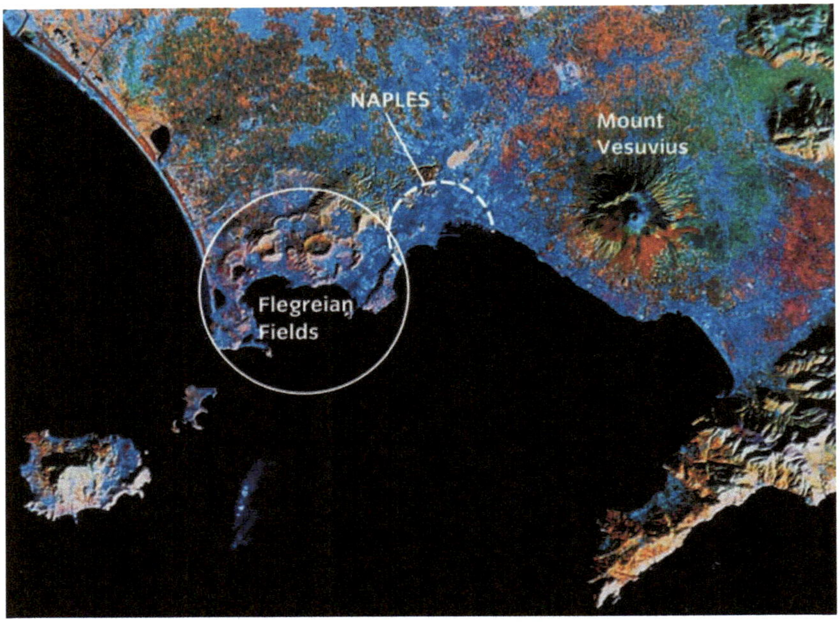

Figure 1.13 A false-color satellite image of the area around Naples, Italy. The locations of two volcanoes, the Flegreian Fields and Mount Vesuvius are located on opposite sides of the city of Naples, which means Naples has a significant volcanic hazard. Scientists have a large number of instruments on both volcanoes so they can warn the citizens of the area of any volcanic activity and advise city officials who may issue evacuation orders. From Murck & Skinner, 2011.

As a result, people who were already impoverished due to natural disaster were vulnerable to the second disaster, the collapse of the landfill that buried over 100 people in rubbish.

1.9 The Role of Government in Vulnerability and Preparing for Disasters

Governments are composed of people. The people in power have choices to make about how to spend the country's money and humanpower. In relatively wealthy countries, disaster preparedness is a function of a network of government agencies, and recovery from disasters is usually fairly rapid and organized. For example, when buildings are built in Japan, they must be engineered to withstand the largest earthquake they are likely to encounter. This is a form of disaster preparedness, and is managed by the construction industry, engineers, architects, and local permitting officials. Community volunteers in some coastal cities are trained to close enormous gates in seawalls to protect communities in the event that a large earthquake generates a tsunami. Government agencies responsible for monitoring the ocean for tsunamis have access to warning sirens and to the media, and they are quick to get the word out about this danger to the public. People are aware of tsunami evacuation routes. The government of Japan has enough money and enough credit with the international community to pay for recovery even after a very large and damaging earthquake. Thus, when an earthquake strikes in Japan, response is rapid. Local and national government agencies set about the work of cleaning up and rebuilding. Any rebuilding that happens occurs to the same (or better) standards as that which existed prior to the earthquake. Though earthquakes can still be deadly and damaging even in a country like Japan, death tolls are typically less than those in developing countries for the same magnitude of event. The country is also capable of recovering quite well, becoming economically stronger than ever within a few years of the event. While every society is vulnerable in some sense to natural disasters, Japan has done everything possible to reduce that vulnerability.

A magnitude 7.0 earthquake struck 7.5 mi (12 km) beneath the city of Kumamoto on island of Kyushu, Japan, in April of 2016. Fifty people were killed and about 3,000 were injured. Water was temporarily disabled in parts of the city, and there were widespread power outages. An estimated 1,000 buildings were seriously damaged, with about 90 buildings completely destroyed. Damage estimates were about $3.2 billion. Disaster preparedness spared many lives and limited damage to buildings and infrastructure. Rapid response meant that people were cared for and sheltered quickly, and efforts were under way very quickly to restore power and water, to clean up rubble, and to assess the structural stability of damaged homes and businesses.

In developing countries, however, disaster preparedness is often less well organized, and even the basic needs of the people during recovery may require the assistance of other nations. Government resources, including revenue that would fund such disaster preparedness and recovery, are meager. Due to high poverty levels, ineffective governments, and poor preparedness, some populations are more vulnerable to natural disasters. Haiti, for example, is not known for having a strong central government focused on the disaster preparedness of its population.

When the magnitude 7.0 earthquake struck 8.1 mi (13 km) beneath the capital city Port-au-Prince in January of 2010, the country was already economically stratified, with many people living in poverty and in substandard housing, and somewhat reliant on aid money from other nations. Over 220,000 people died in the earthquake and about 300,000 more were injured and in desperate need of medical attention. About 1.5 million people were suddenly homeless (Fig. 1.14). At the same time, infrastructure was in shambles, power was out, trees were down, and many of the city's places of business were in ruins. Though people seemed to accept that houses of the poor would be destroyed in such an earthquake, it shook many people that even the National Palace had been destroyed. Immediately, the government requested international aid. Medical teams from many countries assisted in treating the injured. The International Red Cross set up temporary housing and feeding stations. Volunteers searched rubble for survivors and the bodies of those who were not so fortunate. A cholera outbreak hit the island and medical professionals worked to contain it. Billions of dollars were pumped into Haiti in an effort to help its people recover. (Damage estimates are about $14 billion.) Yet even with all of this international help, more than 8 years later people continued to suffer. Many people still lived in tent cities. The Red Cross has been accused of squandering most of its donated funds on internal expenses, failing to help rebuild homes. Hurricanes in 2012 and 2016 killed more people and created additional crises. Though the people of Haiti strive to make their country a better place following such a devastating event, the recovery has been slow and fraught with problems. Assassination of Haiti's president in 2021 further demoralized the public and left the country less prepared than ever to handle another major disaster.

In what other ways can government help or hinder disaster preparedness and recovery? In some countries, there is a general mistrust of government, or at least of agents of the government. Evacuation orders given by corrupt local police forces may be received with suspicion. This results in a failure to convince a population of a very real risk, and therefore a failure to evacuate. In still other cases, mistrust of law enforcement, or perception that law enforcement is absent in a community causes people to stay to protect their homes

Figure 1.14 Utter destruction of homes on a hillside in Port-au-Prince, Haiti, following a magnitude 7 earthquake in January of 2010. In this impoverished country, very few buildings are built to withstand shaking from major earthquakes. As a result, thousands of buildings collapsed, killing over 220,000 people and injuring more than 300,000. From Strahler, 2013.

and property when it would be better for them to evacuate. This is particularly true if crime and corruption are rampant prior to the disaster.

Failure of officials to distribute aid effectively can worsen disasters as well. There was a great deal of criticism of the Ethiopian government in the mid-1980s. A multiyear drought had led to crop failure in many areas of the country as early as 1983. The economy of Ethiopia is highly dependent on agriculture. As a result, much of the tax revenue for the country is derived from agricultural products. Not only did people not have enough to eat, the country's economy was stretched thin from lack of revenue from taxes. Famine began as early as 1983. Unfortunately, the country was in the midst of a civil war and the government made the decision to redirect not only funds, but also food reserves, to the military. This left those in the countryside with no food aid, and no choice but to buy grain in markets. However, grain prices were increasing and food became out of reach financially for many families. The government resettled people in an effort to prevent rebels from gaining a majority in any given area, and cited the famine as a reason for the resettlement. It moved people to large-scale farms and attempted to implement a "famine relief tax" to raise money, but in reality there was little effort to provide food to the rural poor. An international response to the drought did not start until well into the third year of the famine. Large amounts of food were distributed to the Ethiopian government via the United Nations and various international relief agencies. The government, in an effort to quash insurgencies, refused to send food to rebel-held areas. Several countries carried out air drops of food into these areas. Aid money was in many cases redirected to military operations and was used to buy arms instead of food. As a result of the poorly run and callous response to the drought disaster, between 500,000 and 1 million people died of malnutrition and starvation.

1.10 Disasters and Social Change

Natural disasters can be a force for social change. Social changes that are common during natural disasters and their recoveries include dismantling and rebuilding of communities, mass migration of people to new environments, anger over poor responses and recovery efforts, and political change.

In the year 1883, the volcano Krakatau (also known as Krakatoa) located in the Sunda Strait between the two Indonesian islands of Java and Sumatra, erupted. People in Indonesia are no strangers to volcanic eruptions. Indonesia is an island nation and the islands are composed of volcanoes. As a result, volcanic eruptions are commonplace. This eruption, however, was larger and more devastating than any the island had seen in decades. When the eruption began, volcanic ash filled the air and darkened the skies for 3 days. Pyroclastic flows surged down the flanks of the volcano and across the water. Some of these pyroclastic flows reached Java and Sumatra and caused the deaths of hundreds of people unlucky enough to be near the shorelines when the flows arrived. At the peak of the eruption, the volcano tore itself apart and released the loudest sound heard in historic times. The explosion was heard more than 3,000 mi (4,800 km) away in the Indian Ocean. With the destruction of the island came a large tsunami that hit the shores of both Java and Sumatra. Approximately 36,000 people died in the tsunami making it the deadliest such wave in history until the 2004 Boxing Day tsunami in the Indian Ocean.

Indonesia at this time was ruled by a colonial Dutch government. The Dutch were governors of the islands and controlled much of the land and trade in the area. When the eruption was over, the Dutch government saw to the needs of the minority Dutch population, but largely ignored the indigenous Indonesian population. The people became very hostile toward the Dutch governors and began small-scale resistance to the colonialists. This was the beginning of a serious movement to oust the Dutch from Indonesia and regain local control of the nation. Though independence took several decades, anger over the response to the disaster was the catalyst that led to social change and eventually to the Indonesians expelling the Dutch from the islands.

In the mid-1800s, a fungus attacked potato crops in Ireland. Because potatoes were the main food and seed crop in Ireland among the lower classes, the fungus infestation meant that both the food and seed crops were destroyed. A famine ensued, and people were unable to feed their families or replant new crops. They had little choice but to move from their homes to a new environment. Many Irish emigrated to England, Scotland, and the United States. In the United States, they were shunned by many and looked down upon as second-class citizens invading the country. However, as with every wave of immigrants to foreign lands fleeing famine, economic hardship, war, or persecution, they formed communities, began contributing to their new countries, and eventually became an integral and celebrated part of their new societies.

1.11 Community Response

A great deal of research is ongoing into how people behave before and after disasters. We have already discussed people taking calculated risks, but what happens when a natural event is threatening to strike in the short term? There are certain behavioral patterns that many people fit into. Let us examine those.

1.11.1 Before the Disaster

When a natural event is threatening a specific area, it is the job of authorities, often under the coordination of an office of emergency management, to warn people of the dangers they face. Most people living in the Midwest are familiar with tornado sirens, usually tested once each month. When the siren sounds, people generally know that a tornado is in the area and they should take shelter. What people do in reality though may surprise you. Some people do, of course, heed the sirens' warnings and seek shelter. Others will try to call authorities to verify the threat is real. Many more will check the Internet for the most up-to-date reports, weather maps, and other primary sources of information. People will go to their windows to see what is going on outside, to judge for themselves whether the situation is bad enough for them to take shelter (Fig. 1.15). About a quarter of the time, people are warned "unofficially" by friends or family who call them to ask if they know about the danger and if they've taken shelter.

When a dangerous natural event threatens, people overwhelmingly respond with disbelief, or do everything in their power to rationalize why they are not in danger. Some people can be convinced to take shelter, evacuate, or otherwise take appropriate action only after verifying the threat is real or being invited to shelter at a friend or family member's house. Research has led emergency managers to understand that in order to convince people the threat is real, they need to convey specific information to targeted audiences and give express instructions for what they should do. What is it that is threatening? Who is being threatened? Which geographic areas are affected? When is this threat coming? What protective action should be taken? If competing television or radio stations have conflicting information, people will almost always choose to believe those who downplay the dangers.

Despite rhetoric to the contrary and the wishes of emergency managers, there really is no such thing as a mandatory evacuation. If people choose to stay, it is unlikely they will be arrested. In fact, it is unpopular and possibly unconstitutional to do so. It is in the emergency manager's best interest to convince them to leave. In any given disaster, between 10% and 25% of people will refuse to evacuate without

Figure 1.15 A tornado makes its way across the landscape in Nebraska in 2003. Rather than take shelter, the photographer went outside to assess the threat and take photographs. From Murck & Skinner, 2011.

extremely convincing arguments and assistance to those who need it. Prior to Hurricane Katrina, Director of Homeland Security and Emergency Management for St. Bernard Parish tried everything he could think of to get people to leave. He finally told people during a press conference, "We strongly recommend that you leave now because I don't believe I have enough body bags to cover the people that stay." Law enforcement personnel in other parishes of Louisiana used language such as "Well, if you're not going to leave, can I have the contact information for your next of kin?" clearly conveying the gravity of the situation.

In one parish, the evacuation strategy was very effective in the few days before Hurricane Katrina. In Plaquemines Parish, leaders issued a mandatory evacuation warning on Saturday, 27 August, when landfall was expected on Monday, 29 August. Deputies helped direct traffic in town that day during the initial waves of the evacuation. The next day, they went door-to-door to warn people to evacuate and identify those who needed help doing so via public transportation, pickup at home, medical evacuation, or other means. They were able to achieve a 97% evacuation rate. This evacuation rate was unheard of, and it probably saved hundreds of lives.

Some areas of the world have what emergency managers refer to as a *disaster subculture*. In these areas, natural events happen with enough regularity that people have a heightened awareness and almost a routine to what they do when this particular event threatens. This is true in communities threatened by tornadoes. The sirens are tested every month, and usually in the springtime the National Weather Service, local media, schools, Red Cross, and perhaps emergency managers work together to remind people what to do to keep themselves safe when spring storms start rolling through. The same is true of hurricanes in south Florida. Each year in the late spring, newspaper supplements prepared by the National Hurricane Center are dropped on the doorsteps of people across the region, reminding them of what to do if a hurricane threatens. Though many longtime residents know this information, it is always a good idea to reinforce this information with them, and provide it for the first time to new residents.

Other simple measures can help people decide what to do when a disaster threatens. In 1976, there was a flash flood in a steep-walled canyon just outside Loveland, Colorado. The Big Thompson Canyon Flood started as a heavy rainfall event in the areas above the canyon. Water flowed downhill, eventually making its way into the canyon. People were warned at the last minute to evacuate. They did what seemed most logical at the time. They jumped in their cars and tried to head up the road, out of the canyon. Unfortunately, the flood had washed out the road, and many people died when they were trapped in their cars in ravines cut by the floodwaters. More recent education campaigns have instructed residents to climb to safety, directly up the walls of the canyon. By simply climbing upward along the walls of the canyon, they can vertically escape the floodwaters and find safety.

The same strategy is employed in airports across the United States. Since it is impossible to train every person who flies through an airport what to do if a tornado strikes, most airports have prominent signs near the restrooms indicating that these are tornado shelters. By placing signs in plain view, the hope is that if severe weather strikes, people will remember where they saw the signs and seek shelter in the restrooms.

In several countries that have a tsunami risk, signs placed near the shoreline clearly indicate that you are entering a tsunami hazard zone. Nearby signs indicate tsunami evacuation routes. Again, the hope is that displaying these signs prominently on the roads and highways near the coasts will jog people's memories in the case of an actual tsunami, reminding them that they may be in harm's way and can follow a particular road to safety.

The same type of sign is now being posted in areas around Mount Rainier in the state of Washington, except the context is different. Mount Rainier is prone to mudflows after heavy rainfall, and these have the potential to impact populations around the volcano. There are now signs in the area that call attention to volcano evacuation routes.

1.11.2 After the Disaster

Disasters are by their very nature jarring events. People must adjust to a new reality almost instantly, and predictably this causes distress. After disastrous events, there will often be short-term shifts in behavior. Divorce rates may temporarily change. Alcohol and drug use may escalate. Marriage rates may temporarily decrease as people focus on putting their lives and homes back in order. Suicide rates may increase temporarily. One major risk factor for suicide is post-traumatic stress disorder (PTSD). If people do not receive the psychological care they need in the days and weeks after the disaster, PTSD is much more likely, and there is an increased risk of suicidality. Despite the trends discussed above, disasters actually tend to bring families and friends closer together, at least in the short term.

1.11.3 Creating Resilience in Disaster Survivors

Mental health professionals are often called upon to respond to natural disasters. The Red Cross, for example, has an entire Disaster Mental Health corps within the agency's Disaster Cycle Services. In the immediate aftermath of a disaster, it is important to make sure people are treated for their physical injuries, but it is also important to assure they are armed with the strength to begin recovery of their emotional and social lives. A disaster mental health approach focuses on what just happened and what can be done now to look forward into the future. The focus is on short-term help and assisting the individual to achieve a frame of mind that will allow him or her to remain calm, helping in his or her own recovery. An approach to this mental recovery has been developed by psychologists and is known as *Psychological First Aid* (PFA). Psychological First Aid is meant to be a short-term response to emotional trauma, and its goal is to make people feel safe and empowered to help themselves in the aftermath of the disaster. While PFA is not a substitute for more in-depth counseling, it often helps individuals start the recovery process and allows them to be able to assist relief workers in helping them and their families. Often, a brief set of cognitive behavioral therapy sessions after PFA is sufficient to get people past the initial PTSD and anxiety and on the road to emotional recovery. It is not necessary for those who provide Psychological First Aid to be licensed psychologists. Even volunteers can be adequately trained to carry out this vital role in disaster recovery.

The major components of Psychological First Aid (Young & Ruzek, 2012b) are:

1) Contact and engagement. Locate survivors and engage them in conversation.
2) Provide safety and comfort. Explain to a survivor that they are safe. The goal is to communicate to the survivor's brain that it is no longer necessary to be in "fight or flight" mode. You can show them what you've done to make the area safe, explain to them what you've done to remove them from danger, shield them from disturbing scenes, or point out that the danger has passed.
3) Achieve stabilization. Calm yourself first. Speak and act calmly so the survivor can start calming themselves. Emphasize the present, the practical, and the possible.
4) Gather information. Make sure you know who the person is, what family members, friends, pets, and other loved ones were with them, and if they know everyone's whereabouts and status.
5) Practical assistance. Help the survivor to be an active part of their own rescue, and allow

them to help care for themselves and their companions. Help them get cleaned up, get a fresh set of clothes if necessary, get treatment for wounds, and a meal if desired. Involve them in decisions about their care. This helps them overcome feelings of helplessness and gives them a purpose and a "place on the team." Recognize and remind people of their strengths.

6) Create connections. Build a relationship with the survivor. Use their name. Build rapport. Connect the patient with family, friends, and pets, if possible.
7) Provide information on coping and support. Involve the survivor in talking about realistic next steps. Maintain and communicate hope.
8) Link to collaborative services. Help the survivor locate services he/she needs, such as Red Cross for food and shelter; federal and state officials to file paperwork for aid, their insurance company; and a psychologist, counselor, or social worker for additional counseling sessions.

Survivors of disasters are not the only people who need psychological and emotional care following major events. First responders often suffer from similar levels of emotional distress and its effects. In some cases, they may themselves be victims of the same disaster to which they are responding, yet they have to leave their families behind to deal with problems at home while they help the wider community. This separation causes distress both in the first responder and the family, and these people need support as much as anyone else in the community. Being a first responder is a stressful job even under normal conditions, and they are not immune to the stress of responding to a disaster. Psychological First Aid should be practiced with responders as well.

In conclusion, natural hazards are ubiquitous around the world. In any given month, there may be severe weather resulting in a tornado outbreak in Alabama, a hurricane in the Lesser Antilles in the Caribbean, a heat wave in Spain, a major earthquake in New Zealand, flooding in China, and a wildland fire in northern California. Each of these events leads to cascading effects, which may cause these natural events to turn into disasters very quickly. The magnitude of the natural event and preparedness of the population determine whether a natural event turns into a natural disaster. Throughout this book, we will be talking about the causes and effects of natural events, and ways we can mitigate their effects to protect life and property.

1.12 Chapter in Review

Natural disasters strike every country on the planet and billions of dollars in property are lost due to those natural disasters. They can cause loss of life and homelessness as well, affecting a population for years as it struggles to repair, rebuild, and recover. The focus of this book is on the major categories of natural events that occur with some regularity (earthquakes, tsunamis, volcanoes, landslides, floods, severe weather, disease outbreaks, and meteor impacts) and how we can mitigate the effects of those events to try to prevent the events from turning into disasters. We also spend a chapter on climate change. Though climate change itself is the result of human activity, the effects include more and larger natural disasters.

Three terms that will be used throughout the book are *hazard*, *risk*, and *mitigation*. A natural hazard is any natural event that has the potential to impact a population. As an example, floodplains present a hazard of being inundated with water during heavy rainfall. Risk is the likelihood that a natural event will affect a population. The risk to a population depends on the magnitude of a rainfall event and how many people are living in the floodplains. Mitigation is any action we can take to minimize the impact of a natural event. One form of mitigation is to zone floodplains as agricultural or parkland so people do not reside in those areas.

The Earth is composed of four interacting systems. One of these is the geosphere (also known in some circles as the lithosphere), which is the solid Earth. A second is the hydrosphere, which includes the Earth's water: lakes, rivers, glaciers, and groundwater. A third is the atmosphere, which includes the air and water vapor above the Earth's surface. The fourth is the biosphere, which represents all life on Earth. In most cases, an event or disaster in one realm of the Earth system has a ripple effect, meaning that other parts of the Earth system are also affected. There are also extraterrestrial hazards to the Earth, in the form of meteors that fall through the atmosphere, exploding above the Earth's surface, and perhaps impacting the Earth. Even small meteors can cause a great deal of damage.

Natural disasters are usually predictable, but they are not always forecastable. We know, for example, which areas of the Earth can expect earthquakes. However, we have no ability at this point in our scientific understanding of the Earth to forecast these events. We know they are bound to occur sooner or later in those areas, but we as scientists cannot issue a forecast telling a population to prepare for an earthquake of a specific size and depth at a specific time on a specific day. Some events, however, can be forecast with a great deal of accuracy several days in advance. For example, scientists at the National Weather Service can forecast severe thunderstorms in their region, tell members of the public when and where those storms will form, and whether they are likely to produce severe thunderstorms and even tornadoes. As that time approaches and storms start to form, they can use weather radar to determine where tornadoes are forming and give people several minutes of warning before a tornado strikes.

Geologist Wendy Bohon (Associated Press, 2021) stated, "I think it's important to recognize that there's no such thing as a natural disaster. What you have is a natural hazard that overlaps with a vulnerable system." Unfortunately, not all populations have the same level of risk. Some populations are more vulnerable than others because of poverty, the history of the society and culture, and the overall level of preparedness of that population. In general, poorer communities are at greater risk than wealthier ones. This is due to several factors, including lack of enforcement of building codes, location of their communities in less safe areas, lack of government help during evacuation efforts, and general lack of organized preparedness projects and information. This has been demonstrated in many areas of the world where substandard buildings collapse during earthquakes, people in floodplains have drowned because of flooding, and those in areas at risk of being left behind during evacuations because they do not have transportation or money to evacuate. Little will likely change unless governments recognize that these populations are at increased risk and expend resources to help these communities plan and prepare for the inevitable events that will threaten them.

Since natural hazards are everywhere, each family must determine what risks they are willing to live with. Factors that people should consider include whether the hazards in their area are forecastable, and whether they have enough information to keep their families safe if the event occurs. In Florida, a family must come to terms with the fact that hurricanes are common occurrences. They will receive several days of warning before the hurricane strikes, so they can prepare their home to minimize damage and develop an evacuation plan. Other forms of preparation may include hurricane and flood insurance, and a set of friends that live out of harm's way that they can stay with until the hurricane is over and they can return home. The government also helps people in Florida prepare for hurricane season by broadcasting and printing information about hurricane preparedness to the people in that area well before hurricane season.

In some cases, once a natural disaster strikes, it becomes an impetus for social change. In 1883, an eruption of Krakatau volcano in

Indonesia and the Dutch response to that eruption set a series of events in motion that eventually resulted in the Dutch relinquishing their governance of the island nation. The Potato Famine in Ireland led to emigration of people to the United States and other nations. Even the response to Hurricane Katrina in the United States led to the migration of people across the United States. The socioeconomic status of people evacuated to other cities throughout the United States was a large factor in whether they were able to return home to New Orleans.

People may respond to the threat of a natural event in many different ways. Some will prepare diligently. Some will seek shelter as soon as officials advise that course of action. Still others will try to verify that the threat is real before seeking shelter, or will refuse to leave their homes to seek shelter because they worry about the security of their homes or the well-being of their pets. Some will decide to ride out a hurricane at home because they survived a previous storm. Response to threats is generally better if the public is educated and well-informed about the risk and mitigation strategies, if they see reminders often of safe routes or shelters, and if they feel that their homes will be protected. Some members of vulnerable populations may also need help when a threat presents itself because they may not have the financial means or mobility to help themselves.

In the days following a disaster, aid workers from organizations such as the Red Cross and Salvation Army are likely to be called to the scene to help survivors and first responders. One effective method of helping people get on the path to recovery is called Psychological First Aid. This is a set of actions a relief worker can take to reassure a survivor they are out of immediate danger, make sure they have their immediate needs for shelter and food met, and guide them to the resources they need to begin their recovery. First responders also must deal with the trauma of the event, both because their own families are affected, and because they often see terrible things during the rescue and recovery missions they are involved in. As a result, first responders are often in need of Psychological First Aid as well.

Part I

Geologic Events as Hazards

2

Plate Tectonics

Bethany D. Hinga
University of Nebraska at Kearney, USA

2.1 Introduction

Plate tectonics is a framework through which geologists view the world. It is rooted in an idea that dates to the earliest accurate maps of the Atlantic and the continents on either side of that ocean. Plate tectonics was not fully developed as a theory, however, until the early 1960s. As such, it is one of the youngest of the important geologic concepts to be developed. It is no exaggeration to say that it brought about a revolution in the Earth sciences. In broadest terms, plate tectonics explains the existence and location of oceans, mountain ranges, volcanoes, and earthquakes. Plate tectonics tells us that the Earth's outermost layer is divided into several large pieces called plates. These plates move very slowly, colliding with each other, moving past each other, and moving away from each other. It is these interactions between plates that have created the Earth's surface as we see it today, including the current configuration of continents and oceans. Boundaries between plates are where we most often see earthquakes and the creation of magma to feed volcanic eruptions.

2.2 Continental Drift

The idea that the Earth's continents are not stationary but move (or drift) quite substantial distances is not new. As early as the mid-1700s, when the first accurate maps were drawn of coastlines on either side of the Atlantic, people noted that the coastlines of eastern South America and western Africa looked like they would fit together (Fig. 2.1). However, the first serious scientific investigation of this idea was not until the early 1900s, by a German scientist named Alfred Wegener. Wegener was trained as an atmospheric scientist and an astronomer at three prominent universities in Germany. After he graduated from university, he worked with his brother at an astronomical observatory for a number of years. The two of them had an interest in hot air balloons and together completed what was then the world's longest balloon flight, during which they tested an instrument used for navigation. Wegener then moved on to work in the atmospheric sciences, publishing a book on atmospheric physics. He became interested in polar exploration, completing a number of expeditions to Greenland.

In his work as an atmospheric scientist, Wegener spent a great deal of time looking at maps and reading scientific literature. Like many before him, he noticed that the continents of North and South America looked as if they would fit very snugly against Europe and Africa. His keen interest in geology and geophysics led him to start a survey of geologic literature to see if any evidence existed that the continents were

Earth's Natural Hazards and Disasters, Advanced Textbook 7, First Edition. Bethany D. Hinga.

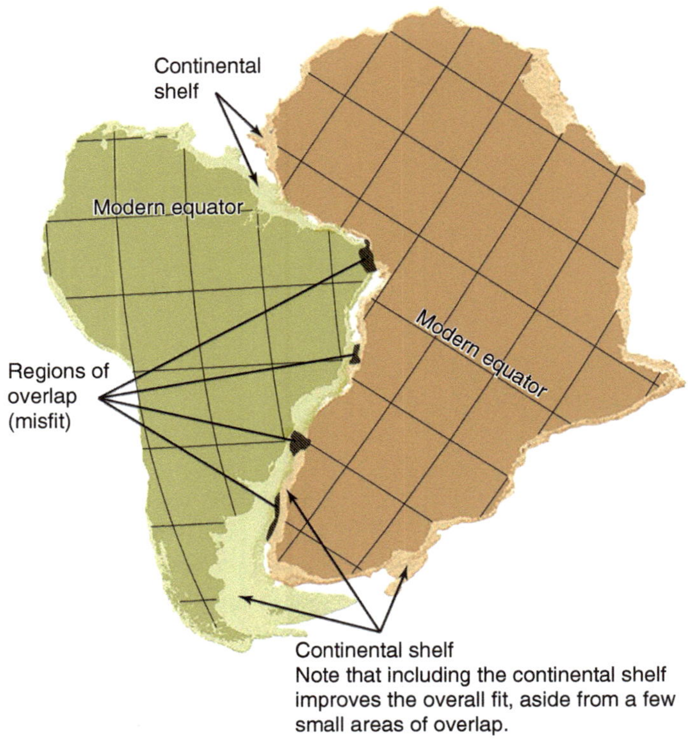

Figure 2.1 Fit of the continental shelves of South America and Africa. From Murck & Skinner, 2016.

in fact joined in this manner at some point in the geologic past. What he found was compelling.

He started with geologic maps as they existed at that time. He found if he used geologic information from all continents, he had evidence they were part of one large landmass, which he called *Pangaea* (Fig. 2.2). Pangaea is Latin for "all lands." He noted if the land masses on either side of the Atlantic converged, mountains of a similar age on North America, Greenland, Europe, and Africa combined to form a single long, continuous mountain range (Fig. 2.3).

The field of paleoclimatology (the study of ancient climates) added evidence to Wegener's case for the existence of Pangaea. Evidence of glaciers can been found in southern Africa, South America, India, Australia, and Antarctica. This evidence consists of grooves worn into rock outcrops by stones on the undersides of glaciers, as well as deposits of glacial debris. Today these regions are separated by thousands of miles, and many are in tropical or subtropical areas. These are not places where we would expect to find glaciers today. Reconstructions of Pangaea show that all of these continents were joined at one time with a continuous ice sheet covering them, and this land mass was centered on the South Pole (Fig. 2.4). This explains why evidence of glaciers from the past are found in these areas today.

Paleontology (the study of fossils) had yet more to contribute to Wegener's reconstruction of Pangaea (Fig. 2.5). Approximately 300 million years ago, life on Earth looked quite different from the life of today. One type of plant that was quite widespread during this time was a seed fern called *Glossopteris*. While today's ferns reproduce with spores, the now extinct seed ferns reproduced with seeds and did not contain spores. The seeds with which *Glossopteris* reproduced were large and heavy, unable to float. *Glossopteris* was found across all of the southern continents in a wide, continuous band across Pangaea. Had Pangaea not been

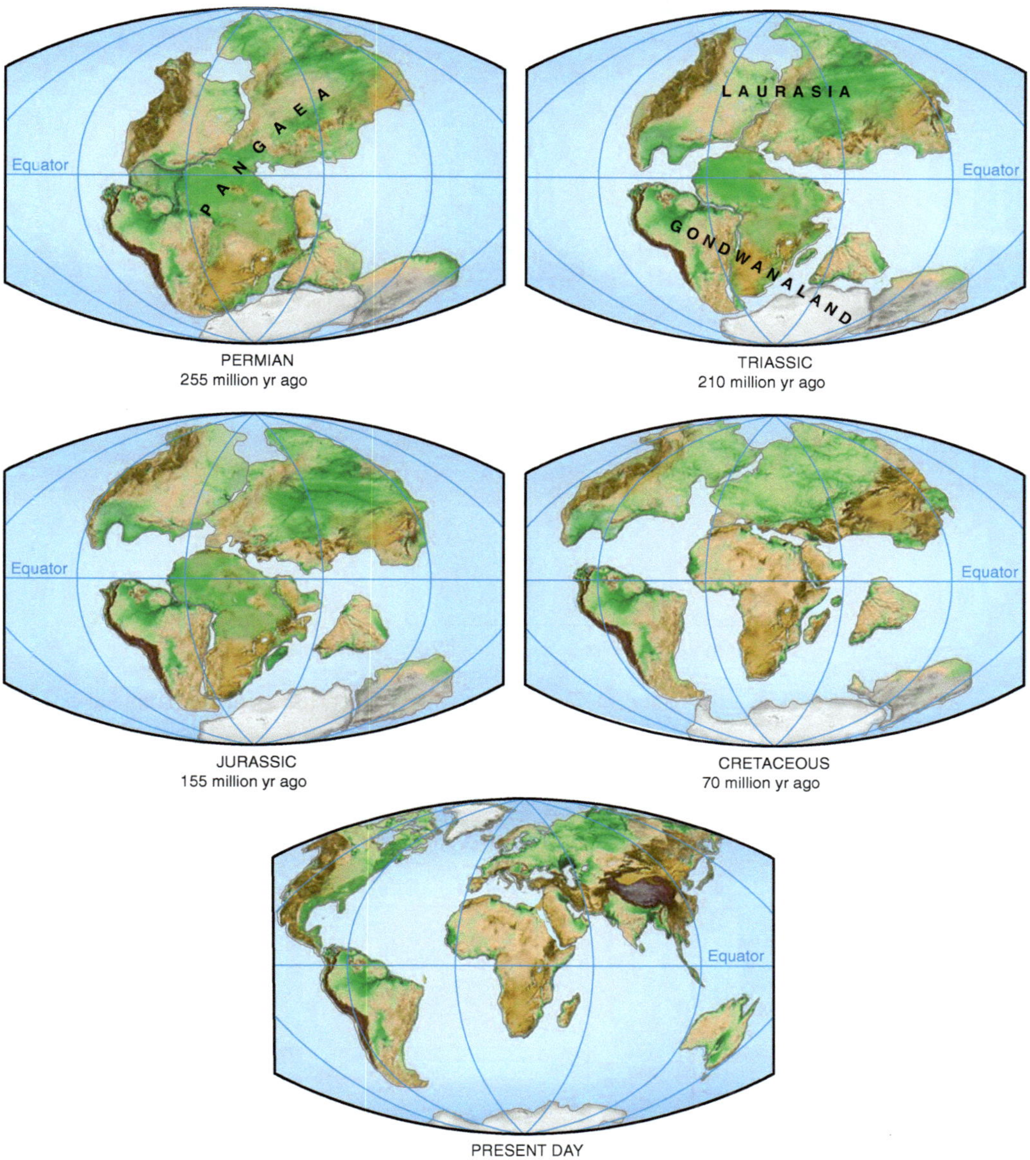

Figure 2.2 Wegener's reconstruction of Pangaea and the drift of continents from 255 million years ago to today. From Fletcher, 2014.

in its presumed configuration, the seeds would have needed to be blown by the wind across an ocean, or floated on ocean currents in order to spread the plant to other continents. Due to the weight of the seeds, scientists do not believe either of these scenarios is possible. The plants must have spread as seeds were dropped on the ground while the continents were part of Pangaea.

Similarly, fossils of two land reptiles called *Lystrosaurus* and *Cynognathus*, and a freshwater reptile called *Mesosaurus* are found across all of the southern continents. These reptile fossils are found in rocks that are approximately

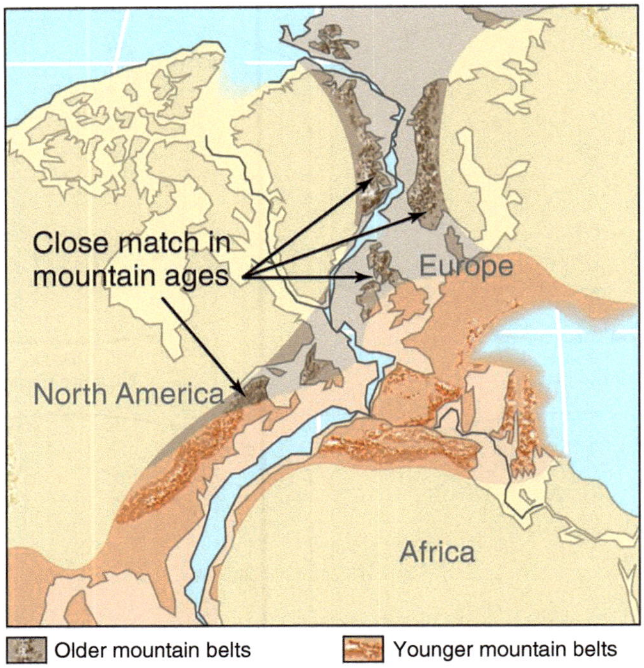

Figure 2.3 Mountain ranges on several continents line up when Pangaea is assembled. From Murck & Skinner, 2016.

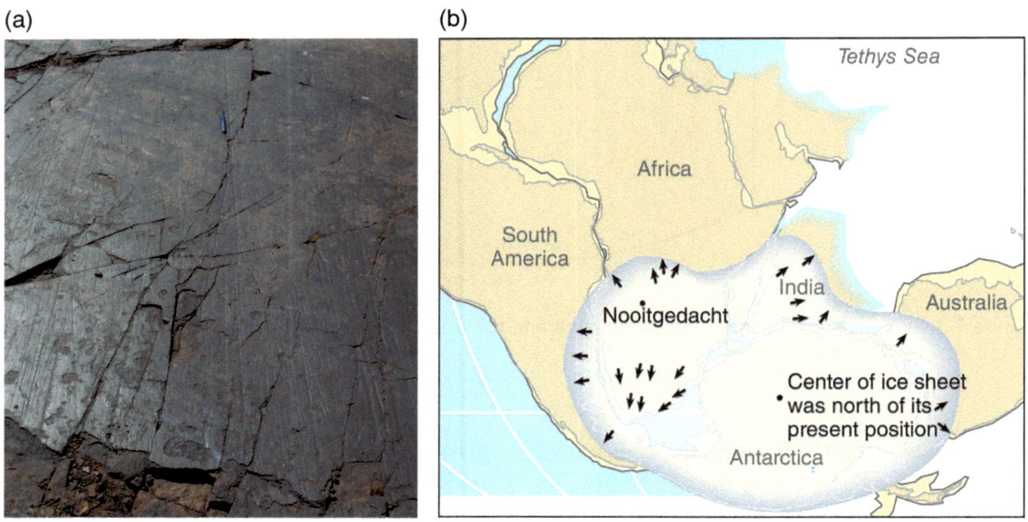

Figure 2.4 (a) Grooves left in rocks in South Africa. These grooves were carved into the rocks when glaciers covered South Africa during the time Pangaea was assembled, roughly 250–300 million years ago. From Murck & Skinner, 2016. (b) Polar ice sheets covered many areas of Pangaea that today would not logically have glaciers. From Murck & Skinner, 2016.

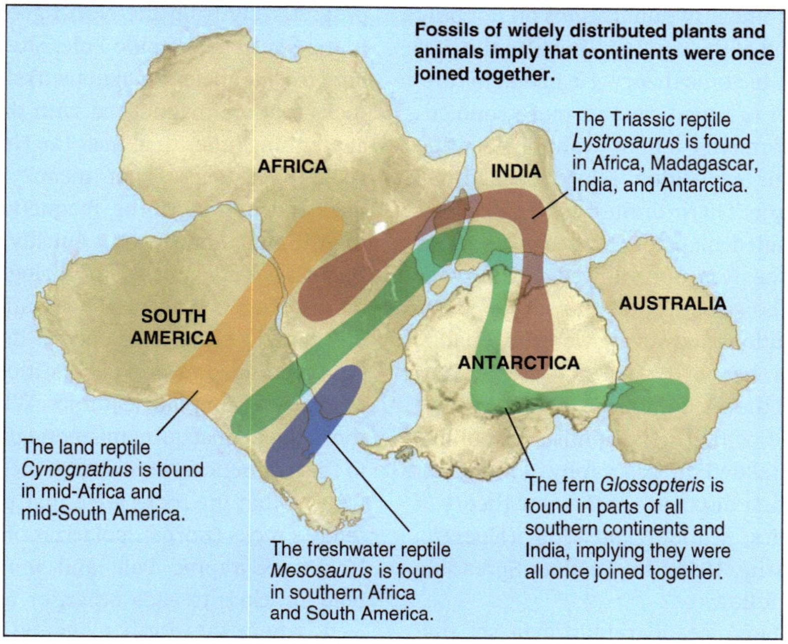

Figure 2.5 Distribution of plants and animals from the time when Pangaea was assembled. From Fletcher, 2014.

300 million years old. As land-based reptiles, they could not swim across oceans to move from one continent to another. Instead, scientists believe that the southern continents were merged together about 300 million years ago and there was no need for migration of these animals across an ocean. Thus, Pangaea must have existed about 300 million years ago.

The first question that likely comes to mind is, "How did we get from a single continent of Pangaea to the seven continents we have today?" This was the same question asked during Wegener's time. Wegener published a book, *The Origin of Continents and Oceans*, in 1915. In his book, he detailed arguments from many areas of geology and geophysics that convinced many that Pangaea had, in fact, existed (Wegener, 1966). However, he could not answer the fundamental questions about mechanisms that might cause a continent to break up and move across the Earth. Wegener kept investigating this idea for the rest of his life, and published four subsequent editions of his book (Wegener, 2011), incorporating an ever-growing body of evidence in support of his idea. However, his critics were not satisfied with unproven hypotheses about how continents could make their way from one location to another, seemingly crossing entire oceans. Unfortunately, Wegener died on an expedition in Greenland in 1930 at the young age of 50. He did not live to see his idea become accepted theory across the geologic community.

2.3 New Data: Magnetism

It was not until the 1950s that additional data were gathered that began to shed light on those elusive mechanisms of breaking up a continent and moving continents across the Earth. The United States military commissioned detailed maps of the ocean floor. Many scientists took part in the surveys, but one of the best known was Harry Hess (Hess, 1960). He commanded several of the mapping expeditions and authored a number of publications on not only the geography of the sea floor, but also the first parts of plate tectonic theory. He had at his disposal a team of scientists and more sophisticated instruments than any scientific oceanographic expedition prior to that time. Scientist Marie Tharp drafted his results into the first detailed maps of the ocean floor. As a woman, she was not allowed to sail with the rest of the scientists on the survey ships. Her contributions, however, were invaluable. Not only was she the first person to create accurate maps of the ocean floor (in 1959), but she was also one of the first scientists to publish articles that related the geography of the ocean floor to the first discovery of the new theory of plate tectonics: seafloor spreading (Heezen., Tharp, & Ewing, 1959); see Seafloor Spreading section that follows).

The first new oceanographic data set that was collected by scientists was that of ocean floor magnetism. The Earth's magnetic field is not stationary. Instead, the magnetic poles are in constant motion. In fact, the Earth's magnetic field changes so quickly that magnetic maps of the Earth are redrawn about every 5 years. In the short term, the magnetic north pole wanders within several degrees of the geographic north pole. In the longer term, and at uneven intervals from a few thousand years to a few million years, the magnetic poles actually flip polarity. This means magnetic north switches from geographic north to geographic south. We do not fully understand how the magnetic poles flip because it has not happened in the modern era. It is possible that the magnetic field loses strength until it is nonexistent, then becomes stronger but in the opposite magnetic polarity. In other words, the magnetic field gets weaker in its current configuration where the north magnetic pole is near the North geographic Pole, then after it fades to zero strength, it reemerges reversed, where the north magnetic pole is aligned with the South geographic Pole. The other possibility is that the north magnetic pole remains the same strength, but moves across the Earth progressively from the North geographic Pole to the South geographic Pole. Studies of *paleomagnetism* (ancient magnetism) show very few rocks that are magnetized with the north pole at anything other than near the North or South geographic Poles. That means if the latter mechanism is how the magnetic poles shift, then it must happen very quickly, geologically speaking. Very quickly in geologic terms is a matter of a few years to a few hundreds of years.

Magnetic maps of the ocean floor are quite enlightening, and provide additional evidence to the theory of plate tectonics. When we study rocks that can acquire magnetic properties due to the magnetic minerals contained in the rock, we find that the rocks are either polarized like today's rocks (normal polarization), where the North geographic Pole and north magnetic pole are close to each other, or reverse polarized, where the north magnetic pole is near the South geographic Pole. In the 1950s, the rocks of the central Atlantic were mapped with respect to both their age through dredging of ocean-floor rock samples and chemically dating the rocks, and their magnetic polarity measured with magnetometers towed behind ships. What the scientists found was fascinating (Fig. 2.6). The youngest rocks were found at the centers of the oceans, and the oldest rocks were found on the edges of the oceans, next to the continents. The rocks on the ocean floor are volcanic, which is an interesting observation. In addition, the magnetic signature of the ocean floor showed that the very youngest rocks in the center of the ocean exhibit normal polarity. On either side of those rocks of normal polarity are slightly older rocks with reverse polarity. Just on the other side of those older rocks with reverse polarity were still older rocks with normal polarity. Both the age and polarity of rocks seemed to be symmetrical about the center of the ocean. Further surveying of the depth of the ocean floor showed that those youngest rocks in the center of the ocean were found at the center of a broad underwater mountain range. The pieces of a new theory were falling into place.

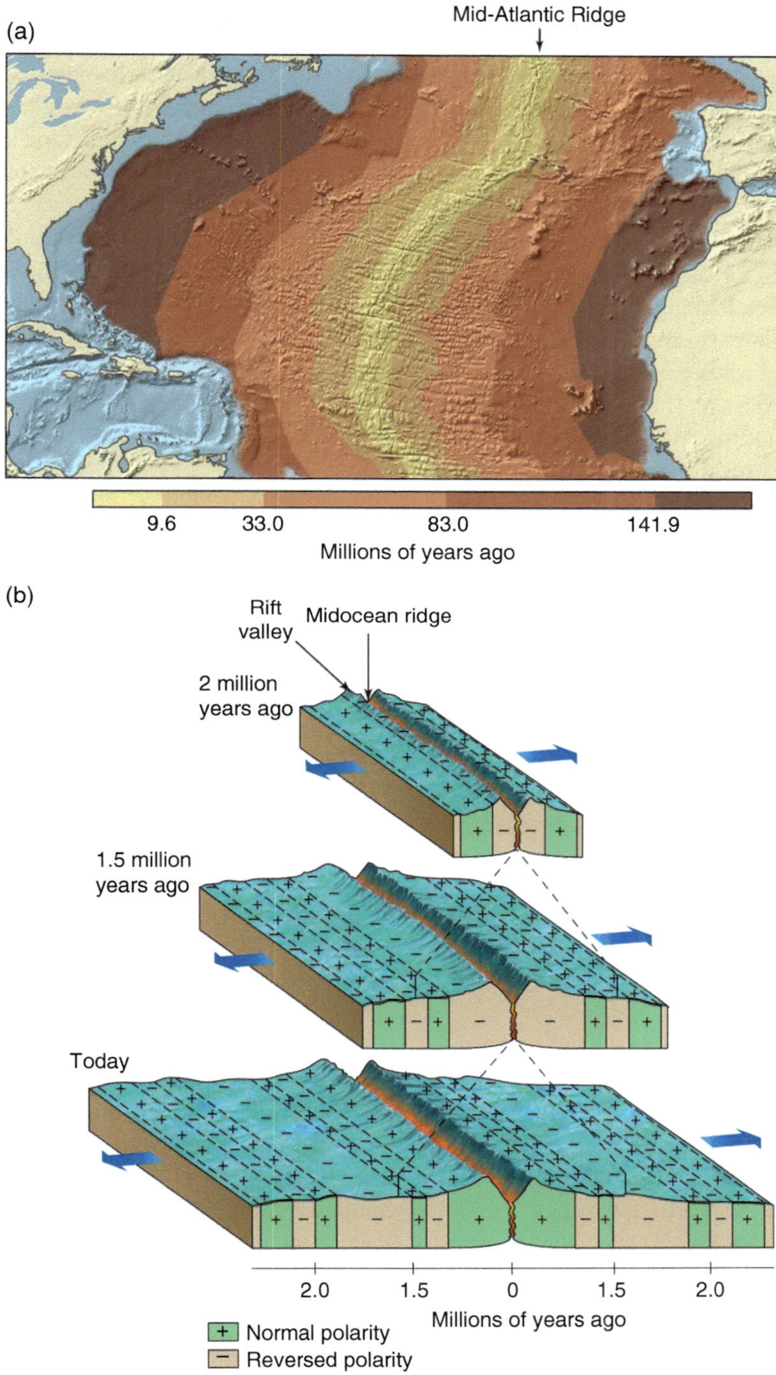

Figure 2.6 (a) The age of the ocean floor is youngest in the center, and oldest on the outer edge. (b) The mechanism for formation of magnetic stripes on the ocean floor due to seafloor spreading. From Murck & Skinner, 2016.

2.4 Seafloor Spreading

When scientists Marie Tharp and Harry Hess analyzed the data they had collected throughout the early 1950s, they brought the information together into a theory called *seafloor spreading*. The first publication about seafloor spreading came out in 1960 (Hess, 1960), and described the mechanism for moving continents that Wegener had been looking for. It explains how continents can move from one part of the Earth to another, and how oceans can form between them (see also Heezen, Tharp, & Ewing, 1959).

In the center of the Atlantic Ocean is a volcanic mountain range. At the center of that mountain range, lava is erupted and flows down both sides of the mountain more or less symmetrically. The two sides of the volcano are in constant motion, pulling away from each other. This means the mountain, and therefore the ocean, is splitting in its center, and getting wider by tiny amounts every year. Today we know that the average rate of seafloor spreading varies greatly depending on the ocean, but about 1/4 in./year (5 mm/year) is a good approximation of the rate of spreading worldwide. If the ocean is getting wider every year, this means the continents on either side of the Atlantic are growing farther apart every year. Thus the continents appear to be moving across the surface of the Earth.

2.5 Plate Tectonics

Seafloor spreading was the first part of the modern theory of plate tectonics to fall into place. While Wegener's evidence supporting the existence of Pangaea is accepted even today, his idea of continental drift did not rise to the level of a theory because he got some parts of the idea wrong. He imagined mobile continents plowing their way through stationary ocean basins. Other proponents of continental drift had equally bizarre ideas about how continents broke apart and moved. Plate tectonics is a theory that not only validates Wegener's ideas about Pangaea, but also expands upon his work to explain processes that have been acting within the Earth since early in its history. From its humble beginnings with Wegener's publication in 1915, to the groundbreaking idea of seafloor spreading, to its full development as a theory in 1963, plate tectonics was revolutionary. Plate tectonics is now a universally accepted theory in the geologic community. In order to fully understand plate tectonic theory, we need to start with Earth structure.

2.6 Earth Structure

The Earth is composed of layers (Fig. 2.7). Some layers are defined based on composition of material. Based on composition, the Earth is composed of three layers. The outermost layer is called the *crust*. In general, minerals that compose the crust are less dense than those that make up the rest of the Earth. There are two types of crust, one continental and the other oceanic. As you might expect from the names, continents tend to be composed of continental crust, and the ocean floors are composed of oceanic crust. Continental crust is a bit less dense than oceanic crust. It is thicker than oceanic crust, and sits at a higher elevation than oceanic crust. The continental shelves, which are shallow areas of the ocean adjacent to continents, are transition zones from one type of crust to the other. Beneath the crust is a layer called the *mantle*. The Earth's mantle is composed of medium-density minerals and is solid but flexible. The third layer of the Earth is called the *core*, which is composed of metals. Geophysical modeling suggests about 94% of the core is composed of iron. Most of the remaining material in the core is nickel.

Another way of looking at the Earth reveals a different structure. Some layers are defined based on how they behave when energy from earthquakes moves through the planet. This energy is called seismic energy, or in more common terms, *seismic waves*. Seismic waves

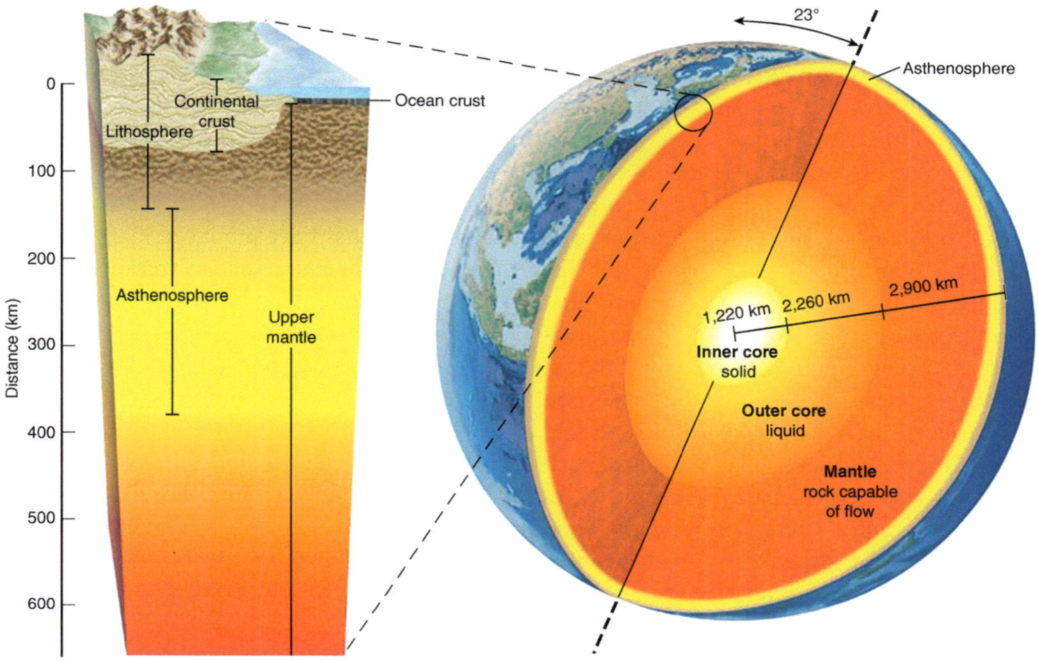

Figure 2.7 Structure of the Earth. The Earth is divided into three layers based on composition: crust (continental and oceanic), mantle, and core. Behavior of the Earth with respect to energy passing through it indicates five mechanical layers: lithosphere, asthenosphere, mesosphere (remainder of the mantle), outer core, and inner core. From Fletcher, 2014.

reveal five mechanical layers to the Earth. The top mechanical layer of the Earth is called the *lithosphere*. It is composed of the crust and the uppermost part of the mantle, which are welded together into a single unit. We describe continental and oceanic lithosphere as it exists beneath the continents and oceans, respectively. Plate tectonic theory tells us it is the Earth's lithosphere that is broken into plates. Some plates are composed primarily of continental lithosphere. Others are primarily oceanic lithosphere. Almost all plates contain some percentage of both.

Beneath the lithosphere is an unusually pliable part of the mantle called the *asthenosphere*. It is convection, or circulation, within this layer that drives the motion of lithospheric plates above. Though it is solid, it circulates slowly. This circulation is a process called *convection*. Hot material from deeper in the asthenosphere rises because it is less dense than the cooler material above it. Cooler material toward the top of the asthenosphere sinks because it is more dense than the hotter material below it. Convection, though it happens slowly, is real and consequential in our story of plate tectonics. The asthenosphere circulates fast enough to keep the plates of the lithosphere in constant motion.

The lower part of the mantle is called the *mesosphere*. It is composed of the same minerals as the asthenosphere in the upper mantle, but it is more rigid than the asthenosphere. There is evidence that it also circulates like the asthenosphere, but at a much slower rate. It is difficult to image this part of the Earth because of its great depth, but computer modeling of the mantle suggests convection below the asthenosphere is necessary to explain the structures we see in the lower mantle.

The final two mechanical layers of the Earth are located within the compositional layer of the core. The *outer core* is easily distinguished from the *inner core* because the two parts of the core exist in different states of matter. The outer core is liquid, and the inner core is solid. Danish seismologist Inge Lehmann discovered that the Earth had a liquid outer and solid inner core in 1936. Certain seismic waves are unable to travel through liquids. Those seismic waves were not traveling through a region of the Earth. This was strong evidence that the Earth has a liquid layer. She also found that some seismic waves traveled a path that indicated they encountered a solid material in the very center of the Earth. She theorized the existence of a solid inner core inside the liquid outer core. She developed a model and calculated travel times and paths of seismic waves through the liquid outer core and the solid inner core, and found they matched the observed times of seismic waves from the largest earthquakes. Therefore, she confirmed the existence of a liquid outer core and solid inner core within the Earth's interior.

2.7 Plate Boundaries

Plate tectonic theory begins with the premise that the Earth's lithosphere is broken into pieces called plates (Fig. 2.8). The plate boundaries correspond very closely with zones of intense earthquake activity (Fig. 2.9). Depending on how you define "large," there about 13 large plates and several smaller ones that form the tectonic framework of the Earth. These plates interact at their boundaries. It is those interactions between plates at those boundaries that have produced many types of geologic phenomena, such as volcanoes, earthquakes, and mountain ranges.

2.7.1 Plate Boundary Types

There are three major types of plate boundaries. *Divergent boundaries* are places where two plates are moving away from each other. At *convergent boundaries*, two plates move toward each other. Two plates move laterally past each other at *transform boundaries*. A fourth phenomenon, known as a *hotspot*, is a place where extra-hot mantle material flows toward the surface and produces volcanoes, often in the interiors of plates.

2.7.2 Divergent Boundaries

The first type of plate boundary to be described by scientists was that of the divergent boundary. At divergent boundaries, two plates move away from each other. Divergent boundaries

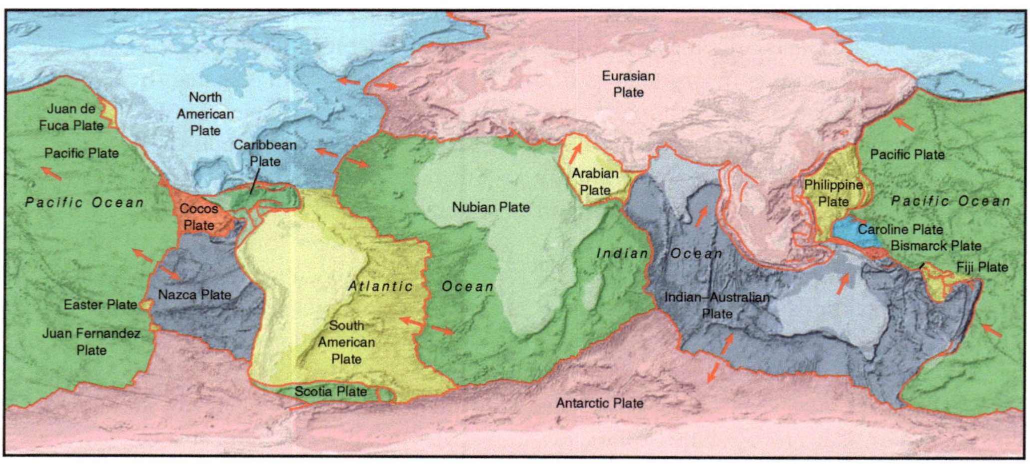

Figure 2.8 A map of the Earth indicating locations of tectonic plates and their boundaries. From Fletcher, 2014.

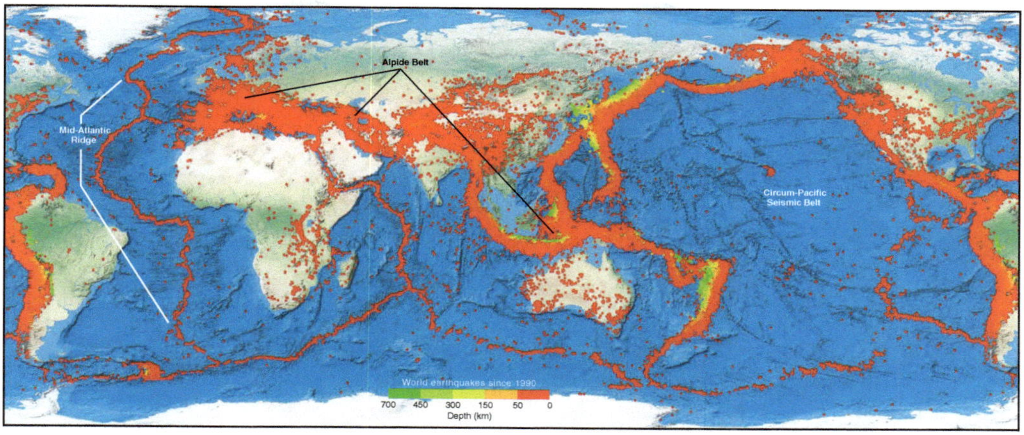

Figure 2.9 Worldwide occurrences of earthquakes for the years 1990 to 2013. Note that high concentrations of earthquakes often coincide with plate boundaries from Figure 2.8. From Fletcher, 2014.

are known by many names. Seafloor spreading, as discussed earlier, was the first term used to describe this action. We also refer to the process of seafloor spreading simply as spreading, or as rifting, or as divergence. Locations where spreading occurs in the oceans are called *mid-ocean ridges*, or simply ridges. Within a continental setting, these boundaries are known as *continental rifts*. We also refer to divergent boundaries as spreading centers in both contexts (continental and oceanic). All of these terms and actions indicate that two plates are moving away from each other (Fig. 2.10).

The Mid-Atlantic Ridge is an excellent example of an oceanic spreading center, or midocean ridge. The Mid-Atlantic Ridge spans the length of the Atlantic Ocean from near the North Pole to near Antarctica. It parallels the coastlines on either side of the ocean, but is found almost directly in the center of the Atlantic. The Mid-Atlantic Ridge is a volcanic mountain range that rises several thousand feet from the deep ocean basin on either side. In the center of the mountain range is a volcanically active region that includes not only active volcanoes that periodically erupt lava, but also hot springs and hot mineral seeps. Some of these hot mineral seeps feature boiling water laden with sulfate minerals, giving the expelled water a roiling black appearance. For this reason, these hot mineral seeps are known as *hydrothermal* (hydro = water, thermal = hot) *vents* or *black smokers* (Fig. 2.11).

Black smokers were discovered in the 1970s and have been a source of incredible new information about life on Earth. These black smokers occur at great depth within the ocean, where no sunlight is able to penetrate. However, around many of these black smokers are vast communities of life including tube worms with white tubes and brilliant red gills, shrimp, crabs, clams, and fish. Upon discovery in the mid-1970s, scientists found that the basis for the food chain in these areas was a bit of a mystery. Up to that point, all life on Earth was believed to have photosynthetic organisms as the basis of the food chain. In other words, prior to discovery of the black smokers, scientists thought the simplest of foods was produced only through *photosynthesis*, which requires sunlight. However, in the areas around black smokers there was no sunlight to help organisms produce food via photosynthesis. What scientists found instead was a form of bacteria that relied on a process called *chemosynthesis* to produce the same type of food. These bacteria use the heat energy from the black smokers to produce food, rather than energy from sunlight. Therefore, the basis of the food chain was chemosynthetic bacteria, which the tube worms

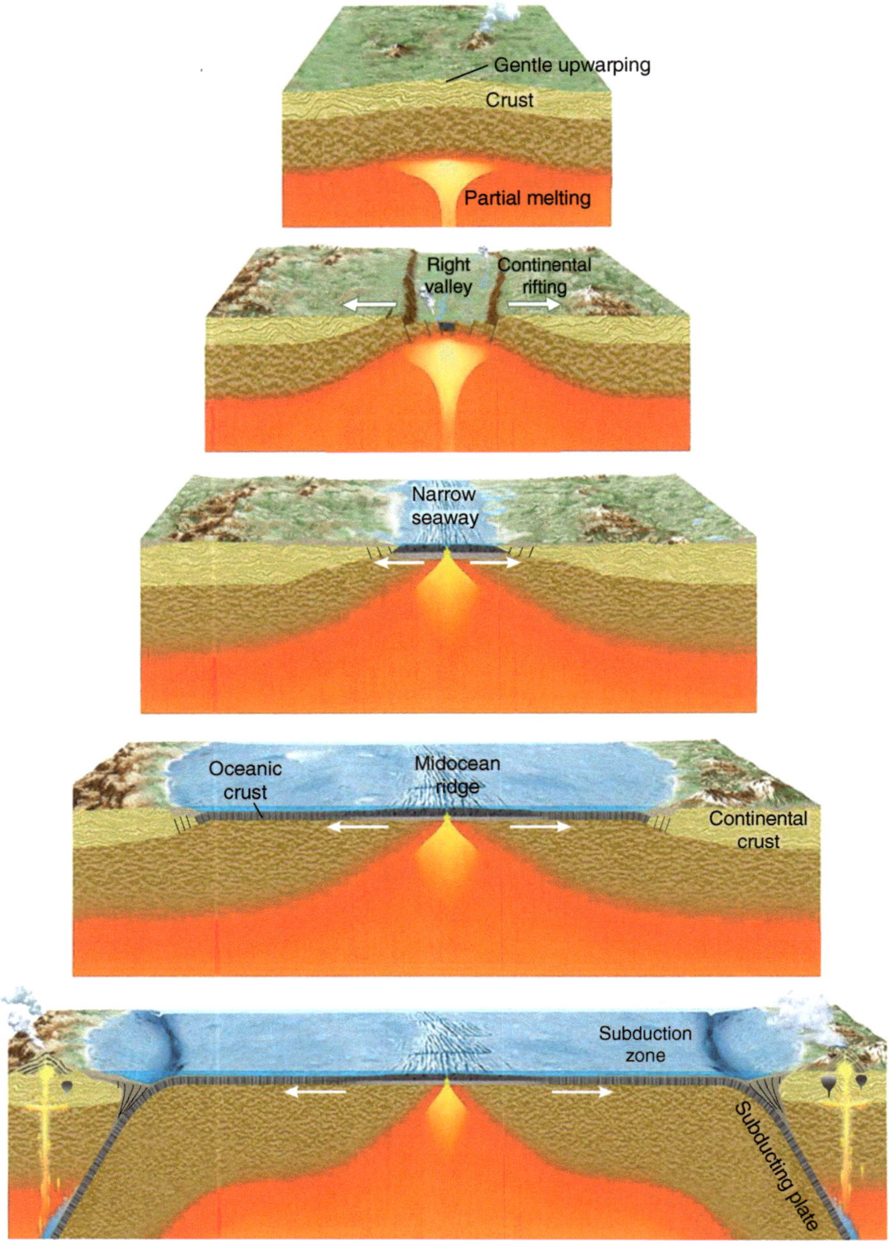

Figure 2.10 Divergent boundary as it may evolve from continental rifting, creation of a narrow seaway, and eventual expansion of the narrow seaway to become an ocean. From Fletcher, 2014.

and other organisms eat. This was a revelation for biologists, who discovered new forms of bacteria and a new process for producing food within the ocean's depths.

Midocean ridges, however, are at their heart volcanic mountain ranges. Figure 2.12 is a typical scene from a midocean ridge. Black volcanic rocks in pillow shapes cover the ocean floor.

Figure 2.11 Sully Vent in the northeast Pacific Ocean in 2004, with monitoring equipment. The white objects topped with red is a bed of tube worms. The tube worms and the rest of the vent community are able to survive because of the chemosynthetic bacteria at the bottom of the food chain. Image courtesy of NOAA; https://www.pmel.noaa.gov/eoi/gallery/smoker-images.html

The pillow shape is unique to lava erupted underwater. Even when these areas, due to tectonic forces, are shoved up onto land, they are easy to recognize as seafloor features because they are covered in *pillow lavas*.

Lava is erupted at the center of a midocean ridge, and lava flows cover the entire *vent* area. As the two adjacent plates are pulled apart, lava of the same age is carried on each plate away from the center of the ridge like the two sides of diverging conveyor belts. Therefore, the youngest rocks are in the center of the ridge and rocks get older with increasing distance away from the spreading center. The oldest oceanic rocks are found on the edges of the oceans next to the continents. These rocks immediately adjacent to the continents can be used to approximately date the time at which the ocean began to form. Volcanic rocks on either side of the Atlantic Ocean are roughly 180 million years old. We know from fossil and rock evidence that Pangaea was formed about 300 million years ago and existed for many millions of years. The age of volcanic rocks on either side of the Atlantic is strong evidence that Pangaea began breaking apart, and the Atlantic Ocean began opening, around 180 million years ago.

What caused the opening of the Atlantic Ocean? If spreading centers create new oceanic lithosphere, what was there before? New oceans form when a continent splits apart through the process of continental rifting. Continental rifting occurs when, for reasons unknown to us at this time, convection cycles in the asthenosphere below a continent reorganize and begin to pull two parts of a continent away from each other.

Figure 2.12 Pillow lavas erupted onto the ocean floor at a midocean ridge. Photo courtesy of NOAA; https://oceanexplorer.noaa.gov/explorations/05galapagosrift/media/687A2327.html

At the location where the rift forms, the continental lithosphere becomes thinned. Initially, the area may rise in elevation because the heat from the rising mantle causes the lithosphere to expand. The crust then begins to fracture, valleys form, and volcanic activity begins throughout the rift. As the two sides of the rift begin to pull away from each other, the elevation of the valley floor drops. Eventually, if the valley connects to the ocean, it fills with seawater. A new linear ocean is formed. The process continues until a wide ocean has developed between what used to be two parts of the same continent. This is the process that occurred in the opening of the Atlantic Ocean when the supercontinent of Pangaea broke apart about 180 million years ago.

The process of continental rifting is occurring today in eastern Africa. The easternmost part of Africa is moving away from the rest of the continent at the East African Rift. Wildlife enthusiasts may know about the Great Rift Valley as a habitat for many of the large mammals on the African savannah. The Great Rift Valley gets its name from its geology and plate tectonic setting. It is a continental rift system. The rift has three arms (Fig. 2.13). One arm has formed a small linear ocean between Africa and the Arabian Peninsula. This is the Red Sea. To the southeast of the Arabian Peninsula is the Gulf of Aden, which is located between the Horn of Africa and the southeastern section of the Arabian Peninsula. These are two of the three arms of this rift system. The third arm is the Great Rift Valley, which extends from Somalia southward through Ethiopia, Kenya, and Tanzania. Here, the eastern part of Africa is moving away from the rest of Africa very slowly, at about 0.04 in./year to 0.16 in./year (1 to 4 mm/year). There are many volcanoes throughout the rift system, including some that are nearly continuously active. Hot springs dot the rift valley as well. All of these are indications that the rift is still active, and eventually, in a few million years, the rifting activity may result in another narrow linear seaway in eastern Africa.

Spreading centers are host to a number of geologic phenomena. They usually include linear valleys bounded by dramatic faults on either side. Earthquakes occur on these faults. Volcanoes dot both the axis of the valley and

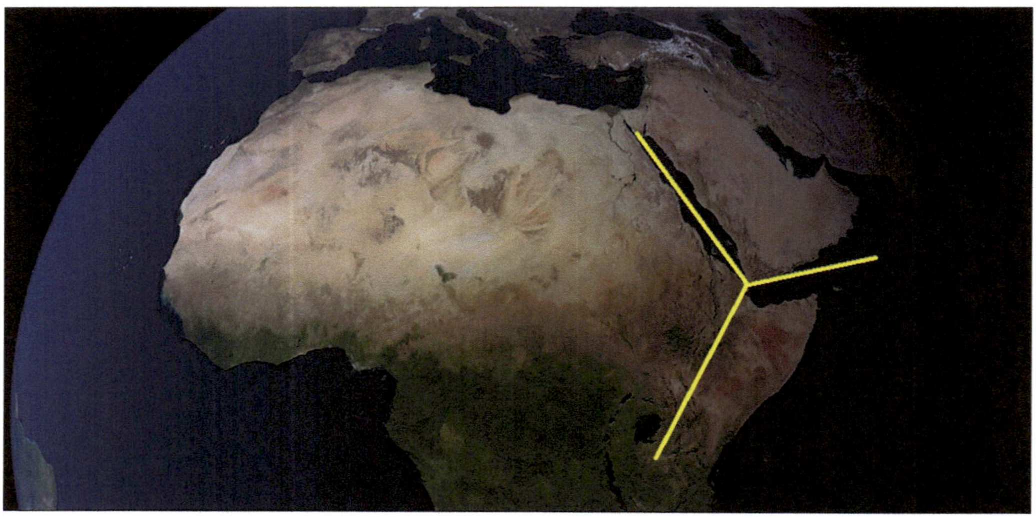

Figure 2.13 Satellite composite of the continent of Africa and the Arabian Peninsula. Note the two linear seas that make up the western (Red Sea) and southeastern (Gulf of Aden) boundaries of the Arabian Peninsula. These are two arms of a rift system. The third arm is located in eastern Africa where a continental rift is splitting eastern Africa away from the rest of the continent. Image courtesy of NASA; https://www.nasa.gov/images/content/338498main_bluemarbleafrica_full.jpg

the edges of the rift. Hot springs are also found in and around the spreading center. At all spreading centers, including continental rifts, new lithosphere is created.

2.7.3 Convergent Boundaries

Since new lithosphere is created at spreading centers, and the Earth is not growing in size, lithosphere must be destroyed elsewhere. This destruction happens at *convergent plate boundaries*. Anywhere two plates are moving toward each other is a convergent plate boundary. Convergent plate boundaries come in two varieties, based on what type of lithosphere is involved and interacting at the boundary. Most commonly, at least one of the plates involved is oceanic. In this case, oceanic lithosphere is being driven downward into the mantle of the Earth as the two plates converge. This is a type of convergent boundary called a *subduction zone*. The other type of convergent plate boundary exists where the two plates are composed of continental lithosphere, and neither is thin or dense enough to dive beneath the other. Instead, the two plates are fused and deformed, creating a tall mountain range at the boundary. This is called a *collision zone*.

2.8 Subduction Zones

Subduction zones are locations in which at least one plate involved is being thrust downward into the mantle. Scientists are still unsure whether plates are pushed from the spreading centers or pulled from subduction zones, or some combination of the two. In any case, the plate that dives into the mantle is called the subducted plate. The other plate that does not subduct is commonly called the overriding plate.

2.8.1 Continental-Oceanic Subduction Zones

Subduction zones most commonly occur where continental and oceanic lithosphere come together (Fig. 2.14). In this case, the thinner, denser oceanic lithosphere dives beneath the thicker, less dense continental lithosphere. Keep

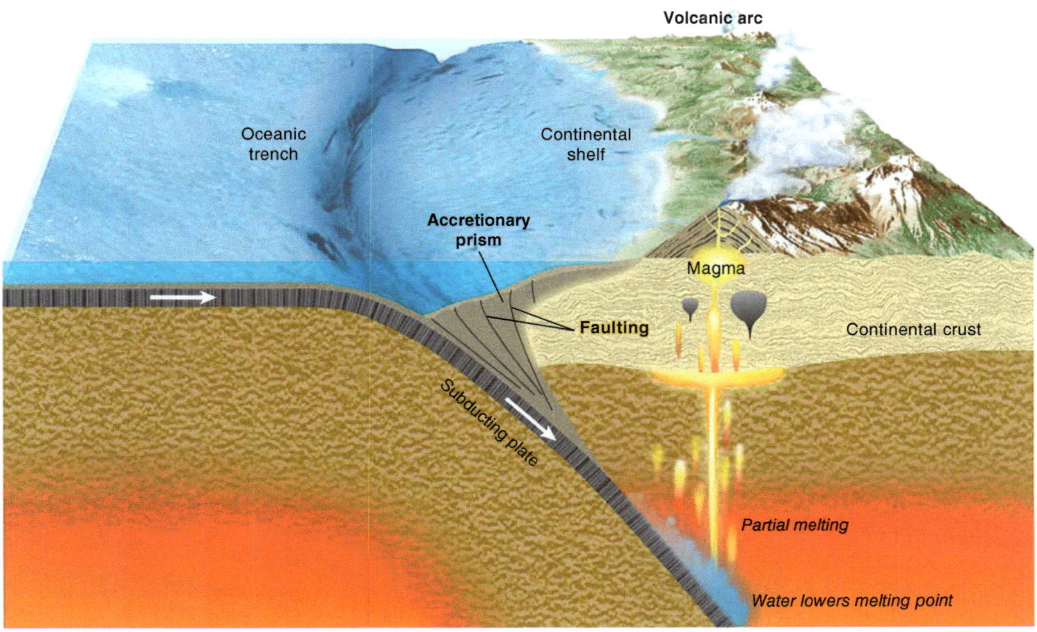

Figure 2.14 Cross section of a subduction zone in which one plate is oceanic and the other is continental. The subduction zone along the west coast of South America is of this type. From Fletcher, 2014.

in mind that the mantle beneath the lithosphere is flexible but still solid, so this is a somewhat violent process. As a result, dramatic effects are seen at subduction zones. Eleven of the 12 largest earthquakes since 1900 have occurred on subduction zones. (The other earthquake occurred at a collision zone, another type of convergent boundary.) The largest earthquake ever recorded was a magnitude 9.6 earthquake near the city of Valdavia, Chile, in May of 1960. Here, the oceanic Nazca Plate is being subducted beneath the western edge of the continental South American Plate. The 1960 earthquake left an estimated 2 million people homeless in the region where it occurred. It also caused destruction around the entire Pacific Rim because it created a set of seismically created ocean waves called a *tsunami* that traveled across the Pacific Ocean. The tsunami caused damage and deaths not only along the west coast of Chile but also in Hawaii, California, Alaska, Japan, and the Philippines.

Volcanoes are also common sights around subduction zones. Nearly the entire Pacific Rim is surrounded by subduction zones, leading some to call this the Ring of Fire (Fig. 2.15). What makes volcanoes such a common occurrence at subduction zones? As the subducted plate reaches a certain depth within the Earth, heat causes some of the water in the subducted plate to be driven out into the surrounding mantle. This lowers the melting temperature of the mantle, causing partial melting (about 5% to 7% of the rock volume) of the mantle rocks. This molten rock, now called magma, rises through the uppermost mantle and crust. Within the crust, the magma may be stored in large amounts in magma chambers, where it can cool and solidify and turn into rock while it is in the crust. Alternately, magma may make its way to the surface from a magma chamber, creating a volcanic eruption.

Along the west coast of South America is a very tall mountain range called the Andes. The Andes are formed from two different mechanisms. One is the pressure of two converging plates pushing rocks upward into mountains. But many of the mountains in the Andes are volcanoes created by the process of subduction. The volcanoes within the Andes present a risk to residents of nations within western South America because they can produce extremely violent eruptions.

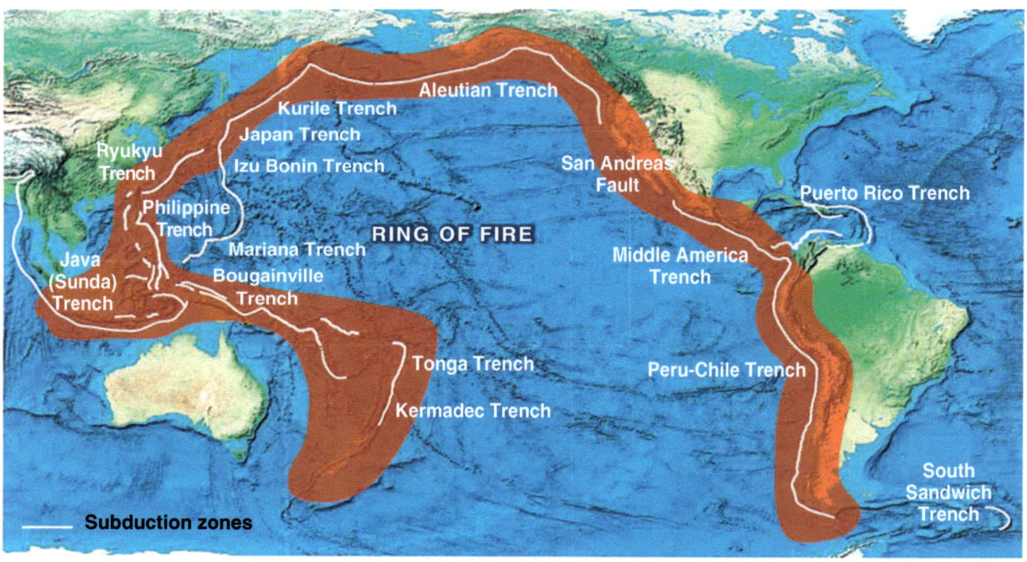

Figure 2.15 The Ring of Fire that nearly encircles the Pacific Ocean. In these areas, subduction zones are the norm. Common features of subduction zones include earthquakes, volcanoes, trenches, and tsunamis. From Fletcher, 2014.

Another feature of subduction zones are oceanic *trenches* (see Fig. 2.14). Trenches are prominent oceanic features and are the deepest places in the world's oceans, and they mark the point of subduction of one plate beneath another. Every subduction zone has an associated trench, though some trenches are much deeper than others. One example of a trench at a subduction zone that involves one continental and one oceanic plate is the Peru-Chile Trench. This trench is the boundary between the continental South American Plate to the east, and the oceanic Nazca Plate to the west. Another example of this type of trench is the Middle America Trench, where the oceanic Cocos Plate is subducting beneath the North American Plate off the west coast of Mexico and Central America.

2.8.2 Oceanic-Oceanic Subduction

Subduction zones also occur in locations where two oceanic plates converge (Fig. 2.16). Though it is difficult to predict which plate will give way and dive into the mantle, one inevitably does, and a subduction zone is born. One area where this occurs is in the western Pacific, where the oceanic Pacific Plate is subducting beneath the primarily oceanic Philippine Plate. Here, the Mariana Trench marks this boundary. The Challenger Deep, a point within the Mariana Trench, is the deepest place in any ocean. The bottom of the Challenger Deep is approximately 36,000 ft (11 km) below sea level. For comparison, the summit of Mt. Everest is approximately 29,000 ft, or 8.8 km, above sea level.

Another example of an oceanic-oceanic subduction zone outside the Pacific Rim is found in the eastern Caribbean. The chain of islands known as the Lesser Antilles is a volcanic island arc created by subduction of the Caribbean Plate beneath the South American plate just north of South America. The island of Martinique is home to the volcano Mt. Pelée. In May of 1902, Mt. Pelée erupted with great force, sending searing hot material into the city of St. Pierre, killing over 28,000 people in an instant. The island of Montserrat experienced a prolonged volcanic eruption in the 1990s that

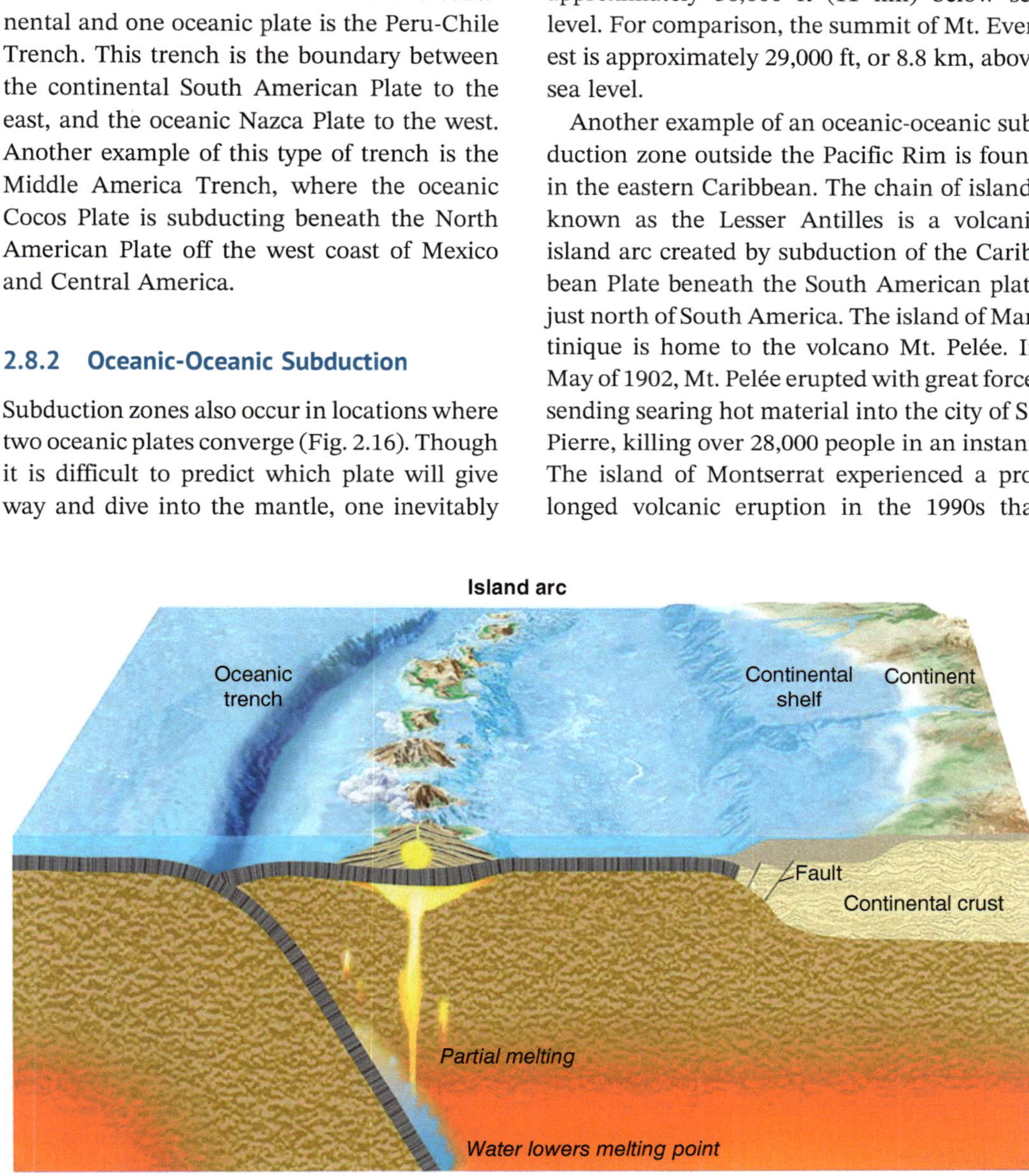

Figure 2.16 A subduction zone in which two oceanic plates converge. One plate dives beneath the other, creating a subduction zone. From Fletcher, 2014.

buried the capital city of Plymouth in several feet of volcanic debris.

Whether two oceanic plates or an oceanic and a continental plate come together, subduction yields the same results. Volcanoes are commonplace in oceanic-oceanic subduction zones, just as they are in oceanic-continental subduction zones. In oceanic-oceanic subduction zones, however, they form a chain of islands in the ocean rather than a chain of volcanoes on land. In the western Pacific region, it is the Mariana Islands that were created by the subduction of the Pacific Plate beneath the Philippine Plate. The Aleutian Islands in Alaska were also formed in this way, where the Pacific Plate is being subducted beneath an oceanic portion of the North American Plate. Each of the Aleutian Islands has a resident volcano, and many of these volcanoes have been active within the last 100 years. The second largest earthquake since 1900 occurred in 1964 at the eastern boundary of the Aleutian Islands, in Prince William Sound near Anchorage, Alaska. The ground shook for 3 min, causing a great deal of damage in and near Anchorage, and produced a large tsunami that took lives as far away as California.

2.9 Collision Zones

When the two plates involved at convergent boundary are both continental plates, the result is not subduction (Fig. 2.17) but collision. Both plates are strong, thick, and buoyant. As a result, neither will yield and sink beneath the other. Instead, they collide with a violent force that causes the rocks in the *collision zone* to be compressed, broken, bent, and deformed. The rocks are pushed upward as long as the two

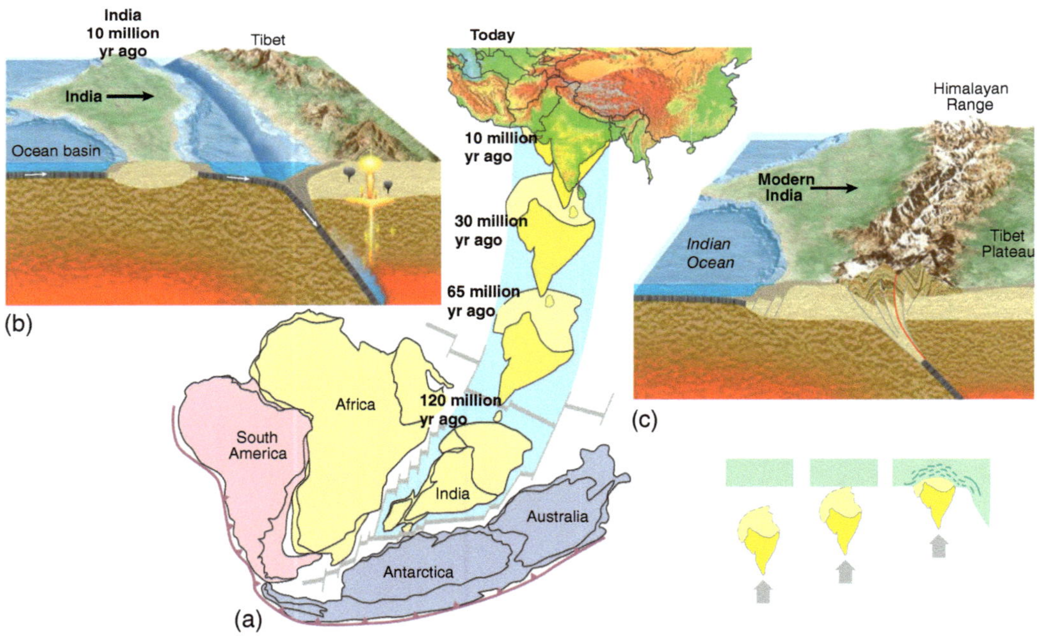

Figure 2.17 (a) The Indian subcontinent used to be a separate land mass from the rest of Asia. It separated from the southern continents about 120 million years ago and began moving northward toward Asia. (b) As India moved toward Asia, a subduction zone was created that consumed oceanic lithosphere between India and Asia. (c) As the ocean between India and the rest of Asia closed, the two continental blocks collided with each other, stopping subduction and creating a collision zone. The boundary between Asia and India became the Himalaya Mountains and surrounding Tibetan Plateau. From Fletcher, 2014.

plates converge. A collision zone is marked not by volcanoes and trenches, but instead by a tall mountain range.

The most frequently cited collision zone on the Earth today is the Himalayas, between the Indian subcontinent and the rest of mainland Asia. India, Pakistan, Tibet, Nepal, and Bhutan are located on this collision zone. The Himalayas began to form about 35 million years ago. Before that time, there was an ocean between India and Asia. The two continental blocks on either side of this ocean moved toward each other, and the ocean between them subducted beneath Asia. About 35 million years ago the ocean between India and Asia closed and the two continental blocks touched. When the two continents came into contact with each other, subduction stopped and collision began. Rocks that had been formed at or below sea level were pushed ever-farther upward, and now the tallest mountain range in the world is composed largely of rock formed on the seafloor when the ocean existed between the two continental blocks.

Collision zones are not known for volcanic activity. The conditions are not right for generating magma, which is necessary to feed volcanic eruptions. Instead, collision zones are known for large earthquakes and rising mountains. Earthquakes are not confined to a narrow zone at the plate boundary. Instead, they occur over a large area that is affected by this collision. In the case of the Himalayas, large and damaging earthquakes occur not only in Nepal, Tibet, Bhutan, India, and Pakistan but also well northward into China and westward into Afghanistan. One of the deadliest earthquakes in recorded history was in China in 1556, which is believed to have killed over 800,000 people.

2.10 Transform Boundaries

A third type of plate boundary is created when two plates do not diverge or converge, but instead move laterally past each other. This is called a transform boundary (Fig. 2.18).

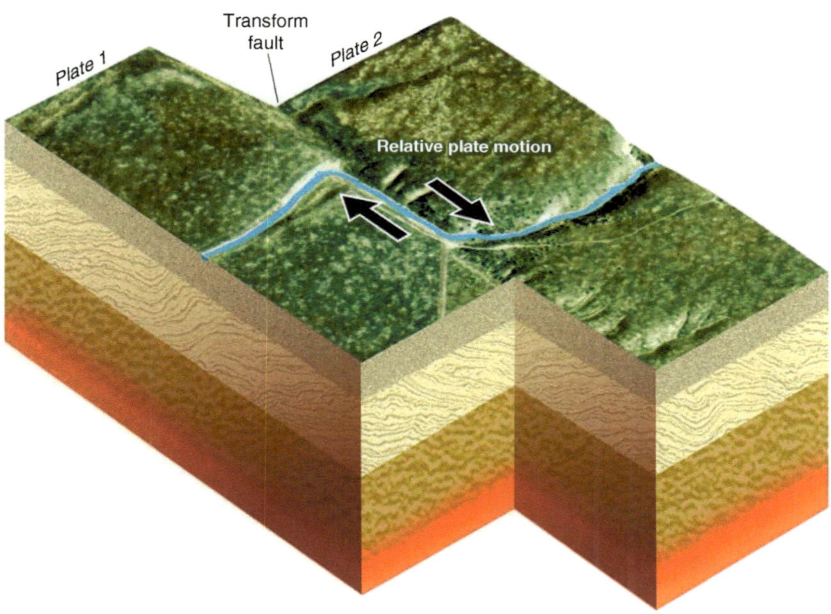

Figure 2.18 Diagram of a transform boundary. The two plates are moving laterally past each other. From Fletcher, 2014.

Transform boundaries often connect *segments* of other types of boundaries, as seen on the west coast of the United States, Canada, and Mexico (Fig. 2.19). For example, most midocean ridges, are broken into several segments. The larger plates on either side of the midocean ridge are moving as one large piece. However the boundary between them has a stair-step appearance. The steps are small segments of divergent boundary offset by transform boundaries. The faults that connect the small segments of divergent boundary (seen as red lines in Fig. 2.19) are called transform boundaries. The Explorer Fracture Zone (upper left corner of Fig. 2.19) is an example of a transform boundary, as is the San Andreas Fault (Fig. 2.20).

The San Andreas Fault is probably the most well-known of transform plate boundaries in the world. Geologist Tanya Atwater first described the formation and growth of the San Andreas Fault in a 1970 journal article (Atwater, 1970). Prior to the formation of the San Andreas Fault, there was a larger plate between the Pacific Plate and the North American Plate, called the Farallon Plate. A portion of the Farallon Plate was subducted beneath North America, putting the Pacific Plate in direct contact with the North American Plate. It is here, where the Pacific and North American Plates are adjacent to each other, that the San Andreas Fault was formed and continues to exist today. The San Andreas Fault is also a transform boundary, one of many along the western margin of North America. The San Andreas connects segments of midocean ridge to each other on both its northern and southern

Figure 2.19 Map of the San Andreas Fault, a transform plate boundary. Note also the other small transform boundaries that connect segments of midocean ridge to the northwest and southeast of the San Andreas Fault. From Fletcher, 2014.

Figure 2.20 View of the San Andreas Fault where it broke to the surface on the Skinner Ranch near Olema, California. Photo courtesy of the U.S. Geological Survey via ScienceBase-Catalog; https://www.sciencebase.gov/catalog/file/get/51dc6b33e4b097e4d3837698?f=__disk__cb%2F14%2Fce%2Fcb14ce4d6d367265db59e2ecc92605257f6eff43&width=4038&height=4011

ends. The tiny Juan de Fuca Plate to the north of the San Andreas is a small remnant of the ancient Farallon Plate. The Nazca and Cocos Plates to the south are also remnants of the Farallon Plate. Atwater's work is key to understanding the relationship between midocean ridge segments and transform faults worldwide, and is the definitive work on the formation and growth of the San Andreas.

At the San Andreas Fault, the area to the west of the fault is moving northwest, relative to the area east of the fault. That lateral motion generates earthquakes, but not volcanoes or trenches. However, note that there are also places on the map in Figure 2.19 called fracture zones that have lateral motion. The Mendocino Fracture Zone, for example, extends from the west coast of California out to sea at the southern end of the Gorda Plate. There, it merges with a spreading center in the Pacific Ocean (the Gorda Ridge) where the very small Gorda Plate is being created off the west coast of northern California, Oregon, Washington, and southern British Columbia in Canada. The Gorda Ridge, Juan de Fuca Ridge, and Explorer Ridge are three segments of the same spreading center. The segments of the spreading center are joined by fracture zones (such as the Explorer Fracture Zone), which also happen to be transform boundaries. Motion along the San Andreas Fault is about 2 in., or 5 cm/year. Los Angeles, on the west side of the fault, is moving ever so slowly toward San Francisco, which is on the east side of the fault. At this rate of motion, the two cities will merge in approximately 20 million years.

2.11 Hotspots

Once plate tectonics was nearly fully developed as a theory in the early 1960s, some scientists working in certain parts of the world wondered why they were seeing phenomena usually associated with plate boundaries in places where it was clear there was no plate boundary. For example, Yellowstone National Park is clearly a vast volcanic region, and has many active *geothermal* features such as geysers, hot springs, and mud pots. However, Yellowstone exists in the interior of the North American Plate, far from any plate boundary. Hawaii has active volcanoes and occasional large earthquakes, yet it is in the center of the Pacific Plate. Iceland is located on the Mid-Atlantic Ridge, but why is it above sea level when the rest of the ridge is below sea level? These questions plagued researchers. Measurements of the flow of heat from the interior of the Earth to the exterior, called *heat flow*, provided a clue as to why these situations occurred. Heat flow was much higher in these regions, even accounting for the volcanic activity, than in other similar areas. Thus there had to be an immensely powerful heat source beneath each of these regions.

The final piece of plate tectonic theory to be described in 1963 was the existence of hotspots, which are areas within the Earth's mantle that, for reasons we still cannot explain, are extra hot. Excess heat in the mantle causes the Earth's crust to be buoyed up a bit, and also causes a small percentage of the upper mantle to melt, forming magma. Magma, or molten rock, from the upper mantle rises through the crust and is either held in magma chambers beneath volcanic regions, or is erupted to the surface via volcanoes. As a result, areas in which hotspots exist are prone to volcanic and geothermal activity.

There are about 60 known or suspected hotspots within the Earth's mantle (Fig. 2.21). By far, the most intensely studied is the Hawaiian Hotspot (Fig. 2.22). Volcanoes produced by this hotspot extend from the island of Hawaii in the central Pacific Ocean northwestward to Midway Island. At that location, the chain turns and proceeds almost due north to the Aleutian Trench where any older material erupted as a result of the hotspot has been subducted. This is called the Hawaiian-Emperor chain. This hotspot has been active for at least 70 million years, as this is the age of the oldest volcano in the chain located at the subduction zone.

Scientists believe that hotspots are more or less stationary. As the lithosphere moves over the hotspot, the lithosphere expands slightly. A small percentage of the upper mantle beneath the lithosphere melts, which feeds volcanic activity at the surface. Over perhaps 500,000 to 2 million years, a volcanic island grows. As the lithosphere moves away from the hotspot, volcanic activity ceases. Erosion takes over as the primary activity at the volcano. In addition, the elevation of that volcano decreases since heat is no longer buoying up the lithosphere in that area. What once was a tall island volcano becomes a *seamount*, or underwater mountain. This has happened countless times in the center of the Pacific Ocean, creating a long chain of seamounts and islands in the Hawaiian-Emperor chain.

Beneath a continent, the effect is somewhat similar. Heat from the hotspot causes the

Figure 2.21 Locations of the Earth's plate boundaries and major hotspots. From Kearey et al., 2009.

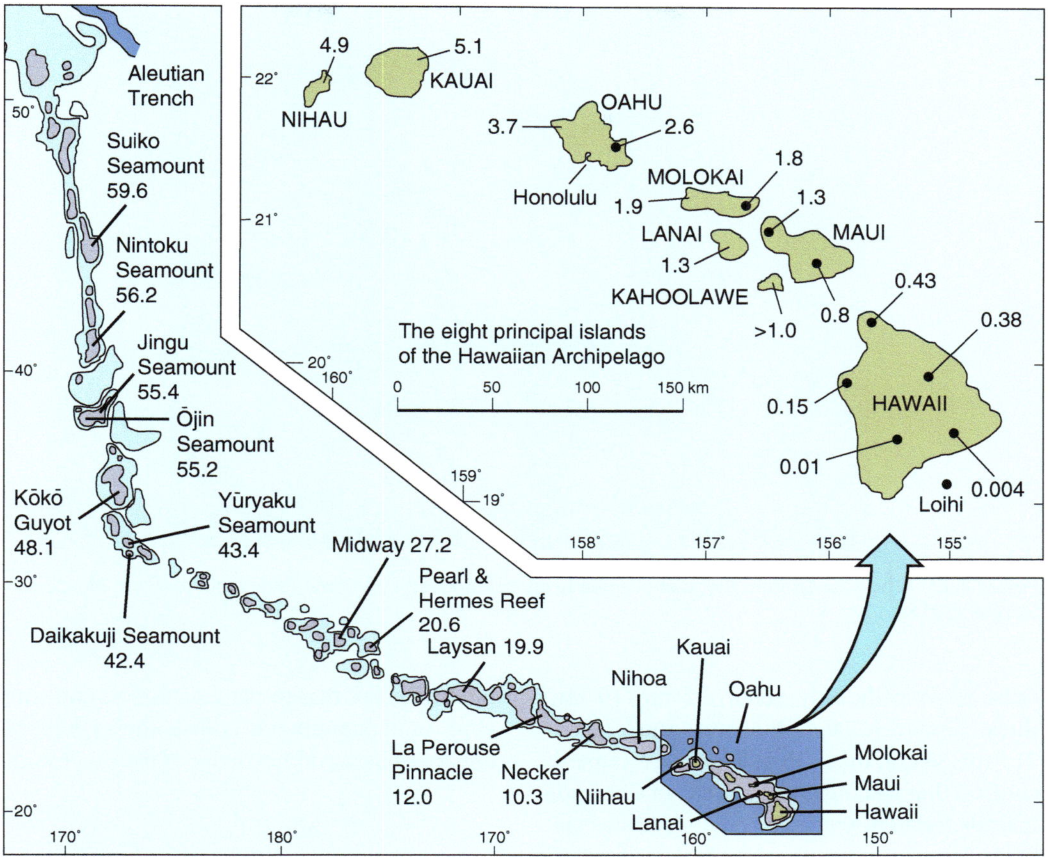

Figure 2.22 Trace of the Hawaiian Hotspot in the Pacific Ocean. From Skinner & Murck, 2011.

lithosphere to expand and rise in elevation. Volcanic activity occurs thanks to the melting of a small percentage of the upper mantle in the location of the hotspot. The Yellowstone Volcano in Wyoming is an example of a volcano on a continent formed by a hotspot. There is evidence of volcanic activity that can be traced through Idaho and Wyoming, indicating that the North American Plate has traveled over the hotspot, and the volcanic rocks in this area, including those in the Snake River Plain of Idaho, can be attributed to hotspot volcanism. Yellowstone is the current location of the hotspot. Heat flow is high over the Yellowstone region, and the immense amount of geothermal activity in the area is evidence the lithosphere is still quite hot. Though Yellowstone has not had a major eruption in about 600,000 years it is possible it may erupt at some point in the future. It could produce another catastrophic eruption as it did 600,000 years ago that affects the entire central United States, or it may (more likely) produce a smaller eruption barely noticed outside of Yellowstone National Park.

Iceland is a rather interesting case of a hotspot located at a plate boundary. Iceland sits atop the Mid-Atlantic Ridge, but why is it above sea level when the rest of the ridge is below sea level? Its location on the ridge is inadequate to explain the great volume of volcanic activity throughout the island and surrounding seafloor as well as the impressive thickness of the crust beneath Iceland. Oceanic

Figure 2.23 The Mid-Atlantic Ridge above sea level in Thingvellir National Park, Iceland. From Murck & Skinner, 2016.

crust is typically only about 6.2 mi (10 km) thick. Under Iceland, however, the crust is 21.7 mi (35 km) thick! Heat flow measurements confirm that Iceland is also the site of a hotspot. It feeds frequent volcanic activity and has created a large plateau on which the island is situated. Iceland is an excellent place to visit for geologists because it is one of the few places one can visit a midocean ridge above sea level. This feature can be found and readily accessed at Thingvellir National Park (Fig. 2.23).

2.12 Conclusion

Plate tectonics was a revolution in the Earth sciences, explaining how continents come together and break apart, why volcanoes and earthquakes occur where they do, and why mountain ranges exist. Interactions of plates at their boundaries are the reason for many of the catastrophic natural events that turn into disasters and human tragedies. Much of the rest of the geologic portion of this book will rely on concepts introduced in this chapter. Earthquakes, volcanoes, tsunamis, and even landslides are tied to plate tectonics, and thus your understanding of those topics will be enhanced by your knowledge of plate tectonics.

2.13 Chapter in Review

Plate tectonics is a relatively young theory in the field of geology. For centuries, people noticed that certain continents looked like they could fit together like jigsaw puzzle pieces, but it was not until the early 1900s that a scientist gathered enough information to make a credible claim that all continents were joined together into one massive supercontinent. This scientist was German meteorologist Alfred Wegener, and he called his idea Continental Drift. Wegener named his supercontinent Pangaea. He produced maps of Pangaea and carefully documented his evidence that the continents had existed in the configuration he proposed. His evidence included the fit of continents' continental shelves, the location of continuous mountain ranges, evidence of glaciers covering a number of continents in a continuous ice sheet, and the ranges of fossil land plants and animals. Although he presented a compelling case, he could not answer questions about why a continent would break apart and how continents moved across the face of the planet.

Wegner's ideas about Pangaea were correct. However, it was not until after his death that technologies existed that allowed scientists to gather information to begin answering the big questions that plagued him. First, scientists determined the structure of the Earth. The Earth is divided into three layers by composition: the crust, mantle, and core. The crust is composed of lighter minerals than in the mantle below, and comes in three main varieties. Oceanic crust is found beneath the oceans. It is thinner yet more dense than the second type of crust, which is continental crust. Continents are composed of continental crust, which is composed of lighter materials and is thicker than oceanic crust. Transitional type crust exists under continental shelves. The mantle is composed of heavier minerals, and the core is composed primarily of iron with small amounts of nickel.

Mechanically, the Earth is composed of five layers. The crust and uppermost mantle form the lithosphere. Lithosphere may be continental or oceanic. Tectonic plates are composed of lithosphere. The rest of the upper mantle beneath the lithosphere is called the asthenosphere. Slow circulation, called convection, occurs within the solid but flexible asthenosphere. Convection keeps the plates in constant motion. The lower mantle is called the mesosphere. Convection likely happens within the mesosphere as well, but it is more dense and likely more rigid than the asthenosphere. The Earth's core is divided into two parts. The outer core is composed of liquid iron and nickel. The inner core is solid iron and nickel.

A survey of the world's oceans revealed that the oceans have large linear volcanic mountain ranges at their centers. The age of the ocean floor is youngest at these long linear volcanic mountain ranges, and oldest near the edges of the continents. In addition, magnetic signatures of rocks at these mountain ranges indicate that the Earth's magnetic field changes polarity and leaves magnetic stripes symmetrical about the axis of these mountain ranges. Analysis of data revealed the existence of spreading centers, which are also known as divergent boundaries. Divergent boundaries are places where two tectonic plates move away from each other and new lithosphere is created. The Mid-Atlantic Ridge, a volcanic mountain range that extends along the center of the Atlantic Ocean, is an example of a divergent boundary. The presence of divergent boundaries explains how individual continents drifted away from each other and how oceans formed between them when Pangaea broke apart.

Further investigation led to the discovery of convergent boundaries. Convergent boundaries are places where two plates come together or converge. They may take one of two forms. In the first form, one plate dives beneath the other. This is called a subduction zone. Subduction zones are marked by earthquakes, volcanoes, and deep oceanic trenches. Tsunamis may be generated by large earthquakes at this type of plate boundary. The west coast of South America is an excellent example of a subduction zone. In the second form, two continental plates come together and neither is heavy or weak enough to subduct beneath the other. This is called a collision zone. Collision zones do not have volcanoes or trenches associated with them, but they do experience major earthquakes and form tall mountain ranges. The Himalayas, found between the Indian subcontinent and mainland Asia, are the world's best example of a collision zone.

The third type of plate boundary is the transform boundary. At this type of boundary, two plates move past each other. Transform boundaries are not typical places for volcanic activity, but they are known for earthquakes, including some that are large and damaging. The San Andreas Fault in California is one example of a transform boundary. Transform boundaries often connect segments of other types of plate boundaries. Midocean ridges are broken into segments connected to each other by transform boundaries.

Hotspots are places in the upper mantle that are unusually hot. They produce a small amount of melt within the upper mantle, which migrates to the surface as magma. This produces volcanic activity over the hotspot. Hotspots are believed to be relatively stationary. A line of volcanoes may be produced as a plate moves over a hotspot as a lithospheric plate moves over the stationary hotspot. The island of Hawaii sits over a hotspot in the center of the Pacific Ocean, but the Hawaiian Hotspot has been active for at least 70 million years. The Hawaiian-Emperor chain of volcanoes stretches from the island of Hawaii northward to the Aleutian Trench. The Yellowstone Volcano is another product of a hotspot.

Plate tectonics explains the existence and location of earthquakes, volcanoes, deep ocean trenches, and mountain ranges around the world. It truly created a revolution in the Earth sciences when the theory was published in the 1960s and early 1970s. Plate tectonics shapes our vision of the Earth and its history.

3

Volcanoes

Bethany D. Hinga

University of Nebraska at Kearney, USA

An erupting volcano is one of the most awe-inspiring sights a person can witness. Some volcanoes erupt with beautiful streams of red lava. Others erupt with billowing clouds of gray ash and avalanches of hot rocks and dust. Still other volcanoes hurl red-hot rocks out of their craters and into the surrounding landscape. All volcanoes pose hazards, and those hazards are linked to the type of volcano and the type of magma erupted from them. In this chapter, we will explore several volcanoes and their related hazards.

3.1 A Lengthy Eruption With an Unexpected End

Kīlauea, a *shield volcano* on the island of Hawai'i (Fig. 3.1), has a history of frequent eruptions and is cited as one of the world's most active volcanoes. As is the case with most volcanoes, the mountain gave many signals it was building to an eruption. These signals include thousands of earthquakes within the summit region of the volcano and along a linear feature called its East *Rift Zone*. The East Rift Zone emanates from the summit and curves initially to the southeast, then to northeast. This is a zone of weakness within the volcano, and a frequent site of eruptions. The surface of the ground in the volcano's summit area and along the East Rift Zone deformed, indicating there was magma, or molten rock, moving toward the surface. Volcanic gases, including water vapor, carbon dioxide, and sulfur dioxide were emitted in increasing quantities indicating that magma was close enough to the Earth's surface that gases dissolved in the magma were released, escaping to the surface. All of this precursory activity led scientists to the conclusion that the eruption would start in the middle East Rift Zone, about halfway between the mountain's summit region and the eastern tip of the island.

The eruption of Kīlauea began early on the morning of 3 January 1983. It started with multiple episodes of lava fountaining (Fig. 3.2) that occurred for 1 or 2 days at a time, roughly once every 3–5 weeks. These fountaining episodes occurred along newly opened cracks in the ground, called *fissures*, on the volcano's middle East Rift Zone on the border between Hawai'i Volcanoes National Park and a forest reserve. Within 6 months, the volcano's eruption focused at a single point. More than 40 fountaining events over the course of 3 years built a *cinder and spatter cone* there that was later named Pu`u`ō`ō (Fig. 3.3).

Pu`u* means "hill" in the Hawaiian language. In Hawaiian legends, Pele is recognized as the goddess of volcanoes. She is said to have traveled to each of the Hawaiian islands in search of a home. She used her *o`o, or "digging stick,"

Earth's Natural Hazards and Disasters, Advanced Textbook 7, First Edition. Bethany D. Hinga.

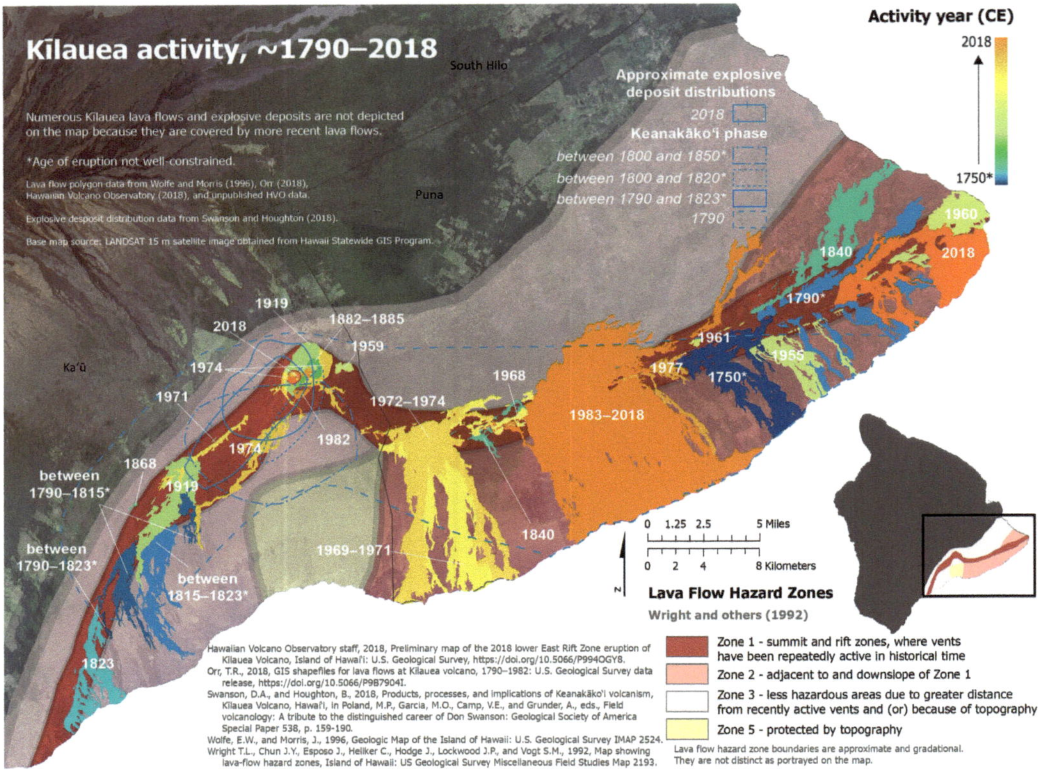

Figure 3.1 Map of the volcano Kīlauea (see inset for location of Kīlauea on the island of Hawai'i). The volcano's geology includes a summit caldera and two rift zones. The summit area is an oblong area just to the left of center on the image. Kīlauea's rift zones are outlined in red. The Southwest Rift Zone extends from the summit area to the southwest. The East Rift Zone extends from the summit area to the southeast, then curves toward the northeast. Eruption dates reveal that much of the volcano has been covered by lava flows since 1790. Map courtesy of the U.S. Geological Survey; https://www.usgs.gov/media/images/map-showing-subaerial-extents-lava-flows-and-explosive-deposits-kilauea-1790-2018

Figure 3.2 Fissure eruption on the volcano Mauna Loa in Hawai'i on 22 November, 2022. Lava is erupting from a long crack, called a fissure.

Figure 3.3 Pu`u`ō`ō, one vent that was active during Kīlauea's 1983–2018 eruption. In this photo, taken during the formation of Pu`u`ō`ō's cone, are high lava fountains that characterized much of Pu`u`ō`ō's early life. Paul Chelsey/NG Image Collection. From Murck & Skinner, 2016.

to create volcanic craters all along her journey. She eventually settled on the island of Hawai`i. The name Pu`u`ō`ō refers to a hill created by Pele's o`o, or digging stick.

During this phase of the eruption, lava erupted from Pu`u`ō`ō flowed downhill out of Hawai'i Volcanoes National Park and into a subdivision called Royal Gardens. There, residents could do nothing but stand by helplessly as lava flows destroyed their homes. Lava covered portions of Highway 130 as well, which was one of the main access roads to the community.

Lava continued flowing both inside and outside Hawai'i Volcanoes National Park for decades. In 1989 and 1990, lava from a secondary vent called Kupaianaha threatened the coastal town of Kalapana and its beautiful black sand beaches. People rallied to move some homes and a small community church. Other homeowners could only move belongings out of their homes as the lava advanced (Fig. 3.4). Kalapana was progressively covered by lava in March of 1990. During the same lava advance, archaeological sites and the seaside visitor center to Hawai'i Volcanoes National Park were also destroyed.

Activity moved back to Pu`u`ō`ō in 1992. For the next 16 years, Pu`u`ō`ō was the sole eruption site on Kīlauea. Lava flows continued to cover land, homes, roads, and archaeological sites on the wide coastal plain below the East Rift Zone. Throughout much of the Pu`u`ō`ō-Kupaianaha eruption, the volcano frequently sent lava flows to areas near the ocean where visitors to Hawai'i Volcanoes National Park were allowed to walk or hike to the lava flows.

In 2008, an additional *vent*, or eruption site, opened within a large crater called Halema`uma`u, located at the summit of Kīlauea (Fig. 3.5). The lava lake, which was situated in a smaller pit within Halema`uma`u, persisted from 2008 until 2018. The surface of the lava lake was usually somewhere in the vicinity of 165 ft (50 m) below the rim of the crater within Halema`uma`u. A handful of times the lava lake overflowed onto Halema`uma`u's crater floor, and at other times the surface of the lava lake fell to more than 300 ft (91 m) below the rim of its crater. The summit area of Kīlauea is part of Hawai'i Volcanoes National Park and is accessible both by car and hiking trails, offering tourists unparalleled opportunities to view an active volcano. Though lava in the crater of Halema`uma`u was usually not directly visible to park visitors, an impressive glow from the crater was present on clear nights from many spots within the summit area of the volcano. Lava is so hot that it glows a bright, brilliant red. This is difficult to see in the daytime because sunlight is brighter than the glow. However, at night the red glow is bright and easily seen from mi away.

Kīlauea's long eruption ended abruptly in late April of 2018. In mid-April, scientists discovered through monitoring of earthquakes

Figure 3.4 A home in Kalapana, Hawai'i, is surrounded by lava from the eruption of Kīlauea in 1989. Ultimately, the house was consumed by fire before lava buried the home site. From Skinner & Murck, 2011.

Figure 3.5 Aerial view of the summit of Kīlauea in Hawai'i. The obvious crater in the photo is called Halema`uma`u. At the time this photo was taken, Halema`uma`u was about 0.6 mi (1 km) across. The small collection of buildings to the right of Halema`uma`u was the Hawaiian Volcano Observatory and Jaggar Museum, within Hawai'i Volcanoes National Park. Stephen & Donna O'Meara/Photo Researchers/Getty Image. From Fletcher, 2014.

and ground deformation that the magma system beneath Pu`u`ō`ō was becoming pressurized. A week later, the lava lake in Halema`uma`u overflowed its crater rim. A few days afterward, the Pu`u`ō`ō cone suffered a collapse, signaling the end of the eruption that began in 1983.

Scientists could tell magma beneath the East Rift Zone was moving down the rift, toward the lower elevations to the northeast. On 1 May 2018, *volcanologists* (the scientists who study the volcano) issued an alert that an eruption was likely in the lower East Rift Zone area, closer to the eastern tip of the island, and unfortunately within populated areas. On 2 May, the first small ground cracks opened in Leilani Estates, a residential neighborhood on the lower East Rift Zone (Fig. 3.6). At 10:30 a.m. local time on 3 May, a magnitude 5.0 earthquake struck the volcano just south of the middle East Rift Zone, and caused additional collapse at Pu`u`ō`ō. At 5 p.m. local time, an eruption at Leilani Estates began. People were forced to evacuate their homes, and the general public was kept away from the dangerous and unfolding situation. Scientists mapped dozens of cracks opening within Leilani Estates subdivision and in surrounding areas. Some of those cracks would later become eruption sites. Others only emitted noxious volcanic gases.

Within the 24 hr of the onset of the eruption, six eruptive fissures opened within Leilani Estates and nearby communities. Low lava fountains formed, tossing *spatter* (blobs of

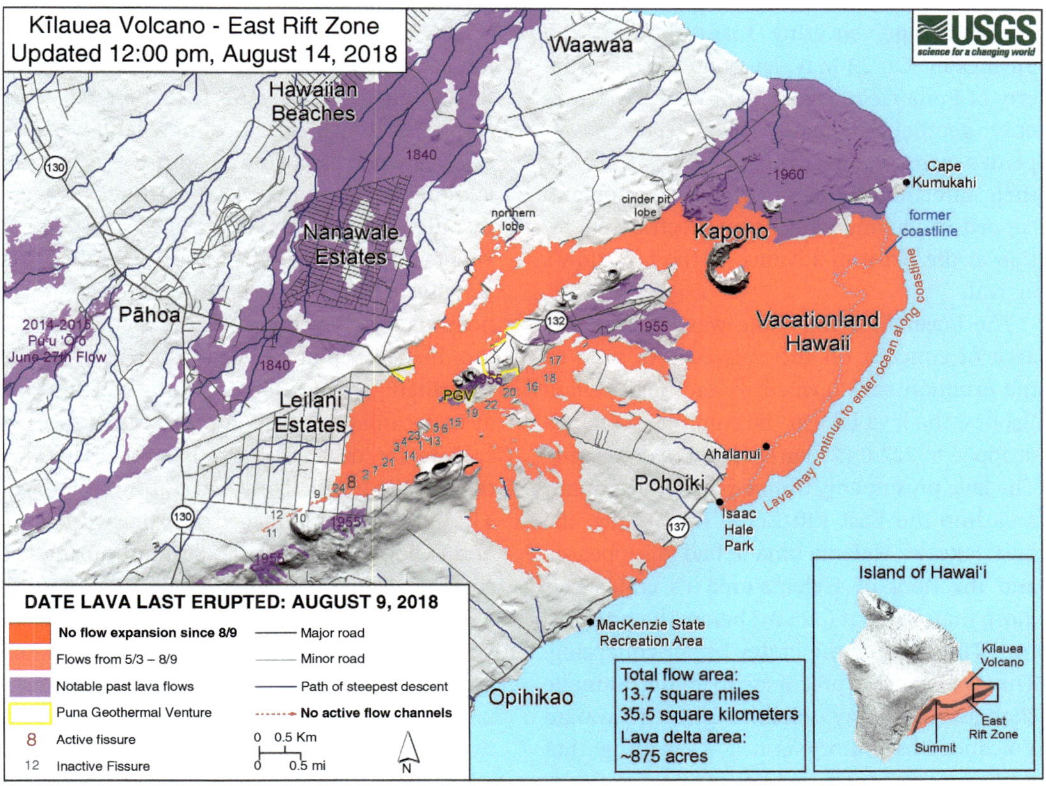

Figure 3.6 Map of lava flows on Kīlauea's lower East Rift Zone (LERZ). Areas in red indicate lava flows erupted between 3 May 2018 and 9 August 2018. U.S. Geological Survey; https://volcanoes.usgs.gov/observatories/hvo/maps_uploads/image-543.jpg/ Public Domain

molten lava) 100 ft (30 m) into the air from several fissures. Lava flows began moving through the subdivision. Three earthquakes of magnitude 5.4, 6.9, and 5.3 occurred near the middle East Rift Zone, further rattling nerves and causing additional collapse at Pu`u`ō`ō. On 5 May, another three fissures opened. By 10 May, 15 fissures were erupting lava into neighborhoods on the lower East Rift Zone. Then, abruptly, the fissures grew quiet. The pause in the eruption was short-lived. Early on the morning of 12 May, Fissure 16 opened. On 13 May, two more new fissures opened, hurling large *bombs*, or pieces of rock, as high as 400 ft (125 m) into the air. Spatter erupted to a height of 300 ft (90 m). As pieces of spatter flew apart, thin filaments of hair-like glass called *Pele's hair* were formed and drifted downwind. The next day fissures began opening nearby, in another residential area. Lava poured onto the landscape, covering roads, homes, and businesses. On 22 May, lava entered the property of Puna Geothermal Venture, the island's only geothermal power plant. The power plant's owners knew the lava flow was headed their direction so they removed equipment, capped wells, and shut down electric generation at the plant in advance of the lava flow's arrival.

The lower East Rift Zone was not the only area of concern on the volcano. As soon as the eruption sites opened in and near Leilani Estates, the lava lake at the summit of Kīlauea dropped to 722 ft (220 m) below the crater rim. The lava presumably drained from the summit area into the East Rift Zone. By 10 May, the lava lake at Halema`uma`u had disappeared and the floor of Halema`uma`u's crater was more than 1,070 ft (325 m) below the previous rim. The sides of the crater began collapsing. This was a mere precursor to more dramatic changes. On 9 May, scientists at the Hawaiian Volcano Observatory issued a warning that explosive events were likely from the Halema`uma`u vent. To protect the public, Hawai'i Volcanoes National Park closed to visitors on 11 May. On 16 May, the first of a series of collapses and accompanying explosive events occurred at Halema`uma`u, sending ash clouds as high as 30,000 ft (9,145 m) above sea level. Beginning on 29 May, these explosive eruptions were accompanied by daily collapse events equivalent to magnitude 5.0 earthquakes. The entire summit area of Kīlauea changed (Fig. 3.7).

Throughout the summer of 2018, Kīlauea's activity continued both at the summit area and in the lower East Rift Zone. Eruptive and explosive activity finally ended in the first week of August. Between May and August of 2018, lava flows on the lower East Rift Zone, most from Fissure 8 (now called Ahu`aila`au), claimed 716 homes, a popular beach park called Ahalanui, which contained hot springs, Isaac Hale Beach Park, which was a popular spot for surfing, and which came within about 110 yd (100 m) of the boat ramp at Pohoiki (Fig. 3.6). In all, 13.1 mi^2 (34 km^2) of existing land and 30 mi (48 km) of road were covered to a depth of up to 167 ft (51 m; Fig. 3.8). Approximately 875 acres of new land was also created when lava flowed into the ocean. In addition, more than 60,000 earthquakes were recorded between the onset of new activity at Kīlauea on 30 April 2018 and the end of that activity in early August 2018. Hawai'i Volcanoes National Park partially reopened to visitors in late September of 2018. However, the Hawaiian Volcano Observatory, which stood at its location on the rim of Kīlauea's summit caldera for decades, was abandoned in May of 2018 and will likely never be reoccupied. The building was damaged by explosive activity and collapse events at the summit during the summer of 2018, and it is no longer structurally sound. Scientists are working from locations in Hilo, the largest city on the island, located approximately 35 mi (56 km) away. The Hawaiian Volcano Observatory will soon be housed in a new building in the city of Hilo.

Kīlauea has been the site of three subsequent eruptions. Those eruptions began in December 2020, May 2021, and September 2021. At this writing (March 2023), the September 2021

Figure 3.7 Aerial view of Kīlauea's summit with the abandoned Hawaiian Volcano Observatory building in the foreground. The large crater is the expanded Halema`uma`u, following months of collapse events and explosive activity. The entire landscape is covered in volcanic ash. Image courtesy of the U.S. Geological Survey; https://home.nps.gov/havo/planyourvisit/upload/PrelimSum_LERZ-Summit_2018_508.pdf

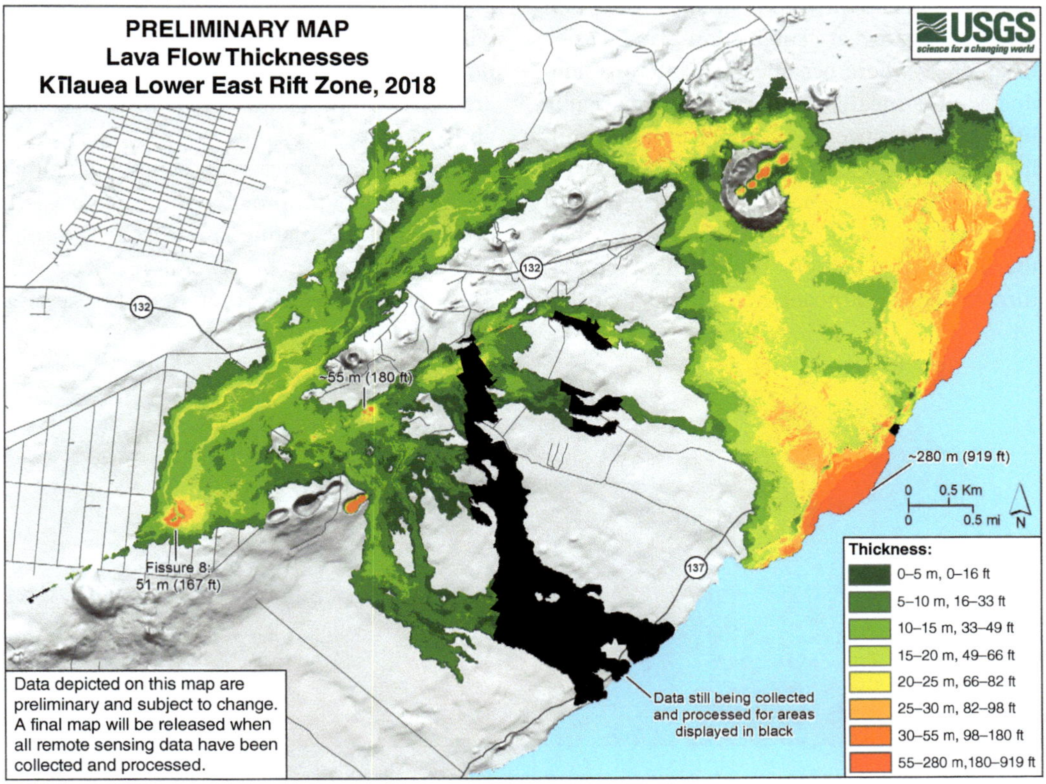

Figure 3.8 Preliminary map of lava flow thickness, compiled by the U.S. Geological Survey. Thicknesses are of lava flows erupted between May and August of 2018 on Kīlauea's lower East Rift Zone. Map courtesy of the U.S. Geological Survey; https://volcanoes.usgs.gov/observatories/hvo/maps_uploads/image-545.jpg/Public Domain

eruption is ongoing. These 2020–2023 eruptions have taken place entirely within the Halema`uma`u crater. As of February 2022, they filled Halema`uma`u's large new crater with more than 850 ft (260 m) of lava, significantly raising the floor of the crater.

3.2 Magma Generation

The word *magma* is used to describe molten rock beneath the surface of the Earth. Magma is created when a small percentage of rock within the upper part of the Earth's solid mantle melts. Melting occurs for a variety of reasons (Fig. 3.9). One reason the solid mantle can melt is if water is introduced into the upper mantle via subducted lithosphere. A second reason is release of pressure, found in places where two plates are diverging. A third condition that can cause melting is the existence of a hotspot.

Melting due to the introduction of water typically happens at subduction zones (see Chapter 2) where oceanic lithosphere is subducted underneath another plate. The oceanic lithosphere has been beneath the ocean for millions of years, and is saturated with water. It moves downward beneath the other plate and into the mantle. The mantle is hotter than the lithosphere because the Earth's temperature increases with depth. As the subducted plate reaches about 62 to 75 mi (100 to 120 km) beneath the surface, it is heated to a high enough temperature that water is driven out of the subducted oceanic plate and into the surrounding mantle. The introduction of water into the solid mantle around the subduction zone lowers its melting point. A very small percentage of the rock melts and produces magma. The molten rock is less dense than surrounding solid rock, so it rises through the upper mantle and into the crust. It heats the crust, which expands slightly and cracks. The cracks can provide a pathway for magma to reach the upper crust and eventually the surface, creating volcanoes. It is common to find a chain of volcanoes on the overriding plate when this occurs. If the overriding plate is continental lithosphere, a chain of volcanoes, called a volcanic arc, will form on land parallel to the subduction zone (Fig. 3.10). The Andes Mountains in western South America are an example of this type of chain. If the overriding plate is oceanic lithosphere, a chain of islands will form in the ocean parallel to the subduction zone (Fig. 3.11). The islands of Indonesia are an example of this phenomenon, called an island arc.

A second reason we may see partial melting of the mantle and production of magma is if pressure on the mantle is released. The solid mantle is overlain by several mi of crustal material. The weight of the crust exerts a

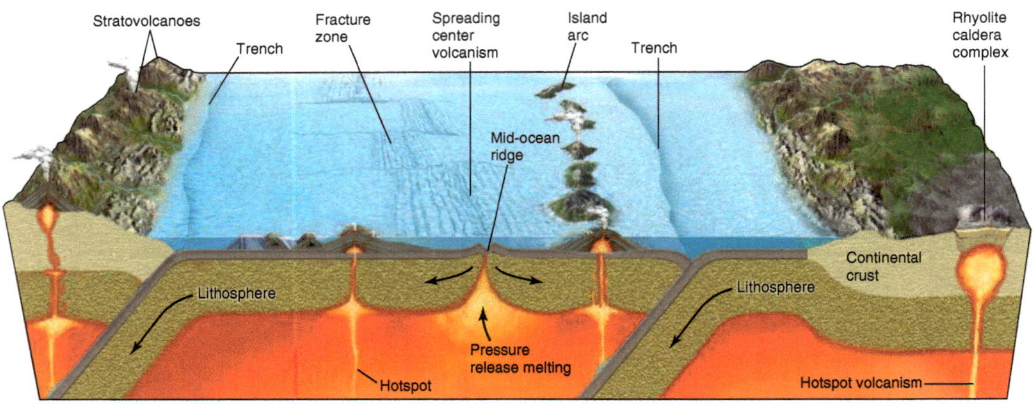

Figure 3.9 Most volcanic activity occurs at one of three plate tectonic settings: divergent boundaries, subduction zones, and hotspots. From Fletcher, 2014.

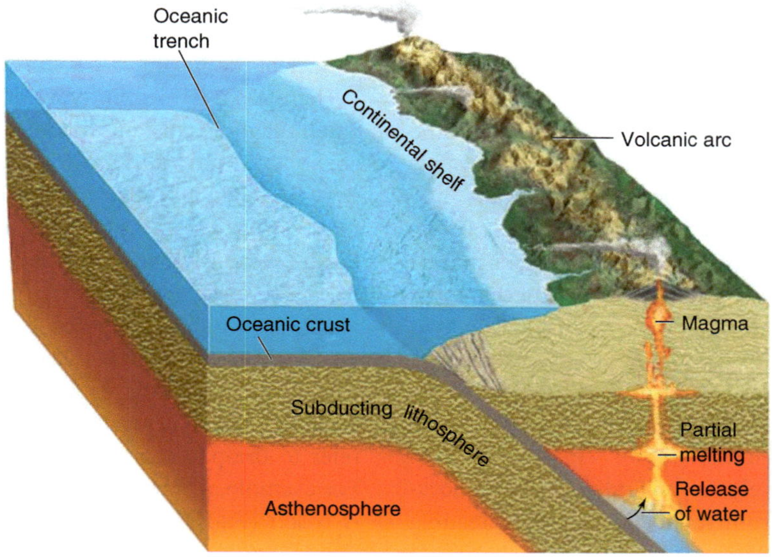

Figure 3.10 Continental-oceanic subduction zone creates a chain of volcanoes near the edge of the continent, parallel to the subduction zone. From Fletcher, 2014.

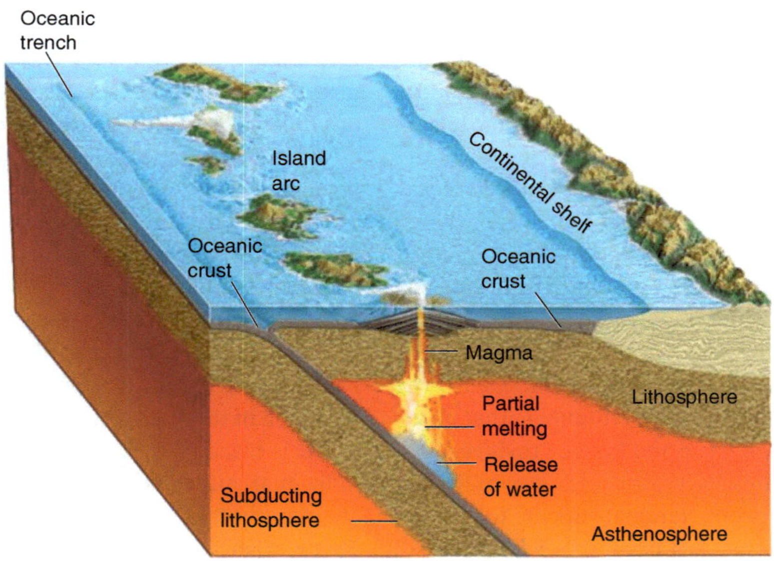

Figure 3.11 Oceanic-oceanic subduction zone creates a chain of volcanoes in the ocean called an island arc. From Fletcher, 2014.

certain amount of pressure on the mantle, which keeps the mantle solid even though its temperature is very high. However, in a location where there is a divergent plate boundary (see Chapter 2), the crust is thinned and hence pressure on the mantle is reduced. The reduced pressure lowers the melting temperature in that location, which in turn allows a small amount of melting to occur within the upper mantle, producing magma. Just as in the case of subduction zones, that magma is less dense than surrounding rocks, and it rises through the upper mantle and crust to find the surface (Fig. 3.12).

The third case in which melting occurs in the mantle is in the presence of a hotspot. Hotspots are areas where, for reasons we do not fully understand, the mantle has a great deal of extra heat. That extra heat within the upper mantle is able to melt a small percentage of the upper mantle rock. Molten rock, or magma, follows the pattern we have established to reach the surface and erupt, forming volcanoes. They form linear chains of volcanoes, such as the Hawaiian Islands.

It is extremely rare to find volcanoes in locations away from subduction zones, spreading centers, or hotspots. This is because it is so difficult to melt mantle rock to create magma to feed those volcanoes (Fig. 3.13) outside of the three previously mentioned conditions that are allow partial melting of the mantle. In the mainland United States, however, there are several locations in the interior of the country with very young (less than 10,000 years old) volcanoes. They are not located at the plate boundaries that mark the edges of the continent to the east and west but are located at hotspots and places called continental rifts where the crust has been thinned. Volcanoes in central New Mexico and Colorado are related to a spreading center in that area of North America, called the Rio Grande Rift. Volcanoes throughout Arizona, Nevada, and Utah are also related to crustal thinning that has occurred in the Basin and Range Province over the last few million years. Volcanoes in Idaho and at Yellowstone National Park are the result of a hotspot that has been active for at least 17 million years.

3.2.1 Eruption of Magma

When magma reaches the surface of the Earth, it is erupted to the surface through a *vent*. A vent is an opening or crack in the Earth. When it reaches the surface, it is *lava*. The only difference between magma and lava is that magma is molten rock found below the surface of the Earth, and lava is molten rock above the surface of the Earth.

There are a few dozen types of rock formed when magma cools and hardens. These rocks are all types of *igneous rocks*. For our purposes, we will divide igneous rocks into two groups: those formed at or above the surface of the Earth (*extrusive* or *volcanic*) and those formed below the surface of the Earth (*intrusive* or *plutonic*). At a fundamental level, igneous rocks are classified based on their texture and composition. *Texture* usually refers to the size of the crystals that make up the rock, but can also refer to the particles that make up the rock. For example, a common type of volcanic rock is *basalt* (Fig. 3.14). Basalt typically is composed mainly of the minerals feldspar, olivine, and orthopyroxene, which are gray, green, and black, respectively. The crystals are microscopic, giving the rock a matte appearance. Its texture is *aphanitic*, meaning it is composed primarily of microscopic crystals.

In Figure 3.15, the rock basalt is pictured next to a colorful circle. The circle is a view of a *thin section*, a thin slice of the rock through a microscope with polarized light shining through it. The colors and shapes in the photo are individual minerals. The rock sample in Figure 3.15 is weathered, which means it has been exposed to the weather at the Earth's surface for many years. Weathering causes the minerals in rocks to break down. This rock has taken on a brown appearance due to the oxidation of iron-rich minerals within it. Fresh basalt is black (see Fig. 3.14). Notice that the individual minerals that make up the rock are tiny, visible only under the microscope. This is what aphanitic texture looks like. This is a texture common in rocks that are formed above the surface of the Earth. Aphanitic rocks are extrusive, or volcanic.

Another texture is glassy, which means there are no crystals. *Obsidian* (Fig. 3.16) is a black glass that forms when lava does not contain much gas (because it has been released earlier in an eruption) and cools extremely quickly. One large and publicly accessible obsidian flow

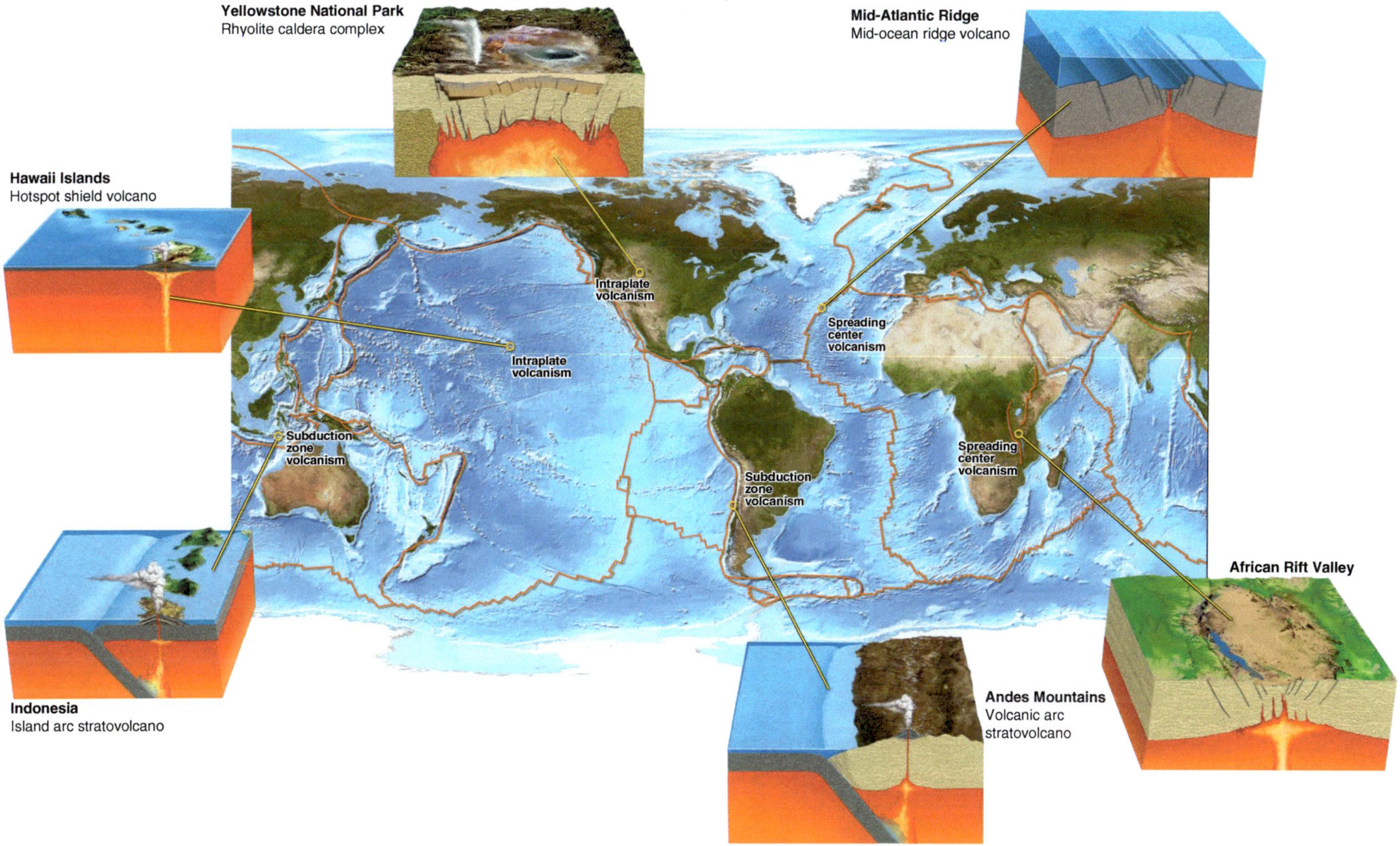

Figure 3.12 Examples of volcanic locations around the world. Most volcanic activity in the world occurs at divergent and convergent plate boundaries. Intraplate volcanic activity is often caused by hotspots. Yellowstone and the Hawaiian Islands in the United States are the result of hotspot volcanism. From Fletcher, 2014.

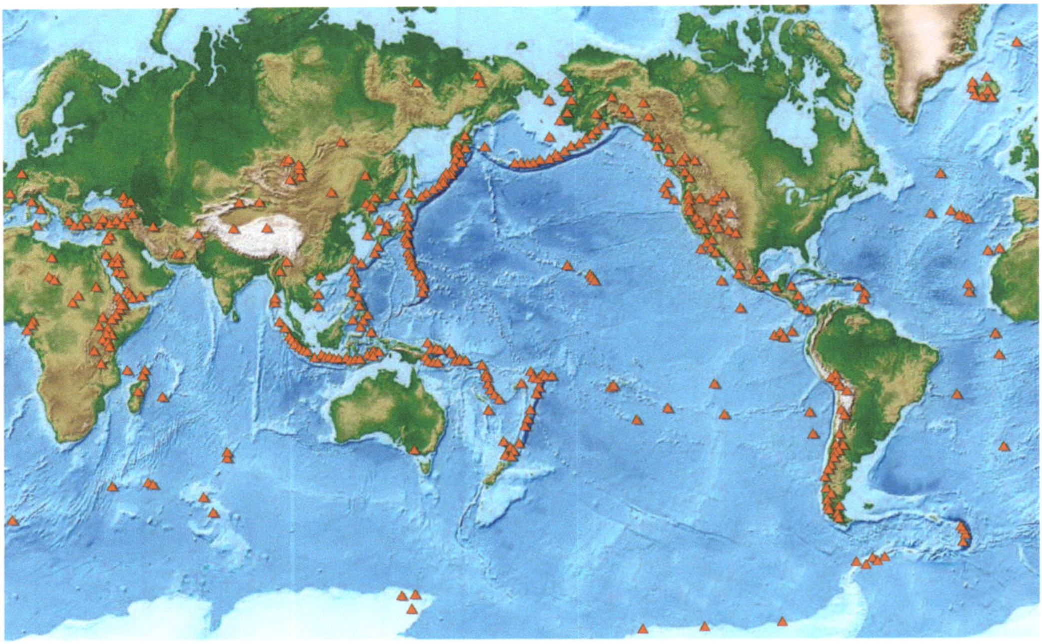

Figure 3.13 Worldwide occurrences of volcanoes. Note that most of the volcanoes are located at plate boundaries. Volcanoes within the interiors of plates are often the result of hotspot activity. From Fletcher, 2014.

Figure 3.14 Basalt lava on the volcano Kīlauea. Basalt is a black extrusive igneous rock. Phillippe Bourseillier/The Image Bank/Getty Images. From Murck & Skinner, 2016.

Figure 3.15 Two views of the rock basalt. The rock in the photograph is a sample of basalt as it appears to the observer. The circular image is a view of a thin slice of basalt as seen through a microscope with polarized light shining through it. Individual minerals that compose the rock are only visible under magnification. PASIEKA/SCIENCE PHOTOLIBRARY/ Photo Researchers, Inc. From Fletcher, 2014.

is located at Newberry Crater, Oregon. It formed when the gases in a magma body escaped through two previous eruptions of the Newberry Crater Volcano, and left gas-poor magma that erupted and cooled quickly. Obsidian forms at the surface of the Earth, making it a volcanic or extrusive igneous rock.

Yet another texture is *vesicular*. This means the rock is full of holes (Fig. 3.17). Holes form

Figure 3.16 The volcanic rock obsidian, which has a glassy texture. Slim Sepp/Shutterstock.com. From Murck & Skinner, 2016.

Figure 3.17 A close-up view of vesicular basalt. This rock has a vesicular texture, which is caused when the rock solidifies around gas bubbles trapped in the lava. Vesicles (holes) may be fairly large, as seen here, or so small they are barely visible. From Fletcher, 2016.

in a rock when gas-rich lava is erupted at the Earth's surface. The rock in the photo is a vesicular basalt, which means it is a basalt that is full of holes. The holes are gas bubbles that were preserved in the rock when it cooled and solidified. A rock called *pumice* exhibits a vesicular texture. The holes in it are tiny but permeate the rock, making it so light it often floats on water (Fig. 3.18)! All of these are common textures for volcanic igneous rocks.

Plutonic igneous rocks form beneath the surface of the Earth, when magma cools slowly over hundreds to millions of years. They are composed entirely of minerals large enough to be seen with the naked eye. The minerals are able to grow this large because the magma in which they formed are kept insulated within the Earth, meaning they cool and form extremely slowly. The slow cooling gives elements in the magma time to arrange themselves into larger crystals. This is a distinct texture, which makes intrusive or plutonic igneous rocks easy to identify. The igneous rock texture in which all minerals are visible to the naked eye is called *phaneritic*. Figure 3.19 is a photograph of the rock *diorite*, which is a particularly striking example of a phaneritic igneous rock. This sample is composed of black and white minerals. The circle in the lower right side of the photograph is a view through a microscope of a thin slice of the rock with polarized light shining through. As you can see, the microscopic view is of large minerals.

Composition is the second criterion we use to identify igneous rocks. Composition is sometimes linked to the color of the rock. When scientists study rocks, they investigate rocks in many ways. As you have already seen, they both visually inspect the rock and look at it under a microscope. A third way they investigate rocks is through chemical analysis. A small sample of the rock is crushed and chemically analyzed, which gives the scientist detailed information about the rock's composition. The component of a rock's composition that is most important for identifying igneous

Figure 3.18 A scientist hefts a large piece of pumice above his head. Photo by Diane Cook, Len Jenshe/ National Geographic Creative. From Murck & Skinner, 2016.

Figure 3.19 Two views of the rock diorite. Diorite is intermediate in composition, and is often composed of half black, half white minerals. Within a diorite, the black minerals are typically amphiboles and pyroxenes, and the white minerals are usually feldspars. The individual minerals are large enough to see with the naked eye. In the microscopic view of the rock (circular image) the individual minerals are both large and easy to see. From Fletcher, 2014.

rocks is the percentage of the rock composed of silica, or silicon dioxide (SiO_2). A rock which contains about 45% to 54% silica is typically black or red in color, and is said to be *basaltic* in composition. The extrusive rock is called basalt (see Fig. 3.14). The intrusive, or plutonic, rock with the identical composition is called *gabbro* (Fig. 3.20). The composition of the next category of rock, which contains 55% to 69% silica, is called *andesitic*. Andesitic rocks are typically a medium gray, brown, or half black and half white. The volcanic andesitic rock is called *andesite*, and the plutonic andesitic rock is called diorite. Rocks with more than 69% silica are *rhyolitic*. They are usually light in color, often appearing as light gray, white, pink, or orange. The volcanic rock that is rhyolitic in composition is called *rhyolite*. The plutonic rhyolitic rock is called *granite*.

3.3 Volcano Behavior

Composition is one of the most important factors in determining the type of volcano and how the volcano is likely to behave. Lower-silica lava, such as basaltic lava, tends to have lower *viscosity* than other types of magma. (Viscosity is a measure of a substance's resistance to flow.) This means lava can flow for impressive distances across the surface of the

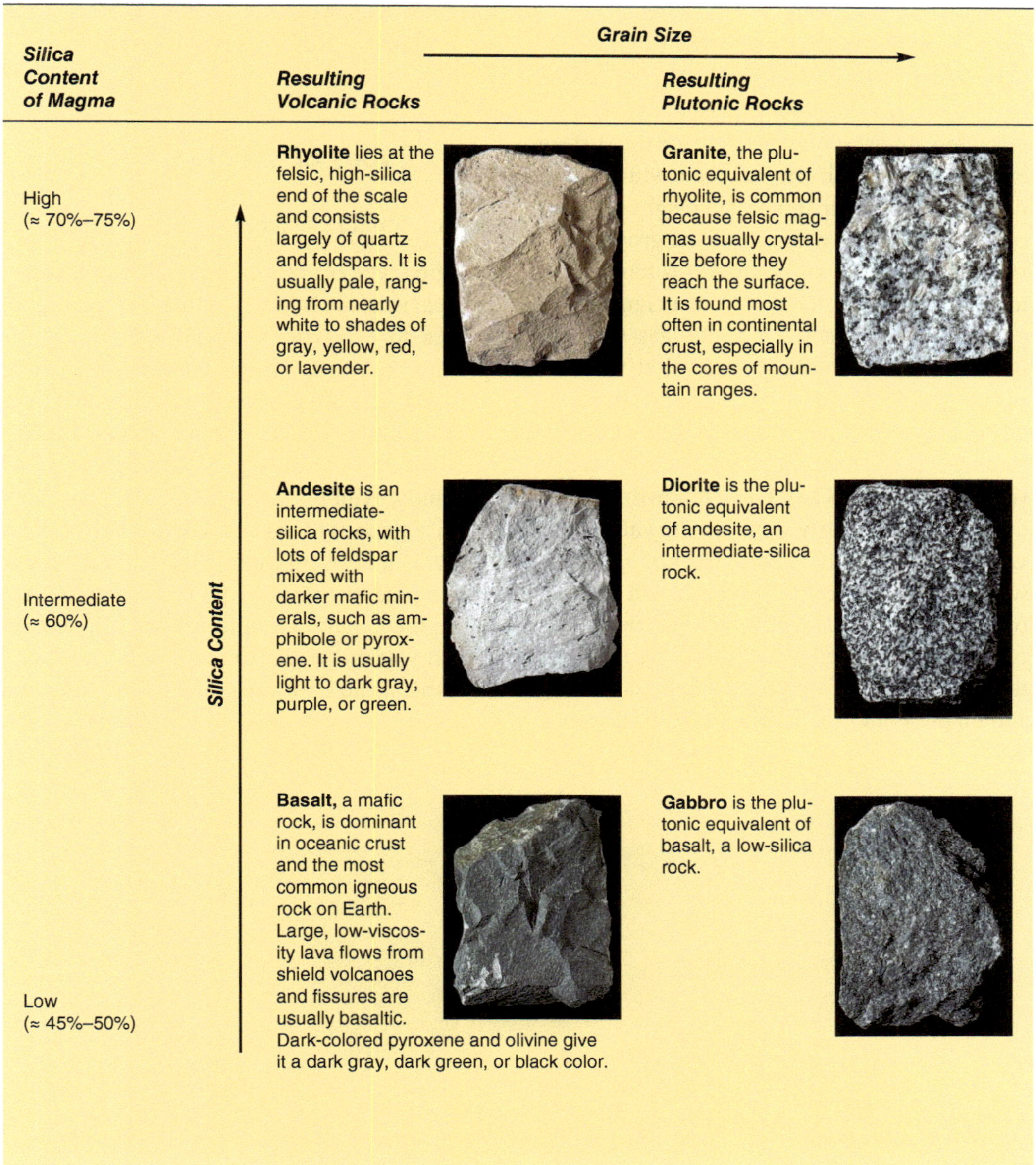

Figure 3.20 Common volcanic and plutonic rocks organized by composition and texture. From Skinner & Murck, 2011.

Earth. The low viscosity allows the gases within the magma to escape easily and pressure can't build up to high levels. The gas leaks out of cracks in the rock near the surface, and the little gas pressure that is built up before an eruption is usually dissipated fairly quickly in fountaining eruptions. This typically happens within a few hours to a few days, but can continue for months if the eruption volume is high. Lava flows are erupted as part of these fountaining episodes and after the fountains dissipate, and eruptions can continue for years, as demonstrated by Kīlauea's 1983–2018 eruption.

Volcanoes with higher silica magmas are much more dangerous and violent than those with lower silica magmas. Higher silica magma, such as rhyolite, is much more viscous. It is often almost solid when it nears the surface of the Earth. Lava, if erupted at all, will produce flows short in length, but tall in height. In some cases, lava simply rises out of the ground and sits on top of the vent because it is too viscous to flow anywhere. This type of magma traps gases very effectively. Once the gas pressure within a volcano builds to a level at which it cannot be contained by the overlying rocks, an explosive eruption occurs. Magma and solid rock are pulverized into tiny particles called *volcanic ash* (Fig. 3.21). Huge billowing clouds of ash can rise nearly 8 mi (12 km) above the volcano, and distributed hundreds of mi downwind from the volcano. Clouds of ash and hot rocks can create hot avalanches of superheated rocks and ash called *pyroclastic flows* that roar down the slopes of a volcano at speeds nearing 100 mi per hour (161 km/hr). We will examine hazards posed by different volcanoes, and the magmas responsible for building those types of volcanoes, in the following sections.

3.4 Types of Volcanoes

3.4.1 Shield Volcanoes

The first type of volcano we will cover is called a *shield volcano*. It gets its name from its resemblance to a warrior's shield laid flat on the ground. It has a large diameter and gentle slopes. These are among the largest volcanoes both on Earth and in our solar system. The largest volcano in the solar system is the shield volcano Olympus Mons, found on the planet Mars. Olympus Mons is 388 mi (624 km) across and 15.5 mi (25 km) high. It is roughly the size of the state of Arizona, and is more than three times the height of Mount Everest. The largest shield volcano on Earth is Mauna Loa, found on the island of Hawai'i. The subaerial (above-sea level) portion of Mauna Loa is 75 mi (120 km) long and 13,677 ft (4,169 m)

Figure 3.21 An eruption of Mount St. Helens (Washington, USA) in May of 1980 produced huge billowing clouds of volcanic ash. Volcanic ash is usually gray in color, and is composed of tiny fragments of pulverized rock and magma. From Skinner & Murck, 2011.

Figure 3.22 Mauna Kea, a shield volcano in Hawai'i. Shield volcanoes tend to have a large footprint, but they are not steep sided. Rocks in the foreground are from Mauna Loa, another large shield volcano. From Skinner & Murck, 2011.

above sea level. Mauna Loa's neighbor Mauna Kea (Fig. 3.22) is slightly taller at 13,803 ft (4,207 m). Both volcanoes are tall enough they often have snow coating their summits, especially during the winter months. Shield volcanoes are also present in the Galapagos Islands, Iceland, and the Canary Islands near Spain, and on Reunion Island in the Indian Ocean.

Typical shield volcanoes erupt basaltic lavas. Basalt is a dark colored rock that, when molten, has a relatively low viscosity. This allows the basalt to flow downhill, often great distances (Fig. 3.23). This creates a large volcano with gentle slopes. In Hawai'i, where shield volcanoes are the norm, individual lava flows can be traced for dozens of miles, from the vent to the ocean.

The main hazards from shield volcanoes are lava flows and the gases that come from the volcano. Lava flows follow the contours of the land as they move downhill. The speed of a lava flow varies greatly with the slope of the land over which it is flowing. Fast moving lava flows on steep slopes may travel at speeds of several feet per second. On flat ground, lava may take several minutes or even longer to

Figure 3.23 A lava flow in Hawai'i in 1983. The lava is organized in a channel, which allows it to flow efficiently. The lava can travel many miles from the vent (in the background). From Skinner & Murck, 2011.

travel a single yard. While we say basaltic lava has low viscosity, this is a relative term. Molten basalt has the consistency of cold honey. On flat ground, this means flows often wander around and settle in low spots. When low spots are filled in, the lava will again begin lateral or downhill movement. In these flat areas, lava flows are quite approachable. In Hawai'i Volcanoes National Park, when lava flows are accessible, tourists are able to walk right up to lava flows and observe them as closely as the intense heat will allow (Fig. 3.24). These lava flows are at least 1,800° F, or more than 1,000° C.

Since basaltic lava flows on flat ground typically do not move terribly quickly (generally just 10 to 20 ft (3 to 6 mi/hr), it is a rare occurrence when lava claims human life, or lives of animals that are able to fly or run away. Most at risk from lava flows are stationary objects such as trees, shrubs, grasses, homes, businesses, churches, and roads. Of course animals that may live within small patches of forest burned or covered by lava flows would also be at risk, as could people trapped between fast-moving lava flows. In the 1983–2018 eruption of Kīlauea in Hawai`i, 8.9 mi (14.3 km) of road within Hawai'i Volcanoes National Park, and 30 mi (48 km) outside the park were covered with lava during the volcano's 35 year eruption. In all, this eruption claimed about 930 structures, more than 900 of which were homes. Lava flows were not responsible for a single death, however.

Basaltic lava flows come in two varieties (Fig. 3.25). One, with a rope-like or smooth surface, is called *pāhoehoe*. This is a common type of lava flow that occurs when the lava is still gas-rich and moving smoothly. A second variety, called *`a`ā*, has a rough and jagged surface. The lava is stretched into sharp pieces of rubble that litter the top and front of the flow. `A`ā is typically formed when gases have escaped the lava flow, making the liquid magma gas-poor and more viscous. This often happens after flowing down a steep slope where the rapid

Figure 3.24 A slow-moving lava flow on the island of Hawai'i. Because the lava is flowing so slowly, this scientist is able to approach the flow and take a sample with his rock hammer. Note that he is wearing leather boots, gloves, and cotton clothes. He is shielding his face with his cotton shirt. These materials provide modest protection from the lava's heat, which would burn exposed skin instantly. University of Hawaii photo by Robert Chinn. From Fletcher, 2014.

(a)

(b)

Figure 3.25 (a) An active pāhoehoe flow. The surface of this lava flow looks like coils of rope. © Martin Rietze/ Westend61/Corbis. From Fletcher, 2014. (b) An active `a`ā flow. The surface of this lava flow is made of jagged chunks of sharp rocks. Photo courtesy of U.S. Geological Survey, Hawaiian Volcano Observatory. From Fletcher, 2014.

movement has allowed gases to escape relatively quickly.

As mentioned earlier, lava has the ability to flow great distances from the vent. One mechanism by which this occurs is through a lava tube. Lava tubes are created when the surface of a lava flow cools and forms a roof over the liquid lava below. The roof insulates the lava inside and keeps it hot so it can flow farther than it could if it were exposed to the much cooler air. Lava tubes are common sights in volcanic landscapes, but nowhere are they more common than on shield volcanoes. Lava tubes form extensive cave networks on many shield volcanoes. Recently, scientists have suggested colonies both on the moon and on Mars could be built within the large lava tubes on these bodies. The roofs of the lava tubes could protect future astronauts from cosmic radiation and micrometeorites, regulate temperature swings, and provide natural barriers against which to build containment structures.

A far more dangerous hazard to human health and life is volcanic gases. Volcanic gases, regardless of volcano type, are the same, though the amounts of gases released varies by type. They include water vapor, carbon dioxide, hydrogen sulfide, sulfur dioxide, chlorine, fluorine, and assorted other trace gases. The gases themselves can be quite deadly. Water vapor can scald. It can also combine with other gases like sulfur dioxide to form corrosive acids. Carbon dioxide is more dense than air. If carbon dioxide settles in a valley, it can suffocate people in the valley because it displaces air that contains the oxygen people need to breathe.

An eruption of carbon dioxide gas from Lake Nyos in Cameroon in 1986 resulted in the death of over 1,700 people and 3,500 head of livestock. A large amount of carbon dioxide had been seeping from cracks underneath the lake for years and was trapped beneath the water. One October night, the carbon dioxide escaped to the surface and burst onto the landscape. People and livestock were killed not from a volcanic blast, but when carbon dioxide displaced the air around the shore of the lake and suffocated them. Since this tragic event, engineers have installed a system to safely remove and disperse gases from the bottom of the lake through a series of pipes and a fountain. Carbon dioxide is no longer able to accumulate in large volumes on the bottom of the lake.

Hydrogen sulfide smells like rotten eggs, making volcanoes very stinky places to be. Sulfur dioxide combines with water vapor in the air to form sulfuric acid. The acid is toxic. If inhaled, it burns the mucous membranes and lungs, and can bring about permanent lung damage. It is also corrosive to anything metal in the vicinity, and is toxic to vegetation. Chlorine and fluorine gas are usually present only in minute concentrations, but they too are toxic to humans and livestock. In areas where fluorine is present in fresh volcanic deposits, livestock have developed fluoride poisoning from eating vegetation coated with volcanic ash. This weakens the teeth and bones and causes a slow, agonizing death.

Shield volcanoes are so large they often have several dozen to several hundreds of individual vents. At shield volcanoes, most vents are located at the summit or along zones of weakness emanating from the summit, called rift zones. It is common for shield volcanoes to have two, or occasionally three, rift zones along which eruptions can occur. Kīlauea, on the island of Hawai`i, has two rift zones. The East Rift Zone has been the center of activity for the last several decades. It also has a Southwest Rift Zone, which is far less active. The Southwest Rift Zone was last active in 1974.

Vents on volcanoes may be long cracks in the ground, which only erupt for a few hours. Cracks in the ground along which eruptions occur are called fissures. Fissure eruptions are spectacular to witness (see Fig. 3.2). These eruptions also go by the rather descriptive name "curtain of fire." Often, a fissure eruption lasts only a short time before it narrows to a single point along the crack. At that point, it may start forming a cone. The hill can be a miniature version of a shield volcano (called a *lava shield*) if lava flows freely from the vent. It could also become a cinder cone or spatter cone if the eruption is slightly more energetic (Fig. 3.26). A *cinder cone* is a hill built from small pieces of lava that have been erupted into the air, cooled while in the air, and come back down as a solid rock (Fig. 3.27). These small pieces of lava are called *scoria*, and are informally known as cinders. They are sometimes mined and used as decorative landscaping stone. A spatter cone is similar, except that

Figure 3.26 Short time lapse photo of a cinder cone in eruption in Kivu, Zaire. The beautiful arcs of red and yellow are the trajectories of individual pieces of spatter and bombs as they are erupted from the crater. From Skinner & Murck, 2011.

Figure 3.27 Satellite image of SP Crater cinder cone in Arizona, USA. A cinder cone is composed of scoria, bombs, spatter, and lava flows. Note the arc-shaped black lava flow that extends toward right of this photo. NASA Earth Observatory image by Jesse Allen and Robert Simmon, using ALI data from the NASA EO-1 team. Public domain, via Wikimedia Commons.

the lava is still only semisolid when it reaches the ground. As a result, the appearance of the cone is more solid, but obviously composed of individual pieces that were soft and deformable as they reached the ground. In both cases, the cone builds up around the vent, and has steeper sides than a lava shield. Both a small lava shield and a cinder or spatter cone can issue lava flows. Both are also capable of erupting *bombs* (Fig. 3.28), which are larger blocks or blobs of lava catapulted through the air. If the lava is liquid, it spins in the air and often takes on a shape that is tapered at the ends and wide in the center. This is called a *spindle bomb*. Another common type of bomb is called a *bread crust bomb*, which has a surface that looks like crusty bread. *Blocks* are pieces of solid rock hurled out of a volcano's crater (Fig. 3.29). Close examination of a cinder cone often reveals the presence of all of these types of materials: cinders, spatter, spindle bombs, bread crust bombs, and blocks.

At times, fissure eruptions continue for months or even years. This is rare, but has happened a number of times in the geologic past. The most recent time was in Iceland over an 8 month period in 1782–1783. The volcano Laki erupted from a number of fissures simultaneously. An estimated 2.5 mi^3 (14.7 km^3) of lava was erupted from 140 vents along a 14.3 mi (23 km) long fissure system. Volcanic gases that escaped from the volcano created volcanic fog, which not only affected Iceland, but also much of western Europe. A famine ensued and, as a result, about 25% of the population of Iceland perished.

Figure 3.28 Small bombs erupted from a volcano. These are spindle bombs, which are thicker in the center and taper toward the ends. Bombs are formed when liquid lava is thrown in the air from an erupting volcano. The bombs cool and solidify before they hit the ground. J.P. Lockwood/Volcano Hazards Program/U.S. Geological Survey. From Fletcher, 2014.

Figure 3.29 Basalt blocks. Blocks are pieces of solid rock that are hurled from an erupting volcano. They were not composed of liquid lava at the time of eruption. Instead, they were solid rock ripped from the edges of the crater and thrown skyward. Red Eckmann, University of Portland.

In the distant past, fissures were responsible for *flood basalts*, catastrophic long-lived eruptions of basalt that covered vast landscapes to thicknesses of over a mile (1.6 km). The Pacific Northwest of the United States is one of these areas. The Columbia River basalts (Fig. 3.30) were erupted over approximately 11 million years, with about 90% of eruptive activity occurring in the 1.1 million year period between 16.7 million and 15.6 million years ago. The total volume of rock in the Columbia River basalts is estimated at 42,000 mi^3 (175,000 km^3). Lava flows cover the eastern two-thirds of Washington and Oregon, and reach into western Idaho.

Another type of feature seen on most shield volcanoes is a *caldera*, or depressed area, at or near the summit. A caldera is different from a *crater* in that a crater is formed by explosions outward, and a caldera is formed by collapse inward. A caldera typically forms when magma drains from a chamber beneath the volcano's

Figure 3.30 Several stacked lava flows from the Columbia River Basalts. Photo was taken from Rowena Crest Viewpoint, Oregon. Photo by Liz Westby/USGS.

surface, leaving the area above unsupported (Fig. 3.31). The unsupported land then collapses, leaving a depression. Though this most commonly happens at the volcano's summit, it can happen at other locations on the volcano as well. The Hawaiian volcano Mauna Loa has three summit calderas, and the volcano Kīlauea has one large caldera. Both of these volcanoes also have dozens of several smaller calderas known as pit craters.

Calderas can also occur on large composite volcanoes. Crater Lake is one example (Fig. 3.32). About 7,500 years ago, Mount Mazama experienced a large and violent eruption. Much of the magma chamber beneath the mountain was emptied, and the summit of the volcano collapsed and formed a caldera. In the time since that eruption, the caldera has filled with water, creating the spectacular Crater Lake. Crater Lake, however, should be called Caldera Lake if we are to accurate describing its origins.

On occasion, shield volcanoes are capable of more explosive eruptions as evidenced in the 2018 eruption of Kīlauea and another famous eruption, which occurred in the same location in 1924 (Fig. 3.33). Examination of deposits around Kīlauea's summit reveals evidence that this happens periodically and is not a terribly rare occurrence. One product of these more explosive eruptions is volcanic ash. There are ash layers in between some of the layers of lava within the walls of the summit caldera. The presence of these ash layers is proof that explosive eruptions are not rare events at the summit of Kīlauea.

3.4.2 Cinder Cones

Cinder cones are often found on shield volcanoes, but also can be found as stand-alone volcanoes. A cinder cone is the smallest of the three volcano types we will examine in this text. Hazards related to cinder cones are similar

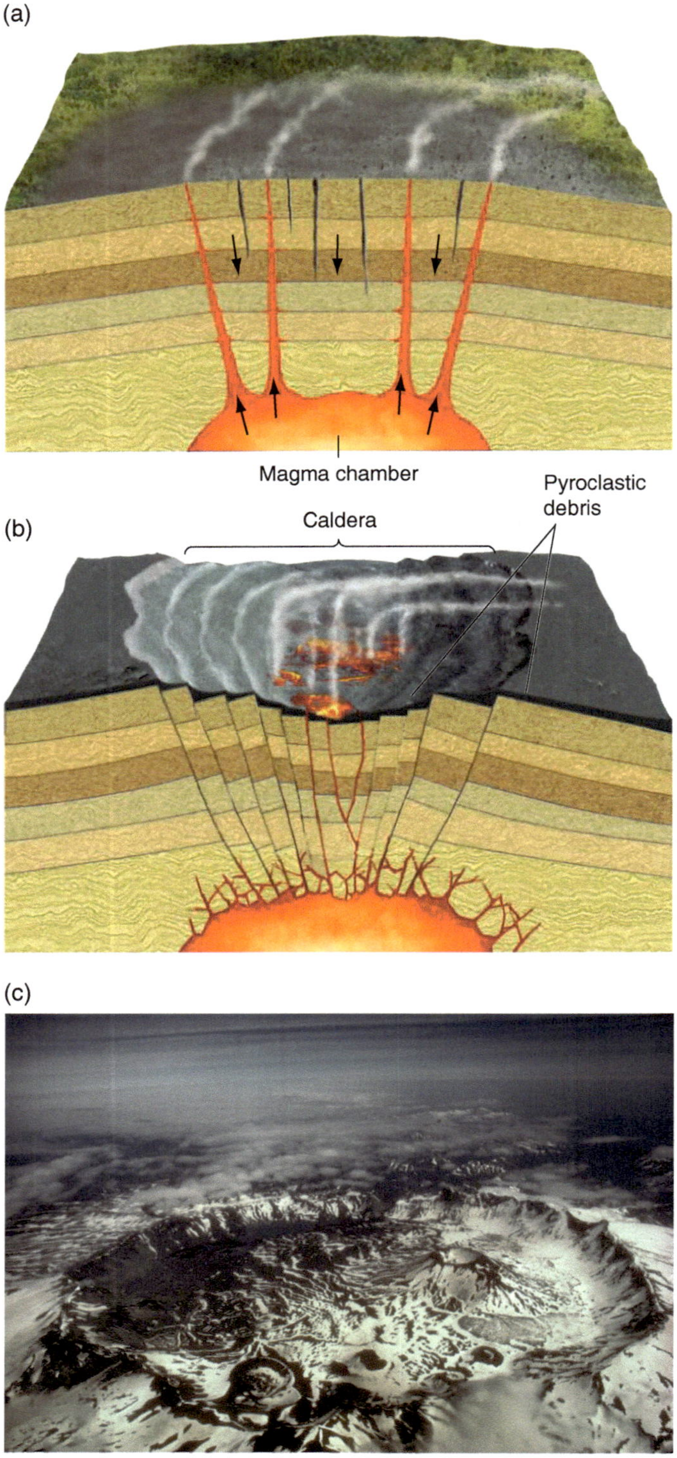

Figure 3.31 Formation of a caldera. (a) A volcano forms above a magma chamber. From Fletcher, 2014. (b) The volcano erupts a large volume of magma, emptying the top of the magma chamber beneath the volcano. The summit of the volcano is left unsupported, and it collapses downward. From Fletcher, 2014. (c) Photo of Aniakchak caldera on the Alaska Peninsula. M. Williams, National Park Service. From Wikimedia Commons.

Figure 3.32 Photo of beautiful Crater Lake, Oregon. Crater Lake is a caldera created roughly 7,500 years ago with a large volume eruption of Mount Mazama. The top of the volcano collapsed, leaving a caldera approximately 5 mi (8 km) in diameter. Photo by Mike Doukas, US Geological Survey.

Figure 3.33 Spectacular explosive eruption at the summit of Kīlauea in Hawai'i in May of 1924. The volcano repeated this style of eruption throughout much of the summer of 2018, drastically changing the appearance of the summit of the volcano. Photo by Kenichi Maehara, courtesy of the Bishop Museum via the U.S. Geological Survey; https://volcanoes.usgs.gov/vsc/images/image_mngr/5000-5099/img5016.jpg

Figure 3.34 Sunset Crater near Flagstaff, Arizona. This young cinder cone erupted in approximately 1085, displacing people who lived and farmed in this area. Sunset Crater is one of many cinder cones in the San Francisco Volcanic Field in northern Arizona. Photo courtesy of the U.S. Geological Survey.

to those of shield volcanoes. Lava flows and volcanic gases are the primary dangers, though cinder cones are perhaps more likely than many shield volcanoes to produce spatter and bombs.

A cinder cone is a hill formed from scoria (cinders) erupted from a single vent (Fig. 3.34). Typically, cinder cones only erupt once (making them *monogenetic*, or formed from a single eruption), though the eruption may go on for several years. In many cases, they exist in *volcanic fields* where there is a dense population of cinder cones of varying ages (Fig. 3.35). In the United States, volcanic fields with cinder cones exist in Arizona (near Flagstaff), Idaho (Craters of the Moon National Volcanic Monument), and New Mexico. One of the best documented and easiest to access is Capulin Volcano in northeastern New Mexico. Capulin is the centerpiece of Capulin Volcano National Monument. This is a classic example of a cinder cone, and is well preserved. In addition, it is accessibly by car, making the volcano easy to study and observe. From a

Figure 3.35 The San Francisco Volcanic Field in Arizona. More than two dozen individual volcanoes are visible in this aerial view of a small portion of the volcanic field. Cinder cones often form in close proximity to other cinder cones. Courtesy Zoltan Sylvester. From Fletcher, 2014.

visitor's viewpoint, it is possible to look down into the crater and imagine bombs being hurled from it. Close observers of the volcano's structure on the way to the top will notice spindle bombs and bread crust bombs in addition to the cinder and spatter that form the bulk of the volcano. Capulin is between 53,000 and 57,000 years old, and is part of the Raton-Clayton Volcanic Field. The entire area is dotted with cinder cones, and Capulin appears to be the youngest of the group.

The volcanoes in Figures 3.34 and 3.35 are found in the San Francisco Volcanic Field in northern Arizona. Sunset Crater is the youngest and best preserved of these volcanoes. Its last eruption occurred in approximately 1085 A.D. A group of farmers lived in the area at the time. The ruins of their homes can be found in the area, burned and filled with volcanic debris. The people moved on to nearby valleys and continued farming. Their descendants, the Hopi and the Zuni, still live in the region, and memories of the eruption have been passed down through the centuries in stories and traditions of the Native Americans in that area.

A cinder cone in Mexico transfixed the volcano community for 9 years beginning in 1943. In a somewhat remote area of Mexico, within a volcanic field already rich with cinder cones, a farmer made his living growing corn. One morning as he went to tend to his crop, he noticed smoke rising from a corner of his field. When he reached the smoke, he found it was coming out of a hole in the ground. The smoke was accompanied by foul-smelling sulfur dioxide mixed with small amounts of volcanic ash and small stones. Within a week, there was a cinder cone over nearly 500 ft (150 m) high where his corn field once stood. By the end of 1943, the volcano was over 1,000 ft (330 m) tall. The volcano was named Paricutin. Over the next several years, the cinder cone Paricutin grew in size and produced a number of lava flows that covered surrounding cropland and eventually buried the nearby town of San Juan Parangaricutiro. Volcanologists from Mexico and the United States kept constant watch on the volcano from the eruption's first days until its last days. They witnessed the eruption of bombs, cinders, `a`ā, and pāhoehoe lava flows, and volcanic ash. The eruption was incredibly well documented and served as a living laboratory for many volcanologists actively working in Mexico at that time. Unfortunately, it also displaced hundreds of people who were unable to return to their homes or their farms, which were covered with many feet of lava. Scientists do not expect Paricutin will erupt again, though it is likely other cinder cones will appear within the same volcanic field in the future.

3.4.3 Composite Volcanoes, Stratovolcanoes, or Composite Cones

A third type of volcano is called a *composite volcano*. This type of volcano also goes by the name stratovolcano or composite cone. A composite volcano is composed of a variety of materials, including lava flows, volcanic ash, mud flows, bombs, scoria, and spatter. The composition of the lava that makes up the volcano is quite variable as well. In many cases, the lower parts of the volcano are composed of basalt, and the upper parts are composed of other types of lava, such as andesite and rhyolite, though any type of lava can erupt from this type of volcano at any time in its history. It may have one style of eruption for years, then switch styles with the next eruption. Composite volcanoes are often tall and steep sided (Fig. 3.36).

3.5 The Case of Mount St. Helens

In early 1980, the volcano Mount St. Helens in southwestern Washington was a serene, symmetrical mountain (Fig. 3.37) that was often compared favorably with the beautiful Mount Fuji in Japan. The mountain was a prominent landmark and a popular recreation spot in both the summer and the winter. Families camped,

(a)

(b)

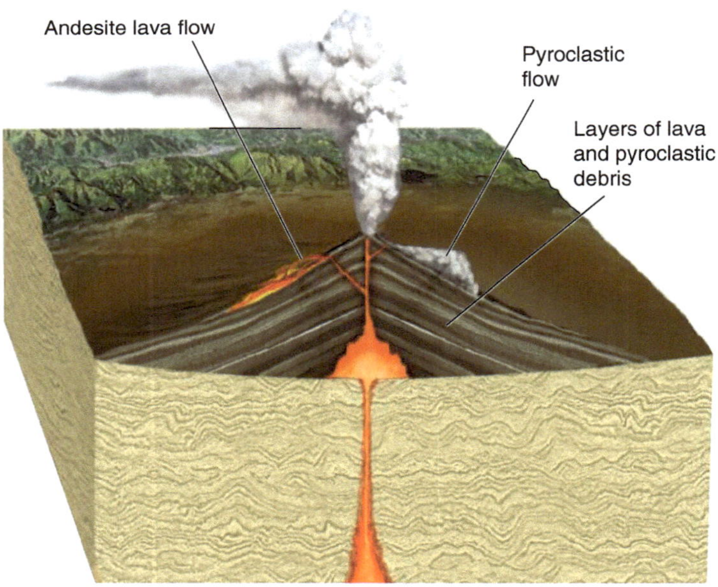

Figure 3.36 (a) Mount. Fuji, Japan. Fuji has the classic shape of a composite volcano. It is a tall, steep-sided conical volcano. Gavin Hellier/Getty Images. From Fletcher, 2014. (b) Diagram of a composite volcano. The cone is composed of layers of lava and pyroclastic debris such as ash, bombs, blocks, and mud flow deposits. These volcanoes may erupt from the summit or flank (side). A pyroclastic flow rushes downward from the summit, carrying toxic gases and hot volcanic ash.

Figure 3.37 Photo of Mount St. Helens prior to the 1980 eruption. Image courtesy of the U.S. Geological Survey; https://volcanoes.usgs.gov/vsc/images/image_mngr/900-999/img995.jpg

swam, hiked, snowmobiled, fished, and hunted in the shadow of the volcano. The mountain was surrounded by pine forests. Logging companies actively worked the area and provided employment for thousands of people.

In March of 1980, the volcano began to stir. Small earthquakes, most less than magnitude 3, rocked the volcano at shallower and shallower depths. On 27 March 1980, the first explosion occurred at the volcano, blasting a hole in the top of the mountain and the glacier covering its summit. Within the next few days, a second crater formed. Shortly thereafter, the two craters merged. Gray volcanic ash covered the formerly pristine white snow near the mountain's summit. Area residents were excited about the volcanic activity, because it is a rare sight in the Pacific Northwest. The last volcanic eruption in the continental United States prior to Mount St. Helens in 1980 was the eruption of Mount Lassen, California, in 1912–1914. The last eruption at Mount St. Helens was about 1860. Native Americans lived in the area at that time, but no written records exist of their observations from that eruption. Citizens of the area in 1980 had little frame of reference for how a large composite volcano was likely to behave. Many people assumed the volcano would erupt much like the Hawaiian volcanoes they had seen on television and on vacation, and they gave excited interviews to the press expressing impatience that the volcano had not put on a good show yet. However, Mount St. Helens is a composite volcano, and behaves very differently from the shield volcanoes in Hawai'i.

As soon as unrest was discovered at Mount St. Helens, volcanologists with the U.S. Geological Survey set up 24 hr monitoring stations on and near the volcano, and convinced the local law enforcement agencies to establish an exclusion zone at a 5 mi (8 km) radius around the volcano. People were evacuated from their homes, and visitors were kept at what they felt was a safe distance, outside the exclusion zone. Scientists installed extra seismometers to better detect earthquakes. They set up stations where they could directly measure the deformation of the volcano as it swelled with magma from within. (Volcanoes swell when they are being inflated with magma, and that deformation can be detected by surveying instruments.)

They monitored sulfur dioxide emissions with specialized instruments. They set up observation posts where geologists could view the volcano directly, then radio information about happenings at the volcano in real time to their base station near Vancouver, Washington. Throughout the spring of 1980, the volcano continued to emit small amounts of volcanic ash and have periodic small explosions, and the crater enlarged. The volcano drew visitors from around the world.

In early April, scientists noted that the north side of the volcano seemed to be expanding. Measurements conducted on that part of the volcano indicated it was expanding outward at a rate approaching 5 ft (just over 1.5 m) per day. Scientists initially thought their measurements were incorrect, but time and further measurement proved the rate of expansion was correct. This activity created a large bulge on the north side of the volcano (Fig. 3.38). By the beginning of May, geologists told people in the area that landslides were likely from the bulge, and urged them to be alert for warning sirens that would warn them of flooding in the Toutle River valley north of the volcano due to landslides on the mountain. Scientists were concerned that a very large landslide might be possible because the north face of the mountain was clearly unstable.

The volcano had experienced a decrease in activity in late April and early May. Emission of volcanic gases was occurring at a low rate, and there were few explosions from the summit crater. But there was the ominous bulge. By 17 May, the bulge had grown outward about 450 ft (135 m). People who lived and owned summer cabins in the exclusion zones viewed the lack of explosive activity as a sign that the volcano was not in a dangerous phase, and were impatient to get into their properties. Volcanologists knew that regardless of its current state of activity, it was possible the volcano was building toward a large eruption.

At 8:32 a.m. on the morning of Sunday, 18 May 1980, there was a magnitude 5.2 earthquake directly beneath the volcano. This triggered a massive landslide of the entire north flank of the volcano where the mountain had been expanding outward (Fig. 3.39). Debris

Figure 3.38 A U.S. Geological Survey scientist surveys the conspicuous bulge on the north side of Mount St. Helens in April of 1980. Image courtesy of the U.S. Geological Survey; https://prd-wret.s3-us-west-2.amazonaws.com/assets/palladium/production/s3fs-public/vhp_img519.jpg

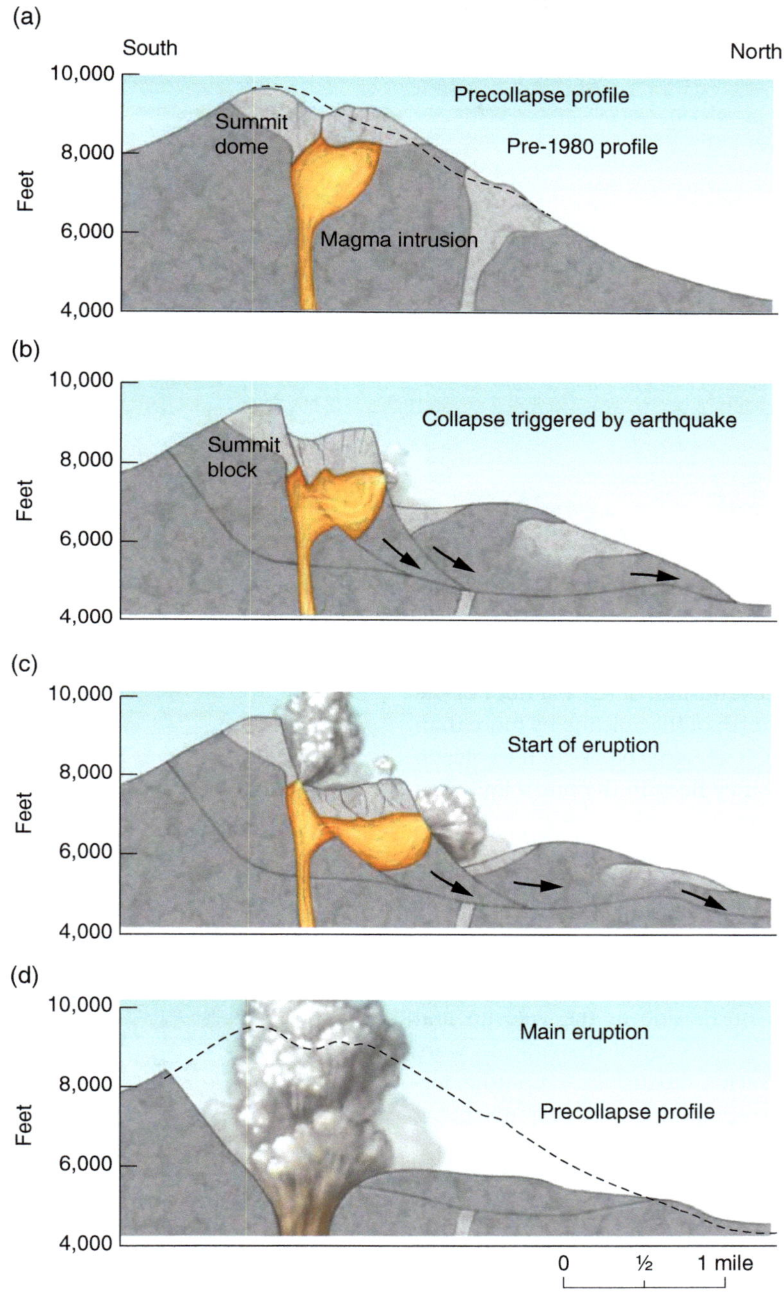

Figure 3.39 Sequence of events in the 18 May 1980 eruption of Mount St. Helens, Washington. (a) The eruption was triggered by an earthquake that caused a catastrophic collapse of the north flank of the volcano. A large landslide caused the north flank of the volcano to move downward and to the north. (b) The release of pressure on the column of magma within the volcano caused both a directed blast toward the north and a vertical Plinian column. (c) During the main phase of the eruption, which lasted about 9 hours, a well developed Plinian column propelled volcanic ash, gases, and other volcanic debris into the atmosphere. From Fletcher, 2014.

Figure 3.40 Trees felled by the lateral blast of Mount St. Helens on 18 May 1980. Photo courtesy of the U.S. Geological Survey; https://volcanoes.usgs.gov/vsc/images/image_mngr/900-999/img985.jpg

from this huge landslide raised the floor of the valley to the north of the volcano by more than 600 ft (180 m). It left large blocks of the volcano littering the valley floor to the north and west of the mountain.

The landslide released pressure from the magma column rising within the volcano and triggered a powerful explosive eruption. The magma conduit in this case was pointed to the north rather than straight up toward the summit. The north side of the volcano and the top 1,350 ft (400 m) of the volcano were blasted away. This created an amphitheater-shaped crater opening toward the north. At the start of this phase of the eruption, a lateral blast rocketed northward at speeds of at least 300 mi/hr (483 km/ hr). Rock, ash, and hot gases moved over the landscape, overtaking ridges and forests. An area of 230 mi^2 (600 km^2) was completely devastated by this lateral blast. Nearly every tree in this area was stripped of its branches, limbs, twigs, and needles, then toppled (Fig. 3.40). It appeared that nearly every living thing within the blast zone was gone.

A few seconds after the lateral blast began, an eruption column of volcanic ash and hot gases, known as a *Plinian column*, rose vertically from the volcano (Fig. 3.41). It shot 15 mi (24 km) into the sky. The eruption column was sustained for approximately 9 hr, further testament to the amount of pressure that

Figure 3.41 Plinian column rising from Mount St. Helens on 18 May 1980. Dark streaks on the sides of the volcano are individual mudflows. Image courtesy of the U.S. Geological Survey; https://volcanoes.usgs.gov/volcanoes/st_helens/st_helens_gallery_23.html

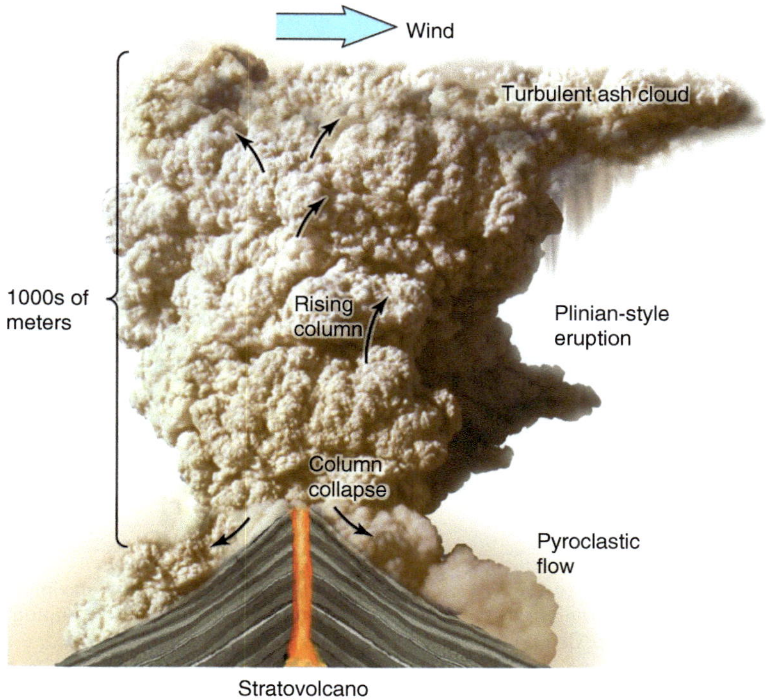

Figure 3.42 Diagram of a Plinian-style eruption of a composite volcano. Ash is propelled thousands of feet into the atmosphere. Wind catches the upper part of the ash cloud and blows it downwind. Closer to the volcano, ash rushes down the sides of the volcano in pyroclastic flows. They are usually deadly to anyone caught in them, and can destroy even buildings made of reinforced cement or masonry. From Fletcher, 2014.

had built up beneath the volcano. Upper level winds caught the ash and blew it eastward (Fig. 3.42). This turned a bright, sunny morning in eastern Washington as dark as midnight. Ash fell like a blanket on the landscape.

Hot (1,475° F or 800° C) avalanches of volcanic ash and rocks rolled down the sides of the volcano, many approaching speeds of 100 mi/hr (160 km/hr). These avalanches are called *pyroclastic flows*. These are flows of volcanic ash and rocks mixed with hot gases escaping the volcano. Pyroclastic flows are among the deadliest of volcanic hazards. They cannot be outrun, and they are deadly if a person is caught within one. Not only is the ash and rock moving at hurricane speeds, it is mixed with toxic gases.

Ash mixed with melting water from the volcano's snow pack and glaciers, creating vast rivers of mud that rolled down the sides of the volcano in all directions. Mud flows, also called *lahars* (Fig. 3.43), coalesced in the larger rivers leading away from the volcano. Some of these larger lahars were up to 60 ft (20 m) thick and traveled at speeds of more than 70 mi/hr (115 km/hr).

Lahars can be incredibly dangerous because they are in many cases a hidden threat. Composite volcanoes tend to be tall mountains, and they often have many forested ridges and river systems extending outward from the volcano. The largest of mudflows are found in river systems leading away from volcanoes, so the communities at highest risk from mudflows are found in river valleys some distance from the volcano.

Volcanologist David Johnston, who worked for the U.S. Geological Survey, was killed in the eruption. He was sitting at an observation post just a few miles north of the volcano on

Figure 3.43 Lahar deposits from Mount St. Helens' eruption on 18 May 1980. The maximum height of the lahar can be discerned by measuring to the level of mud coating trees near the river. Geologist in photo for scale. Photo by Lyn Topinka, U.S. Geological Survey.

the morning of 18 May 1980. He felt the earthquake at 8:32 that morning, then saw the beginning of the massive landslide that followed. He radioed a short message to the geologists in Vancouver. His final words were "Vancouver, Vancouver, this is it!" Unfortunately, he was one of 57 people who died that day. Only three of those people, including David Johnston, were inside the exclusion zone. The rest were outside the exclusion zone in areas scientists had believed were safe. While a landslide from the deformed north flank of the volcano was anticipated, the lateral blast that followed was not. The blast zone extended 17 mi (27.3 km) from the volcano, which was 12 mi (19.3 km) outside the exclusion zone on the north side of the volcano. Scientists learned a great many hard lessons with the eruption of Mount St. Helens.

After a summer of explosive eruptions, the volcano began a new phase of eruption. A lava dome began growing in the crater (Fig. 3.44). A *lava dome* is a lava flow that covers the vent of a volcano. The dome grew to over 0.6 mi (1 km) in diameter over the next several years. The eruption was declared over in 1986. In 2004, however, the volcano surprised many by renewing its dome-building activity. The volcano was again active between 2004 and 2008, and almost tripled the size of the dome in the volcano's crater.

3.6 Hazards of Composite Volcanoes

Mount St. Helens exhibited many of the major hazards associated with composite volcanoes, and some that are less common. Certainly large landslides and lateral blasts are rare occurrences. However, the vertical Plinian eruption column is a common feature at composite volcanoes, as are lahars and pyroclastic flows.

Volcanic ash (Fig. 3.45) is a major hazard of composite volcanoes. Volcanic ash is created when magma is torn apart in a violent explosion, or when rapidly expanding gas within a rock causes it to shatter. As a result, when viewed under a microscope, the particles of ash look like small shards of glass or tiny rock fragments. The ash particles are irritating to the

Figure 3.44 A lava dome formed in the crater of Mount St. Helens, 1984. A lava dome is a tall lava flow that sits over the vent in the crater of a volcano. Photo by the U.S. Geological Survey; https://volcanoes.usgs.gov/vsc/images/image_mngr/500-599/img538.jpg

Figure 3.45 Volcanic ash blankets a farm in Oregon following the May 1980 eruption of Mount St. Helens. Volcanic ash is composed of tiny particles of pulverized rock and lava. Volcanic ash is a major hazard of composite volcanoes. From Skinner & Murck, 2011.

eyes, nose, throat, and lungs. Ash poses a significant suffocation risk, as particles are small and can be easily inhaled. Ash is incredibly abrasive in machinery as well, and particularly hazardous in engines. Volcanic ash is recognized as a major hazard to aviation. There are observatories dedicated to detecting volcanic ash hazards as volcanoes erupt, particularly around the Pacific Rim where composite volcanoes are most common. These observatories were established following an alarming occurrence in 1984 in which a British Airways passenger jet flew through an ash cloud over Indonesia and all four of its engines stalled. The plane lost all power and fell thousands of feet. The pilots were able to get the engines restarted once the plane fell beneath the ash cloud, and the plane was able to make an emergency landing. The front windshield of the plane had been so abraded by the volcanic ash that the pilots had to look out of side windows to see clearly enough to land the plane. More recently, the volcano Eyjafjallajokull in Iceland (Fig. 3.46) erupted in 2010 and disrupted flights to, from, and within Europe for a period of several days. Improved ash warning

Figure 3.46 The April 14, 2010 eruption of Eyjafjallajokull in Iceland. Ash from the volcano brought air travel in Europe to a standstill for a week. Photo by Bjarki Sigursveinsson, via Wikimedia Commons.

systems prevented any aviation disasters during its eruption.

Ash is also a hazard as it combines with water to form *lahars*. In 1985, a composite volcano called Nevado del Ruiz (Colombia) erupted on a November night. Though the eruption was not terribly large, it did unleash pyroclastic flows, which moved across snowfields in the upper parts of the volcano, partially melting them. Volcanic ash from pyroclastic flows mixed with the melted snow and formed lahars. Mudflows started in the small rivers on the upper slopes of the volcano and grew larger each time smaller streams joined into larger ones. People in communities along the rivers were many miles away from the volcano and, in fact, could not even see the volcano from their homes. However, that did not mean they were safe from danger. The city of Armero was located on a floodplain of the Lagunillas River, whose headwaters were located high on Nevado del Ruiz. A lahar swept down the Lagunillas River and an estimated 25,000 people died that night when the city was overtaken by mud (Fig. 3.47). Armero was located 44 mi (72 km) from the volcano. Though scientists correctly forecast the location and extent of lahars, the residents of Armero had no monitoring system to tell them the lahar was on its way, and no warning system to tell them to get to higher ground. After this tragedy, scientists vowed to step up monitoring and alert networks worldwide to prevent similar disasters from occurring elsewhere.

One of the most tragic occurrences of a pyroclastic flow (Fig. 3.48) was in the island community of Saint-Pierre, on the island of Martinique in the Caribbean. In the spring of 1902, the volcano Mont Pelée, which loomed above the city, showed alarming signs of increasing activity. What began as a few muffled explosions from the crater soon grew to ever-increasing ash emissions that began to kill

Figure 3.47 The town of Armero, Colombia, was buried by a lahar in 1985. Approximately 25,000 people died when the lahar struck. They received little reliable advance warning and were unable to escape the fast-moving wall of mud when it arrived. From Strahler, 2013.

Figure 3.48 A pyroclastic flow from Mayon Volcano in the Philippines, 1984. These ground-hugging turbulent clouds of ash and gases are deadly. Bullit Marquez/©AP/Wide World Photos. From Fletcher, 2014.

Figure 3.49 Aniakchak Caldera, Alaska. The caldera was formed approximately 3,450 years ago when the volcano erupted approximately 12 mi^3 (50 km^3) of magma. Later activity produced lava domes within the caldera. M. Williams, National Park Service/U.S. Geological Survey. From Fletcher, 2014.

wildlife and livestock on the island. Then the lahars began. On 5 May 1902, a rum distillery and sugar plantation were overrun by a fast-moving lahar and about a dozen people were killed. The increasing intensity of volcanic activity drove terrified people from the countryside into the closest city, which was Saint-Pierre. On 8 May of 1902, Mont Pelée erupted violently, sending a large pyroclastic flow down a valley and directly into the city of Saint-Pierre, whose population was swollen with refugees from the surrounding countryside.

The pyroclastic flow killed all but a few inhabitants of the city, about 29,000 people in total. Even many people on ships in the city's harbor were killed as the pyroclastic flows moved over water as easily as they did over land. It tore roofs from buildings, knocked down masonry walls and buildings, and left people in the direct path of the flow burned to charcoal. It was a horrific event. Scientists from the United States, Great Britain, and France were first on the scene to investigate. They documented the damage they saw and marveled at the incredible force the volcano unleashed upon this community. At that time, pyroclastic flows were largely unknown to science. Saint-Pierre became a valuable laboratory in which the world's first true volcanologists learned about these phenomena. One of the scientists, Thomas Jaggar from the Massachusetts Institute of Technology, was profoundly moved by what he saw. He recognized that in order to prevent deaths from volcanic eruptions in the future, scientists needed to study active volcanoes to learn their signs and signals that foretell an eruption. He went on to found the Hawaiian Volcano Observatory in 1912, an entity that even today is a world-class facility for studying active volcanoes.

Pyroclastic flows have certainly taken a number of lives since 1902, but thankfully no further losses have occurred on the magnitude of Saint-Pierre. A tragic episode in 1991 illustrated just how unpredictable and deadly they can be. Three world-renowned volcanologists, Harry Glicken from the United States and Katia and Maurice Krafft from France, were in Japan observing and filming small pyroclastic flows from the volcano Unzen in Japan. The flows had been small and confined to a single valley for quite a long time. The Kraffts and Glicken felt they were safe, observing and documenting the pyroclastic flows from a distance. The Kraffts had seen more volcanoes in eruption than any other people at that point in history, living or dead. However, they misjudged this volcano. A pyroclastic flow left the valley they had been watching and overtook them. The Kraffts, Glicken, and a number of journalists with them died instantly. A recent documentary entitled *Fire of Love* follows the Kraffts through their lives and careers.

Like other types of volcanoes, composite volcanoes are perfectly capable of producing spectacular, glowing lava flows. These lava flows are often composed of andesite and do not typically flow as far as basalt lava flows. Mount Etna, a composite volcano on the island of Sicily, is also well known for its spectacular lava flows.

Like shield volcanoes, it is also possible for composite volcanoes to develop calderas. Calderas on composite volcanoes are usually found at the summit of a volcano. In 1991, the volcano Pinatubo in the Philippines erupted violently, in the second largest eruption of the twentieth century. During the volcano's catastrophic June 1991 eruption, a caldera was formed. The caldera is approximately 2.5 mi (4 km) in diameter, encompassing what used to be the summit of the volcano. Some volcanoes, however, can produce giant calderas. The Valles Caldera in New Mexico is located near Los Alamos. Voluminous eruptions between 1.1 and 1.4 million years ago left a caldera 13 mi (21 km) across. Subsequent volcanic activity has resulted in lava domes on the floor of the caldera.

Another example of a large caldera is that of the Yellowstone volcano in Wyoming. The actual volcano at Yellowstone is so large it cannot be seen in its entirety from any vantage point, except from space. It does not have the typical shape of a single mountain. Instead, it is a complex landscape that has been shaped by hundreds of volcanic eruptions over the last 3 million years. In that time, many dome-forming eruptions have occurred, and three large caldera-forming eruptions have happened as well. The latest caldera formed by the Yellowstone volcano occurred following the eruption of more than 240 mi^3 (1,000 km^3) of material. The caldera formed in that eruption 640,000 years ago is an incredible 28 mi by 53 mi across (45 km by 85 km). Yellowstone ash is found across many of the midwestern and southern states. Since the last caldera-forming eruption, there have been a number of smaller eruptions of rhyolite lava flows. The area is still underlain by a great deal of hot rock and magma, as evidenced by the active geothermal system, which includes geysers, hot springs, and mud pots. Thankfully, huge eruptions such as the one that formed this caldera at Yellowstone, are extremely rare, occurring infrequently. The last eruption even approaching a tenth of this size was of Toba in Indonesia about 74,000 years ago. The odds of a caldera-forming eruption occurring at Yellowstone in the next several centuries are very, very small.

Another hazard presents itself with the largest of eruptions. Following the 1815 eruption of Tambora in Indonesia (the largest of the nineteenth century), ash and sulfur droplets in the atmosphere blocked heat from the sun from reaching the surface of the Earth. As a result, the following year was known as the "Year Without a Summer." Snow fell in June across the United States and Europe. Crop production was low, and many people found it difficult to make it through the following winter due to food insecurity. It was during this time (the summer of 1816) that Mary Shelley wrote her book Frankenstein. She was likely influenced by the dreary cold weather that resulted from the 1815 eruption of Tambora.

After 1991 the eruption of Pinatubo in the Philippines, global temperature decreased by about 1° F, or 0.5° C for more than a year. We know from these two examples, and several others throughout history, that large volcanic eruptions can disrupt temperature for months afterward. They can also create colorful sunrises and sunsets as the sun's rays bend through particulates in the atmosphere. Edvard Munch's paintings *The Scream* and *Anxiety*, painted in the 1893 and 1894, respectively, depict brilliantly colored skies. It is likely that these skies were modeled on those that occurred in Europe after the 1883 eruption of Krakatau in Indonesia.

3.7 Mitigation

Like all dangers in the natural world, there are a host of actions that can be taken to mitigate volcanic hazards. First, scientists have a number of tools at their disposal to monitor the status of unrest at volcanoes (Fig. 3.50). Assessing the status of a volcano is accomplished through a variety of activities, including through review of satellite imagery and images from remote cameras, tracking the magnitude and locations of earthquakes, monitoring the type and quantity of gas being emitted by the volcano, thermal imaging, and monitoring of ground deformation.

Scientists use seismometers to detect earthquakes that occur beneath volcanoes. An increase in earthquake activity can be an indication magma is on the move. By determining how many earthquakes of various types are occurring at a range of depths, scientists have a good idea of whether magma is moving through the volcano or toward the surface.

When installed in river valleys, seismometers may also be used to measure ground vibrations that indicate a lahar is moving down a valley. Lahars occur during eruptions of volcanoes, but can also occur months or even years after an eruption has blanketed a landscape with ash. Any time there is a large rainfall event, or an unusually large amount of snow melt, the excess water can mix with volcanic ash and trigger a lahar. Many countries have installed lahar detection systems in valleys near major volcanoes. When a lahar is detected, a siren warns residents downstream to seek higher ground. Since lahars can occur even in the absence of volcanic eruptions, this warning system is a critical piece of safety infrastructure around composite volcanoes worldwide.

Another technique that has been used to monitor volcanoes is measuring the subsonic sound waves that travel through the air. This is called infrasound monitoring. The instrument itself is an array of several

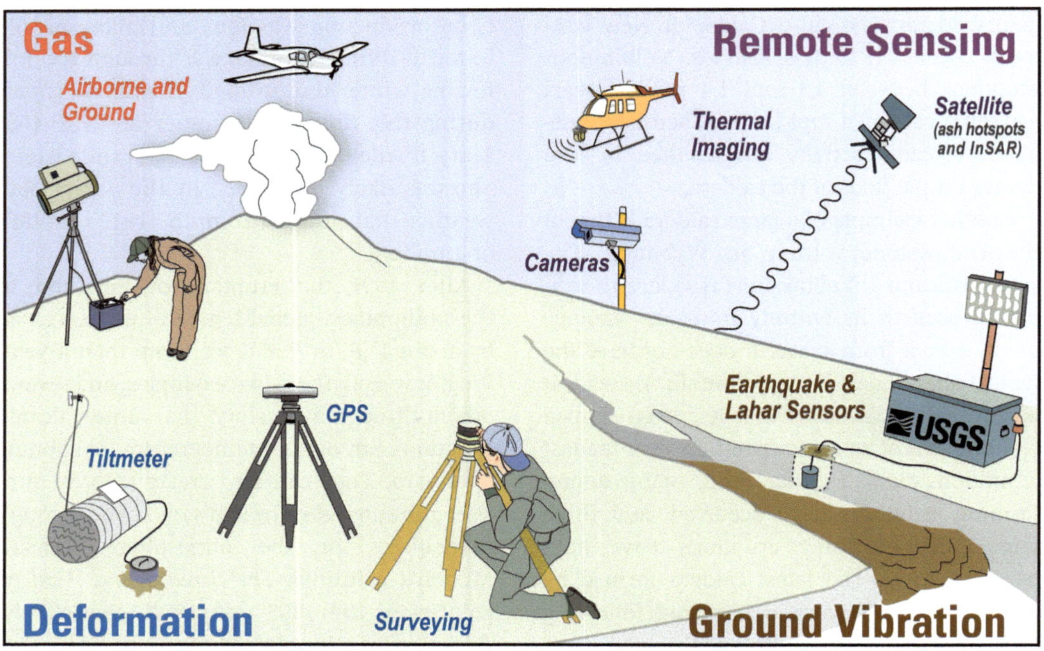

Figure 3.50 Scientists have a number of tools at their disposal to monitor the status of an active volcano. Image courtesy of the U.S. Geological Survey; https://volcanoes.usgs.gov/vsc/images/image_mngr/1100-1199/img1138.jpg/ Public Domain

microphone-like structures that extend many meters away from a central point. Infrasound monitoring is very helpful in detecting explosions, both from volcanoes and from other sources. This can be incredibly valuable to scientists when clouds obscure volcanic mountains and they are unable to view the mountain directly.

Satellites and ground-based tools such as tilt meters and ultrasensitive GPS instruments measure deformation of the volcano. As a volcano fills with magma, and that magma moves toward the surface, the volcano swells like a balloon. These instruments help detect the changing tilt of the volcano's slopes and changes in elevations. Scientists also monitor volcanic gases. An increase in emissions of volcanic gases signals magma is near the surface of the Earth.

Satellites and ground-based instruments are also used to monitor the temperature and eruptive status of volcanoes. Especially in remote areas where volcanoes are inaccessible, remote cameras and satellite sensors can be extremely helpful in determining whether a volcano is either close to or in eruption. Thermal sensors on satellites have detected eruptions in both remote areas and cloud-covered areas before people could see them. Webcams are deployed on many dangerous volcanoes around the world.

All of these monitoring techniques have a singular goal, which is to protect people from the dangers of volcanic eruptions. The ultimate objective is to forecast when a volcano will erupt, within a narrow window of time, so people can stay in their homes and jobs as long as possible until the threat of an eruption becomes real and present. Alarming a population too early creates resentment from lost wages and loss of access to homes and businesses. Alarming a population too late results in mass panic and confusion around evacuation, or in the worst case, not enough time for an evacuation and great loss of life. Alarming a population when no volcanic eruption occurs at all creates distrust of scientists and public officials, and a tendency to see any future warnings as false alarms. Pressure is on scientists to get it right every time.

In late 2017, a volcano called Sinabung in Indonesia showed signs of unrest. Scientists monitored the volcano and were confident enough in forecasting an eruption a few days before it occurred to convince public officials to issue an evacuation order. Just days before the volcano began its violent eruption, thousands of people and many of their livestock were evacuated and saved from pyroclastic flows and heavy ash fall. This is a success story. There have been many other success as well since the tragedies of the 1980s eruptions of Mount St. Helens and Nevado del Ruiz. Scientists have learned a great deal by studying active volcanoes around the world, and are applying that knowledge to every new volcanic threat near populated areas.

In addition to scientific monitoring and warnings from public officials, education and planning are important. Land-use planning around dangerous volcanoes can be incredibly helpful in reducing risk to a population. Leaving the slopes of a volcano and valleys leading from a volcano open for recreation and cropland, and restricting residential areas within those regions, means fewer people are sleeping in the most dangerous places every night. Planning streets and highways so they are wide enough to accommodate evacuation traffic is important as well. Educating the public not only about hazards they face from a volcano, but also about what an evacuation order will look like and the evacuation routes they should take are critical to allow for the safest and most efficient evacuation possible.

3.8 Chapter in Review

Magma is molten rock beneath the surface of the Earth. Magma is created by one of three processes, which allows melting to take place within small pockets of the upper mantle. The first process, which can create magma, is removing weight and pressure off of the mantle. This happens at divergent plate boundaries or in other places where the crust is thinned.

The second process, which can generate magma, is introducing water into the mantle. This happens at subduction zones. The subducted oceanic plate is saturated with water after sitting underneath the ocean for millions of years. As it moves downward into the upper mantle and is heated, some of that water escapes into the surrounding mantle. Water lowers the melting temperature of mantle rocks, creating a situation in which small pockets of the mantle melt to form magma. The third situation in which magma is created is melting around a hotspot. A hotspot is a place in the upper mantle where the temperature is higher than the surrounding mantle. This allows for a small amount of melting of the upper mantle, creating pockets of magma.

Igneous rocks are those formed when magma cools and solidifies. This can happen beneath the surface of the Earth, creating plutonic or intrusive igneous rocks. When magma is erupted to the surface of the Earth, it is lava. When lava cools, the result is volcanic or extrusive igneous rocks. Magma with a specific composition can create both intrusive and extrusive igneous rocks. The characteristics of the rocks are different, resulting in different rock names. Rocks are grouped into types based on their composition. The three categories are low-silica (45%–54% silica), intermediate (55%–69% silica) and high-silica (more than 69% silica). Basalt is a low-silica volcanic rock. Andesite is an intermediate volcanic rock. Rhyolite is a high-silica rock. Other volcanic rocks of note are pumice, obsidian, and scoria.

The largest of all volcanoes are shield volcanoes. Shield volcanoes are found in Hawai'i, the Galapagos Islands, the Canary Islands, and Reunion Island. They resemble a warrior's shield laid flat on the ground. Shield volcanoes have large diameters and gentle slopes. The most common type of lava erupted from them is basalt. A shield volcano may have several vents, including fissures (cracks through which lava is erupted), craters, and cinder or spatter cones. Many shield volcanoes also have at least one caldera. The surface of a lava flow can be smooth or ropy, which is called pāhoehoe, or it can be rough and jagged, called `a`ā. Lava flows, volcanic gases, and occasional explosive eruptions are the major hazards of shield volcanoes. Basalt lava is typically more than 1,800° F (over 1,000° C) so it can burn forests, grasses, and structures. It can also cover roads and neighborhoods. Lava is not usually a threat to human life. Its major threat is to objects such as human-built structures and any form of life that is unable to fly or run away. Volcanic gases include water vapor, carbon dioxide, sulfur dioxide, hydrogen sulfide, and various other gases in trace quantities. Carbon monoxide and carbon dioxide can suffocate. Hydrogen sulfide causes gastric distress. Sulfur dioxide combines with water vapor to create sulfuric acid, which can burn skin, mucous membranes, and lungs.

The smallest of the volcanoes discussed in the text is the cinder cone. Cinder cones may occur on other types of volcanoes, or they may occur as single volcanoes within volcanic fields, which are provinces with many individual volcanoes. Cinder cones can be found in Hawai'i, Idaho, Arizona, and New Mexico. These volcanoes are composed primarily of scoria (a type of basalt lava that is red or black in color and filled with holes), as well as basalt or andesite lava flows, spatter, and bombs. The hazards of cinder cones are identical to those of shield volcanoes. Lava flows, volcanic gases, and occasional explosive activity are the major hazards of this type of volcano.

The third and most dangerous type of volcano is the composite volcano, also known as a composite cone or stratovolcano. These volcanoes are composed of lava flows, scoria, volcanic ash, and other volcanic materials. Prominent composite volcanoes are Mount St. Helens in the state of Washington, Mount Lassen in California, Mont Pelée on the Caribbean island of Martinique, Mount Etna on the island of Sicily, and the Yellowstone Volcano in Wyoming. Hazards associated with composite volcanoes are lava flows, volcanic gases, explosive activity, volcanic ash, lahars, and pyroclastic flows.

Volcanic ash is pulverized magma, which has the consistency of fine sand or dust. It presents a suffocation hazard for people and animals because it is easy to inhale. Volcanic ash can also clog engines and abrade metal and glass. Lahars are mudflows, which form when volcanic ash mixes with water from snowmelt, rainfall, or rivers. They can travel tens of miles from a volcano along stream valleys, and may travel at speeds in excess of 50 mi/hr (80 km/hr). A particularly deadly lahar in Columbia in 1985 killed approximately 25,000 people. Pyroclastic flows are deadly fast-moving clouds of hot gases and volcanic ash, which may move at 100 mi/hr (161 km/hr). They can destroy masonry structures, set wooden structures on fire, and usually kill anyone in their path. A pyroclastic flow from Mont Pelée in Martinique killed almost all 29,000 inhabitants of the city of Saint-Pierre.

Composite volcanoes can also form calderas. Examples of volcanoes that have formed calderas include Mount Mazama, which is now called Crater Lake, and Mount Pinatubo in the Philippines. Large calderas include the Valles Caldera in New Mexico and the Yellowstone Caldera in Wyoming. Calderas form when a great volume of magma erupts from a chamber beneath the volcano, leaving the roof of the chamber insufficiently supported. The surface of that part of the volcano collapses, forming a depression. Later eruptions can leave behind volcanic domes.

There are many strategies for mitigating the hazards of volcanic activity. Scientists monitor volcanoes for many types of activity, including gas emissions, earthquakes, ground deformation, and thermal action. The goal is to forecast an eruption accurately and convince local officials to evacuate people from areas of danger. Early warning systems are an important step forward in keeping populations aware and safe. Education of the public about what the early warning system detects, what a warning looks like, and evacuation routes they should take are also critical pieces of the mitigation strategy. Land-use zoning, which keeps the slopes of the volcano and valleys leading from it free from permanent residences, can also be helpful.

4

Earthquakes

Bethany D. Hinga

University of Nebraska at Kearney, USA

4.1 Introduction

The early 1900s was an exciting time to live in San Francisco. The city was booming, and the economy was growing by leaps and bounds. The city had seen fast population growth and rapid building since the late 1840s when gold was discovered nearby. People poured into California by horse and wagon and by sea. New construction was everywhere. The city's population was a diverse melting pot of sophisticated socialites and hardy working-class citizens, which included large and vibrant communities of Asian immigrants. San Francisco was one of the shining jewels of the west coast. The city was looking forward to a bright future.

At 5:12 a.m. on 18 April 1906 a large *earthquake* struck and shook the city for approximately 45 s. After a lull of 10 to 12 s, a second, larger shock occurred and shook the area for a full minute longer. Residents were jolted from their sleep, and eyewitnesses describe the second shock as being strong enough that they were unable to stay on their feet. Buildings were destroyed. Brick and stone walls that formed the facades of many buildings in the city crumbled and fell onto the streets below (Fig. 4.1). Streets, sidewalks, and street car rails bent and buckled. Damage was swift and significant. The worst, however, was yet to come.

As many people were struggling to understand what had just happened, dangers were mounting. In 1906, much of the city's indoor and outdoor lighting was provided by gas lamps. In the violent shaking of the earthquake, many of the city's gas lines ruptured. Gas streamed out of the broken gas lines and was ignited by open flames and sparks from downed electrical lines. Hundreds of small fires started simultaneously. The extreme fire risk in San Francisco had been acknowledged in an official report to city leaders only a year before. The city's fire stations were not positioned strategically to most effectively fight fires within the city, and the city's water distribution system was outdated, inadequate, and in disrepair. The city's fire chief knew that any outbreak of fires would be a challenge for his firefighters.

Firefighters nonetheless rushed to their posts and with their horse-drawn fire engines began navigating the city streets that were choked with debris. However, when they arrived at the scene of a fire and tried to hook into the water lines, they found the water lines dry. Water lines, like the gas lines, had broken in the earthquake. There was no water to fight the fires. Fires grew, combined, and raged for 3 days (Fig. 4.2). Untrained but well-meaning firefighters made fire breaks by destroying buildings in the path of the fires with dynamite. This created many hard feelings in the city, but

Earth's Natural Hazards and Disasters, Advanced Textbook 7, First Edition. Bethany D. Hinga.

Figure 4.1 Ruins of City Hall following the 1906 earthquake in San Francisco. From the National Archives, Photo 92-ER-24. Photo courtesy of the National Archives.

Figure 4.2 Two women having their portrait taken overlooking the city of San Francisco as it burns following the 1906 earthquake. From the Library of Congress; https://www.loc.gov/item/agc1996000206/PP/

in the end it was the only thing that prevented even more of the city from being consumed by fire. Ultimately, fires burned an area of 4.7 mi^2 (12.2 km^2) and caused even more damage to San Francisco than the earthquake itself. Estimates of losses were approximately $400 million, and only 3% to 10% of that damage was due to the earthquake. The rest was due to the resulting fire.

This was not San Francisco's first experience with fire. Prior to the 1906 earthquake, significant parts of the city had burned multiple times due to a combination of an abundance of fuel (in the form of cooking oil, fuel oil, and flammable building materials), buildings situated immediately next to each other allowing for easy spread of fires, and fierce winds.

In some areas of the city, devastation was nearly total (Fig. 4.3). Entire sectors of the city had been leveled. Fully half of the population was homeless. Few parts of the city were untouched by the earthquake and fire. The city needed strong leadership to organize the population after the earthquake and to start the recovery process. People needed shelter and food, and the injured needed medical care. Two men immediately took control. One was the mayor of San Francisco, Eugene Schmitz. The other was a U.S. Army officer, Brigadier General Frederick Funston. Funston was the highest-ranking military officer in San Francisco at the time.

Immediately following the earthquake, Funston summoned all available troops and ordered them to make themselves available to city officials, including the mayor, police chief, and fire chief. He mobilized troops to secure federal buildings and government assets. He requested tents, food, and medicine from Washington, D.C. Within a few days, all of the tents the military owned were in San Francisco, and the largest hospital train in U.S. history was en route from Virginia. Soon after, 10% of the U.S. Army was in San Francisco as well. The U.S. Navy and Revenue Cutter Service (which later became the U.S. Coast Guard) sent fireboats and helped with evacuations.

Mayor Schmitz issued several proclamations. He declared that looters would be shot on sight. He also ordered a curfew, closed saloons, banned "naked candles," and informed the public that the electric and gas companies would remain closed until it was safe for them to resume operation. These two leaders worked together to set up tent camps for those left suddenly homeless by the earthquake and fire. A feeding program was immediately begun. Within short order the camps had running water, working sewers, and bath houses. The U.S. Post Office started delivering mail to the tent camps a mere 2 days after they were established. The postmaster even accepted letters without stamps so people could "tell their friends and loved ones of their condition and their needs." Small cottages eventually replaced some of the tents, and people occupied them for years. The city was functioning again within a very short time due largely to the leadership of these two men.

We will never know the exact death toll for this earthquake, but the best estimates put the number around 3,000. Census numbers for San Francisco recorded the number of households, but not the number of people within the city. Even those scant records were first buried in the rubble of City Hall, then burned by subsequent fires that swept through the area. Many people who perished in the earthquake were buried beneath the rubble. When fires raged in the city, it is suspected that many bodies were incinerated. After the fires were out, people went through the city to collect the dead, and buried many of them in mass graves with little effort to determine their identities or even to count the number of people in each grave. It is even suspected that city officials misrepresented the number of dead to make the earthquake and resulting fire seem less severe to the outside world. They wanted

Figure 4.3 Near total destruction occurred in some parts of San Francisco as a result of the earthquake and ensuing fire. Image by Waldemar Lindgren/NG Image Collection. From Murck & Skinner, 2016.

to encourage continued investment in the city, and were afraid that news of a massive earthquake would scare away investors. In fact, many started referring to the event as the Great San Francisco Fire to minimize the role of the earthquake that led to the fires.

We will also never know the *magnitude* of this earthquake, as it occurred before the invention and installation of modern *seismometers*. Seismometers are modern instruments used to record earthquakes. Records from seismometers are required to calculate magnitudes of earthquakes. Circumstantial evidence based primarily on the damage done by the earthquake, the length of time the ground shook, and measurements taken along the fault leads scientists to believe the earthquake was in the range of a magnitude 7.7 to a magnitude 7.9. However, that is only the best estimation scientists have for the event.

Following the 1906 earthquake and fire, the city was finally forced to design a new water distribution system and place fire stations in locations that made sense for providing rapid and reliable service to the city's growing population. This was a positive outcome of the earthquake.

Some poor decisions were made during the recovery period as well. For instance, much of the debris from destroyed buildings was dumped into San Francisco Bay, near the shoreline. The result was several acres of new, *made land*. Since it was waterfront property, this became high-value real estate and many of the city's most prominent citizens built on this new land. This is the Marina District of San Francisco, and it is still home to some of the most valuable real estate in the city. One hallmark of the Marina District is that the homes were built extremely close together. This created a fire hazard by having buildings in such close proximity to each other. In addition, the risk of severe damage from extreme shaking in the next major earthquake is high. Made land is composed of loose material that is saturated with water. This type of material is known to experience enhanced shaking during earthquakes, so these homes on some of the most expensive real estate in the city are primed for destruction.

In a number of other locations around the city, building began at a furious pace. Part of what made buildings vulnerable to earthquakes prior to 1906 was that they were constructed quickly and of poor-quality materials. In addition, they were not engineered to withstand shaking from major earthquakes. People were repeating the same mistakes with new construction. It was not until decades later that stringent building codes were put in place that were meant to make buildings better able to survive severe shaking.

In 1989, the San Francisco area experienced another earthquake, this time of magnitude 6.9. This became known as the Loma Prieta earthquake. With its epicenter near Santa Cruz, roughly 60 mi (97 km) southeast of San Francisco, one might have expected damage to be minimal within the San Francisco area. Unfortunately, damage was considerable in areas where thick soil, mud, and sand deposits existed. Sixty-four people lost their lives. Many homes in the Marina District were damaged or destroyed (Fig. 4.4). Part of the Cypress Structure, a two-tiered viaduct that ran through Oakland, California, collapsed during the shaking (Fig. 4.5). Many people died when the upper deck collapsed onto the lower deck, crushing cars and their occupants on the lower deck of the highway. Portions of the highway that collapsed were built on soft mud substrate.

Many people in California are expecting a large, damaging earthquake they refer to as "the big one." The Loma Prieta earthquake was not the big one northern California was anticipating. The big one is expected to be about a magnitude 8 event. Even so, as the Loma Prieta earthquake demonstrated, a smaller

Figure 4.4 Damage to homes in the famous Marina District of San Francisco following the Loma Prieta earthquake of 1989. These homes were built on debris from the 1906 earthquake that was used to fill in part of San Francisco Bay. This is some of the most expensive real estate in the San Francisco area, but it is also some of the most unstable. The homes are situated close to one another, making spread of fire more likely than if they were spaced farther apart. Many homes also have soft first stories, which means the first floors of these buildings are not adequately braced to withstand horizontal forces of an earthquake. The first stories of these buildings, used for parking, collapsed. Loma Prieta Collection, Image LP0514. NISEE-Berkeley; https://nisee.berkeley.edu/elibrary/Image/LP0514

(a)

(b)

Figure 4.5 (a) A collapsed portion of the Cypress Structure (Cypress Street Viaduct), an elevated stretch of highway with two decks. In areas, the structure's upper deck collapsed on the lower deck, trapping people in crushed cars. The greatest loss of life in the earthquake was in the Cypress Structure. From NISEE-Berkeley Loma Prieta collection; Photo by Robert E. Kayen, Image LP0465; https://nisee.berkeley.edu/elibrary/Image/LP0465 (b) View from the top deck of a collapsed section of the Cypress Street Viaduct following the Loma Prieta earthquake in 1989. Photo by Caltrans; Photo 2009082118 from the Jim Roberts Talk Collection, NISEE-Berkeley; https://nisee.berkeley.edu/elibrary/Image/2009082118

event centered several miles away can still do considerable damage within San Francisco, and of course to surrounding communities closer to the epicenter. In a study published in 2014, scientists estimate there is a 72% chance of a magnitude 6.7 or greater earthquake that will occur on one of the faults in the San Francisco Bay region before the year 2043 (Fig. 4.6).

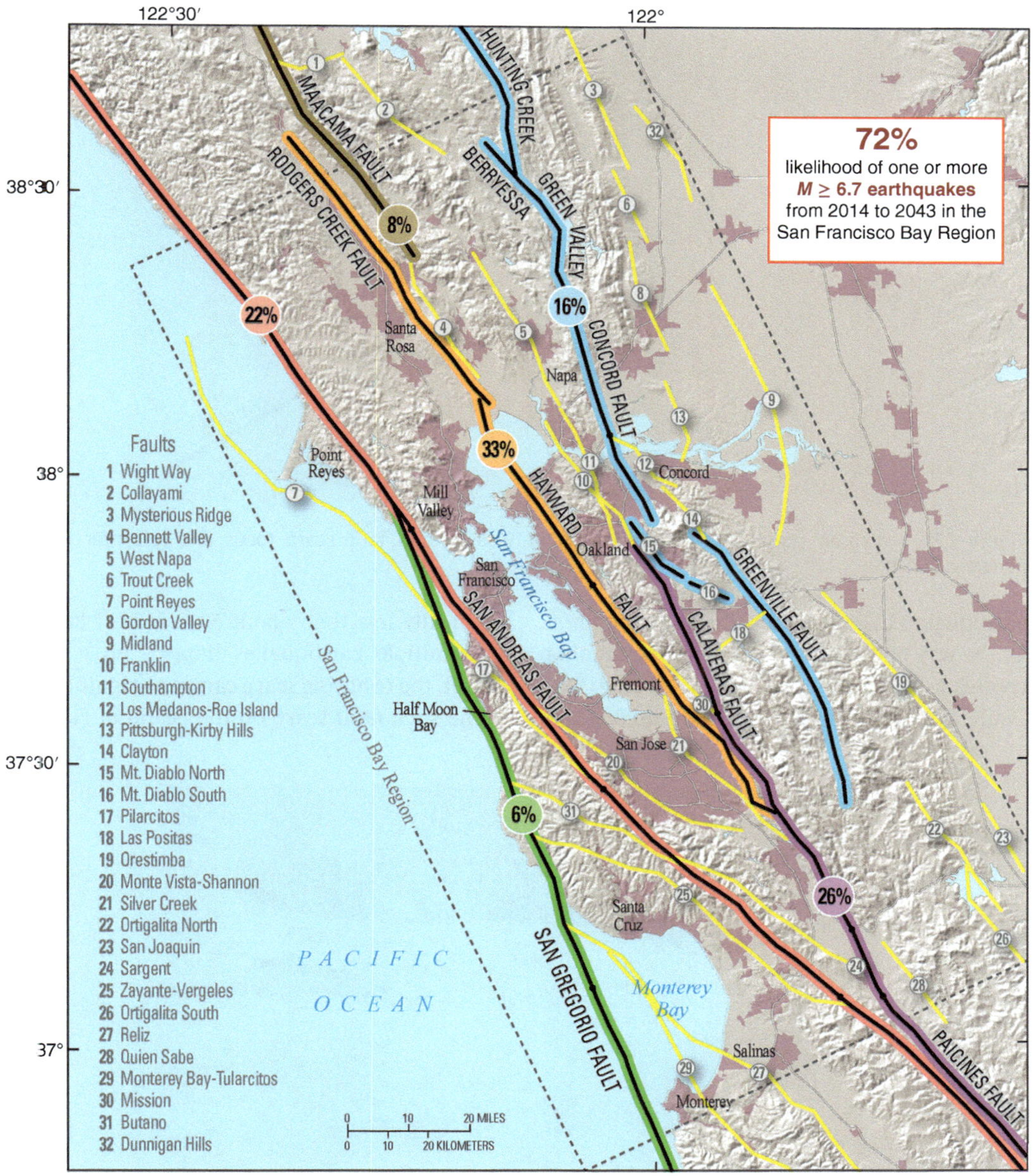

Figure 4.6 Probability of a major earthquake (magnitude equal to or greater than 6.7) in the San Francisco Bay region before the year 2043. Major faults in the area are depicted with bold black lines and highlighted, with their specific probabilities for a major earthquake occurring on each of those faults. Minor faults are numbered, with names indicated in the list on the left side of the map. The probability that an earthquake of this size will occur on a lesser-known minor fault is 13%. From Aagard and others, 2014 / https://pubs.usgs.gov/fs/2016/3020/fs20163020.pdf /Public Domain.

4.2 What Is an Earthquake?

An earthquake is sudden and rapid movement along a *fault*. A fault is a break in the Earth's crust (Fig. 4.7). The term *earthquake* is commonly used to describe the resulting shaking of the Earth's crust, which results from slippage along the fault, as well as the corresponding energy that spreads out in all directions from the starting point of the

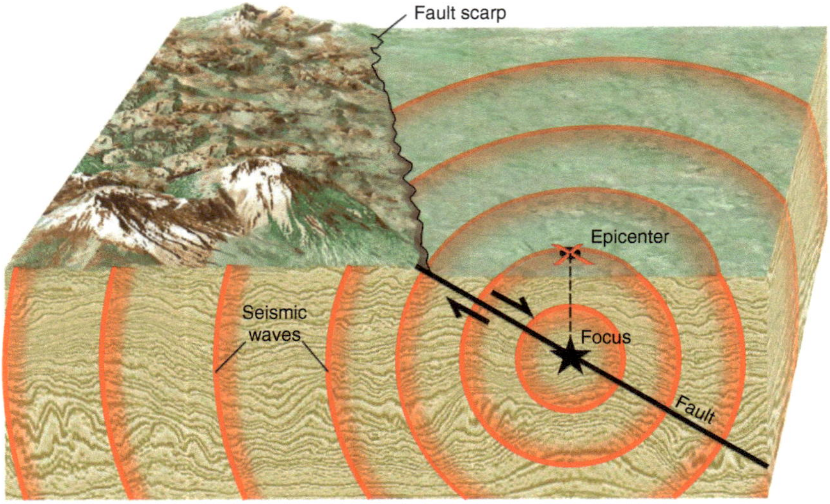

Figure 4.7 A diagram portraying the relationship between a fault, fault scarp, focus, and epicenter. From Fletcher, 2014.

earthquake. In some cases, the fault can break all the way to the surface. When it does, it often creates a slope, called a *fault scarp* (Fig. 4.8). Fault scarps for individual earthquakes are typically less than about 50 ft (15 m) tall, but if multiple earthquakes happen on a single fault, the resulting scarp can grow significantly. One such fault scarp on the island of Hawai'i,

Figure 4.8 Fault scarp from the magnitude 7.5 Hebgen Lake earthquake in 1959 with a geologist and vehicle for scale. Photo courtesy of the U.S. Geological Survey.

Figure 4.9 Aerial photo of the Hilina Fault System on the island of Hawai'i. The fault scarp (steep slope in this image) has a maximum height of about 1,475 ft (400 m). It was formed over thousands of years from multiple earthquakes. Photo courtesy of the U.S. Geological Survey.

which formed over thousands of years, is 1,475 ft (450 m) tall (Fig. 4.9).

The location on a fault where an earthquake begins is called the *focus* of the earthquake. A focus also goes by the name *hypocenter*. The focus or hypocenter of an earthquake is always beneath ground level. The point on the surface of the Earth directly above an earthquake's focus is called the *epicenter* of the earthquake. Most people are concerned primarily with an earthquake's epicenter, but geologists can tell a great deal about an earthquake by knowing the depth of the earthquake as well. For instance, earthquakes with a shallow focus tend to cause more damage than those with a deep focus. Deep focus earthquakes occur primarily along subduction zones (Fig. 4.10). Subduction zones were first mapped using the foci (focus locations) of earthquakes.

Energy that emanates from the focus of an earthquake is called seismic energy. We usually refer to this seismic energy as *seismic waves*. Seismic waves radiate outward from the focus in all directions. If an earthquake is large enough, seismic waves from that earthquake can travel around the entire planet and can be detected anywhere in the world. For instance, when the magnitude 9.2 Boxing Day earthquake occurred near the island of Sumatra in 2004, scientists all over the world were able to observe seismic waves on their instruments.

The instruments used to detect and record seismic waves are called seismometers. Seismometers are at their core quite simple machines. They are composed of a mass suspended on a spring or wire (Fig. 4.11). When the Earth moves, the mass attempts to stay in place while the rock around it moves. The result is squiggly lines on a paper or a computer screen that illustrates how the Earth is moving relative to the mass. The size of the jagged line on the paper or computer screen is dependent on how close the instrument is to the earthquake, the size of the earthquake, the amplification settings on the instrument, and the design of the instrument itself. There are dozens of models of seismometers in use today, but modern seismometers are digital. The ground shaking is translated into an electrical

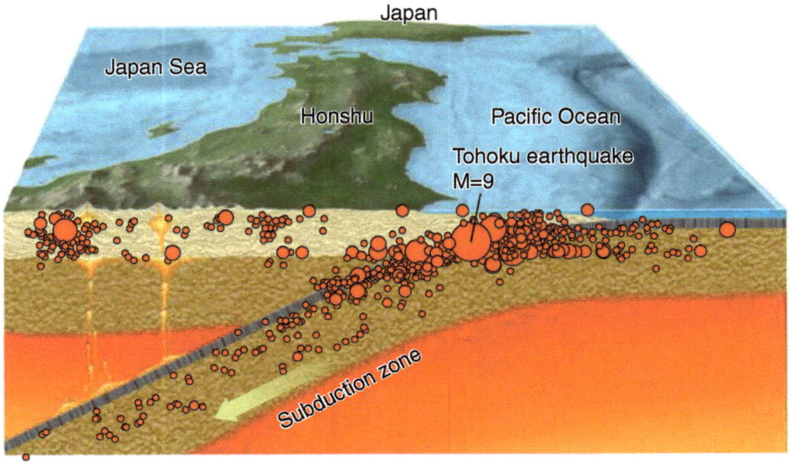

Figure 4.10 Cross section of a subduction zone off the coast of Honshu, Japan. Earthquakes occur in and near the lithosphere that has been subducted. The size of each circle indicates the magnitude of an earthquake. The largest earthquake, the 2011 Tohoku earthquake, had a magnitude of 9. Note also that earthquakes occur in areas where magma is rising through the lithosphere (top light brown layer and medium brown layer), but earthquakes rarely occur in the mantle beneath the lithosphere. From Fletcher, 2014.

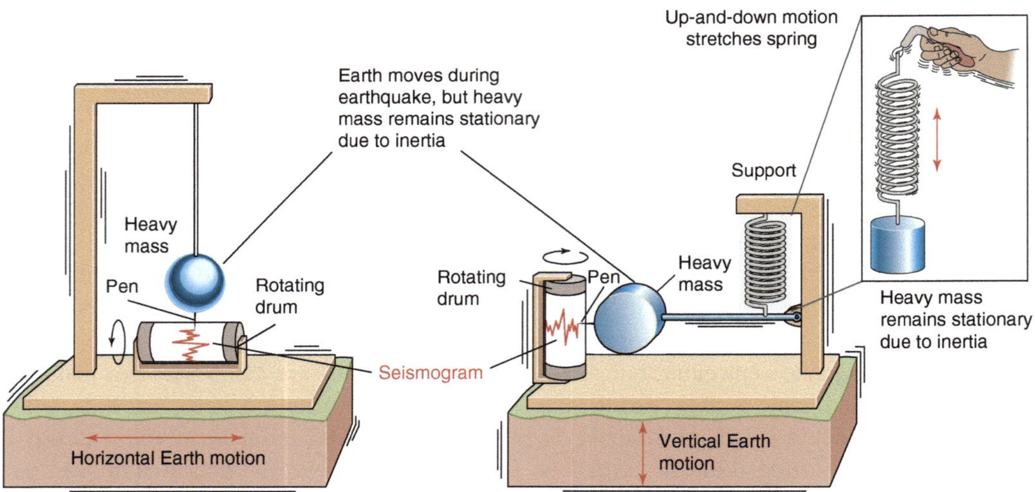

Figure 4.11 Simplified design of seismometers. A heavy mass is suspended on a spring or wire. When an earthquake occurs, inertia causes the mass to remain stationary while the instrument moves around it. A pen recorded the motion. Most current instruments are digital, where ground shaking is converted to electrical signals that are then plotted on a computer screen. From Skinner & Murck, 2011.

signal. The strength of the electrical signal is represented in the peaks and valleys seen on a computer screen that indicate seismic waves. The electronic or paper record of seismic waves is called a *seismogram*.

The solid Earth is in constant motion. Even when there is no earthquake in the vicinity of a seismometer, it is always recording motion called *background noise*. Background noise may be due to the wind, thunder, cars on a nearby highway, waves crashing on the shore, or cows rubbing on guy wires holding up the seismometer's radio antenna. Usually it is some combination of these. When an earthquake occurs that is large enough to be recorded by the seismometer, there is a jump in the height, or *amplitude*, of the seismic waves on the seismogram. This is how *seismologists*, or earthquake scientists, determine that they have recorded an earthquake. A great deal of information is then gleaned from the seismogram that allows scientists to determine a host of parameters about the earthquake. In order to intelligently discuss those parameters, we need to become familiar with the nature of seismic waves.

4.3 Seismic Waves

Seismic waves come in four varieties (Fig. 4.12), and those are divided into two major types. The first major type of seismic wave is one that travels through the interior body of the Earth. These are known as *body waves*. Body waves come in two types, *P waves* and *S waves*. P stands for primary. P waves are the fastest of all seismic waves. Because they are the fastest waves, they are the first to arrive at any seismometer from an earthquake. Therefore, they are the first, or primary, waves a seismologist sees when examining a seismogram. P waves are also known as compressional waves. They are formed when the Earth contracts and expands in the direction of wave movement. Imagine that you have a Slinky toy stretched out on a table. If you gather some coils toward you, this means you are compressing them. Other coils are dilated, or stretched apart. If you let the compressed coils go suddenly, you'll see that an area of compressed coils travels along the Slinky parallel to the direction of the wave. Compressional waves can travel through any type of material: solids, liquids, and gases. In fact, sound waves are a type of compressional wave. Sound waves can travel through the air, through water, and through solids. They may travel at different speeds through all of those materials, but travel they do.

The second type of body wave is called an S wave. S stands for secondary or shear, and

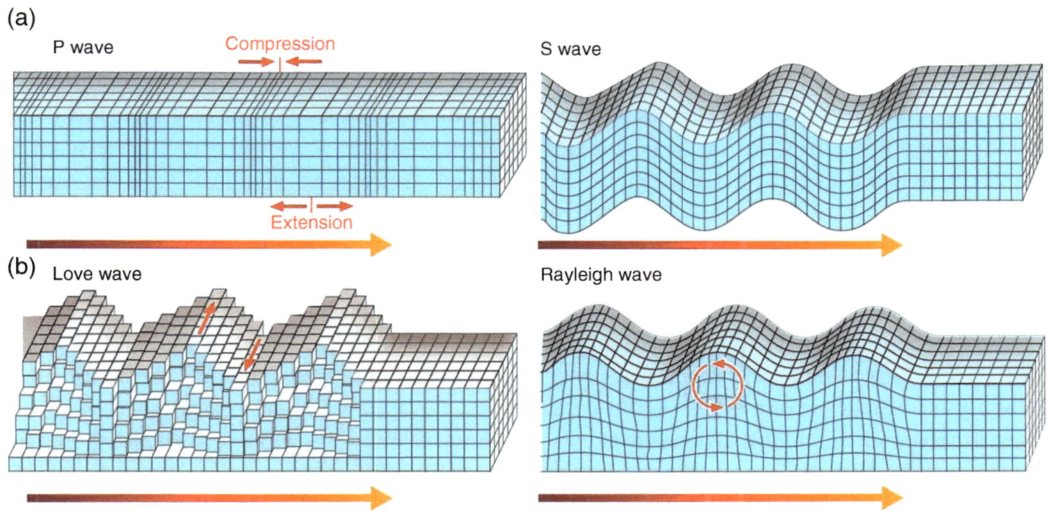

Figure 4.12 Seismic waves. (a) P waves and S waves are body waves. They move through the body of the Earth. (b) Love and Rayleigh waves are surface waves, confined to the surface of the Earth. Note that the movement of Love and Rayleigh waves dies out with depth, but P and S waves do not. From Fletcher, 2014.

these are the second fastest type of seismic waves. Since they are the second fastest type of seismic wave, they are often the second wave to arrive at a seismometer after an earthquake has occurred. If an earthquake occurs very close to a seismometer, it may be difficult to see where a P wave ends and an S wave begins since the travel time for each wave is so short. However, the greater the distance between an earthquake focus and a seismometer, the greater the separation between the P wave and the S wave. The separation between the arrival of these two waves can be several minutes if an earthquake is thousands of miles away from the seismometer recording it.

As previously noted, S also stands for shear, which refers to the motion of this type of wave. Shear motion means that the Earth moves back and forth perpendicular to the direction of motion. Imagine a rope tied at one end to a wall. If you take the rope and jerk it upward quickly, you will see an S shape form in the rope. The wave you've created in the rope will move toward the wall, but the rope itself moves in an S shape perpendicular to the direction the wave is moving. Shear waves can travel only through solid materials. They cannot travel through liquids or gases.

In the 1930s, Danish scientist Inge Lehmann used properties of P and S waves to determine that the Earth had a liquid outer core and solid inner core. She observed that P waves traveled through the entire Earth, but S waves could not move through one layer of the Earth (Fig. 4.13). In Figure 4.13a, you will see that S waves that originate in the Earth's lithosphere curve through the Earth until they hit the liquid core. The paths curve because the Earth generally becomes more dense with increasing depth in the Earth. This causes the waves to curve. When the S waves encounter the liquid core, they simply stop. Some of their energy is converted in to P waves, but S waves are no longer produced or transmitted. This results in something called an S wave shadow zone. Seismometers located within the S wave shadow zone simply do not see S waves from even the largest earthquakes (Kölbl-Ebert, 2001).

In Figure 4.13b, P waves (in blue) have been added to the S wave paths (in black). As you can see, the P waves are able to move through the entire Earth. They are bent because of the change in density of the Earth. Notice that the angle of the P wave paths changes dramatically at the boundary of the mantle and liquid outer core. This is because the outer core is less dense than the mantle above it. This results in a ring-shaped P wave shadow zone around the Earth. When P waves pass from the outer core to the solid inner core, they speed up again. As the P waves exit the solid inner core and move into the liquid outer core again, they temporarily slow down again. The waves then speed up in the lower mantle. The resulting wave paths are illustrated in Figure 4.13b.

The second major type of seismic wave is one that stays close to the surface of the Earth. This type of seismic wave is called a *surface wave*. There are two varieties of surface waves. The first is called a *Love wave*. Love waves are somewhat similar to S waves in their type of motion, in that they tend to move in an S shape perpendicular to the wave's motion. However, the motion is horizontal rather than vertical. They are like S waves lying on their sides. Love waves are also slower than S waves. Love waves have a higher amplitude, or height, near the surface of the Earth and decrease in amplitude with depth.

The second variety of surface wave is called a *Rayleigh wave*. Rayleigh waves have what is called "retrograde elliptical" motion. Retrograde means opposite the direction of wave motion, or backward. Elliptical means oval shaped. So the motion of a Rayleigh wave is almost circular but the circle direction is opposite the direction of wave motion. Rayleigh waves also move slower than S waves. Rayleigh and Love waves in fact travel at close to the same velocity, so without specialized knowledge and different types of seismometers, it is difficult to tease them apart. For that reason, when we look at a seismogram, we will often point out P waves, S waves, and surface waves without differentiating between Love waves and Rayleigh waves.

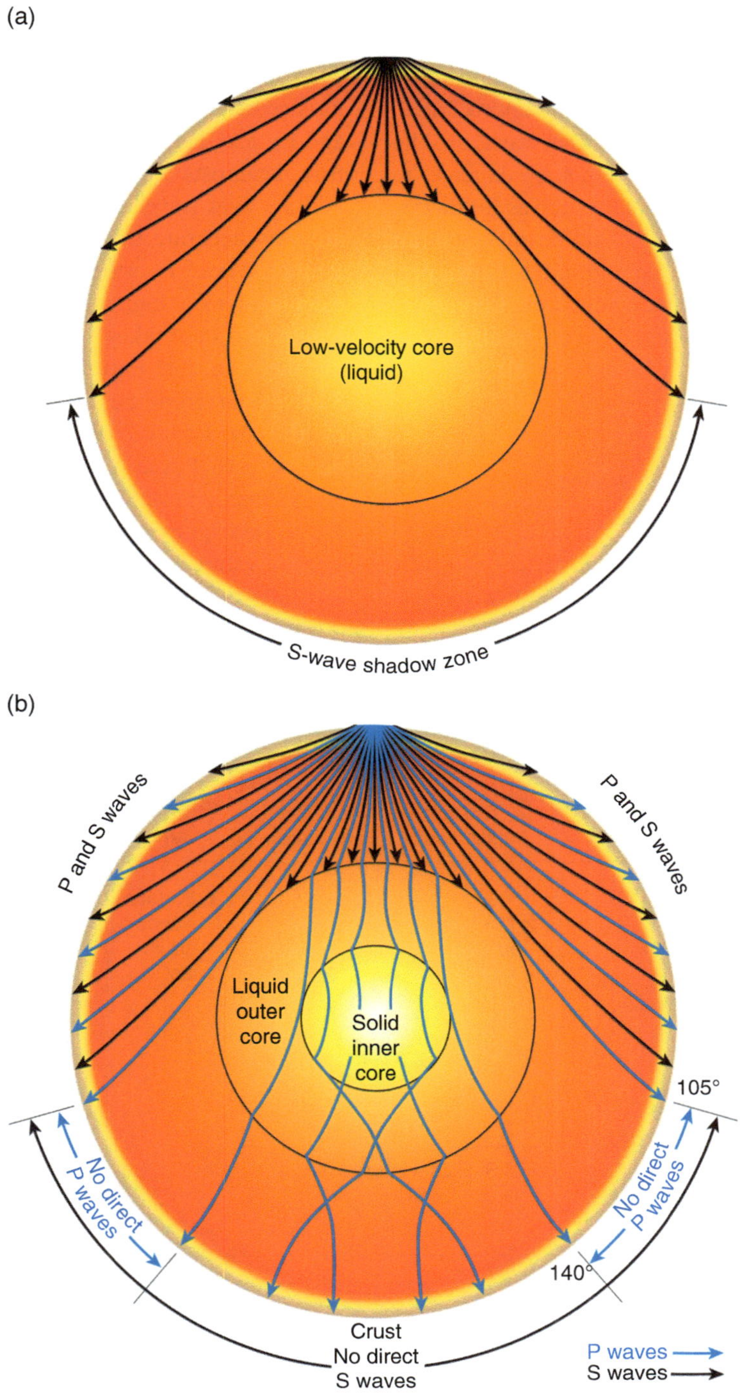

Figure 4.13 Generalized paths of S and P waves through the Earth. (a) S waves travel from the earthquake focus through the interior of the Earth. They stop when the encounter the liquid outer core because S waves cannot travel through liquids. This results in a large S wave shadow zone where S waves are not seen on seismograms. (b) P waves travel through the entire Earth. However, they slow down in the liquid outer core because the outer core is less dense than the mantle above and inner core beneath it. This results in a ring-shaped P wave shadow zone. From Fletcher, 2014.

If an earthquake is shallow, meaning that it occurs only a few miles or kilometers beneath the Earth's surface, surface waves tend to be the highest amplitude waves that are recorded. In shallow earthquakes, surface waves are the most damaging waves, causing damage to buildings and infrastructure such as highways, bridges, overpasses, and train tracks. If an earthquake occurs deep within the Earth, perhaps a few hundred miles or kilometers beneath the surface of the Earth, the amplitude of surface waves may be quite small.

4.3.1 Use of Seismic Waves

As stated earlier, seismic waves can be used to determine a host of parameters about an earthquake. They can be used to determine the size of an earthquake, the distance between an earthquake and the seismometer that detects it, and, through calculation, the amount of energy released by an earthquake.

Earthquake Location

The first two pieces of information most people want to know when they hear an earthquake has occurred are the location of the earthquake and its size. Let us tackle the location matter first. Scientists know how quickly P waves and S waves travel through the Earth. There are very accurate travel-time charts for both P and S waves. In other words, scientists have calculated to a high degree of accuracy the length of time it takes P and S waves to travel distances in the Earth. Both P waves and S waves start from the same location, the focus of the earthquake. When the earthquake occurs, both P and S waves race outward in every direction from the focus. P waves are fastest, so they quickly outpace the S waves. Both seismic waves arrive at a seismometer. However, the P wave arrives some time before the S wave (Fig. 4.14). Scientists record the arrival time of each type of wave, then find the difference between the arrival times as precisely as possible. In many cases with current digital

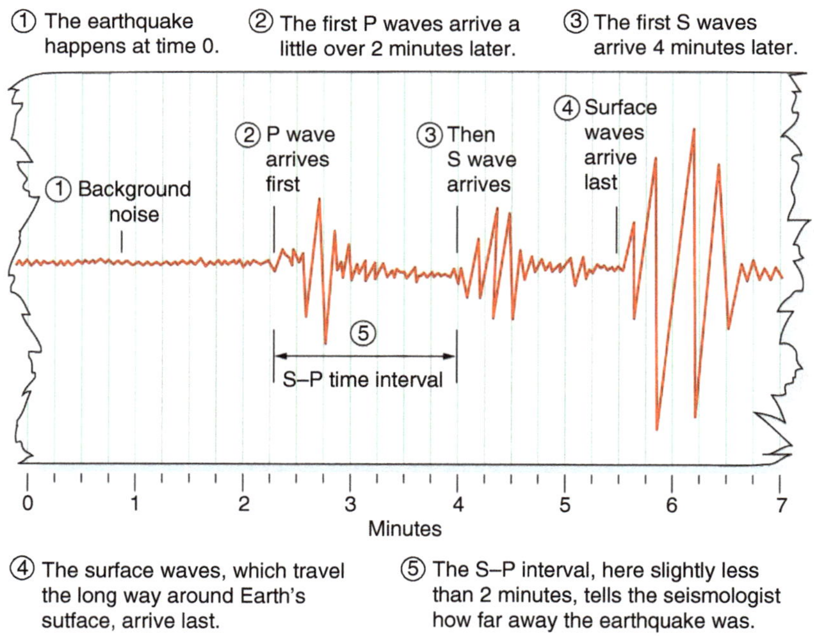

Figure 4.14 A seismogram of a typical earthquake. P waves arrive first, followed by S waves and finally surface waves. The time of P wave and S wave arrivals are noted. The S-P time interval is the time between the arrival of the two waves. It is used to calculate the distance from the seismometer to the earthquake's focus. From Murck & Skinner, 2016.

technology, this can be determined to the tenth or even hundredth of a second. The difference between the arrival times of the two waves corresponds to a specific distance between the earthquake's focus and the seismometer that recorded those waves. Using compass, pen, and paper (or an application on a computer that does this automatically), the seismologist can draw a circle around the location of the seismometer that corresponds to the distance between the epicenter of the earthquake (the location of the earthquake projected onto the on the Earth's surface above the focus) and the location of the seismometer (Fig. 4.15). The seismologist then repeats this exercise for another seismometer. The two circles will intersect at exactly two points. Repeating this exercise a third time will produce a single point at which all three circles cross. This is the epicenter of the earthquake. Today, seismologists do not calculate earthquake locations in this way. They obtain data from many seismometers and computer algorithms determine the focus (both the epicenter and depth) of an earthquake.

Earthquake Size

Once seismologists know how far away an earthquake occurred, they can calculate the magnitude, or size, of the earthquake. A familiar term used in relation to earthquakes is *Richter Scale*. The Richter Scale was the first

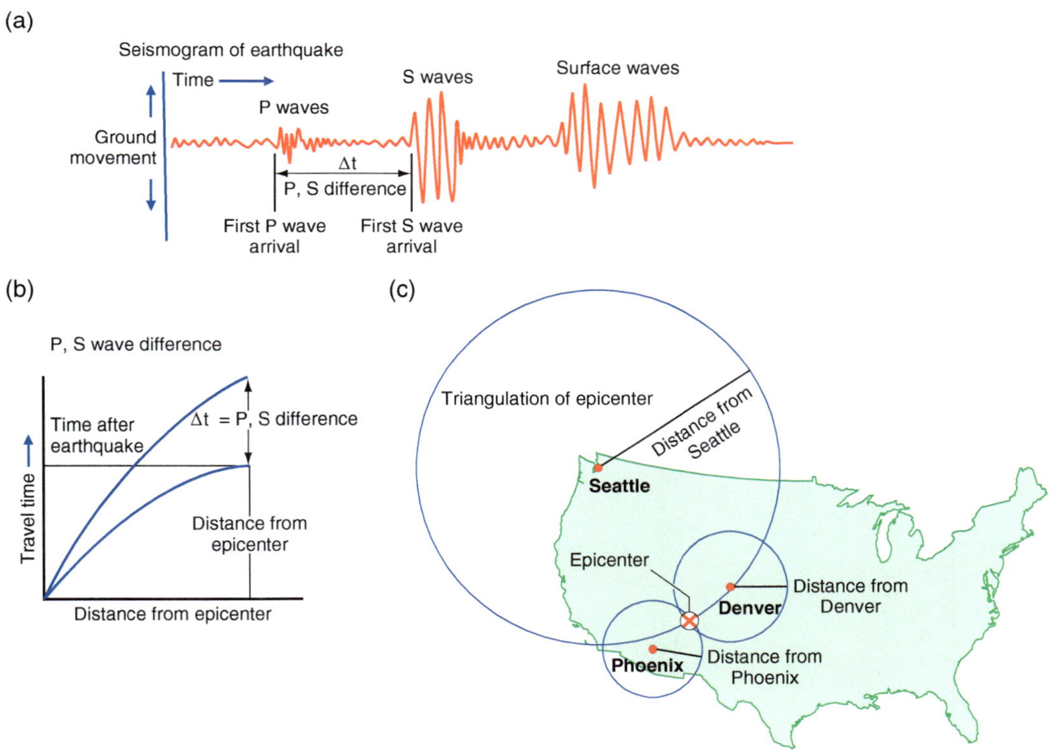

Figure 4.15 Triangulation is used to determine the location of an earthquake. (a) Seismograms from three earthquakes are examined, and P-S time intervals are determined for each. (b) P-S time intervals are used to determine distance from the seismometer to the earthquake's epicenter. (c) A circle with a radius of the distance from the seismometer to the earthquake's epicenter is drawn with the center on the seismometer's location. When this is completed for three individual seismometers, they overlap at a single point. This point is the epicenter of the earthquake. From Fletcher, 2014.

magnitude scale developed, published in the mid-1930s. It is named for seismologist Charles Richter, who was employed at California Polytechnic University, or Caltech, at that time. He worked with seismologist Beno Gutenberg to devise the very first magnitude scale. Because Richter was the person most interviewed by the news media about this scale, his name became forever associated with it.

The Richter Scale is not a physical scale you place on a table or hang on the wall. It is a set of equations used to determine earthquake size, established by Gutenberg and Richter. The equations describe the size of an earthquake mathematically and objectively. Before this time, the only measure of the size of an earthquake was how much damage it did. However, damage is dependent on depth of an earthquake, construction practices, and type of rock or soil on which buildings sit. What about earthquakes that occur in sparsely populated areas or under the ocean? There was no way to determine earthquake size in those cases. The scientific community needed an unambiguous way to determine the size of each earthquake, and the ability to objectively describe the size of earthquakes relative to each other.

The method used to determine the magnitude of an earthquake begins with examination of a seismogram. The highest amplitude of seismic waves from an earthquake is measured in millimeters. Then a set of equations correct the amplitude for the distance between the earthquake and the seismometer. Through another set of equations, a magnitude is calculated. In theory, a single earthquake recorded anywhere in the world should have the same calculated magnitude. In practice, different sets of equations work better for different sets of conditions. We will come back to that idea in a few paragraphs. A simple way to estimate the magnitude of an earthquake is shown as Figure 4.16. This is a *nomogram*, or a diagram that allows you to plot two values to determine

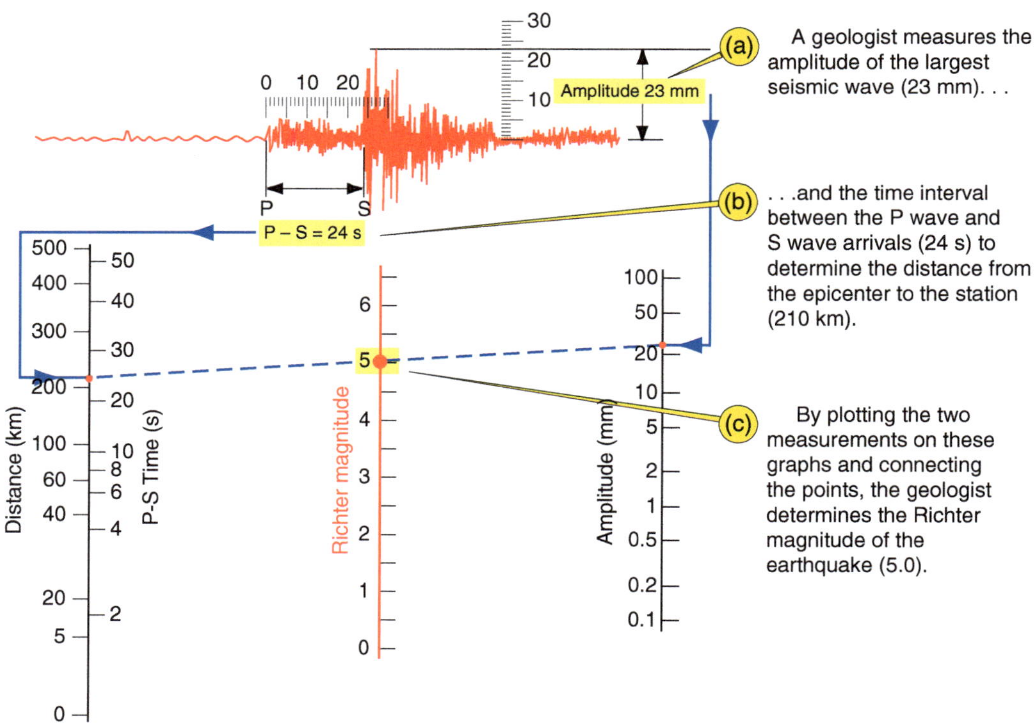

Figure 4.16 A nomogram, used to estimate the magnitude of an earthquake. If two points on the nomogram are known, the third can be determined by drawing a straight line through the two known points. Where the line intersects the third scale is the value of the third parameter. From Fletcher, 2014.

a third value. In this case, if you know the earthquake's distance from the epicenter to a specific seismometer, and the greatest amplitude of the tallest wave on the seismogram, you can estimate the Richter magnitude of an earthquake.

Magnitude scales are logarithmic, which means that every increase of one on this scale (for example from a magnitude 4 to a magnitude 5) corresponds to a 10-fold increase in the amplitude of ground shaking. There is also a roughly 32-fold increase in energy released. That means that an increase of two on the scale, such as from a magnitude 4 to a magnitude 6, corresponds to a 100-fold increase in the amplitude of ground shaking. There is also a roughly 1,024-fold increase in energy released. Because of the logarithmic nature of the equations, a magnitude scale can include the smallest earthquakes that are too small to be felt, all the way up to the largest earthquakes that are capable of destroying entire cities. Contrary to popular belief, magnitude scales do not run from 1 to 10. Earthquakes can actually have negative magnitudes, and there is in theory no upper limit to the scale. However, the largest earthquake recorded was a magnitude 9.6, which occurred in Chile in 1960.

Though Richter and Gutenberg were the first to work out these types of equations, and they (especially Richter) became famous for doing so, their equations were specific to the relatively small, shallow earthquakes they recorded in southern California during the period between about 1910 and 1930. This is where their network of seismometers was located, and these were the types of earthquakes they recorded. When other seismologists attempted to use their equations to determine earthquake size, they found the equations and correction factors did not work so well. The equations were modified through the decades, and now new sets of equations that are more widely applicable are used instead of the equations worked out by Richter and Gutenberg. Thus, we no longer discuss earthquakes and their magnitude on the Richter Scale. We instead refer to the size of an earthquake as being on the *Magnitude Scale.*

To find the most accurate locations for earthquakes, there are even different equations used depending on the estimated size of an earthquake and its depth. A magnitude scale used for small earthquakes (less than magnitude 5) within close proximity (62 mi or 100 km) of a set of seismometers is called the local magnitude, delimited by the symbol M_L. The equations used to calculate local magnitude are most similar to those used by Richter and Gutenberg. For deep earthquakes, body-wave magnitude, which is more dependent on P wave amplitudes, is most appropriate. It is denoted by the symbol m_b. It tends to work best for earthquakes up to a m_b of 7.0. For shallow earthquakes, surface-wave magnitudes are often calculated. Surface-wave magnitudes use the symbol M_S. They are calculated using the maximum amplitude of the surface waves. The surface-wave magnitude equations work particularly well for large, shallow earthquakes. Ultimately, the goal of many seismologists is to determine how much energy was released during an earthquake. An additional magnitude scale was formulated with this goal in mind. It is called the moment magnitude, denoted by the symbol M_W. The moment magnitude seems to be the most accurate scale for the largest of earthquakes.

Is that the only way we express the size of an earthquake? No, there is another way that is more subjective. The *intensity* scale is used to describe the size of an earthquake based solely on the type of damage done. As you will find out later in this chapter, the damage done by an earthquake is dependent on a number of factors. These include not only the size of an earthquake and the amount of energy released when it occurs, but also construction practices, population density, and the type of ground beneath buildings. Thus, intensity is not a reliable measure of earthquake size, but instead is best used to describe damage done during an earthquake.

The end result of an intensity study is a map of severity of ground shaking. The map is constructed by gathering damage reports and witness statements from various locations where the earthquake was felt. For each location at which an observation was made, an intensity is determined. Those intensities are plotted on a map, and the result is a map showing zones of ground-shaking severity. The U. S. Geological Survey collects data from individuals to create intensity maps, which they call ShakeMaps. To view a ShakeMap for a large earthquake, go to the U.S. Geological Survey's earthquakes page (https://www.usgs.gov/programs/earthquake-hazards). Find the list of recent earthquakes and click on one of those earthquakes. You should be given the opportunity to view a ShakeMap. If you experience an earthquake, visit this page on the U.S. Geological Survey's earthquake site, and click on that earthquake in its records. A link called "Did You Feel It?" will allow you to contribute your observations. User responses are compiled to create ShakeMaps.

Even though it does not give us mathematical precision in determining the size of an earthquake, intensity still has its uses. Insurance companies are very interested in intensity maps, or ShakeMaps. Some areas of cities shake more severely during earthquakes than others, and insurance companies like to have this information when they are making decisions about providing earthquake insurance to a homeowner. People looking for a new home should be just as interested in this information because it is one measure of how well their potential new home is likely to perform in an earthquake. The intensity scale is also useful as a way to measure the size of historic earthquakes, those that occurred before seismometers were common and widespread. Intensity is the best size information we have for earthquakes that occurred before about 1910.

The intensity scale has gone through a number of revisions over the years, but the most widely used scale is called the *Modified Mercalli Intensity Scale* (Table 4.1). This scale uses Roman numerals and goes from I (barely felt by anyone) to X (extreme damage). In general, when examining historic earthquakes, we expect the epicenter to have been located in the zone of worst damage. In reality, however, epicenters could be outside of populated areas or in a zone of less-severe damage than the most serious that occurred. The MMI scale is therefore quite unreliable for determining epicenter location. It is also unreliable as a measure of earthquake magnitude.

Magnitude should, in theory, correlate quite well to the intensity scale. However, as stated before, it is highly dependent on a number of variables including the depth of the earthquake, variations in building materials and construction practices, and the makeup of ground materials beneath buildings. All of these factors can have huge effects on the type of damage done during an earthquake.

Fortunately, large and damaging earthquakes are relatively rare events (Table 4.2). Even the largest earthquakes must occur close to the surface of the Earth to cause widespread damage, so earthquakes that are large but occur more than about 35 mi (60 km) deep, produce far less damage than those located closer to the surface.

4.3.2 Additional Capabilities of Seismology

Seismology, or the science of studying seismic waves, can be useful not only in studying earthquakes, but also in detection of explosions. The United States is party to a number of treaties, including the Comprehensive Nuclear Test Ban Treaty. Nations that have signed onto the treaty may not test nuclear weapons. This means they may not test the weapons anywhere-- under the Earth, at the surface of the Earth, in the atmosphere, in space, or under the oceans. As with any treaty, it is meaningful only if it is verifiable.

There are a number of methods by which the Earth is monitored for nuclear blasts. One is by satellite, where the Earth's atmosphere is monitored for bright flashes that could be nuclear

Table 4.1 Modified Mercalli intensity scale.

Intensity	Shaking	Description/Damage
I	Not felt	Not felt except by a very few under especially favorable conditions.
II	Weak	Felt only by a few persons at rest, especially on upper floors of buildings.
III	Weak	Felt quite noticeably by persons indoors, especially on upper floors of buildings. Many people do not recognize it as an earthquake. Standing motor cars may rock slightly. Vibrations similar to the passing of a truck. Duration estimated.
IV	Light	Felt indoors by many, outdoors by few during the day. At night, some awakened. Dishes, windows, doors disturbed walls make cracking sound. Sensation like heavy truck striking building. Standing motor cars rocked noticeably.
V	Moderate	Felt by nearly everyone; many awakened. Some dishes, windows broken. Unstable objects overturned. Pendulum clocks may stop.
VI	Strong	Felt by all, many frightened. Some heavy furniture moved; a few instances of fallen plaster. Damage slight.
VII	Very strong	Damage negligible in buildings of good design and construction; slight to moderate in well-built ordinary structures; considerable damage in poorly built or badly designed structures; some chimneys broken.
VIII	Severe	Damage slight in specially designed structures; considerable damage in ordinary substantial buildings with partial collapse. Damage great in poorly built structures. Fall of chimneys, factory stacks, columns, monuments, walls. Heavy furniture overturned.
IX	Violent	Damage considerable in specially designed structures; well-designed frame structures thrown out of plumb. Damage great in substantial buildings, with partial collapse. Buildings shifted off foundations.
X	Extreme	Some well-built wooden structures destroyed; most masonry and frame structures destroyed with foundations. Rails bent.

Source: Courtesy of the U.S. Geological Survey; https://www.usgs.gov/media/images/modified-mercalli-intensity-scale/Public Domain
Note: The intensity scale is a subjective scale, based on the severity of effects from the earthquake, including severity of ground shaking and damage done. Reports are collected from the area in which the earthquake was felt. For each location, an intensity value is assigned. The intensities are plotted on a map, and the map is shaded to represent zones of damage. The result is an intensity map, or ShakeMap.

Table 4.2 Historical occurrences of earthquakes and their magnitudes.

Descriptor	Magnitude	Average number per year
Great	8 and higher	1
Major	7–7.9	18
Strong	6–6.9	120
Moderate	5–5.9	800
Light	4–4.9	6,200 (estimated)
Minor	3–3.9	49,000 (estimated)
Very minor	Less than 3.0	More than 1,000 per day (estimated)

Source: After Fletcher, 2014.
Note: In general, larger earthquakes occur much less frequently than smaller earthquakes. Earthquakes of magnitude 8 or greater occur at a rate of less than one time per year. In contrast, the smallest of earthquakes (less than magnitude 3) occur more than a thousand times per day.

detonations at the Earth's surface or in the atmosphere. Another is by air testing. By testing the air for distinctive elements created when a nuclear bomb is detonated at, near, or above the Earth's surface, scientists can state definitively whether an explosion was due to a nuclear blast. The third method used to monitor the Earth in support of this treaty is through a worldwide network of seismometers. These seismometers are found in many countries around the world as well as under the oceans. Scientists can determine with relative ease the difference between an earthquake and an explosion. When a large explosion is detected, particularly near a known nuclear testing site, the explosion is further investigated using air sampling and imagery from satellites. For example, when North Korea has tested its nuclear weapons, it has rarely been a surprise. The West has satellites trained on North Korea's known nuclear test site, and they typically detect increased activity at the site in the weeks prior to a nuclear test. When a large explosion is detected, air sampling typically finds distinctive evidence that the blast was nuclear, and shortly after the explosion, it is not uncommon from North Korean media to broadcast news of a successful nuclear test. North Korea is one of the very few countries in the world that has not signed onto the Comprehensive Nuclear Test Ban Treaty.

Seismology has been used in other unconventional capacities as well. When terrorists crashed planes in the World Trade Center buildings in New York City on 11 September 2001, the buildings burned for a short period of time, then collapsed. There was speculation that terrorists had planted large bombs within the buildings that detonated and brought the buildings down. Seismologists from nearby Columbia University recorded the events of that morning on their seismometers (Fig. 4.17). They detected the impacts of the airplanes with the towers, but did not detect an explosion immediately before either building collapsed. It was therefore definitively proven that there were no explosives in the towers.

In 2000, ocean-floor seismometers were used to determine the fate of the Russian submarine *Kursk*. Upon examination of seismograms from instruments on the seafloor, it was determined that the *Kursk* experienced two explosions on board, which led to the sinking of the submarine and loss of all sailors on board. In yet another unconventional use of seismometers, soccer fans gathered in Mexico City and literally jumped for joy in the summer of 2018 after its World Cup team advanced. The jumping was energetic enough to be picked up by seismometers within the city. The news media erroneously reported that the fans had caused an earthquake. In reality, there was no motion

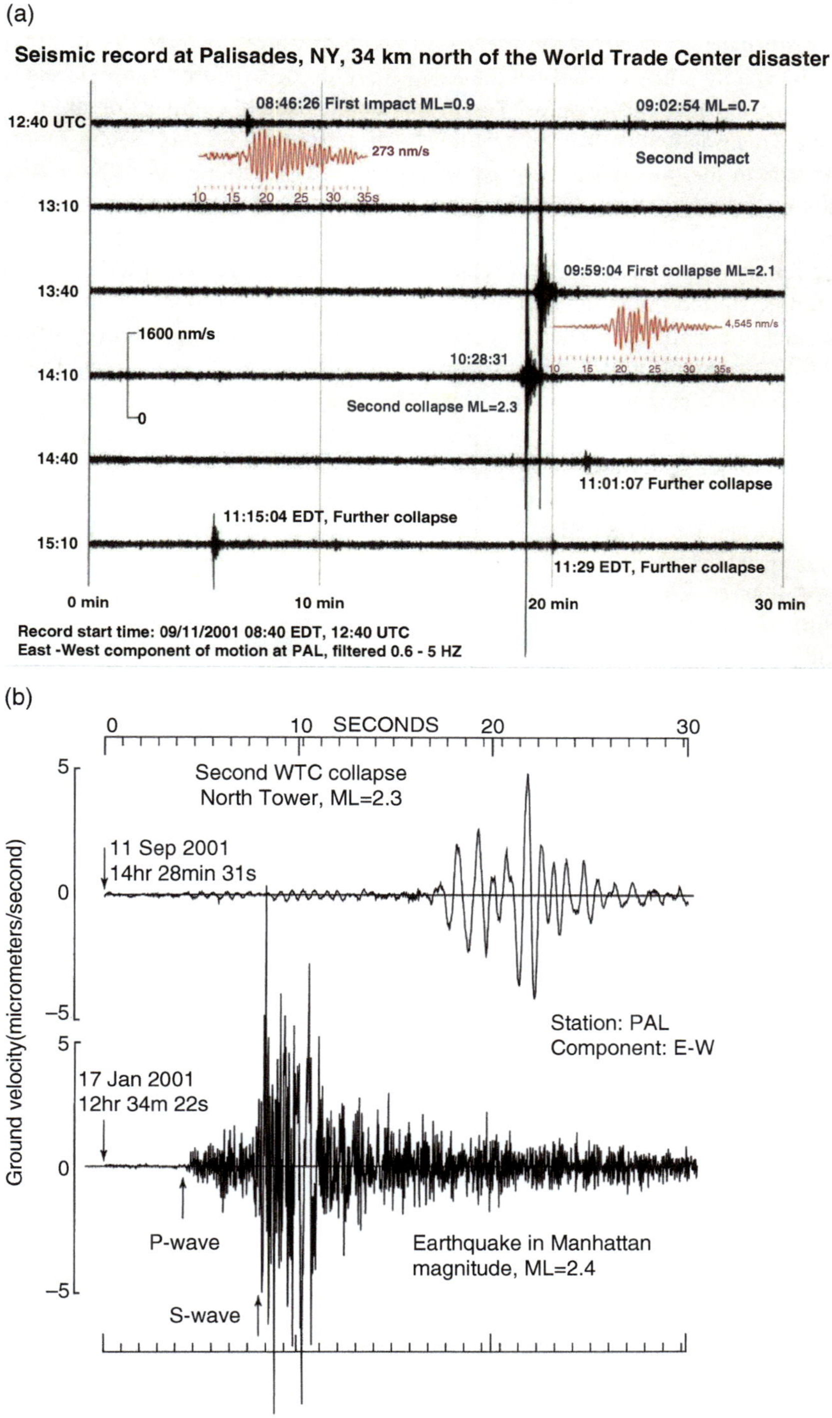

Figure 4.17 (a) Seismic records from Columbia University's seismic station PAL from 11 September 2001. The first image shows the seismic record for three continuous hours starting at 8:40 a.m., which is 12:40 UTC. Small seismic signals mark the impacts of the two planes (the first is expanded in red), and the larger signals are from the collapse of the two towers (again, the first is expanded in red). Smaller signals at 11:01 and 11:15 a.m. are from further collapses that day. (b) The second image is a comparison of the tower collapse signal and an earthquake that produced similar local magnitudes. From Kim et al., 2001.

on any known fault in the area, so it was merely a ground disturbance picked up by seismometers in the vicinity and not an earthquake.

Can humans ever cause earthquakes? Yes, in fact they can. This was first confirmed around the Denver area in the mid-1960s. Disposal of waste fluids by injecting them deep into the ground changed the stresses around preexisting but stationary faults. When the stresses changed, faults started moving, and earthquakes began happening. When people stopped the practice of injecting fluid into deep wells, the earthquakes stopped. More recently, oilfield practices have led to thousands upon thousands of earthquakes in areas where they are drilling for oil and natural gas. A relatively new method of oil and gas extraction, called fracking requires that large volumes of fluids be pumped into the ground very quickly in order to fracture the rocks. Occasionally, this can cause movement on a fault. Perhaps more common, oil and gas companies are pumping large amounts of saltwater, a common byproduct of oil and gas drilling, into the ground in deep disposal wells. If there are any faults in the area, the change in the pore pressures, or pressure in between grains in the rock, causes those faults to become mobile. Due to the injection of fluid waste into the ground, we are seeing more human-induced earthquakes than ever before (Fig. 4.18). Oklahoma has seen the greatest increase in these earthquakes, with dozens every year in the magnitude 3 range or above. The largest earthquake that has occurred in Oklahoma in recent memory was a result of these practices. It was a magnitude 5.8 earthquake near the city of Pawnee in 2016. It was felt in surrounding states and caused moderate damage within the city.

Earthquakes can be used also to investigate the interior of the Earth. In a small-scale example (Fig. 4.19), geophysicists used locations of earthquakes and travel times of seismic waves to map the plumbing system of Steamboat Geyser and Cistern Spring in Yellowstone National Park. The technique they used is called *tomography*, which measures density differences in the Earth that cause seismic waves to move at different speeds than a model of a uniformly

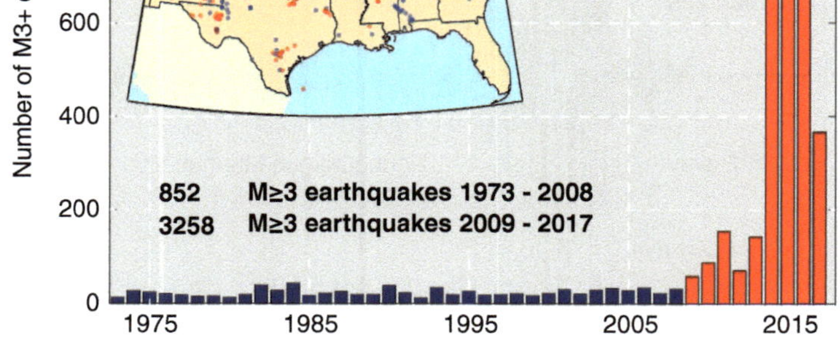

Figure 4.18 Number of earthquakes of magnitude 3 and above that have struck the central United States from 1973 to 2017. Oklahoma has seen the greatest increase in numbers of earthquakes. This is due to fracking and injection of wastewater into deep wells in the state. Courtesy of U.S. Geological Survey; https://Earthquake.usgs.gov/research/induced/overview.php/Public Domain

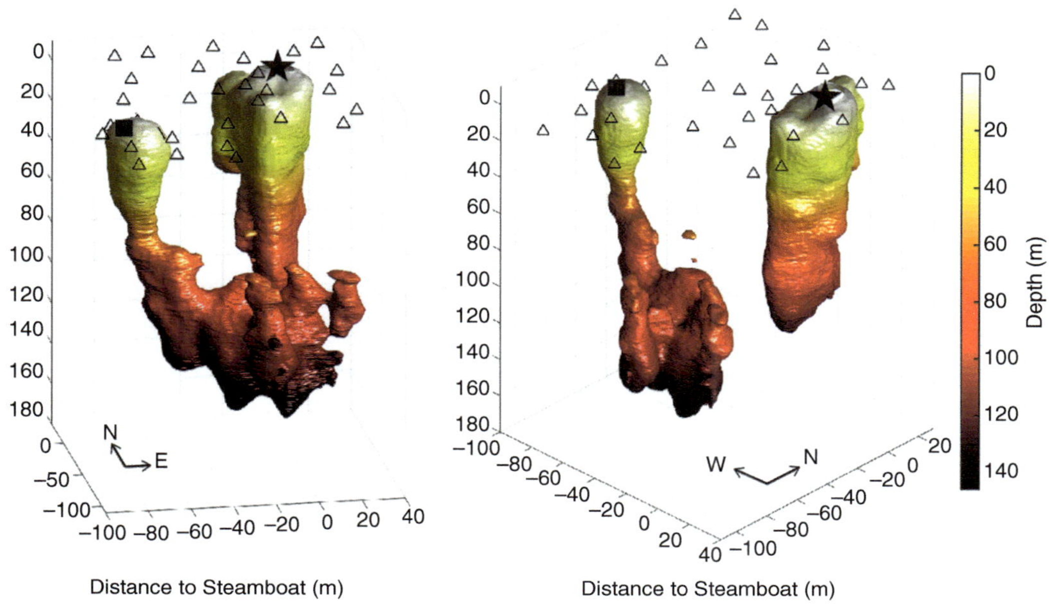

Figure 4.19 Three dimensional map of the underground plumbing system of Steamboat Geyser (black star) and Cistern Spring (black square) created through the process of seismic tomography. The triangles at the top of the diagram indicate locations of seismometers used in the study, and color is an indication of depth, with the lighter colors indicating shallower depths. https://www.usgs.gov/observatories/yvo/news/complex-plumbing-systems-steamboat-geyser-and-cistern-spring /Public Domain

layered Earth would predict. By tracing paths of seismic waves from source to seismometer, a series of slices of the Earth are constructed. When those slices are combined and imaged in three dimensions, underground structures can be mapped. This is similar to how a CT scan of the human body is produced. Instead of seismic waves creating slices through the Earth, narrowly focused x-rays produced by the CT scanner create image slices of the human body. When those x-ray slices are processed and added together, they produce a detailed three dimensional image of the body.

Scientists can also use an artificial source to create seismic waves in order to map faults and other structures in the Earth's subsurface. This is commonly used by oil companies to find structures underground that might trap oil or natural gas (Fig. 4.20), creating an economically viable reservoir for those products. In addition, processed seismic images like that in Figure 4.20 can indicate where faults exist, which helps engineers as they plot the drill's course through the rock.

4.4 How Do Earthquakes Work?

Earthquakes occur when there is movement on a fault. How and why does that movement occur? In most cases, faults that create earthquakes are found at or near plate boundaries (Fig. 4.21), which are always in motion. The motion may be slow, but it is moving nonetheless. The relative motion of two plates can create built-up energy at the boundary. Energy that builds within a body is referred to as *potential energy*. Rocks are deformed, and that deformation is called *strain*. As strain accumulates in the rocks on either side of a plate boundary, the rocks are held in place because they are stronger than the strain that has accumulated. Once the strain exceeds the strength

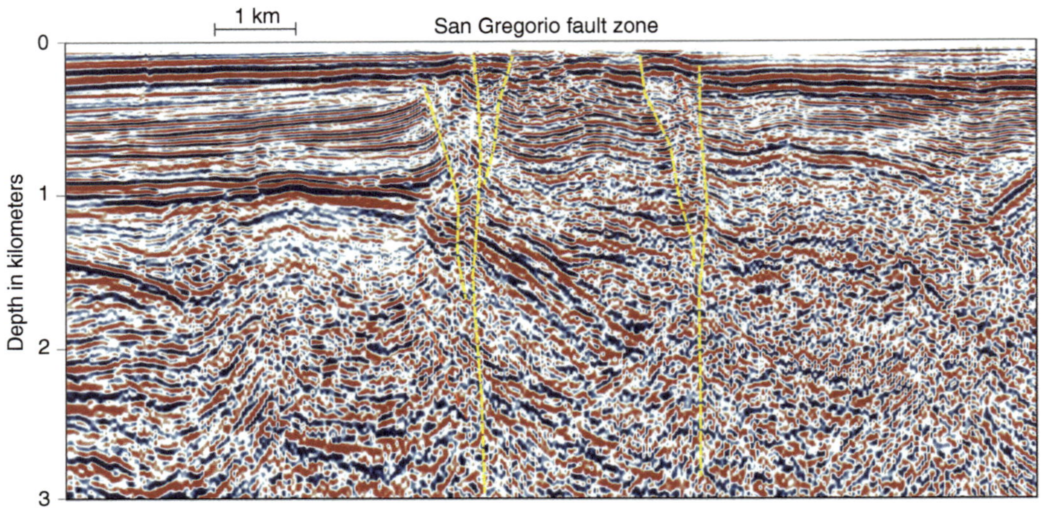

Figure 4.20 Cross section of the San Gregorio Fault Zone off the coast of California, created when human-made seismic waves were bounced off of layers in the subsurface. Faults are indicated by yellow lines, and red and blue linear features indicate layers of rock in the subsurface. Cross section courtesy of the U.S. Geological Survey.

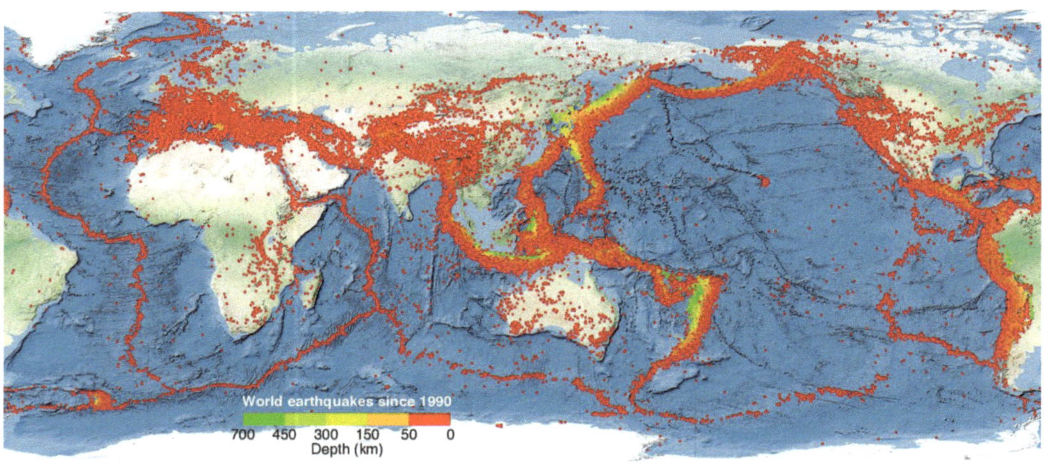

Figure 4.21 Worldwide earthquake occurrences since 1990. The individual dots on the map represent earthquakes, and they are color coded for depth. Most earthquakes are shallow (less than 31 mi or 50 km deep). Deeper earthquakes occur at subduction zones. Subduction zones are found in locations where the colors range from red to yellow to green. From Fletcher, 2014.

of the rocks, the rocks break and move rapidly. This sudden break and movement of rocks on either side of a fault is an earthquake. In an earthquake, the rocks move as strain is released. Potential energy, or stored energy, is converted to energy of motion, or *kinetic energy*.

Imagine that you have a small stick. If you simply hold the stick in your hands, the stick is at rest. It has no outward *stresses*, and it has not been strained or deformed. If you take the stick in both hands and move your hands together, you begin to bend it. You are putting stress on the stick and it is becoming strained,

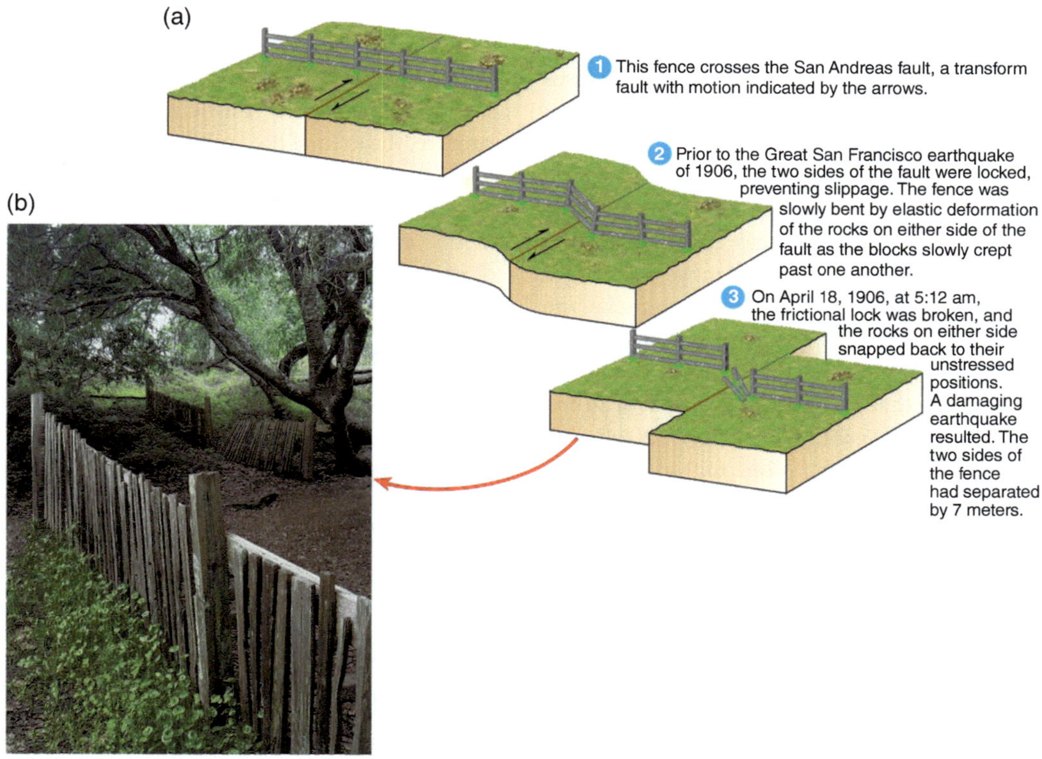

Figure 4.22 Elastic rebound as illustrated by events around the 1906 San Francisco earthquake on the San Andreas Fault. (a) A fence was built in the late 1800s that crossed the San Andreas Fault. Over time, strain accumulated in the fence as it was bent from movement on either side of the fault. When the 1906 earthquake occurred, the stored energy in the fence was converted to kinetic energy (energy of movement). The fence broke with the fault and was separated by several meters. (b) Reconstructed fence built in the location of the 1906 fence that broke. It illustrates the offset of the two sides of the fault as a result of the 1906 earthquake. From Murck & Skinner, 2016.

or deformed. It is also storing energy, which is potential energy. Eventually, if you keep increasing the pressure on the stick, you will break the stick. At this point, the strain is reset because the stick is no longer bent. It is in two undeformed pieces. The act of breaking the stick took the potential (stored) energy and turned it into energy of motion (kinetic energy). This process of increasing the stress and strain on either side of a fault, until it ruptures and recovers to an undeformed state, is called *elastic rebound*. Scientists first recognized the process of elastic rebound following the 1906 earthquake in San Francisco.

To clarify, rocks can behave in both brittle and plastic ways. Brittle deformation means that rocks break. This is what creates faults. Rocks can also respond to high stresses with plastic deformation, in which rocks are bent and twisted but do not break, and that deformation is permanent and nonreversible. Elastic deformation is temporary.

Geologists who observed changes in the landscape north of San Francisco described it in the scientific literature shortly thereafter (Reid, 1911; Fig. 4.22). A fence north of San Francisco crossed the San Andreas Fault. In the years prior to the 1906 earthquake, the part of the fence located directly over the fault began to deform because strain was building on the fault. When the earthquake occurred, rapid movement of the fault caused the fence to break. The fence on either side of the fault was separated by several meters.

Earthquakes are typically rapid processes that begin at a single location on the fault, which is the focus. A break along a fault may expand and move in one direction or both directions away from that location along the length of the fault. In general, the larger the area of a fault that breaks and moves, the larger the earthquake. The larger the earthquake, the longer ground shaking occurs as well.

As an example, the magnitude (M_W) 9.2 earthquake that occurred off the coast of Sumatra on Boxing Day (26 December), 2004, took 10 min to rupture from one end to the other. The length of the rupture was approximately 746–808 mi (1,200–1,300 km), and the width of the rupture was about 62 mi (100 km). One side of the plate boundary moved 66 ft (20 m) relative to the other side. Using the conservative number for the length of the fault rupture, the area of the fault surface that broke was 746 mi long by 62 mi wide, or 46,252 mi^2 (1,200 km x 100 km, or 120,000 km^2). The speed of the rupture was approximately 1.24 mi/s (2 km/s).

4.4.1 Types of Faults

Faults come in three basic varieties, and roughly correspond to the type of motion you might expect to see at the three types of plate boundaries (Fig. 4.23). Each is defined based on the relative movement of the two sides of

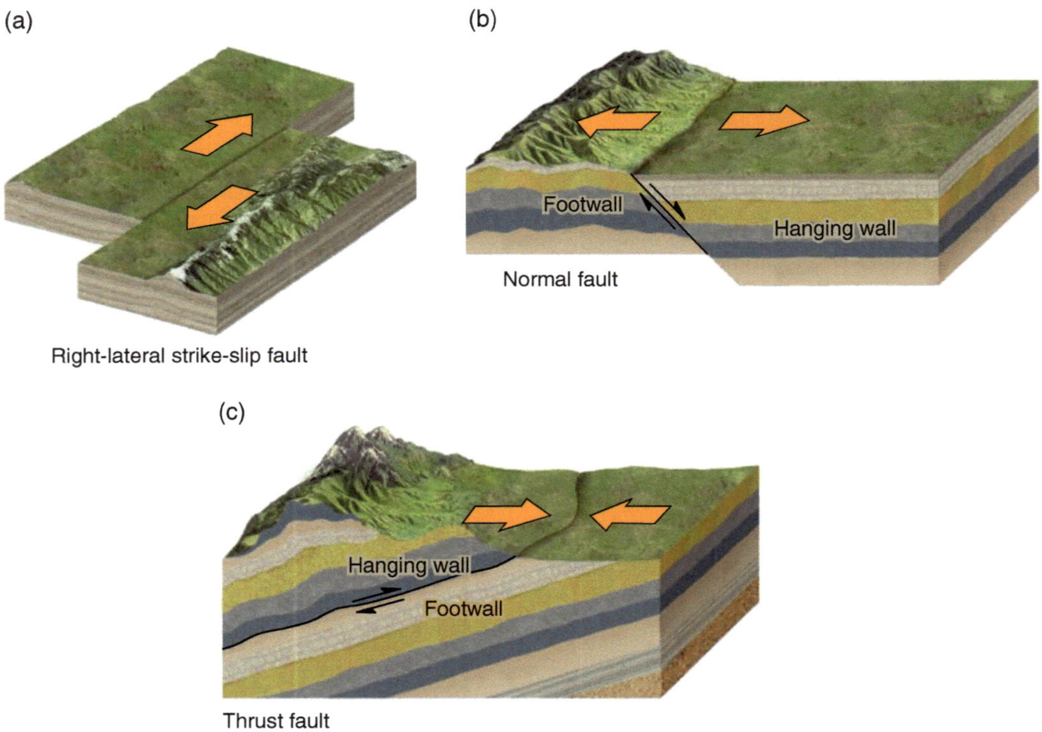

Figure 4.23 Types of faults on which earthquakes occur. (a) A right-lateral strike-slip fault. This is a right-lateral fault because if you were straddling the fault, the right side would move toward you. Alternately, if you were on one side of the fault looking across to the opposite side, the opposite side would move to the right. This type of fault is common at transform plate boundaries. (b) Normal fault where the block above the fault (the hanging wall) moves down relative to the other side of the fault. This is typical of divergent plate boundaries. (c) Thrust fault, where the hanging wall (area above the fault) moves up and over the other side of the fault. This type of fault is often seen at convergent plate boundaries. Images from Fletcher, 2014.

the fault. For the first type of fault, the boundary between the two sides of the fault is usually nearly vertical. The two sides of the fault move laterally, or side-to-side. This is called a *strike-slip fault* (Fig. 4.23a). It mimics the movement seen at transform-plate boundaries, but can be found in association with any type of plate boundary and within plates. We define the type of motion for this type of fault based on which side of the fault moves toward you if you were to straddle the fault. For example, the San Andreas Fault is both a transform plate boundary and a strike-slip fault. If you were to straddle the San Andreas Fault, you would find that the right side of the fault moves toward you. That means it is a right-lateral strike-slip fault. This is true whether you face the north or the south, similar to the idea that we in the United States always drive on the right side of the road. Whether going to work or coming home from work, you are always driving on the right side of the road. The Garlock Fault, which splits from the San Andreas just north of Los Angeles and extends to the east through the desert, is a left-lateral strike-slip fault. That means if you straddled the fault, the left side would be moving toward you. That also means the block of land south of the Garlock fault is moving east, away from the San Andreas. Strike-slip faults can produce large and damaging earthquakes. What makes them particularly dangerous is that earthquakes along strike-slip faults are typically shallow. That means most of the earthquake's energy goes into the crust, and results in severe shaking at the Earth's surface. The 1906 earthquake in San Francisco was on the San Andreas Fault, a strike-slip fault. A 1999 earthquake in Izmit, Turkey, that killed 17,000 people, was also on a strike-slip fault, the North Anatolian Fault (Fig. 4.24).

The other two types of faults usually have a fault surface that is inclined, or tilted. We refer therefore to the side above the fault and below the fault. When the area above the fault (the *hanging wall*) moves down relative to the area below the fault (the *foot wall*), this is called a *normal fault* (Fig. 4.19b). This is the type of motion that might be typically found at a divergent plate boundary, but like a strike-slip fault can be associated with any type of plate boundary, and can be found within plates as well.

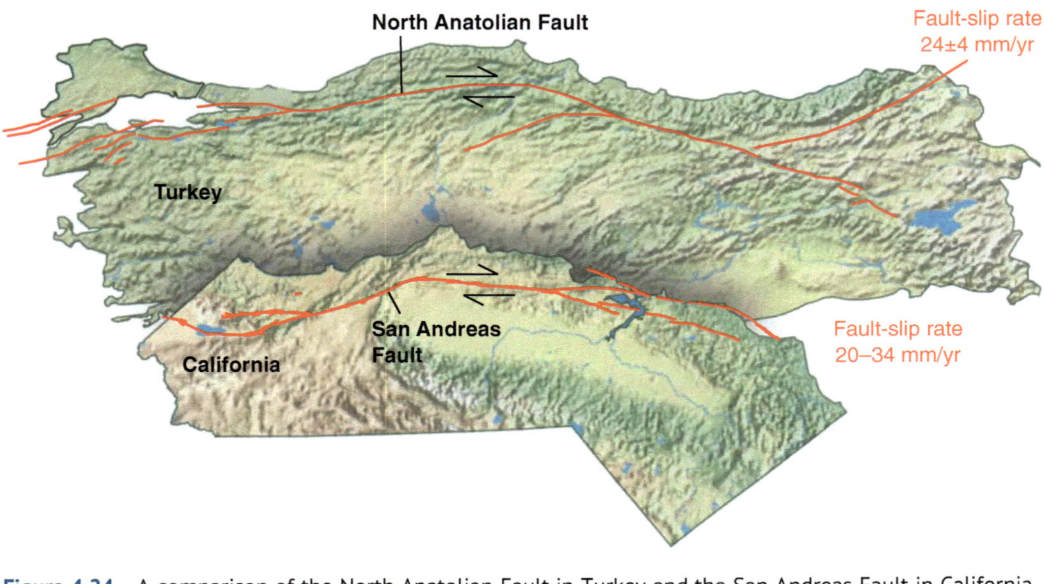

Figure 4.24 A comparison of the North Anatolian Fault in Turkey and the San Andreas Fault in California. Faults in both locations are outlined in red, and the sense of motion is indicated next to the fault. Since the two sides of each fault are moving laterally past each other, these are both strike-slip faults. From Fletcher, 2014.

The third type of fault is called a *thrust fault* or *reverse fault*. This is the type of fault we find when the area above the fault moves up relative to the other side (Fig. 4.19c). This fault type is found most often at convergent plate boundaries, both subduction zones and collision zones, but can be found at any type of plate boundary and within plates. Of the 16 largest earthquakes that have occurred since 1910 when instrumental records began (all magnitude 8.5 and above), all occurred on this type of fault. Fifteen occurred at subduction zones around the world, and 1 occurred at a collision zone where India is colliding with mainland Asia.

Subduction zones are the only locations where deep earthquakes (deeper than 186 mi or 300 km) occur with any regularity or predictability. Deep earthquakes simply do not occur with other types of plate boundaries because earthquakes must occur in the lithosphere. The lithosphere is only about 78 mi (125 km) thick even under the continents. Therefore, earthquakes in other settings do not occur below those depths. However, when subduction occurs, the lithosphere is violently shoved down into the mantle of the Earth. When this happens, the lithosphere remains cold and rigid for millions of years. That means earthquakes can still occur in subducted lithosphere until the slab reaches a depth of up to about 435 mi (700 km). Earthquakes have not been recorded below this depth.

Each day, thousands of earthquakes occur around the globe. Most are too small to be felt. About 75 earthquakes per year occur that are large enough to be felt and do some amount of damage. The largest earthquakes, those above a magnitude 8, occur roughly once every 1 to 3 years. Earthquakes of or above magnitude 8 are known as *great earthquakes*. They are called great not because they are impressive or outstanding or more interesting than smaller earthquakes. It is because they have the ability to completely reset the stress regime in the area in which they occur.

Do earthquakes occur away from plate boundaries? Absolutely they do. These are called *intraplate earthquakes*, meaning they occur within the interiors of plates. In the United States, there are a number of these earthquakes every day. One region that experiences a large number of earthquakes each year is the New Madrid Fault Zone. This is found on the border of Missouri, Kentucky, Illinois, and Tennessee. In 1811 and 1812, there was a series of three powerful earthquakes along this fault zone. All three of the earthquakes, based on descriptions of ground shaking and damage done, were likely of magnitude 8 or above. This is one area in the United States that has an unknown earthquake risk. We have not seen large earthquakes in this area since the early 1800s, and we have no idea what the recurrence interval (how often these earthquakes occur) is for earthquakes on the New Madrid Fault Zone. It could be thousands of years, or just few hundreds of years. This fault zone is still a bit of a mystery to scientists in that sense. However, we do know that earthquakes continue to occur on the fault zone every day, and we know that when a large earthquake in the New Madrid area occurs, it will affect a wide area.

In 1886, a large earthquake occurred near Charleston, South Carolina, causing great damage to the city. Estimates based on damage put this earthquake somewhere near a magnitude 7.5. In 2011, a magnitude 5.8 earthquake struck Virginia. It shook the surrounding states and caused damage to the National Cathedral and Washington Monument in Washington, D.C. (Fig. 4.25). The Washington Monument was closed to visitors for many months afterward as officials worked to repair it.

Earthquakes on the New Madrid Fault Zone and in other parts of the eastern half of the United States are particularly hazardous. Since the eastern United States has thick soil cover, surface waves travel great distances and can be amplified in areas of thick sediment. In the western United States, earthquakes are felt

Figure 4.25 Damage at the Washington National Cathedral following a magnitude 5.8 earthquake in Virginia in 2011. Photo by John Stuhldreher; Courtesy Washington National Cathedral. (Private email/correspondence)

over much smaller areas and damage is limited to regions closest to the earthquake (Fig. 4.26). However, in the eastern United States, damage can extend far from the earthquake's epicenter because of amplification of surface waves (see section 4.5, Earthquake Hazards).

4.5 Earthquake Hazards

Although at this time it is impossible to accurately forecast an earthquake, it is possible to predict where earthquakes will occur. Given a particular location, it is even possible to predict what type of damage is likely to occur in a large earthquake. Earthquake hazards anticipated at sites around the world are well known. They are predictable based on the magnitude and location of each earthquake.

The most obvious and recognizable hazard during an earthquake is ground shaking. The ground can move quickly up and down, back and forth, or even in a circular motion. People describe earthquake motion differently, and

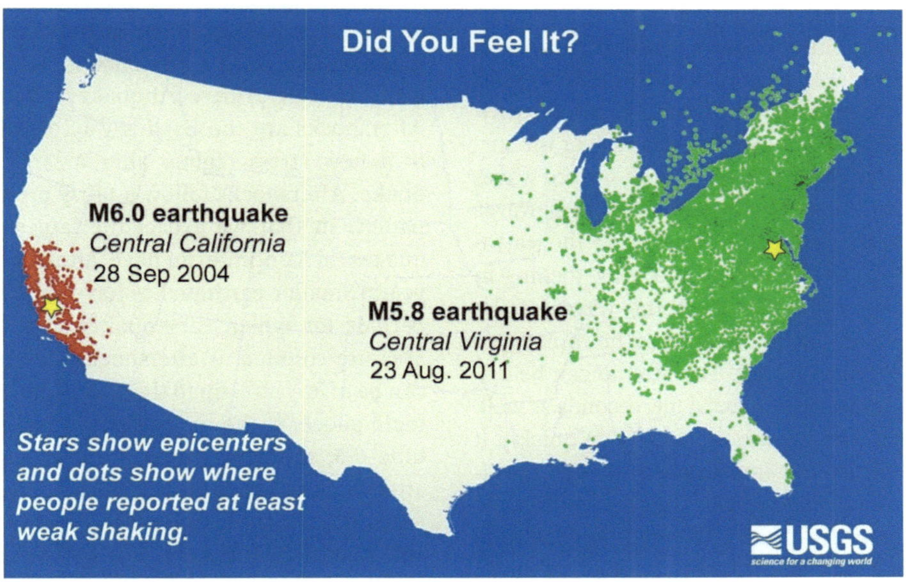

Figure 4.26 Red and green dots indicate locations in which these two earthquakes were felt. The California earthquake was slightly larger in magnitude than the Virginia earthquake, but it was felt over a much smaller area. This is because much of California has solid rock close to the surface of the Earth, so surface waves do not travel very far or at great amplitude in that material. Much of the eastern United States is covered in deep soil. This allows surface waves to be amplified and travel great distances through the thick soil. Courtesy of U.S. Geological Survey; https://blogs.agu.org/wildwildscience/files/2015/02/felt-comparisons.jpg/ Public Domain

Figure 4.27 Collapse of a fireplace façade and chimney inside a residence as a result of the Parkfield, California, earthquake in 2004. © AP/Wide World Photos. From Murck & Skinner, 2016.

these differences are largely due to the type of motion that occurred on the fault and what types of waves are largest in that particular earthquake. If the earthquake occurs on a strike-slip fault, back and forth motion might be the most prominent type of motion experienced. An earthquake that occurs near the surface of the Earth may send large surface waves to communities above. Large Rayleigh waves may make those who experience the earthquake feel as if they are on a boat, rocking in a circular motion as if on the waves of the ocean. In any given earthquake, multiple types of motion may be felt. The ground can be displaced several feet in just a few seconds as well, meaning that the ground moves so quickly it can be difficult or impossible to remain standing. Ground shaking can trigger a number of other hazards. In the local area, this includes objects falling from shelves, out of cabinets, and off of building facades, as well as chimney collapse (Fig. 4.27). It can also include structure collapse, fires, *liquefaction*, and amplification of ground motion. Across a broader area, avalanches, landslides, dam failure, tsunamis, and seiches may be seen.

4.5.1 Aftershocks

Aftershocks are smaller earthquakes that occur after a main shock or major earthquake. (If the first earthquake in a sequence is not the largest, we call prior earthquakes *foreshocks*.) Aftershocks are the Earth's way of adjusting to a new stress regime after a large earthquake. Aftershocks follow a fairly predictable pattern in that we expect the largest earthquakes to continue to have aftershocks for years. Smaller earthquakes have shorter time periods for which earthquakes in the same area are counted as aftershocks. Aftershocks can be of any size, up to the size of the original main shock. If the main shock was a magnitude 7.5, there will usually be at least one aftershock in the magnitude 6 range, a handful of aftershocks between magnitude 5 and 6, and increasing amounts of aftershocks of smaller magnitudes. Aftershocks can be particularly dangerous because they can bring down buildings that were damaged during the main shock. This puts the injured and rescue workers at risk. They can also further frighten an already traumatized population because these can be large, damaging earthquakes in their own right.

4.5.2 Structure Collapse

Structure collapse is a major hazard in areas where ground shaking is severe, especially where construction is poor. In 2003, an earthquake occurred near the city of Bam, Iran. Though it had a magnitude of only 6.6, the city's construction was largely composed of adobe, or mud bricks. The adobe bricks crumbled in the earthquake, flattening homes, businesses, and historic sites including a citadel and UNESCO World Heritage Site that was 2,000 years old. Over 26,000 people died. Most were crushed when their homes, also constructed with bricks of adobe, collapsed.

Poor construction practices were also cited in the enormous death toll of the 2010 earthquake in Port-au-Prince, Haiti (Fig. 4.28). Over 220,000 people died. Like the earthquake in Iran, many of those deaths were caused when people's homes collapsed and buried them under the rubble. The most common type of injury for those who survived was a crushing injury, again from falling debris as buildings collapsed.

On 6 February 2023, a sequence of large earthquakes occurred in southeastern Turkey, near its border with Syria. A magnitude 7.5 earthquake occurred early in the morning of 6 February (Fig. 4.29), and was followed 9 hr later by a magnitude 7.8 earthquake, 59 mi (95 km) from the previous earthquake. The earthquake happened on the strike-slip East Anatolia fault zone near the boundary of the Anatolia, Arabia, and African plates. More than 58,000 people were confirmed dead from this earthquake, 50,783 in Turkey and 8,476 in Syria (Wikipedia, 2023b). Most died as a result of building collapse. Nearly 130,000 people were injured as well. More than 160,000 buildings collapsed in the earthquake, leaving millions of people homeless in both Syria and Turkey. Most of these buildings had not been built to the proper earthquake codes and were structurally unable to withstand the severe shaking from these powerful earthquakes. Many of the collapsed buildings had been built

Figure 4.28 Building collapse was an all-too-frequent cause of death in Port-au-Prince, Haiti, when a magnitude 7.0 earthquake struck in 2010. © Ron Haviv/VII/Corbis. From Murck & Skinner, 2016.

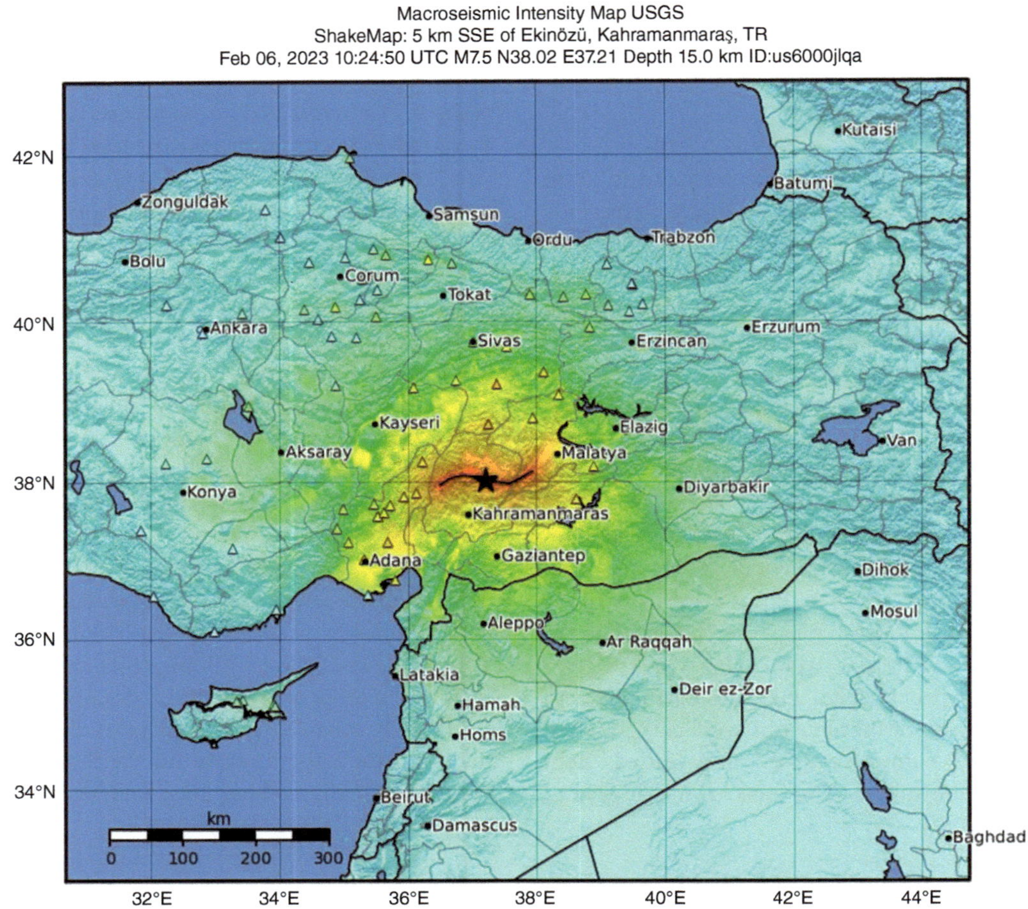

SHAKING	Not felt	Weak	Light	Moderate	Strong	Very strong	Severe	Violent	Extreme
DAMAGE	None	None	None	Very light	Light	Moderate	Moderate/heavy	Heavy	Very heavy
PGA (%g)	<0.0464	0.297	2.76	6.2	11.5	21.5	40.1	74.7	>139
PGV(cm/s)	<0.0215	0.135	1.41	4.65	9.64	20	41.4	85.8	>178
INTENSITY	I	II–III	IV	V	VI	VII	VIII	IX	X+

Scale based on Worden et al. (2012)   Version 9: Processed 2023-03-08T17:21:31Z
△ Seismic Instrument ○ Reported Intensity ★ Epicenter ▭ Rupture

Figure 4.29 ShakeMap for the magnitude 7.5 earthquake that occurred in Turkey. From the U.S. Geological Survey; https://earthquake.usgs.gov/earthquakes/eventpage/us6000jlqa/shakemap/intensity

since the government of Turkey enacted new earthquake codes in 1999. Several engineers, building inspectors, and contractors in Turkey were arrested following the earthquake when buildings they were involved in building were shown to have been built without the proper legal earthquake engineering. For months after the earthquake, many earthquake survivors were housed in tents but others lived out of vehicles, train cars, university residence halls, container homes, and even hotel resorts. There is no estimate for how long it will take for these people to be housed in proper homes, but more sturdy shelters are being built to house people through the winter months in late 2023-early 2024. It will take years for Turkey and Syria to recover from this event.

Perhaps the largest death toll from any natural disaster in history was that of an earthquake in China in the year 1556. The earthquake occurred where many of the people lived in caves dug out of the soft Earth. The earthquake caused the caves and area buildings to collapse, killing over 800,000 people. Another earthquake in Tangshan, China, in 1976 killed approximately 250,000 people. That city was known for its brick factories. As a result, many of the buildings in the city were built from bricks. While bricks are relatively strong, the mortar that holds them together is not. When earthquakes strike, the mortar crumbles and brick structures collapse. Most of those who died in Tangshan died due to collapse of brick structures.

4.5.3 Fire

As illustrated in the case study of San Francisco's 1906 earthquake in the opening of this chapter, fire can also be a major hazard when large earthquakes strike. Gas lines rupture, electric lines break, and the two together mean many small fires can be ignited in short order. Damage to water lines can mean that very little water is available to fight fires, so fires can spread out of control very quickly. In modern cities in industrialized nations, fire is somewhat less of a hazard than it is in developing countries because many utilities in earthquake-prone areas have automatic emergency shutoff valves that operate when an earthquake strikes. In developing nations, many people still cook with open fires, and gas lines do not have automatic emergency shutoff valves. However, fires still plague even the most modern of cities after major earthquakes (Fig. 4.30).

4.5.4 Amplification of Ground Motion

Some materials, such as the made land in San Francisco, are naturally prone to enhanced ground shaking. One might expect loose, soft material to cushion buildings from the severe shaking of an earthquake, but the opposite is true. Loose material such as thick soil, mud, sand, and made land amplifies, or increases, the amplitude of surface waves during earthquakes making the shaking more severe. If

Figure 4.30 Fires are a major hazard after earthquakes. Earthquakes can rupture gas lines and break power lines, and the combination of these two can cause fires to ignite. Ground shaking can also disrupt water distribution systems, making fires difficult to fight. AFP/Getty Images, Inc. From Fletcher, 2014.

an earthquake occurs at a shallow depth, surface waves are already expected to be the highest amplitude waves people experience. If a building sits on loose, soft material such as deep soil or sand, those surface waves are amplified. This causes greater amounts of damage in areas underlain by this material. In the same earthquake, damage in two different parts of a city may look entirely different depending on the type of material that underlies those parts of the city.

An example of this phenomenon occurred in 1985. A magnitude 8.5 earthquake occurred near the coast of Mexico in the state of Michoacan. While there was some damage in the area near the epicenter of the earthquake, the majority of damage occurred inland near the center of the country in Mexico City. Structures near the epicenter were built on solid rock, and there was little damage as a result. However, Mexico City is built on a dry lake bed, and is underlain by sand, silt, and clay. This material amplifies surface waves, and the result is enormous damage any time a large earthquake strikes within a few hundred miles of the city. In 1985, the earthquake was felt by almost 20 million people and at least 9,500 were killed. More than 400 buildings collapsed, including a hospital and several apartment buildings (Fig. 4.31).

4.5.5 Liquefaction

In addition to the amplification of surface waves in soft material, Mexico City had a problem that surfaced during this particular earthquake. The soft material under parts of the city went through a process called liquefaction. Liquefaction can occur in two different forms. In both, the underlying cause is that shaking causes the particles in the soil and sediment to compact. That means the spaces between particles in the soil begin to close. This forces water and air that existed between the particles to be expelled. The result can either be forceful eruption of water and sand from the ground, known as a "sand blow," or the soil can begin to act more like a liquid than a solid (Fig. 4.32a). In the second case, heavy buildings begin to sink into the ground. In Mexico City, many buildings were left relatively intact, but became tilted and partially buried as a result of liquefaction. The same phenomenon was

Figure 4.31 Rescue workers look for survivors in the rubble of a collapsed building in Mexico City in 1985. From Skinner & Murck, 2011.

Figure 4.32 Two results of liquefaction. (a) Sand blows discovered after the Loma Prieta earthquake in 1989. Image LP0429 from the Loma Prieta Collection; https://nisee.berkeley.edu/elibrary/Image/LP0429. (b) A relatively intact building in Izmit, Turkey, was overturned by liquefaction of the soil beneath it. Image IZT-62 from the Izmit Collection; https://nisee.berkeley.edu/elibrary/Image/IZT-62 Both photos from NISEE-Berkeley.

observed in Izmit, Turkey, during an earthquake in 1999 (Fig. 4.32b). Liquefaction and amplification of surface waves often occur together because both occur in similar types of materials.

4.5.6 Landslides and Other Earth Movements

Avalanches and landslides fall under the broader category of "mass wasting" or "Earth movements," the subject of Chapter 6 in this book. Mass wasting occurs when material moves downhill under the influence of gravity. Though gravity may ultimately be what brings material downhill, it can be shaken loose by earthquakes. Earthquakes have triggered some of the largest and deadliest mass wasting events in history. One of the most tragic occurred in 1970 in a mountainous area near Chimbote, Peru.

On 31 May 1970, a magnitude 7.9 earthquake occurred at a shallow depth just off the coast of Peru. In the city of Chimbote, nearest the epicenter, about 80% of the city's buildings received severe damage or completely collapsed. About 3,000 people lost their lives. In the nearby city of Casma, about 90% of the structures were severely damaged but there was little loss of life. Water, sewer, phone, and power lines were damaged throughout the region, as were roads, railroads, and bridges. Though damage was considerable in these coastal cities, the majority of deaths occurred inland, in the mountainous area known as the Cordillera Blanca and the neighboring valley of the Rio Santa. A debris flow started high in the mountains and sped downhill, destroying and covering two villages (Fig. 4.33). The debris flow continued down the valley, overtaking a train and destroying the rail line, roads, and homes. In all, more than 70,000 people were killed. Most victims died in debris flows generated by the earthquake.

4.5.7 Dam Failure

Dam failure following earthquakes is a fairly rare phenomenon, but is not without precedent. During a 1971 earthquake in the San Fernando Valley of California, the San Fernando Dam on the Lower Van Norman Reservoir partially failed. Shaking caused a landslide of loose material from the reservoir side of the dam. It lost 30 ft (9.1 m) of its height and stood at its

Figure 4.33 Path taken by the debris flow that buried the cities of Yungay and Ranrahirca, Peru. Following an earthquake in 1970, the landslide broke loose from Huascaran Norte, the peak to the upper left in the photo, and raced down the mountain and into the valley below. © George Pflaker, U.S. Geological Survey. From Murck & Skinner, 2016.

highest only 4 ft (1.2 m) above water level when the shaking stopped. If shaking had continued for another few seconds, no doubt the dam would have completely failed and released 476.6 million ft^3 (13.5 million m^3) of water onto the area below the dam, populated by 80,000 people.

4.5.8 Tsunamis and Seiches

Tsunamis and *seiches* are types of water motion that occur, often as a result of earthquakes. Tsunamis are the subject of Chapter 5 in this book, but an introduction here is in order. A tsunami is a water wave created by a disturbance of the ocean floor. The most common cause of tsunami generation is an earthquake at a subduction zone. They can also be caused by landslides in the ocean and volcanic eruptions. The deadliest tsunami of which we are aware occurred following the 26 December 2004 Boxing Day earthquake off the coast of Sumatra, Indonesia. Displacement of the ocean floor created a tsunami that reached heights of 33 ft (10 m) in many areas around the rim of the Indian Ocean. In all, more than 230,000 people died in 11 countries (Fig. 4.34). The tsunami claimed lives as far away as the coast of Africa, 2,500 mi (4,100 km) west of the earthquake's epicenter. Tsunamis travel at speeds similar to those of jet aircraft and have the ability, as demonstrated in the 2004 tsunami, to travel across entire ocean basins. In the open ocean, they may present as a swell 3 ft (1 m) tall. It would rarely be noticeable among the normal waves on the ocean. However, as the tsunami approaches land, the shallowing water causes the wave to slow down and grow in height. Many tsunamis hit shore at 50 mi/hr (80 km/hr) or faster. This is too fast to outrun. Those caught unaware are taken by the wave and battered by not only the water but the debris within the water. Few caught up in the water are lucky enough to survive.

A seiche is also a phenomenon related to water movement. They are sloshing motions found in closed basins after large earthquakes.

Figure 4.34 Ruins in the Banda Aceh area of Indonesia, one of the hardest hit areas in the tsunami that followed the 26 December 2004 earthquake off the coast of Sumatra. Very few buildings were left standing, and all of those that were still standing were heavily damaged. Most of the people who lived in this community died during the tsunami. From Associated Press, in Fletcher, 2014.

The closed basins can include bays, estuaries, lakes, swimming pools, and even water wells. Though seiches are less damaging than tsunamis and hardly ever result in loss of life, they are less expected and can occur far from both coastal regions and the epicenters of earthquakes. In this way, they are a surprise hazard to much of the population and can cause millions of dollars in damage.

4.6 Prince William Sound Earthquake

We will focus on one example of an earthquake that included all of the previously mentioned hazards. A magnitude 9.2 earthquake occurred in the vicinity of Prince William Sound, Alaska, near the city of Anchorage on Good Friday (27 March), 1964. The earthquake occurred at a depth of about 15 mi (25 km) in a sparsely populated area. The rupture area was approximately 500 mi (800 km) long and 155 mi (250 km) wide. It occurred on the subduction zone in the northern Pacific where the Pacific Plate is subducting beneath the North American Plate.

This earthquake was the second largest of the twentieth century. Ground shaking went on for several minutes and was felt over an area of 700,000 mi^2 (1.8 million km^2) that included most of Alaska and Canada's Yukon Territory, as well as parts of the Northwest Territories and British Columbia. Most severe earthquake damage was limited to the 80,000 mi^2 (207,000 km^2) around the communities of Anchorage, Valdez, Seward, Kodiak, and Cordova, Alaska. Several prominent buildings were destroyed, including the control tower at the Anchorage Airport, the J.C. Penney building in downtown Anchorage, an elementary school, several buildings with unreinforced masonry walls or facades, and those that did not have all elements of the building tied together firmly. Buildings taller than three stories also fared worse than shorter buildings.

Ground shaking triggered landslides in the Anchorage metropolitan area. Five major slides and a number of minor ones were documented. One of the most spectacular was that of the Turnagain Slide. It involved an area adjacent to Cook Inlet on the outskirts of Anchorage. The Turnagain Heights Subdivision was largely destroyed by this landslide, which occurred because there was a tilted layer of clay beneath the neighborhood. Clay becomes slippery when it becomes wet or shaken. In this case, the shaking caused the clay to become a sliding surface. The ground above this clay layer broke into pieces and slid with the violent shaking. In all, 75 homes were lost when large cracks opened and blocks of Earth subsided. The ground was topsy-turvy with very few horizontal surfaces left by the time the shaking stopped.

Several coastal communities in south-central Alaska have ports used not only for the oil industry but also for fishing boats and other commercial shipping. Many of these communities had oil stored in large tanks near the shore. When the earthquake struck, fires began in many of these port cities. Waves distributed burning oil along the waterfront in some cities as well, spreading fires far from their ignition sources.

The displacement of rock on the bottom of Prince William Sound also produced a tsunami. Not only did many coastal communities have to deal with severe ground shaking, destruction of buildings, landslides, and fires, they also had to deal with a tsunami. All along the shores of south-central Alaska, communities were hit by tsunami waves, some with heights over 100 ft (33 m). Several people lost their lives in these waves.

Seiches also occurred in several small lakes throughout the United States, and even in water wells and swimming pools. Seiches were created as far away as the Gulf of Mexico and coastal bays in Louisiana, 2,500 mi (4,000 km) from the earthquake's epicenter. The seiches caused several boats and ships to break loose of their moorings, flooded Bayou Lafourche in Louisiana, and damaged boat houses, docks, and boats near Denham Springs, Louisiana. Seiches did hundreds of thousands of dollars' worth of damage throughout the gulf coast region.

4.7 Mitigation Strategies

Mitigation in earthquake zones is imperative. Earthquakes will occur, and both communities and individuals need to prepare for them. The number one reason people die in earthquakes is because of collapse of structures. In fact, there is a saying: "Earthquakes don't kill people, buildings do." Collapse of homes, businesses, government buildings, and churches, as well as bridges, overpasses, elevated freeways, tunnels, and other public infrastructure are very real dangers in large earthquakes.

There are a number of steps that can be taken by individuals to ensure their family's safety at home in an earthquake. First, be sure that your home is built to the most recent and stringent earthquake codes available. This means that a number of safety features should be built into the home to ensure that it will remain standing during a large earthquake. Some of these safety features include:bolts that tie the building to its foundation; brackets that tie horizontal and vertical elements of the building to each other (Fig. 4.35); *shear walls*, which are solid boards used to tie studs together within walls; braces, which are diagonal supports within the walls of the structures (Fig. 4.36); and blocks, which are horizontal studs in the walls. In addition, it is possible to construct a building so it is isolated from ground shaking. This can be done either

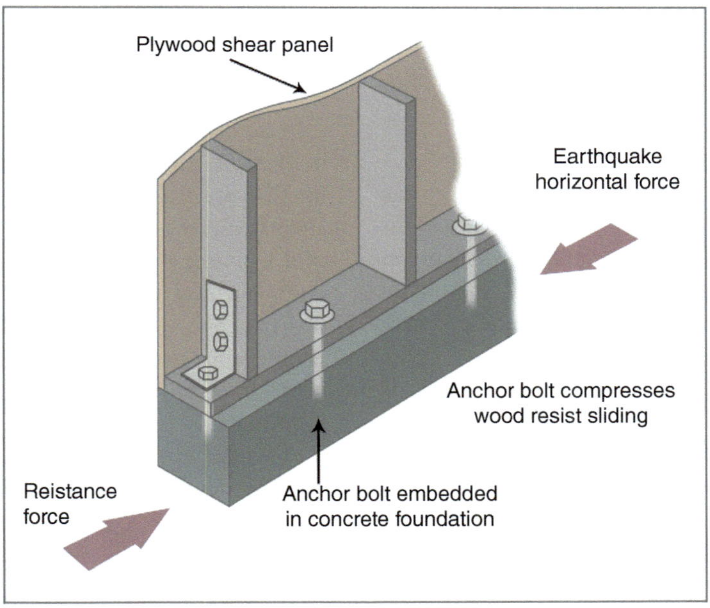

Figure 4.35 Earthquake mitigation strategies in buildings. Anchor bolts should attach the frame of the building to its foundation. Brackets tie horizontal and vertical elements of the building together. A plywood shear panel can reinforce walls against cracking and breaking in an earthquake. From Yeats, 2004; https://oregonstate.edu/instruct/oer/Earthquake/13%20chapter%2011_color.html

Figure 4.36 Bracing added to the exterior of a building in Auckland, New Zealand, to make it more resistant to shaking during an earthquake. From the James M. Kelly collection, NISEE-Berkeley; https://nisee.berkeley.edu/elibrary/Image/NM0126

with *base isolation* bearings or with large casters or roller bearings (Fig. 4.37). If a base isolation system is used, the building does not sit on a solid foundation. Instead, it sits on the base isolators, and is somewhat insulated from ground motion as a result.

Families should also put together an earthquake kit, which consists of water, nonperishable food, a flashlight, a battery-operated radio, and any other necessities such as extra medication and blankets for family members. In addition, they should know what to do when an earthquake strikes. If you are inside, find a sturdy table for shelter (Fig. 4.38). By sitting underneath a table, any falling debris will hit the table instead of your body. For family members in wheel chairs, make sure their wheels are locked and that they have something sturdy to hold over their heads. If you are outside, move away from tall buildings where glass from windows or rock from the building façade could fall down on you.

Communities can prepare for future earthquakes by adopting and strictly enforcing strong building codes. Insist on proper building materials for any new buildings. Employ engineering firms that know how to build earthquake-resistant infrastructure. Upgrade old or vulnerable infrastructure such as aging dams and poorly reinforced support columns for bridges and overpasses. Hold disaster drills so all segments of the population involved in emergency response (city, county, and state officials, police, fire, hospitals, transit authorities, Red Cross and other volunteer organizations, etc.) know what to expect from each of these entities and what their roles will be when a large earthquake strikes. Offer financial assistance to homeowners and businesses that would like to retrofit their homes and offices. Retrofitting means that they are bringing an older building up to the most modern codes available.

In the case of coastal communities, be aware of the tsunami risk and prepare the population for that eventuality. Post tsunami evacuation signs to direct people to safe zones in the event of a tsunami (Fig. 4.39). Install and test tsunami warning sirens. Have community meetings to explain the tsunami risk and make sure people understand that it is imperative they move to a higher elevation if a tsunami is imminent.

The most obvious mitigation for mountainous areas prone to earthquake-induced landslides is not to build in those areas. In places where cities exist that are at risk, specify safer zones of the city, usually on top of hills, as destinations during evacuations. Emphasize to people that if they feel an earthquake, they should make their way to these safe zones rather than wait for word from authorities.

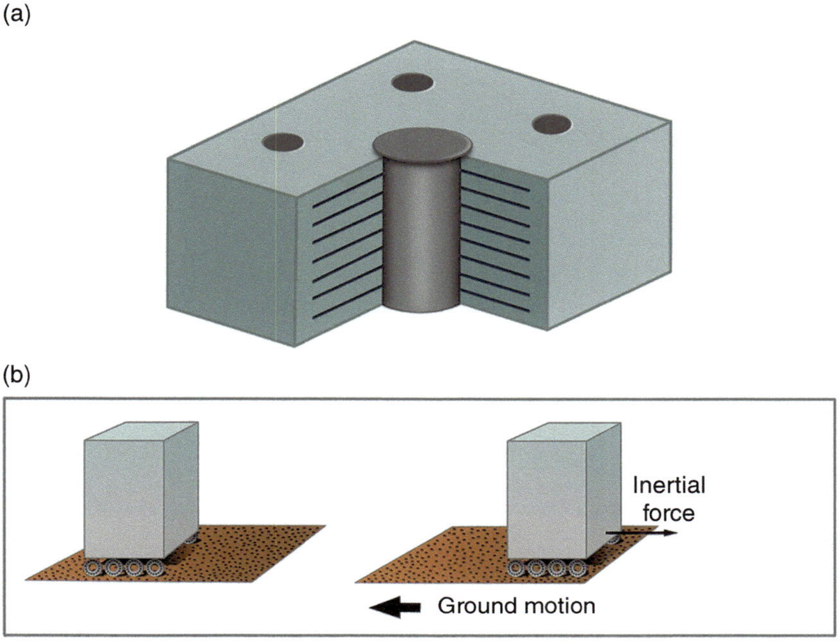

Figure 4.37 Two methods of isolating the building from ground motion, called "base isolation." (a) A base isolation bearing. This is composed of a lead core surrounded by alternating layers of rubber and steel. It acts somewhat like a shock absorber in a car. The energy of the earthquake will be partially absorbed by the lead core. (b) Another method of isolating the building from ground motion is to put the building on large casters or roller bearings. When the ground moves, inertia causes the building to stay put. The ground moves around the building, rather than the building moving as a result of ground motion. From Yeats, 2004; https://oregonstate.edu/instruct/oer/Earthquake/14%20chapter%2012_color.html

Figure 4.38 What to do in case of an earthquake. (a) Those who are able should find a sturdy table and shelter beneath it. From https://www.shakeout.org/messaging/. (b) Those using mobility aids such as walkers and wheelchairs should lock wheels and find something sturdy to hold over their heads. https://www.rentonwa.gov/city_hall/executive_services/emergency_management/hazards_in_renton/earthquake/Public Domain

Figure 4.39 Tsunami evacuation route sign. Photo courtesy of the U.S. Geological Survey.

4.8 Conclusion

Earthquakes are among the most powerful of the Earth's processes. Most earthquakes occur in well-known areas close to plate boundaries or in historically active regions. Earthquakes are somewhat predictable because we know where earthquakes are likely to happen, and how large those earthquakes may be. However, earthquakes are not forecastable. We do not yet have the ability to determine the specific day or time an earthquake will occur, or how large a given earthquake will be. We can do a great deal to protect ourselves from the effects of earthquakes. Engineers can design buildings that will withstand even severe ground shaking. We also know how to recognize areas in which the ground is likely to shake more violently than others, so it is possible to avoid building on the most vulnerable locations.

Seismology, the study of earthquakes, has led to a greater understanding of our planet. Scientist Inge Lehmann discovered that the Earth had a liquid outer core and solid inner core by observing the behavior of seismic waves from the largest earthquakes. Plate boundaries are outlined by zones of earthquake activity. Subduction zones were originally mapped based on the locations of thousands of earthquakes within the subducted lithosphere. The earthquakes occurred at increasing depths beneath the Earth's surface, and the earthquake zone defined an inclined plane that stretched into the asthenosphere. We are now able to determine the size of an earthquake and the amount of energy it produces thanks to the work of seismologists. Seismology also allows us to monitor the Comprehensive Nuclear Test Ban Treaty and investigate phenomena that involve ground shaking in a number of different scenarios.

4.9 Chapter in Review

Earthquakes are sudden movements along breaks (called faults) in the Earth's crust. The location on a fault where an earthquake begins is called the earthquake's focus. Directly above the focus, on the surface of the Earth, is the earthquake's epicenter. Most earthquakes occur entirely on parts of faults that are underground, but occasionally the faults can break all the way to the Earth's surface. When they do, they produce a fault scarp.

Seismometers are modern instruments that record seismic waves that radiate from an earthquake's focus. There are four types of seismic waves, in two different categories. The first category is the body wave, which travels through the interior body of the Earth. One type of body wave is the P wave. P stands for primary. It is the fastest of all seismic waves, and travels through solids, liquids, and gases. Since P waves are faster than all other types of seismic waves, they are the first to arrive at any given seismometer. The second type of body wave is the S wave. S stands for secondary. It is the second fastest of all seismic waves. It travels only through solids. S waves are the second seismic waves to arrive at any given seismometer after an earthquake. The second category of seismic wave is the surface wave, which travels near the surface of the Earth. Surface waves travel slower than body waves, so are the last waves to arrive at a seismometer after an earthquake. For shallow earthquakes, surface waves are the largest waves and most damaging. For deep earthquakes, surface waves may be very small in amplitude.

Location and magnitude of an earthquake are calculated using measurements from seismograms. The location of an earthquake is calculated using the difference between the arrival times of P and S waves at the same set of at least three seismometers. Magnitude is calculated using a set of equations appropriate to the presumed location of the earthquake. For deep earthquakes, a body wave magnitude is calculated. For large, shallow earthquakes, a surface wave magnitude is usually most accurate. For small earthquakes, equations for local magnitude give the best results. For the largest of earthquakes, scientists calculate a moment magnitude. Scientists can also construct an intensity map, or ShakeMap, after an earthquake. An intensity map is constructed with data from hundreds or thousands of observations of damage and ground shaking. Each observation point is assigned an intensity value that corresponds to the severity of ground shaking. The intensity values are plotted on a map, and the end result is a map that defines areas with various intensities of ground shaking and damage. Information about where ground shaking is most severe is helpful to city planners, insurance companies, home owners, and home buyers.

Scientists can use seismometers for purposes other than monitoring, locating, and determining magnitudes for earthquakes. Seismometers have solved mysteries about the sinking of submarines and the collapse of buildings, and are used in monitoring the world for compliance with the Comprehensive Nuclear Test Ban Treaty. Seismic data can be used to map the subsurface as well. Seismometers can provide information to scientists about any event that shakes the ground, whether it is an earthquake or other occurrence. People have the ability to create earthquakes. This happens most commonly when fluids are pumped underground. Fracking, a process of pumping fluids underground to fracture rock to allow for natural gas extraction, is one instance in which artificial earthquakes can occur. Wastewater, such as from oil and gas companies, is sometimes pumped underground for disposal. Pumping this wastewater underground also causes earthquakes. Oklahoma has seen a dramatic jump in earthquakes over the last several years due to this practice of wastewater disposal.

Earthquakes occur on faults. There are three basic types of faults. Strike-slip faults have lateral motion. At normal faults, the hanging wall moves down relative to the footwall. In reverse or thrust faults, the hanging wall moves up relative to the footwall. Any type of fault may exist at any type of plate boundary. However, strike-slip faults are often associated with transform boundaries. Normal faults are commonly found at divergent boundaries, and reverse or thrust faults are most numerous around convergent boundaries.

Earthquake hazards are numerous. The basis for most hazards is ground shaking. Aftershocks are smaller earthquakes that occur in the same general area, but after the main shock. Structures can collapse. Earthquakes can rupture gas lines and electric lines or disturb open fires, causing widespread fire. Loose material such as sand, silt, clay, or mud can amplify ground shaking causing disproportionately more severe damage than in other areas within an earthquake-affected region. Ground shaking can expel water and air from loose material causing liquefaction. Liquefaction causes the ground to act more like a liquid than a solid. Severe ground shaking can displace loose rock, soil, or snow causing landslides and avalanches. Old and poorly engineered dams are in danger of failing during large earthquakes. Tsunamis and seiches can also be produced by large earthquakes. Tsunamis are water waves created by disturbances of the ocean floor. Seiches are sloshing motions in closed basins after an earthquake occurs. Both tsunamis and seiches have the ability to reach populations thousands of miles away from the earthquake's epicenter.

5

Tsunamis

Bethany D. Hinga

University of Nebraska at Kearney, USA

5.1 Introduction

A *tsunami* is a water wave generated by a disturbance of the ocean floor. It can be caused by fault rupture during an earthquake under the seafloor, eruption of a volcano on the seafloor or near the shoreline, underwater landslides, landslides generated on land that flow into the sea, meteor impacts, and even atmospheric phenomena. *Tsunami* is a Japanese term, from the word *tsu*, which means "harbor" and *nami*, which means "wave." As happens so often, those who see or describe a phenomenon most are those whose names typically stick in the international scientific community. The Japanese have been experiencing tsunamis on their shores for the entire duration of their long recorded history. As a result, the best early descriptions we have of tsunamis and their effects have come from the Japanese people.

Tsunamis are not unique to the Japanese Islands, however. Tsunamis have been observed throughout the Pacific Rim, in the Indian and Atlantic oceans, and in the Mediterranean and Caribbean seas. Because tsunamis can be generated by a variety of events and travel across entire ocean basins, no shoreline is without some level of tsunami risk.

Tsunamis have often been called tidal waves. This name is incorrect because tsunamis are not caused by tides. Tsunamis are independent of the Moon and Sun, which create the tides we see every day at the shoreline. However, this misnomer was probably started because it describes the feeling of a tsunami. It comes ashore rapidly much like a very fast-moving high tide. It is an incoming bore of water with the power to devastate coastal communities.

5.2 Unimak Island Earthquake and Tsunami, 1946

The Aleutian Islands are a chain of volcanic islands that extends from mainland Alaska westward toward the Kamchatka Peninsula of Russia. They were created by subduction of the Pacific Plate beneath the North American Plate. Unimak Island is one of the first islands in the Aleutian chain, nearest the Alaska Peninsula. In 1946, it was the location of one town called False Pass, on the eastern part of the island, and two Coast Guard installations on the western tip of the island. One of the Coast Guard installations was the Scotch Cap Lighthouse (Fig. 5.1a), a sturdy reinforced concrete building that was 98 ft (30 m) tall, with its base 46 ft (14 m) above sea level. The other was a D/F (direction finding) station located on the bluff above the lighthouse. Both were manned by Coast Guard personnel.

Earth's Natural Hazards and Disasters, Advanced Textbook 7, First Edition. Bethany D. Hinga.

(a)

(b)

Figure 5.1 Before and after photos of the Scotch Cap Lighthouse on Unimak Island, Alaska. (a) Photo of the lighthouse taken in 1946, just 2 months prior to the earthquake. Courtesy of NOAA/National Geophysical Data Center); https://www.ngdc.noaa.gov/hazardimages/#/tsunami/29 (b) Photo of the lighthouse taken just after the 1946 earthquake and tsunami. Courtesy of the U.S. Coast Guard. The lighthouse was torn from its base and destroyed, killing the crew of five men who were inside at the time. Waves reached the top of the cliffs, a height of about 138 ft (42 m).

At approximately 1:30 a.m. local time on 1 April 1946, a large earthquake struck. The men in both buildings reported that the shaking lasted 30 to 40 s. Though unnerved, they found no damage to either structure as a result of the earthquake. About 30 min later, a larger earthquake occurred, this one a magnitude 8.6. It triggered an underwater landslide and a massive tsunami. The epicenter was about 90 mi (145 km) offshore.

Just 21 min after the larger second earthquake, the men in the D/F station at the top of the bluffs heard a deafening roar. A wave hit the station with tremendous force. The building was severely damaged. The men evacuated to higher ground. They noticed that the lighthouse's light had been extinguished and that the foghorn had gone silent. Additional waves pounded the shoreline at regular intervals for a few hours.

At dawn, the D/F station crew members went to the edge of the cliff and looked down to the site of the lighthouse. What they saw shocked them. The lighthouse had been completely destroyed (Fig. 5.1b). All that remained was twisted steel girders and a few blocks of concrete. The lighthouse had been staffed by five men. All five died that night, but only two bodies were recovered. The other three were presumably washed out to sea with the wave.

The tsunami sped across the Pacific Ocean at an average velocity of 490 mi/hr (790 km/hr). It arrived in the Hawaiian Islands just after sunrise. Maximum wave heights were 33 ft (10 m) on Maui, 36 ft (11 m) on O`ahu, and 55 ft (17 m) on the island of Hawai`i. Therefore, the island of Hawai`i sustained more damage and loss of life than the other islands. On the eastern side of the island of Hawai`i, a coastal railroad existed that connected the sugar plantations on the northeast coast to the port in Hilo. The tsunami tore train cars from the tracks and left the tracks a twisted mess. The railroad was never rebuilt.

The city of Hilo is situated on a picturesque bay. A significant breakwater stretches across the opening of the bay. Because the breakwater was there, it protected the bayfront area from the worst of the tsunami. The business district was flooded, as were several neighborhoods. However, the docks at the main port of Hilo were not protected by the breakwater. Terrible waves overtook the harbor, sweeping away warehouses, docks, and the people who were working there that morning (Fig. 5.2).

The date of the tsunami contributed to the death toll. Because it occurred on 1 April, many people who warned of the tsunami were met with skepticism because it was April Fool's Day. The first instance of this happening was when the wave reached O`ahu just after 6:30 a.m. A pilot in a patrol plane radioed Kaneohe Air Base that he saw a line or small wave on the surface of the ocean. He was instructed to descend for a closer look, but he replied that the phenomenon had outrun his plane and was gone. The radio operator's friends assured him it was an April Fool's joke.

Perhaps the most tragic of the incidents surrounding the tsunami occurred in the small community of Laupahoehoe, about 25 mi (40 km) north of Hilo. The Laupahoehoe School had been built on a low, narrow plain about 20 ft (6 m) above sea level. Most families whose children attended the school lived on the bluffs above near the sugar plantations, and children descended into the valley to school each morning. The children were arriving for school when the tsunami hit. The first part of the wave to hit was the *trough*, which left the sea floor exposed. Some children rushed out to pick up stranded fish, unaware that the *crest* of the wave was on its way. As the crest started moving toward shore, some children shouted "Tidal wave!" only to be met with "Ha, April Fool!" Unfortunately, several children were caught up in the initial wave, which pushed upward and into the area of the school yard and nearby teachers' cottages. Many of the teachers were recruited from the mainland and had never heard of a tsunami so were unaware of the hazard. As the water retreated, a number of people were carried out to sea and never seen again. The second

(a)

(b)

Figure 5.2 (a) Image of tsunami hitting the docks in the harbor at Hilo, Hawai'i, in 1946. The man in the left side of the photo was a dock worker who did not survive the tsunami. There were 159 deaths in Hawai'i as a result of this tsunami. (b) Damage along Kamehameha Avenue, along the bayfront in Hilo, Hawai'i. Photos from the National Geophysical Data Center; https://www.ngdc.noaa.gov/hazardimages/#/tsunami/29

wave was even higher and caught more people in the wave. In all, 46 people died at Laupahoehoe. This included school children, teachers and their spouses, and children of teachers. A memorial still exists at the site. The school was rebuilt at the top of the bluffs, and the former site of the school and teachers' cottages is now a park.

No lives were lost to the earthquake itself. The death toll for this tsunami included the five Coast Guard employees in the Scotch Cap Lighthouse, 149 people in the Hawaiian

Islands, and one person in Santa Cruz, California. All deaths were a result of the tsunami. The tsunami also produced waves up to 30 ft high in French Polynesia, destroyed fishing boats in Chile, and caused damage all along the west coast of North America.

Ultimately, this tsunami and its ocean-wide damage and large death toll in the far-field led to the formation of the Pacific Tsunami Warning Center, which was established in 1948. It is now an installation of the National Oceanic and Atmospheric Administration located in Ewa Beach, on the island of O`ahu. The Pacific Tsunami Warning Center monitors the entire Pacific Rim and Hawai`i for major earthquakes that could potentially cause a tsunami. The Pacific Tsunami Warning Center was the first warning center of its kind. Now tsunami warning centers serve all of the world's coastlines.

5.3 Wave Creation

A tsunami is a water wave created by displacement of water in the ocean. By far the most common cause of a tsunami is a very strong earthquake in the Earth's crust below the seafloor (Fig. 5.3). An earthquake that breaks to the surface (which is under the ocean) can significantly raise or lower the floor of the ocean and displace a large volume of water while doing so. A megathrust earthquake occurs when strain at a subduction zone becomes too great and a fault at the plate boundary ruptures. As it does so, the subducting plate moves downward and the overriding plate snaps upward, creating a vertical disturbance of the ocean floor. This disturbance displaces water upward, creating a tsunami. This happens only when the earthquake is shallow (less than 9 mi or 15 km, in general), and typically occurs only if an earthquake is of magnitude 7 or above. Deep earthquakes do not break to the surface, and therefore do not vertically displace large volumes of water. Earthquakes smaller than magnitude 7 generally do not displace enough water to create a tsunami.

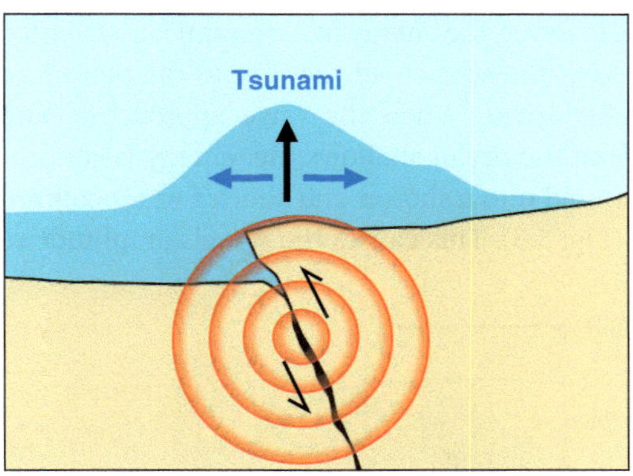

Figure 5.3 A tsunami is generated when there is a disturbance of the ocean floor. Seawater over the disturbance is displaced, creating a long wave. In this instance, an earthquake has disrupted the ocean floor, creating a tsunami above. From Fletcher, 2014.

Landslides are responsible for the largest tsunamis. Often, however, energy from the tallest tsunamis dissipates rapidly. A tsunami resulting from a landslide therefore may have incredible height close to the site of a landslide, but with just a little distance, the wave becomes much smaller and may disappear entirely. Landslides are also sometimes suspected when tsunamis caused by other forces become larger than one would naturally expect. As an example, a magnitude 7.0 earthquake is just large enough to generate a tsunami. The resulting tsunami is usually no more than a meter or two tall. However, in 1998 a magnitude 7.0 earthquake located just off the north shore of Papua New Guinea produced a tsunami 49 ft (15 m) tall. The tsunami washed completely over several communities located on a low-lying sand bar, killing over 3,000 people. Scientists and residents alike were shocked at the size of the waves. Further investigation and computer modeling of the event revealed that an underwater landslide had accompanied the earthquake, adding to the dimensions of the tsunami.

Volcanic eruptions are rare causes of tsunamis, but they have occurred. The best known was a tsunami caused by the collapse of the volcano Krakatoa in Indonesia in the year 1883. A tsunami was generated that swamped the coastlines of the nearby islands of Java and Sumatra, killing 36,000 people. The tsunami killed many more people than did ash and pyroclastic flows from the volcano.

Meteotsunamis are those that are caused by atmospheric phenomena. In 2022, the eruption of the Hunga-Tonga eruption (see Chapter 1) caused a meteotsunami that traveled across the Pacific Ocean. The tsunami was created primarily by the atmospheric blast from the volcano, making this a meteotsunami rather than a typical tsunami generated by a volcanic eruption.

5.4 Wave Characteristics

Tsunamis are waves, and therefore have a number of parameters that we can define that are common to any type of wave (Fig. 5.4). For example, waves have a *wavelength*, measured as a distance from one crest to the next. They also have an amplitude, or height. They have a velocity. And they have a *period*, or time that elapses, from the arrival of one crest to the next. As we will see, regular wind-driven waves are very different from tsunamis in almost all of the parameters listed above.

Tsunamis have very rapid velocities. In the open ocean, tsunamis can travel the speed of a jet plane, or approximately 375 mi/hr to 560 mi/hr (600 km/hr to 900 km/hr). Common wind-driven waves a few miles offshore might have a velocity of 6 mi/hr (10 km/hr). Thus, tsunamis are much faster than wind-driven waves, particularly in the open ocean. In the 2004 Boxing Day tsunami, the first wave traveled from the coast of Sumatra to Sri Lanka (about 1,100 mi or 1,770 km) in just under 2 hr (!), for a velocity of about 550 mi/hr (885 km/hr).

Tsunami wavelengths are also extremely long as compared with wind-driven wave wavelengths. Tsunami wavelengths may be as great as 300 mi (482 km) or more. A wind-generated wave a few miles offshore has a wavelength of perhaps 200 ft (60 m). The wave's length and velocity allow us to calculate the period, or length of time that passes between arrivals of successive wave crests. The relationship between these parameters is period = wavelength / velocity.

Using this relationship, we find that the period of a tsunami is close to 1 hr in the open ocean. This makes them very different from every day waves in the ocean, generated by the wind, which have periods often measured in seconds, rather than several minutes to an hour.

Wave heights, or amplitudes, of tsunamis depend largely on the magnitude of disturbance that created them and the depth of the water. In the open ocean, even the largest tsunami amplitudes may be only 3 ft (1 m) or so, making them impossible for passengers on a ship to detect as they pass beneath the vessel. However, the energy in a tsunami is distributed from the surface all the way to the bottom of the ocean. When the wave approaches land and the ocean shallows, the energy is concentrated into a shorter and shorter water column (Fig. 5.5). This causes the wave's amplitude to

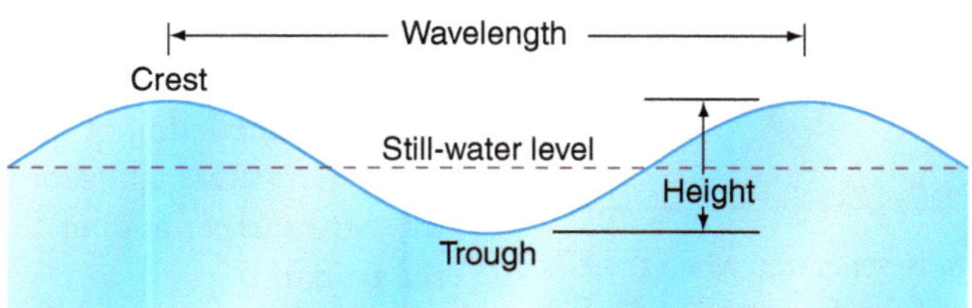

Figure 5.4 Anatomy of a wave. Amplitude is the height of the wave, and the wavelength is the distance from one crest to the next, or one trough to the next. From Strahler, 2013.

Figure 5.5 In the open ocean, a tsunami has a very long wavelength and small amplitude. As it approaches the shoreline, the wavelength shortens and the amplitude grows as the energy of the tsunami is forced into a smaller space. Credit: Wickerwood/Shutterstock.

increase dramatically and its wavelength to decrease. Wave heights for large tsunamis close to their sources can top 165 ft (50 m). Though wind-driven waves may also see an increase in amplitude as they approach shore, they rarely reach heights approaching 50 ft (15 m). When they do, they break offshore. They do not reach the shoreline with an amplitude of 50 ft. Many coastlines around the world are famous for their large waves, sometimes used in surfing competitions. Large wind-driven waves can be found off the coast of Hawai`i, California, Australia, and other coastal areas with steep ocean floors.

Though wave heights for tsunamis can be incredibly large, they do not necessarily have to be. There is no minimum height defined for a tsunami. Rather, a tsunami is defined by its origin, the distribution of energy throughout the water column, and the wave's parameters such as wavelength, velocity, and period. Tsunamis can be just a few centimeters tall. Even small amplitude tsunamis can cause disturbances in coastal communities because they move the water so quickly. They mimic a strong current, and may drag marine life and other items found on the ocean floor, causing damage and disruption.

Tsunamis cause water movement that is very different from that of wind-generated waves. Wind-driven waves generate circular motion within the water column (Fig. 5.6). Each individual water molecule near the surface of the ocean moves in a circular motion. This motion is greatest at the surface and decreases with depth. In relatively calm water on a day with a gentle breeze, all wave motion stops at a depth between 30 and 60 ft (10 and 20 m), called wave base. In a tropical storm, wave base may extend to a few hundred feet deep. Tsunamis move in a straight line, moving water from one location to another like in a swift current. Tsunamis also involve the entire water column, from the surface of the ocean down to the ocean floor unlike wind-driven waves, which are confined to the top several feet of the ocean.

Tsunamis are also different from wind-driven waves in that tsunamis can travel across entire ocean basins to cause deaths on distant shores. This characteristic of tsunamis makes them the farthest-reaching natural disasters on the planet. No other type of natural disaster has the ability to strike a distant shoreline on the opposite side of an ocean. Before rapid transocean communications became routine, those in a tsunami's path far from the wave's source were always caught unaware. An earthquake might occur off the coast of South America and cause deaths in other parts of South America as well as in Mexico, California,

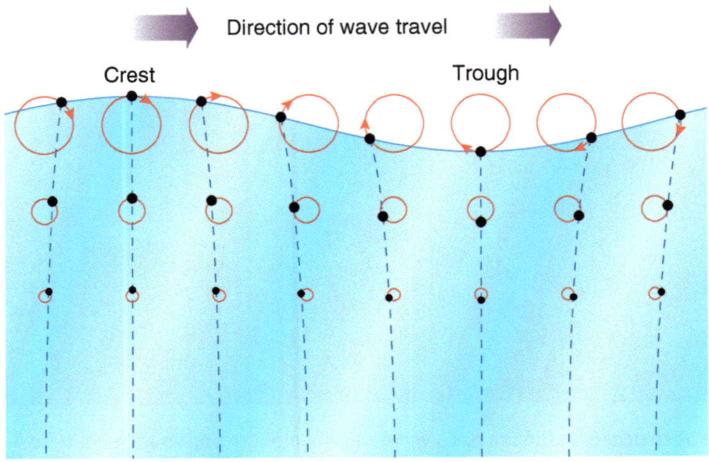

Figure 5.6 Wind-drive wave motion in the ocean. Water molecules oscillate in a circular motion, more or less in place. Wave motion dies out downward in the water column and stops at a depth equal to half of the wavelength. This depth at which water motion stops is called wave base. From Strahler, 2013.

Alaska, Hawai'i, and Japan. People in these locations far from the earthquake epicenter were caught completely unaware because people could not possibly have felt the earthquake and had no warning the tsunami was headed their way. Thankfully, we now have tsunami warning systems in place in all of the world's oceans. These systems provide countries on the world's coasts with advance notification that a wave is expected, allowing them to warn people and move individuals to safety.

As a tsunami approaches shore, it is important to know that it is not always the crest of the wave that arrives first. If the crest arrives first, it only rarely appears as a large crashing wave. Most of the time it arrives as a rushing torrent of water that grows higher and higher for several minutes before it pauses and reverses course, draining not only water but any debris, people, and animals it has picked up back out to sea. If the trough of the wave arrives first, the sea will recede far beyond the normal low tide level, stranding fish and other sea life in the open air. This is a very odd sight to people who live in the area and are used to seeing the ocean look a certain way. This should be an instant warning to people to move inland and uphill. Unfortunately, in some cases it has actually lured people out to the newly dry land to catch fish or just investigate out of curiosity. The wave rushes back as quickly as it receded, so people who do this are often caught off guard and drowned in the wave.

Even though they slow considerably once they hit land, tsunamis move fast enough that they are impossible to outrun. A typical velocity for a tsunami striking land is about 15 mi/hr (7 m/s). The water will slow down even more as it moves across the land because of friction with the Earth, trees, grasses, buildings, and other obstacles. It is powerful enough, however, to move those obstacles and incorporate them into the wave. Though most deaths in tsunamis are due to drowning, a significant number of people are killed by debris in the water. The best defense is to know that a tsunami is coming and to get to higher ground, out of reach of the water.

5.5 Wave Parameters

Tsunamis are often described by the maximums encountered during the passage of a wave. First, it is important to note that a tsunami is typically a series of waves rather than a single wave. Multiple waves hit the same stretch of shoreline one after the other at fairly regular and predictable intervals, if data from the seafloor are collected and analyzed, and forecasts are included in tsunami warning statements. As tsunamis reach shore, their amplitudes increase, wavelengths decrease, and they slow down. Wave parameters vary widely, but it is not uncommon for waves to come ashore with a period, or time between arrival of successive crests, of between 5 min and an hour. It may take several hours for all of a tsunami's waves to pass, and the first wave is not always the largest. In a 1946 earthquake that occurred in Alaska, the first wave was the largest. The same is true of a 2004 tsunami in the Indian Ocean on the island of Sumatra. However, in Sri Lanka, the third wave was the largest. During a tsunami generated by the twentieth-century's largest earthquake, a magnitude 9.6 event off the coast of Chile, the third wave was largest. And in a 1964 Alaskan earthquake, the fifth wave was largest.

The area of land that is covered by the waters of the tsunami is called its *inundation zone*. Tsunamis can move several miles inland if the wave is powerful enough and the land is flat. They can also create a *tsunami bore*, which is a steep, turbulent water flow typically found in an estuary or in the mouth of a river. The bore moves quickly up the river valley like a fast-moving tide and can both erode and deposit debris far inland along the course of the river. In an earthquake off the coast of South America in 1868, the United States warship *Wateree* was carried 1.9 mi (3 km) inland and deposited there by a 69 ft (21 m) tsunami. Since no subsequent wave reached that far inland and the river was not navigable by a ship the size of the *Wateree* under normal conditions, the warship had to be abandoned where it came to rest.

The maximum height reached by a wave as it rushes onshore is called the *run-up*. Run-up of a tsunami is different at every location. In the same tsunami event, some areas may see very high run-ups while other areas see only modest run-ups. In general, run-ups are highest near the source of the tsunami and become smaller with increasing distance from the source. There are exceptions to this general rule. Ocean floor topography is known to influence the height of waves. The coast off Crescent City, California, tends to focus and amplify tsunami waves as they travel toward the shoreline. Areas to the north and south of Crescent City typically have smaller run-ups than are observed within the vicinity of the city.

5.6 Case Studies

5.6.1 Earthquake-Generated Tsunamis

The Cascadia Earthquake and Tsunami, 1700

On the night of 27 January 1700, a major tsunami hit the shores of Japan. There had been no earthquake in the Japanese Islands, no indication that anything dangerous was headed their way. A tsunami arrived around midnight, taking residents by surprise. Thankfully there was little loss of life, but it was an unnerving occurrence. Because nobody knew the source of this tsunami, it was dubbed an "orphan" tsunami. It was not until the mid-1990s that the earthquake that generated this tsunami was identified.

People had long noticed "ghost forests" along the west coast of the United States. These ghost forests are groves of trees that are still upright, but are dead, drowned in salt water. Through carbon dating, scientists found that the trees died about 300 years previously. Since the tree rings indicated they were healthy right up until their death, they surmised the trees were drowned in a sudden event that left the coastline about 1 m (3 ft) lower than it had been when the trees were healthy and alive. They also studied and dated tsunami deposits along the coasts of Washington and Oregon and found they were the same age as the trees, about 300 years old. In addition, they studied Native American legends about a killer wave that struck one winter night after a large earthquake, killing hundreds in coastal villages. Scientists and historians were able to tie all of this information together into a well-researched theory about an earthquake that struck about 9 p.m. local time on 26 January 1700 along the Cascadia Subduction Zone, which lies off the coast of Oregon, Washington, and British Columbia. The earthquake had an estimated magnitude of 9.1, and created a Pacific-wide tsunami. The tsunami struck Japan approximately 8 hr later on the night of 27 January, on the other side of the International Date Line.

Indonesia Earthquake and Tsunami, 2004

The 2004 Indonesian tsunami was the most deadly tsunami on record. It occurred as a result of a magnitude 9.1 earthquake, sometimes referred to as the Boxing Day earthquake or the Great Sumatra-Andaman Island earthquake. This was the world's third largest earthquake of the last 100 years, and the largest since a magnitude 9.2 quake in Alaska in 1964. The earthquake struck at 7:58:47 a.m. local time. The epicenter was nearest the municipality of Banda Aceh on the northernmost tip of the island of Sumatra. Though large earthquakes are common in this part of the world, powerful and deadly tsunamis are not. The last deadly tsunami in the region occurred in 1883 during the eruption of the volcano Krakatoa. Therefore, many people were less concerned about a major tsunami than they perhaps might have been after an earthquake of that size in other coastal areas of the world.

The earthquake that occurred that morning was on a megathrust, in which one plate is thrust beneath another plate at a subduction zone (Fig. 5.7). The strain that had been building at the plate boundary was immense. The portion of the plate boundary that broke is estimated, based on aftershock locations, to be about 750 to 800 mi long (1,200 to 1,300 km). Vertical displacement of the ocean floor was on the order of about 65 ft (20 m). Thousands of cubic miles of seawater were displaced quickly upward when the fault broke, generating a large and terrifying tsunami. It took only 10 min for the tsunami to reach the nearest shoreline, in the municipality of Banda Aceh on the island of Sumatra (Fig. 5.8). At the town of Labuhan, the tsunami reached the dizzying height of 160 ft (48.86 m). The wave penetrated 3.1 mi (5 km) inland, wiping away almost all human-made structures and carrying away people, animals, trees, shrubs, grasses, and even soil. Run-ups varied considerably by location (Table 5.1). In general, run-ups were greatest in areas nearest the earthquake's epicenter.

The provinces of Banda Aceh, Aceh Besar, and the Aceh Jaya on the northwestern tip of Sumatra experienced the greatest mortalities. In this region, crude mortality rates were approximately 23%. That means that of the people who lived in those areas before the tsunami, approximately 23% lost their lives. It is impossible to separate the number of fatalities due to the earthquake from those caused by the tsunami because the tsunami washed away almost all buildings and debris from the area (Fig. 5.9). All evidence of earthquake damage was removed once the tsunami moved through the region. In all, 167,736 people were reported dead or missing in Indonesia alone. Those who were reported missing were likely swept out to sea.

Indonesia was hardest hit by the tsunami that day (Fig. 5.10), but it was not the only country that suffered mass casualties. The tsunami spread throughout the Indian Ocean and affected shorelines more than 5,000 mi (8,000 km) away. The second country to be affected by the tsunami was Sri Lanka (Fig. 5.11). There were 35,322 deaths in Sri Lanka, including more than 1,500 passengers aboard the train *Queen of the Sea*. The train was traveling along its coastal route at the time the tsunami struck, washing it from the tracks. Almost all passengers and crew died.

The beaches of Thailand were hit at nearly the same time as parts of Sri Lanka. There,

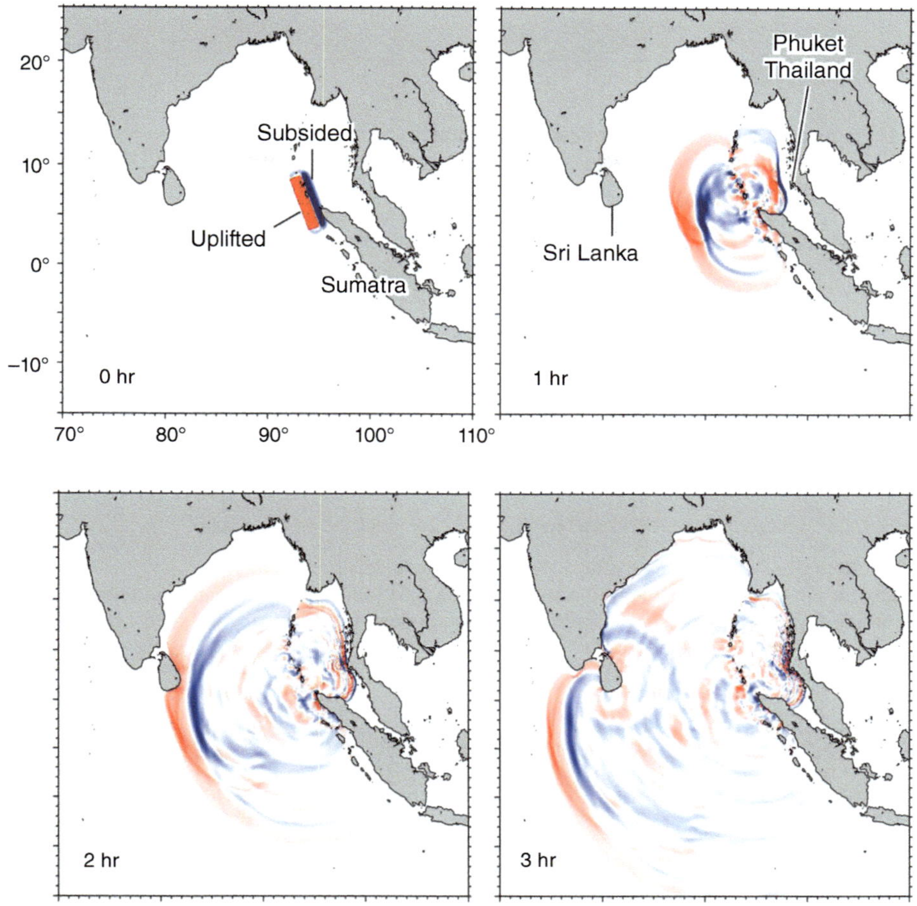

Figure 5.7 The 26 December 2004 Indonesian tsunami was unleashed by a magnitude 9.1 megathrust earthquake. Just off the coast of Sumatra, the Indo-Australian Plate is subducting beneath the Burma-Sumatra Plate. From Murck et al., 2008.

Figure 5.8 An aerial photo of Banda Aceh, Indonesia, following the tsunami. Most structures were destroyed and swept out to sea. In addition to human-made structures, soil and vegetation were stripped from this area of the island, causing total devastation. From Fletcher, 2014.

Table 5.1 Tsunami run-ups, 2004 Indonesia earthquake.

Location	Maximum run-up	Maximum run-up	Death toll
Indonesia	48.86 meters	160.3 feet	167,736
Thailand	19.96 meters	65.5 feet	8,212
Andaman-Nicobar Islands	16.50 meters	54.1 feet	Included with mainland India
India	11.46 meters	37.6 feet	18,045
Sri Lanka	10.87 meters	35.7 feet	35,322
Malaysia	7.40 meters	24.3 feet	75
Myanmar	6.70 meters	22.0 feet	61
Maldives	4.43 meters	14.5 feet	108

Source: Data from Choi, et al., 2006; Bryant, 2008; and Facts and details, 2022.
Note: Maximum run-ups for the hardest-hit countries in the 2004 Indian Ocean tsunami. Additional deaths were recorded on the shores of some African countries but run-ups there were generally less than 1 m (3 ft).

(a) (b)

Figure 5.9 Satellite imagery of Banda Aceh, Indonesia, from before and after the tsunami. (a) This image was taken on 23 June 2004. (b) Image taken just 2 days after the tsunami on 28 December 2004. Most buildings were destroyed, leaving only their foundations behind. Parts of the island were eroded by the tsunami, leaving water where there once was land. In addition, vegetation was stripped from the island so the lush green spaces on the island no longer existed. They were replaced by bare sand and temporary lakes. From DigitalGlobe. From Strahler, 2013.

thousands of Europeans had been enjoying a tropical holiday with their families. This was the country with the highest number of foreigner deaths. Among those dead were hundreds of Swedes, Germans, Finns, British, Swiss, and Austrians. The total death toll in Thailand was 8,212.

The tsunami moved on to India, where it was Full Moon Day. Many Hindus were along the shoreline, ritually bathing in the sea. Several worshippers became victims of the tsunami. In all, India lost 18,045 people. The Maldives are a group of small islands within the Indian Ocean. One hundred and eight residents lost their lives. Additional people were killed in Myanmar, Malaysia, Somalia, Tanzania, the Seychelles, Madagascar, Bangladesh, and Kenya. The tsunami traveled across the entire Indian Ocean and caused fatalities thousands of miles from the earthquake's epicenter. The final estimated death toll from this powerful earthquake and far-ranging tsunami was 229,867. Prior to 2004, the largest known death toll from a tsunami was that of the 1883

Figure 5.10 Ground-level photo of damage in western Sumatra from the 2004 Indian Ocean tsunami. Debris was washed out of homes and into streets. When buildings were torn apart, the debris ended up in the water, and was spread far and wide. Debris left by the tsunami makes streets impassable and can crush and kill anyone unfortunate enough to be caught in the wave. From Skinner & Murck, 2011.

Krakatoa tsunami, which killed approximately 36,000 people, primarily in the areas adjacent to the Sunda Strait between the largest Indonesian islands of Java and Sumatra.

Cell phone and video camera footage of the tsunami and the unbelievable damage it left poured into news agencies and found its way onto the Internet and television. Hair-raising tales of survival and tragic descriptions of deaths were abundant all around the rim of the Indian Ocean. One 10 year old British citizen, Tilly Smith, became famous for saving the lives of all who happened to be on a particular beach in Thailand that morning. She remembered a lesson from her geography class in which she had learned the warning signs of an impending tsunami. In the area in which Tilly and her family were vacationing, the sea receded much farther than the low tide level. Tilly remembered that this was one signal of the arrival of a tsunami, when the trough of the wave arrives first. She warned her parents of the meaning of this phenomenon, and together they urgently advised all of the other people on the beach to evacuate to higher ground. The crest of the tsunami followed shortly after everyone had evacuated the beach, and all of those Tilly and her family had warned survived the tsunami.

Tohoku Earthquake and Tsunami, March 2011

On 11 March 2011, a magnitude 9.0 earthquake struck off the east coast of the island of Honshu, Japan, near the shores of the prefectures of Iwate, Miyagi, and Fukushima. This was one of the five largest earthquakes to occur since 1900, and the largest in recorded history to strike off the shore of Japan. Christened the Tohoku Earthquake and Tsunami, the earthquake was a result of a megathrust

(a)

(b)

Figure 5.11 Before and after satellite images of the Sri Lankan coast. (a) Before image, showing a calm coastline and abundant vegetation on the island. From Digital Globe/Zuma Press/NewsCom. From Skinner & Murck, 2011. (b) After image, showing swirling refracting waves in the water. Vegetation was stripped from the island within the inundation zone as well. From HO/AFP/Getty Images/NewsCom. From Skinner & Murck, 2011.

earthquake on the subduction zone just off the coast of the island of Honshu (Fig. 5.12). During the earthquake, many sturdy and well-constructed buildings were damaged and a handful of buildings collapsed. Gas lines snapped, as did electrical lines. This sparked fires within several cities.

A large tsunami was generated by this earthquake. The first waves hit the coast of Honshu about 30 min after the earthquake. Tsunami warnings had gone out to the general public, but scientists had somewhat underestimated the size of the expected waves due to an original magnitude calculation of 8.0 rather than the true magnitude of 9.0. When waves came ashore, they were much higher than expected. They topped seawalls that had been erected as tsunami barriers and swept away people, cars, and boats (Fig. 5.12). Within minutes, many coastal cities were in great turmoil as water churned through the streets, destroying buildings in its path. Before long, the streets were clogged with watery debris and escape was nearly impossible. People were forced to ride out the tsunami in their homes and businesses, some of which were on fire (Fig. 5.13). As is common with tsunamis of

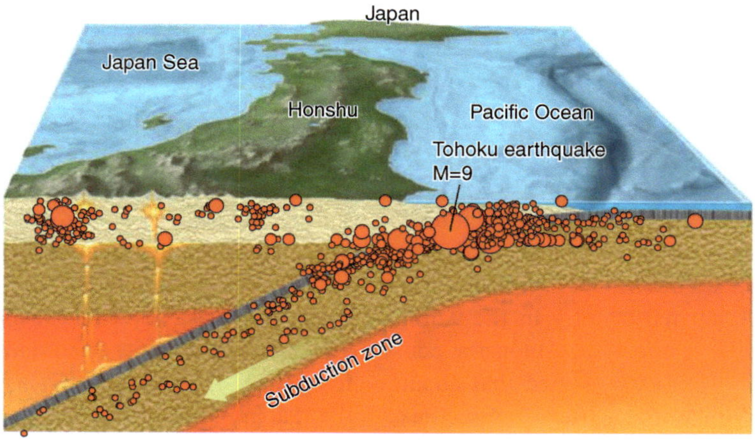

Figure 5.12 A cross section of the subduction zone off the coast of Honshu in Japan. Here, the Pacific Plate to the east is subducting beneath the Okhotsk Plate to the west. From Fletcher, 2014.

(a)

(b)

Figure 5.13 (a) Tsunami overtopping the seawall following the March 2011 Tohoku earthquake in Japan. The seawall had been built years before to mitigate the tsunami hazard in this area. However, this tsunami was larger than the average tsunami, so it easily overtopped the seawall. Water rushed into the city causing widespread destruction and tragic loss of life. Image by AFP/Getty Images. From Fletcher, 2014. (b) Aerial photo of the aftermath of the tsunami. From Fletcher, 2014.

this size, many areas were stripped of vegetation and debris littered the landscape.

The death toll from this earthquake and its resulting tsunami was 18,480. This is a shocking number of fatalities in Japan, which is perhaps the best-prepared country in the world when it comes to tsunamis. Autopsies performed on the deceased indicated that about 90% of the victims drowned, 4.4% of victims were crushed either in the earthquake or by tsunami debris, 1.1% died of burns. For 2%, no cause of death could be identified.

One structure that commanded international attention in the aftermath of this earthquake and tsunami was that of the Fukushima Daiichi Nuclear Power Plant, located in the coastal area of the city of Okuma. The power plant had a number of safety features built into its systems, but the tsunami caused a chain of events to occur that resulted in a meltdown of the reactors' cores. During the earthquake, all of the region's nuclear reactors were shut down automatically. Had the earthquake been the only event that day, the reactors would have been inspected and likely returned to service almost immediately. However, the tsunami knocked out power to the cooling system at four of the reactors at the Fukushima Daiichi power plant. Generators kicked in to continue the cooling system function. However, they were lost to subsequent waves, meaning that the reactors had no way to cool the rapidly heating fuel. Reactor cores became superheated, and seawater was introduced into the reactors in an attempt to cool them. Water and radioactive waste started leaking out of the reactors and into the surrounding ocean shortly thereafter. The situation was brought under control within about 2 weeks. There were no injuries or loss of life from the nuclear accident, and all of the reactors were decommissioned.

Maximum run-up for this tsunami was 128 ft (38.9 m). Some scientists believe that the tsunami was much larger than expected because it was accompanied by an underwater landslide, which occurred during shaking from the earthquake. The combined water displacement from an earthquake and a major landslide, they believe, could account for the much larger than expected tsunami waves.

Other Earthquake-Generated Tsunamis

Tsunamis are hazards around all coastlines of the world. Though typically associated with the Pacific Rim, tsunamis can occur anywhere, on the shores of any ocean or sea. Earthquakes are by far the most common cause of tsunamis, and earthquakes certainly occur outside of the Pacific Rim, as a quick review of history will confirm. The most deadly tsunami in history occurred in 2004 in the Indian Ocean, as did the second deadliest tsunami, which occurred during the eruption of Krakatoa in Indonesia. However, tsunamis have occurred in the Caribbean, the Mediterranean, and the Atlantic as well as the Indian and Pacific Oceans.

The deadliest tsunami in the Caribbean occurred in the year 1692. A large earthquake struck near Jamaica, just offshore near the city of Port Royal. Though Jamaica had experienced many earthquakes before, they were typically quite small. This large earthquake, estimated to be about a magnitude 7.5, struck about 11:30 a.m. local time. A local tsunami followed shortly thereafter. Together, the earthquake and tsunami killed about 2,000 people outright and another 3,000 in the following days from injury and disease. The population density at that time was quite a bit lower than exists today, so if the same size event happened today, the death toll could be much higher.

Another lethal earthquake and tsunami occurred in the area of Lisbon, Portugal, in 1755. The earthquake, estimated to be over a magnitude 7 event, occurred on 1 November, All Saints Day. Because it was a holy day, many people were in church. Several stone masonry churches and other buildings collapsed, crushing those inside. Frightened survivors of the earthquake ran away from collapsing buildings, toward the open spaces in the city near the river and ocean. Shortly after the ground stopped shaking, a tsunami struck the city, washing away many who had sought safety near the shore. In all, about 30,000 people died in the earthquake and tsunami.

5.7 Landslide-Generated Tsunami

Landslides are the second most common cause of tsunamis. Landslides can occur for a number of reasons. They may happen as a result of earthquakes shaking rocks and soil loose, because the soil has become saturated and moves downhill under the force of gravity, or with seemingly no discernable trigger at all. Though tsunamis created by landslides can be large or small, it is interesting to note that the largest run-ups recorded throughout history and in the geologic record appear to be those caused by large landslides.

The largest run-up seen in a tsunami in recorded history was 1,720 ft (524 m). This occurred in 1958 after an earthquake-induced landslide sent rocks cascading into Lituya Bay, Alaska (Fig. 5.14). When the landslide hit the water, it created a giant wave that traveled across the bay and swept everything from the cliffs on the opposite shore up to a run-up of 1,720 ft (524 m). A smaller 100 ft–165 ft (30 m–50 m) tsunami traveled the length of the bay and toward the ocean. A very lucky father and son were on a boat that encountered this wave within the bay and survived to tell their story to scientists investigating the incident. Others on the water that day were not as lucky. Two other boats were swamped by the tsunami and their crews perished as a result.

Note in Fig. 5.14a that the wave was not able to sustain its original height of 1,720 ft as it swept down the length of the bay. The tallest wave was localized to the area directly across from the landslide itself. Landslide-generated tsunamis often create very tall local tsunamis that dissipate quickly with distance from the source. They are rarely ocean-wide phenomena as they tend to be when created by large earthquakes.

5.8 Mega-Tsunamis

There is a class of tsunami that have been dubbed "mega-tsunamis." These produce run-ups much greater, at least in the local area, than those found in earthquake- or volcano-generated tsunamis. They typically involve large portions of land cleaving off of islands. These may have produced run-ups in the local area of over 2,000 ft (610 m). One such area where mega-tsunamis are believed to have occurred is in the Hawaiian Islands. Offshore mapping indicates that all of the Hawaiian Islands are surrounded by immense landslide deposits where large sections of the islands have slid into the Pacific Ocean. With each of these landslides, a mega-tsunami was likely produced. Evidence for these mega-tsunamis includes coral heads and other marine debris found at relatively high elevations on the islands adjacent to the slide areas. Since the islands are formed volcanically, there is no way for coral to reach high elevations on the islands unless people carry them there, which has been ruled out, or they are deposited by large waves. The Canary Islands of the north Atlantic, also volcanic islands, have been the site of mega-tsunamis as well.

5.9 Volcano-Generated Tsunamis

One of the few cases, and certainly the most prominent example, of a tsunami generated by volcanic activity is that of Krakatoa in 1883. The volcano Krakatoa was an Indonesian island located in the Sunda Strait between the islands of Java and Sumatra, the two largest islands in the country. Krakatoa began its eruption in May of 1883 and produced low level activity for months. It became somewhat of a popular curiosity and tourist destination. The volcano's activity level increased throughout

(a)

(b)

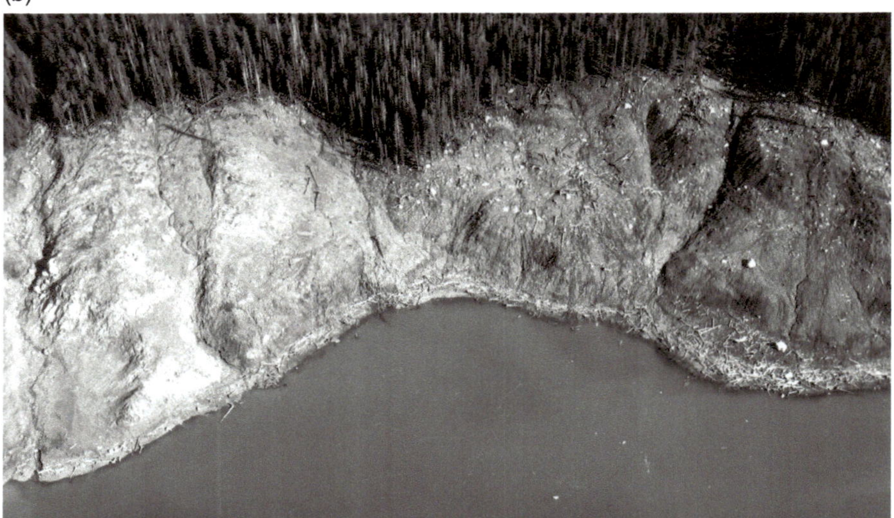

Figure 5.14 (a) Aerial view of Lituya Bay, Alaska, following a 1958 tsunami caused by an earthquake-induced landslide. The white areas along the shoreline were scoured by the tsunami. The greatest wave height of 1,720 ft (524 m) occurred directly across from the landslide area. (See upper center of photo.) Photo by USGS; https://www.sciencebase.gov/catalog/file/get/51dd9157e4b0f72b4471ccdf?f=__disk__64%2Fe1%2Fe8%2F64e1e8ef4dea3f625d6af97c19355125d48d511a&width=5640&height=4436 (b) Close view of the coastline of Lituya Bay. Full-sized trees were uprooted and defoliated, and logs deposited along the shoreline. Soil was stripped from the area, leaving bedrock exposed. Photo by USGS; https://www.sciencebase.gov/catalog/file/get/51dd910de4b0f72b4471cca9?f=__disk__7c%2Fe2%2F1a%2F7ce21a07c6523c4195c1e5663382218f4178850d&width=5544&height=3428

August of 1883. Its cataclysmic eruption began on 26 August 1883. The volcano began its eruption with a strong earthquake early in the afternoon of 26 August. Telegraph operators exchanged communications about their observations, which included a darkening sky and continuous ground shaking. Ships moving through the strait endured deafening explosions. On land, the explosions shook the air and ground. Activity continued through the night of 26 August and into the day on 27 August. At night, a red glow was visible above the volcano and the ground was shaking so severely people were afraid their homes would collapse. The volcano's eruption could clearly be heard by ship captains 70 mi (112 km) away. The sky became inky black even during the day from massive output of volcanic ash.

Krakatoa's eruption reached a climax at 10:02 a.m. on Monday, 27 August. The loudest sound ever recorded issued from the volcano at that time. Scientists believe this sound was the result of the collapse of Krakatoa's summit. The eruption had been so voluminous, it emptied much of the magma chamber beneath the volcano. Krakatoa's summit collapsed into the ocean, creating a caldera. A large volume of water rushed in to fill the void. A tsunami was generated that rushed outward toward the coastlines of the two larger islands. Within an hour, the waves had done their damage. Homes, businesses, piers, customs houses, and other structures on the islands' shorelines were washed away. Pyroclastic flows, which had to travel more than 10 mi to reach the shorelines, killed approximately 2,000 people on Sumatra who managed to escape the tsunami. In all, approximately 36,000 people lost their lives in the eruption. Most died in the tsunami in the coastal areas of Java and Sumatra.

About 30 years after the 1883 eruption of Krakatoa, a small volcanic peak rose above the waves within Krakatoa's caldera. This new vent, named Anak Krakatoa (Child of Krakatoa) has been quite active over the last 120 years. Anak Krakatoa has survived the currents and waves to become a permanent fixture in the Sunda Strait. In 2018, the volcano was the site of yet another tsunami, this time generated by the collapse of approximately 158 acres (64 hectares) of the volcano's southwest flank. The volcano, whose height had been 1,108 ft (338 m) above sea level before the collapse, was reduced to an elevation of 360 ft (110 m) above sea level after the collapse (Fig. 5.15).

No warning went out to the populations on either side of the Sunda Strait. As a result, people were caught completely off guard when the tsunami struck the shoreline. Government officials explained that the tsunami warning system had been out of order since 2012 due to lack of funds, vandalism, and technical faults. Even if it had been in working order, it is likely that it would not have been triggered in time to alert people on Java and Sumatra. The warning system is designed to activate after a large earthquake. This tsunami was triggered by a landslide rather than a large earthquake. By the time ocean buoys had detected a tsunami passing beneath them, the tsunami would have been well on its way to shore. Still, even a minute or two of warning would have been enough to allow people to reach safety.

Coastal areas of both Java and Sumatra, the two large islands on either side of the Sunda Strait, were affected by the tsunami. The popular band Seventeen was playing to a small crowd at the Tanjung Lesung Beach Resort on Java's shoreline. The wave crashed through the venue with no warning, killing three members of the band and more than 100 concertgoers and other resort guests. Similar scenes were replayed in other coastal areas facing the Sunda Strait. People were caught unaware, with no chance to run to shelters or higher ground. At least 426 people were killed and more than 14,000 were injured. An additional 16,000 were displaced from their homes.

Another instance in which a tsunami was produced as a result of a volcanic eruption and caldera collapse was in the Mediterranean around the year 1613 B.C. The island of

(a)

(b)

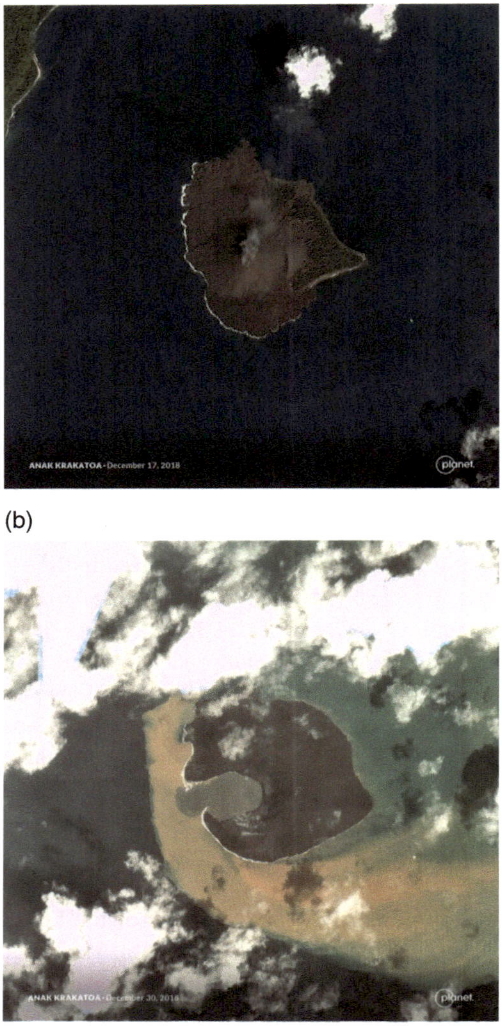

Figure 5.15 Satellite images of Anak Krakatoa taken (a) before and (b) after the collapse on 22 December 2018. Images from Planet Labs, Inc.; https://blogs.agu.org/landslideblog/2019/01/03/anak-Krakatoa-3/

Santorini was the site of a volcano called Thera. Thera erupted violently, and like Krakatoa, the center of the volcano collapsed into the sea. Though there are no contemporary reports of the size of the tsunami, it was likely taller than that of the Krakatoa eruption's tsunami. Many people believe the eruption and resulting tsunami were responsible for the disappearance of the advanced Minoan civilization in the region. Today, Santorini is a beautiful horseshoe-shaped island with sun-drenched white cliffs (Fig. 5.16). It takes a trained eye to spot the violent history of this beautiful and photogenic island.

5.10 Mitigation of Tsunami Hazards

As we have seen, tsunamis are a danger on all coastlines of the world, whether from local earthquake or landslide hazards, volcanic eruptions, or earthquakes that occur far away. The primary defense against tsunami deaths is to leave low-lying coastal areas uninhabited. However, this is not practical in most areas. Millions of people depend on the ocean for their livelihood, and it is unreasonable to ask them to abandon these areas as possible home sites. The second defense is to educate people about tsunamis and their possible signs. People who live in areas where earthquakes trigger tsunamis should know that if a large earthquake occurs, they need to immediately seek shelter on higher ground. People should also be aware that tsunamis can come ashore first as a trough, which will expose the ocean floor well beyond the low tide line. If this occurs, it is a signal to head to higher ground because the crest of the wave will soon follow. A related mitigation strategy that works well in tsunami-prone areas is to use signs to warn people that they are in a tsunami hazard zone, and to point the way to evacuation routes and higher ground.

The Japanese are perhaps better prepared than any other populace for the effects of a tsunami. Not only do they keep many low-lying coastal areas free of development, they also have signage pointing people to evacuation routes and safe shelters. In addition, the Japanese have been proactive in building tsunami defenses in the form of seawalls. Seawalls are typically about 40 ft high (12.5 m) and are designed to protect coastal communities from

Figure 5.16 The caldera of Santorini in Greece. The caldera was created in the sixteenth century B.C. by a massive eruption of the volcano Thera. The collapse of the center of the volcano into the sea generated a large tsunami. Credit: JeniPhoto/Shutterstock.

most tsunamis that are likely to strike the coast. However, they can be overtopped by extraordinarily high run-up tsunamis such as the one that struck the Japanese island of Honshu in March of 2011. Though most people appreciate the security these seawalls provide, they do have their drawbacks. The walls are not attractive. They radiate heat in the summer making people uncomfortable. They block ocean views, and make some people feel as if they are confined. As unattractive and confining as these walls are, most residents in seaside communities agree that the walls make them feel safe, and they are happy to have their protection.

Tsunami warning systems are another line of defense against loss of life due to ocean-wide tsunamis. All populated areas are covered by at least one tsunami warning system. The oldest of the tsunami warning systems is in the Pacific Ocean. When an earthquake occurs anywhere around the Pacific Rim, scientists quickly calculate an earthquake magnitude. If the earthquake's magnitude is above a magnitude 7, they issue a Pacific-wide alert within minutes. Scientists gather data from ocean buoys and landfalls to determine if a tsunami has in fact been produced. If no tsunami has been created, the tsunami alert is canceled. If a tsunami has been generated, subsequent warnings include estimated arrival times for various locations around the ocean basin. Tsunami warning sirens urge people to move inland and to safer ground at higher elevations prior to the approximate arrival time. Tsunami warning systems have been incredibly effective in reducing deaths at great distances from earthquakes around the edge of the Pacific. The Pacific Tsunami Warning Center is run by the United States National Oceanic and Atmospheric Administration, or NOAA.

Other areas of the world are also covered by tsunami warning systems, administered by various international consortia. The northeast Atlantic is covered by a warning system administered by UNESCO, which is the United Nations Educational, Scientific, and Cultural Organization. This monitors European shores and those of the Mediterranean. The Caribbean is covered by a third warning system. Like the Pacific Tsunami Warning Center, the Caribbean Tsunami Warning Program is administered by NOAA. A fairly new warning system is in place in the Indian Ocean as well, though unfortunately it came too late to save hundreds of thousands of lives in the December 2004 tsunami. Like the northeastern Atlantic Ocean and Mediterranean Sea, the Indian Ocean is monitored by a UNESCO consortium. The warning center is staffed by scientists in Australia, India, and Indonesia.

While tsunami warning centers are credited with saving thousands of lives over the decades for which they have been in place, tsunamis move so quickly that those nearest the source will likely receive no formal notification that a tsunami is on the way before that tsunami strikes. Some of the largest ocean-wide tsunamis occur when an earthquake occurs just a few miles offshore at a subduction zone. Those nearest the earthquake's epicenter may experience ground shaking for a period of 3 to 4 min. They may have just enough time to wait out the earthquake and check on the well-being of those in the same building as themselves before a tsunami comes ashore. Tsunami waves can hit shorelines in as few as 7 or 8 min from the time ground shaking ends. In order to escape a tsunami, those in coastal areas need to immediately proceed to higher ground. Those who do not think quickly, or who are unable to leave the area due to injury or infirmity, are then left vulnerable to a tsunami. It takes time for scientists to calculate a magnitude for the earthquake to determine if it is large enough to produce a tsunami, then to send out an appropriate message. In those critical minutes, those closest to the earthquake's epicenter should already be on the move.

Because tsunami energy is distributed throughout the entire height of the water column including along the ocean bottom, scientists can use this property of the waves to detect tsunamis in the open ocean using ocean-bottom sensors known as DART systems. DART stands for Deep-Ocean Assessment and Reporting of Tsunamis. The tsunami waves exert pressure from their strong current on sensors located on the ocean floor. These sensors transmit wave velocity, height, and wavelength data from the ocean floor to a surface buoy, and then from the surface buoy to warning centers around the world. Data are then used as input for models that allow scientists to forecast tsunami arrival times at locations around the ocean basin. Wave heights, however, can be very difficult to predict. Scientists routinely issue tsunami alerts and then downgrade or cancel them because tsunami amplitudes are incredibly small when they are finally detected and measured at the coastline. Still, even low-amplitude tsunamis can be dangerous to those swimming or scuba diving in the ocean. Even though the waves may not be very high, they still move at extremely fast velocities and can result in strong near-shore currents. People, boats, marine life, and debris can get caught in these currents and swept out to sea.

A number of research efforts are underway in laboratories around the world to learn how to make coastal structures more resistant to tsunamis. Scientists use large-scale flumes to create simulated tsunamis in the laboratory. They measure impact force of a tsunami as it overtops a building or bridge within their scale model. Data are analyzed and computer models are built that will allow scientists and engineers to work together to determine ways coastal structures could perhaps be designed to survive undamaged, saving lives and reducing economic losses.

5.11 Conclusion

Thankfully, large and deadly tsunamis are fairly rare events. We are better equipped now for their occurrence than at any time in history. Recent occurrences of large and deadly tsunamis in Indonesia and Japan have reminded the world that this is a major hazard, and new warning systems are in place to help keep people safe. Scientists continue to conduct tsunami research using computer models, wave tanks, and data from actual tsunamis. The hope is that, with time, we will be able to accurately forecast tsunami arrival time and wave heights with great accuracy, and build tsunami-resistant structures that will save lives.

5.12 Chapter in Review

Tsunamis are water waves created due to disturbances of the ocean floor. The most common cause of tsunamis is shallow earthquakes at subduction zones. These large earthquakes cause the subducting plate to be thrust under the overriding plate. The overriding plate snaps upward displacing millions of gallons of seawater and creating a tsunami. Tsunamis are also caused by landslides and volcanic eruptions as well as atmospheric phenomena. The largest of all tsunamis are caused by landslides. These landslide-generated tsunamis have very high local run-ups but often dissipate quickly away from the location of the landslide. Volcanic tsunamis occur most often when an island volcano erupts and collapses, causing seawater to rush into the void. This displacement of the ocean floor and large-scale water movement can generate a tsunami.

Tsunamis are fundamentally different from wind-driven waves in all characteristics. Wind-driven waves only affect the uppermost parts of the ocean. Tsunamis affect the entire water column, from the surface to the ocean floor. Wavelengths of wind-driven waves are fairly short, perhaps only a few hundred feet long. Wavelengths of tsunamis are hundreds of miles long. Wind-driven waves travel at a speeds on the order of a few kilometers per hour. Tsunamis travel as fast as a jet plane, up to 500 mi/hr. Periods of wind-driven waves are just a few seconds, compared with 5 min to 1 hr for tsunamis.

An inundation zone is an area that is covered with water when a tsunami comes ashore. The tsunami's maximum height when it comes ashore is called its run-up. Run-up is measured in many locations, and can be different at every location where it is measured. Run-up is determined by a number of factors including: the size of the disturbance that generated the tsunami, distance from the tsunami source, and topography of the ocean floor. Some areas of the ocean floor are known to focus and amplify tsunamis.

An earthquake along the Cascadia Subduction Zone off the coast of California, Oregon, Washington, and British Columbia was the cause of a Pacific-wide tsunami in January of 1700. Historic records from Japan record an earthquake there on the night of 27 January. Corresponding stories of a great wave occur in Native American folklore from the same time period. Scientists put the two together after studying and carbon dating tsunami deposits and "ghost forests" of dead trees that were drowned suddenly about 300 years ago.

A magnitude 9.1 earthquake off the coast of Indonesia on 26 December 2004 was responsible for generating an ocean-wide tsunami in the Indian Ocean. This was the deadliest tsunami in history, killing nearly 230,000 people. Deaths were reported in 11 countries, with the most deaths occurring in Indonesia. At the time of the earthquake, there was no tsunami warning system in place in the Indian Ocean, so there was no formal process for warning countries around the rim of the Indian Ocean that a tsunami was headed their way. After that time, UNESCO funded a new tsunami warning system in the Indian Ocean, which is staffed by scientists from that region of the world.

The largest earthquake ever to strike Japan caused a major tsunami on 11 March 2011. The tsunami struck the island of Honshu with great force, creating a tsunami with a maximum run-up of 128 ft (38.9 m). More than 18,000 deaths occurred in Japan as a result of this earthquake and tsunami. Autopsies revealed that the vast majority of people drowned, while small numbers of people died in fires or were crushed by debris, either during the earthquake or the tsunami.

A major earthquake off the coast of Lisbon, Portugal, generated a tsunami in 1755. The earthquake and resulting tsunami killed more than 30,000 people. Another major earthquake off the coast of Port Royal, Jamaica, occurred in 1692. The earthquake damaged many buildings nearby and generated a tsunami. The earthquake and tsunami killed 2,000 people outright and another 3,000 people within the next week or two because of injury and disease.

The largest of all tsunamis are caused by landslides. The largest run-up from a tsunami in recorded history occurred in 1958 in Lituya Bay, Alaska. An earthquake-induced landslide sent debris into the bay. On the shoreline opposite the landslide, the run-up reached 1,719 ft (524 m) above sea level. A smaller wave raced along the length of the bay before dissipating a short distance outside the bay's entrance. A class of large tsunamis generated when sections of islands slide into the ocean is called mega-tsunamis. Two locations where scientists believe mega-tsunamis have been produced include the Hawaiian Islands and the Canary Islands.

Volcanoes generate tsunamis as well. The volcano Krakatoa, an island in the Sunda Strait between the two large Indonesian islands of Java and Sumatra, erupted in August of 1883. The eruption was incredibly violent, producing a large volume of volcanic ash and pumice. After erupting for a full day, the summit of Krakatoa collapsed into the water, generating a tsunami that engulfed the shorelines of both Java and Sumatra. The tsunami was responsible for most of the estimated 36,000 deaths from the eruption of Krakatoa. In December of 2018, the small island of Anak Krakatoa, located within the 1883 caldera of Krakatoa, suffered a collapse that generated a tsunami. About 426 people were killed in the tsunami, which came ashore with no warning.

The island of Thera in the Greek Isles had a massive eruption in 1613 B.C. The eruption emptied the magma chamber beneath the volcano, causing collapse of the volcano's summit. The collapse of the volcano's summit into the ocean generated a tsunami that affected nearby islands. The island of Santorini is the only remnant of Thera. Many scientists and archaeologists believe that the Minoan civilization was destroyed by this eruption of Thera.

There are many ways to mitigate tsunami hazards. Countries may zone land use such that people may not build permanent structures within tsunami hazard zones. Education about what a tsunami is and what it looks like is critical for people who live and work in coastal areas. The only way for a person on shore to escape a tsunami is to move to higher ground. Governments can erect signs that notify people when they are entering a tsunami hazard zone. Complementary signs can direct people to areas of higher elevation. Some countries have built seawalls to protect coastal communities from average sized tsunamis. Unfortunately, seawalls are not effective in the largest tsunamis, as was demonstrated in the 2011 Tohoku earthquake and tsunami where the seawalls were overtopped by the tsunami.

International coalitions have worked together to establish tsunami warning systems in most of the world's oceans. Warning centers are staffed by scientists who monitor earthquake activity and data from ocean-bottom sensors. If a tsunami is generated, these warning centers send out alerts to coastal communities. Scientists in many government agencies across the world also use computer models to predict run-up from hypothetical earthquakes in various locations.

6

Earth Movements

Bethany D. Hinga

University of Nebraska at Kearney, USA

6.1 The Role of Gravity in Shaping the Earth

Gravity is an important agent of landscape change. Landscapes are altered by a number of natural events. Streams erode downward through rocks and soil, creating deep river valleys. Stream valleys also widen over time, and it is the work of gravity that allows that valley widening to happen. Water that finds its way into cracks expands when it freezes, making the cracks wider. This creates a process called *ice wedging* in which pieces of rock are forced away from the bedrock by expanding ice in those cracks. Gravity pulls that weathered rock downhill. Heavy rainfall can saturate steep soil-covered slopes and cause debris flows of soil and rock to move quickly downhill. Rocks can fall away from steep slopes in massive rock falls. Such Earth movements caused by gravity are often referred to collectively as *mass wasting*.

Mass wasting is the downslope movement of soil, weathered material called *regolith*, and rocks under the influence of gravity. Mass wasting may happen slowly or quickly, in large volume or small. Common types of Earth movements include landslides, avalanches, slumps, debris flows, rock slides, and soil creep. Slope instability can happen anywhere, even in areas with only gentle slopes. In the chapter that follows, we will examine the various types of mass wasting that continue to shape the world we live in.

6.2 Venezuela, 1999

The state of Vargas is a mountainous region on the northern coast of Venezuela. Because the mountains extend to the ocean, there is very little coastal land available for development. As a result, cities are built on features at the base of the mountain called *alluvial fans*. Alluvial fans (Fig. 6.1) are deposits of debris left behind at the base of a steep slope, typically at the mouth of a narrow canyon. Some of these narrow canyons are occupied by streams. Streams that flow down steep slopes move quickly. As a result, there is enough energy in the water to carry a great deal of sediment. At the bottom of the steep slope, the stream must transition to flow more slowly, over flat land. Sediment that was carried easily downhill in the high velocity water of steep streams can no longer be carried by slower-moving streams on flatter terrain. Sediment drops out of the water, creating features of deposited sediment called alluvial fans. Alluvial fans are composed of boulders, cobbles, sand, and silt.

Some alluvial fans are composed primarily of debris-flow deposits. This means that slope

Earth's Natural Hazards and Disasters, Advanced Textbook 7, First Edition. Bethany D. Hinga.

Figure 6.1 Aerial view of an alluvial fan in Death Valley, California. The alluvial fan formed when a stream in the mountains encountered the flat area at the base of the mountains, and the sediment within the stream was deposited. From Skinner & Murck, 2011.

instability led to the rapid mobilization of material in the narrow canyon that leads to the alluvial fan. The result is a debris flow, which is a chaotic downhill flow of rock, soil, and other items on the slope. This has a great deal of importance in the case of Caraballeda, Venezuela, presented here.

The city of Caraballeda had a population of about 300,000 in 1999. It is situated on an alluvial fan where the Rio San Julian meets the coastal plain leading to the Caribbean Sea. The city is built on relatively flat ground located between the Caribbean and steep mountains. The steep mountain slopes behind the city are composed of thin soil and weathered rock. The route to the capital city of Caracas is a steep road that leads up the face of the mountain to the plateau above. A coastal highway hugs the steep mountain front along the shoreline, and leads to other cities built in settings much like Caraballeda.

The rainy season in this part of Venezuela typically lasts from May to October. However, 1999 was not a typical year. In mid-December, a somewhat unusual winter storm moved into the area, reaching Caraballeda on 14 December 1999. Over the next 3 days, 36 in. (91 cm) of rain fell on the city of Caraballeda and the surrounding mountains. People started noticing flooding on the evening of 15 December. Heavy rainfall caused thin soil on the slopes of the mountains behind the city to become saturated. This is a dangerous situation because steep slopes saturated with water are prone to *debris flows*, mudflows, and other forms of Earth movements.

Debris flows are masses of mud and rocks that move downhill under the influence of gravity. Rocks within these flows can range in size from less than 1 in. in diameter to the size of a small house. Debris flows can travel at speeds ranging from very slow to more than (sometimes MUCH more than) 20 mi/hr (32 km/hr). They also have the ability to gather other debris as they move downhill, including trees, shrubs, cars, parts of homes, and portions of roads.

Initial reports of debris flows reached first responders about 8:30 p.m. on 15 December. Initial debris flows were small and confined to minor canyons on the mountain slope. However, residents knew that if heavy rains continued, the city of Caraballeda could be flooded and perhaps overrun by major debris flows. Many in Caraballeda decided to evacuate inland. People became understandably alarmed when these flows began to damage the road along the mountain front that led from Caraballeda inland to Caracas. Many people who attempted to flee along that route were prevented from doing so. Some were stranded on the mountain slope for the duration of the storm because the road was washed out above and below them.

In addition to the larger debris flows in the major river valleys, hundreds of smaller debris flows also occurred along the mountain front next to the coast. Some of these flows damaged and closed coastal roads. All land routes into Caraballeda were cut off, leaving the Caribbean Sea the only route into and out of the city. This assured that those who lived in these coastal communities had no easy route of escape.

Debris flows became larger during the night of 15 December. A series of deadly and devastating debris flows moved down the Rio San Julian and into the city of Caraballeda that night and overwhelmed the city. Flows continued unabated until late in the afternoon of 16 December. They moved at speeds approaching 31 mi/hr (50 km/hr). Vast areas of the city were covered with mud and rocks to a depth of 20 ft (6 m) or more, well above the second story of buildings (Fig. 6.2). Boulders larger than cars barreled through the city, knocking holes in buildings and crushing other items in their paths. Damage was extensive, and many buildings were completely demolished. Hardest hit were one and two story structures that included single family homes and apartment buildings. Many were destroyed or buried by the fast-moving flow.

When the storm was over and debris flows stopped, approximately 67 million ft^3 (1.9 million m^3) of material had been moved from the mountains onto alluvial fans in the area. Caraballeda was not the only city affected. The coastal communities of Naiguita and La Guaira were also inundated with debris flows. Damage to both mountain and coastal highways impacted search and recovery efforts in the short term and economic recovery in the long term because there was no easy route for rescue workers into the city, nor a way to transport building materials into the city or to road repair infrastructure and buildings.

Initial estimates of deaths in Caraballeda were between 5,000 and 50,000, though the

Figure 6.2 Aerial view of damage within the city of Caraballeda. Photo by Lawson Smith/U.S. Army Corps of Engineers. From Murck & Skinner, 2016.

number of fatalities is now considered to be somewhere between 19,000 and 30,000. This is as much as 10% of the prestorm population of 300,000. There is great uncertainty in the number of fatalities because many bodies were never recovered. Some people were buried under several meters of rocks and mud, and others were washed out to sea by the debris flows and floods. Entire families died, and when communities were washed away, people had no way of checking to see if neighbors were alive and well. Only about 1,000 bodies were recovered so many families have no idea what became of their loved ones and neighbors. This event is a tragic example of how mass wasting can affect vulnerable populations in areas with significant hazards.

6.3 Types of Earth Movements

There are six main types of mass wasting defined by experts (Fig. 6.3). These are creep, falls, slides, slumps, flows, and subsidence. Each type of mass wasting may have several subtypes. For example, flows can be composed of snow and air, water and soil, water and volcanic ash, boulders and mud, or any combination of the above materials. The materials involved may dictate what we call the specific mass wasting event, but the type of mass wasting is based on the nature of the movement of the material.

It is also important to mention that mass wasting happens on a continuum, and it is rarely a simple or straightforward event. One type of

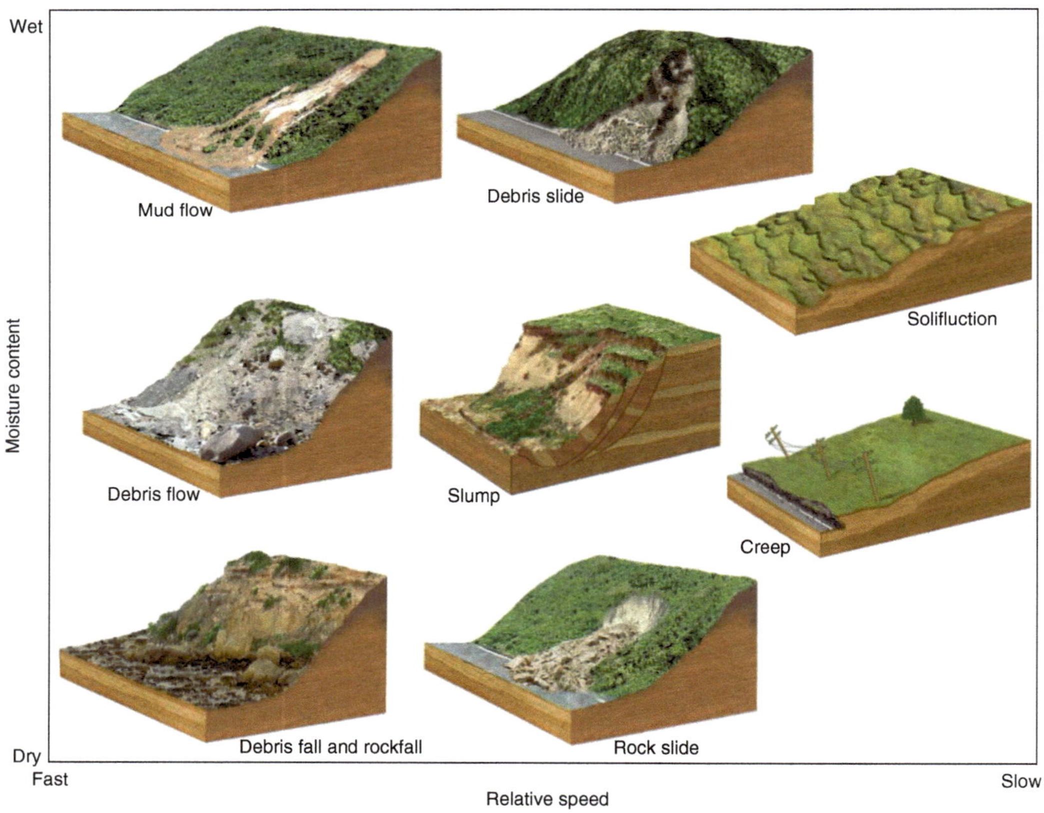

Figure 6.3 Major types of mass wasting, classified by relative speed and moisture content. From Fletcher, 2014.

mass wasting may turn into another type. Or a single mass wasting event may have characteristics of two or more types. Therefore, while the mass wasting classification system offered here is instructive, the real world is complex and harder to classify into neat categories. You will find a number of cases in the following pages in which these complexities come to light.

6.3.1 Creep

The slowest of all Earth movements is *creep* (Fig. 6.4). Creep affects the uppermost layers of the Earth, typically just the soil and the weathered material, or regolith, beneath. Creep is a slow movement of material in a downslope direction that may occur over months, years, decades, or centuries. Creep is rarely dangerous to humans or animals since it occurs so slowly, but can cause a great deal of damage to structures and infrastructure. Over time, creep can displace anything built on top of or within the regolith and soil, including roads, building foundations, sidewalks, retaining walls, power line poles, fences, pipelines, water lines, and sewer lines. Creep

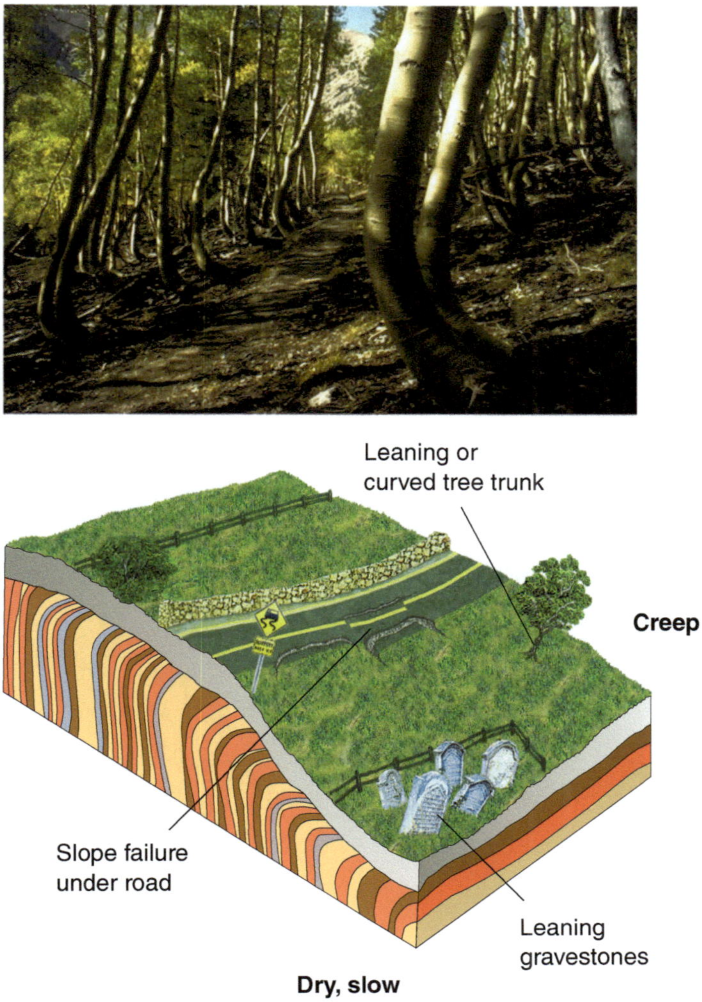

Figure 6.4 Creep, a slow form of mass wasting. Soil and regolith are pulled downhill under the influence of gravity, which can damage and displace human-made structures. Photo by Marli Miller. From Fletcher, 2014.

can even cause the trunks of trees to become bent. Because creep causes movement in any materials found on or within a modest slope, those items may need to be repaired or replaced on a regular schedule. That means that while it is decidedly undramatic, creep can be extremely expensive. Creep does not require steep slopes in order to occur. Even a modest slope of only 10° to 15° is sufficient to cause creep that is disruptive to infrastructure over the span of several years (Fig. 6.5). For this reason, creep is also the most pervasive type of mass wasting on the planet.

What causes creep to occur? Creep occurs because gravity causes soil to shift downhill. Slippery clay minerals in the soil swell and slide with rainfall, then dry and shrink as the soil dries out. Even microscopic changes in a single shrink-swell cycle can cause creep to occur. If those microscopic changes take place for years on end, they become noticeable. Creep can also occur in colder or more temperate climates due to the freezing and thawing of the soil and water within it. In the cold season, water in the soil and regolith freezes and expands. In the warm season, the water thaws and contracts. The expansion and contraction, together with the force of gravity, pulls the soil and regolith downslope. A special case of creep, called *solifluction* (Fig. 6.6), occurs in Arctic climates. Soil that is frozen all year is called permafrost. In the summer months, it can get warm enough to melt a thin layer of permafrost at the surface of the Earth. When that happens, water in the soil causes the soil and overlying vegetation to flow downhill in lobes, making the slope appear to be melting. This happens in permafrost areas in Alaska and other polar regions.

6.3.2 Rock Falls

A *rock fall* is an event that occurs when rock free-falls, or falls through the air, due to gravity (Fig. 6.7). This happens when a piece of rock becomes dislodged from a rock face and is not adequately supported from beneath. It can occur in places where a stream or glacier has undercut a rock face, or where a vertical cliff exists. Falls also occur when water invades cracks in rock faces. In winter, the water freezes and expands the cracks. At some point,

Figure 6.5 The wooden posts in this photograph are tilted downhill due to creep on this hillslope. Photo by Ralph Lee Hopkins/Photo Researchers. From Arbogast, 2017.

Figure 6.6 Solifluction is a form of creep in which the top layer of permafrost melts in summer months and causes the saturated soil to move downhill. It is common in polar regions. © Marli Bryant Miller. From Fletcher, 2014.

Figure 6.7 In a rock fall, rocks fall directly from the face of an outcrop. The rocks at the base of the cliff were deposited there as a result of rock falls and have formed talus. Photo is from Kaibito Canyon in Arizona. Photo by Bill Hatcher/NG Image Collection. From Murck & Skinner, 2016.

gravity takes over and the rock falls to the valley floor beneath. A fall may be triggered by a heavy rainfall, ice wedging, weathering, thermal stress on exfoliation joints, vibrations in the ground from heavy machinery, or an earthquake. When the rock falls to the ground, it may bounce or slide until it finds a place to land where it will be more stable. The deposit of rocks at the base of a slope that have been deposited there as part of a fall event is called *talus*, or a *talus slope* (Fig. 6.8).

6.3.3 Slides

Slides are mass wasting events in which material slides downhill along an inclined plane (Fig. 6.9). Often, this happens when layers of rock are tilted in a downhill direction. The boundaries between layers can serve as planar sliding surfaces along which material slides downhill. Slides can occur with rocks (rock slides) or debris such as soil, regolith, and rock (debris slides). The main characteristic of a slide is that it must move downhill along a planar surface. It may require very little to trigger downhill movement, so there may be very little or no warning that this movement is about to take place.

Figure 6.8 Ice wedging of these cliffs near the shore of Lake Louise in the Canadian Rockies has led to rock falls. The large light gray deposits are talus slopes created from material involved in rock falls. Photo by A. N. Strahler. From Strahler, 2013.

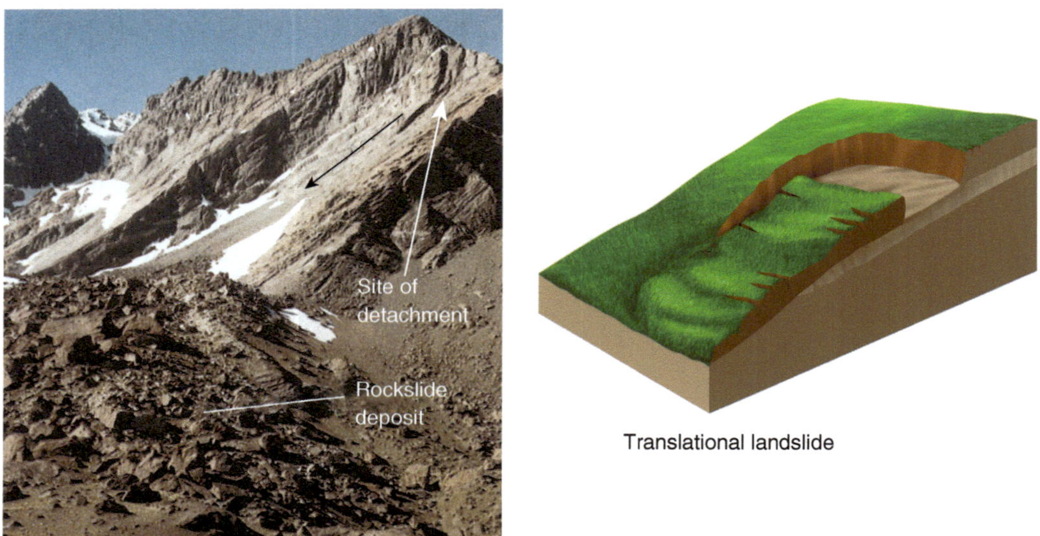

Figure 6.9 A slide. Note that the boundary between the top rock layer and the one underneath is serving as a planar sliding surface. From Fletcher, 2014. The photo is of a rock slide in the Andes Mountains in Argentina. Photo by Stephen C. Porter. From Fletcher, 2014.

One of the modern era's most memorable mass wasting events was a rock slide that occurred in Italy in 1963. In the Vaiont River valley in the Italian Alps, engineers designed and constructed a high dam for the purpose of creating a reservoir and generating hydroelectric power for the communities in the area. However, they did not pay close enough attention to the geology of the area when they undertook this construction project. The rocks on one side of the valley tilted toward the valley floor. The sequence of rocks included a clay layer. Clays become slippery when saturated with water. The shape of clay particles also contributes to their tendency to facilitate slippage. Clays are shaped like sheets of paper (Fig. 6.10). They tend to be very thin and flat. That means the surfaces between clay particles form convenient sliding surfaces.

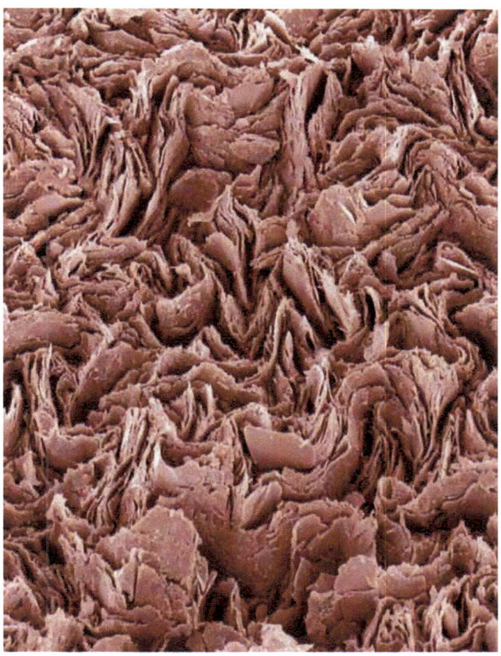

Figure 6.10 Photomicrograph of clay particles. Note their thin structure. Clays shrink and swell with changing moisture levels. Clay layers make excellent sliding surfaces, especially when the clay is wet. From Fletcher, 2014.

When the Vaiont dam was complete, the reservoir it created began filling with water, and the rocks beneath the reservoir became saturated. The water table in the area rose, and soon the clay layer became wet and slippery. Engineers realized their mistake soon after the reservoir filled with water, less than a year after construction was completed. They began monitoring slippage of the slope above the reservoir and at one point realized it was sliding downhill at a rate of 16 in. (40 cm) per day. They tried to slow the motion of the slope by drawing down the water level within the lake, but to no avail. At 10:15 p.m. on 9 October 1963, a large block of rock and soil slid downhill into the reservoir (Fig. 6.11). A wave about 100 m (328 ft) high washed over the dam and into the valley below. It struck the town of Longarone with no warning and killed about 2,600 people. The reservoir near the dam was completely filled with debris. No attempt was made to move the landslide deposit and restore the dam to its original use. The landslide deposit is still visible today.

Another tragic slide occurred on 13 January 2001 in El Salvador. That day, a magnitude 7.6 earthquake struck near the southern coast of El Salvador. The earthquake was strong enough that it was felt as far away as Mexico City. The shaking destabilized a steep slope above the city of Santa Tecla. A large slide was unleashed that turned quickly into a debris flow. It flowed into the Las Colinas neighborhood, destroying several blocks of homes and killing approximately 1,200 people (Fig. 6.12).

An earthquake was to blame for yet another notable landslide, this one in the United States. On the night of 17 August 1959, a magnitude 7.1 earthquake struck near the western edge of Yellowstone National Park. Known as the Hebgen Lake earthquake, it was the largest earthquake in that area in many decades. Three major events happened in quick succession. First, the earthquake triggered a landslide in a canyon along the Madison River. A piece of the south wall of the canyon about 2,000 ft (600 m) long and 1,000 ft (300 m) tall broke

Figure 6.11 Photo of the Vaiont Dam area in Italy. Here, a massive landslide from Mount Toc slid into the reservoir. The landslide created a large wave that swamped the town of Longarone, washing it away in an instant and claiming 2,600 lives downstream. The source of the slide is visible in the left half of the image, on the steep slope. Areas of bare rock indicate the source of the landslide. Photo © CNES 2010, distribution Astrium Services/Spot Image S.A. From Fletcher, 2014.

Figure 6.12 A debris flow tore through the Las Colinas neighborhood of Santa Tecla, El Salvador, on 13 January 2001. A landslide, which turned into a debris flow, was triggered by a magnitude 6.7 earthquake. Image by U.S. Geological Survey Landslide Hazards Program. Public domain via Wikimedia Commons.

away from the canyon and descended about 0.3 mi (0.5 km) down to the Madison River. It moved at approximately 100 mi/hr (160 km/hr). Second, the force of the fast-moving landslide displaced air in front of it, creating gale force winds. Third, the landslide blocked the Madison River, creating a natural dam that caused the water to rise incredibly quickly (Fig. 6.13).

Many people had been camping in the valley that evening, and received the fright of their lives. Several witnessed the terrifying landslide as it overtook the lake and some campsites. Travel trailers were picked up and tossed about by the high winds. Horrified campers also had to outrun rapidly rising water when the slide blocked the river and created a new natural dam on the river (Fig. 6.14). Unfortunately, 28 people lost their lives due to this event. The Madison Slide, as it was later named, created a permanent new lake that is now approximately 330 ft (100 m) deep. It is aptly named Earthquake Lake.

6.3.4 Slumps

A *slump* is a mass movement in which material slides downhill along a curved sliding surface (Fig. 6.15). The sliding surface is steep at the top (head) and shallow near the bottom. Another term for a slump is a *rotational slide*. They may happen on any scale, from a small area on the side of a highway to massive blocks in the mountains. Slumps occur most often when a slope is saturated with moisture and begins to mobilize (Fig. 6.15b). Slumping is often seen during rainfall events where there is thick soil on a slope. It is also seen in banks of streams, in water-saturated sand or soil deposits, and in fractured rocks. Large slumps on stream banks

Figure 6.13 Photo of the Madison Slide in Montana, which occurred on 17 August 1959. The landslide was triggered by a nearby magnitude 7.1 earthquake. The slide blocked the Madison River and created a natural dam. The resulting lake is named Earthquake Lake. Photo by A. N. Strahler. From Strahler, 2013.

Figure 6.14 A photograph of the newly formed Earthquake Lake. The cottages in the photo were spared by the earthquake and landslide, but were destroyed and floated away by flooding that occurred when the Madison Slide dammed the Madison River, creating the lake. Photo courtesy of the U.S. Geological Survey. From Strahler, 2013.

Figure 6.15 Slump or rotational slide. (a) A slump is characterized by a curved sliding surface and rotation of the block above. (b) Photo is of a slump in central California. Photo © Marli Bryant Miller. From Fletcher, 2014.

are not uncommon, particularly after copious amounts of rainfall. Slumping is one method by which stream valleys get wider. Slumps can also include thicker sections of regolith and rock. Large slumps also often occur along coastlines with bluffs or cliffs (Fig. 6.16a) or in the mountains. Major slumps along the coastline in California have landed streets and multi-million-dollar homes in large, broken heaps at the bases of sea cliffs (Fig. 6.16b).

(a) (b)

Figure 6.16 (a) Erosion of the shoreline beneath these cliffs has led to slumping along the cliff face. Photo © Marli Miller, used with permission. (b) Homes have been destroyed as the land they were built on slumped into the ocean below. Photo by Aerial Archives/Alamy. From Fletcher, 2014.

6.3.5 Slump to Flow

In many cases, one type of mass wasting can turn into another quite abruptly. One example of this is the Oso Landslide, which occurred in 2014. In Snohomish County, Washington, a moderate slope of glacier deposits was located, about 200 m (650 ft) high, composed of soil and regolith. The slope led downward toward a state highway and the nearby North Fork Stillaguamish River. A middle-class community of 35 houses called Steelhead Haven was nestled among the trees on this slope. The month of March 2014 was much wetter than usual, thanks to 3 weeks of intense rainfall. On the morning of Saturday, 22 March, a landslide in the form of a slump, or rotational slide, hit the area. Many residents who were home at the time described a crashing, roaring sound. Some had time to put on their shoes and alert family members of the need to evacuate, while others were overtaken too quickly to escape the area. Forty-three people lost their lives that day, including both adults and children. About 283 million ft^3 (8 million m^3) of material was involved. The elevation of the hill at the head of the slide was reduced by about 88 m (290 ft). At the base of the slope where material accumulated, there was 75 ft (23 m) of elevation increase. Further investigation showed that this entire area along the river, which included the Oso Landslide area, was one large landslide complex. This indicated not only that landslides had occurred in the past, but that they are likely to occur well into the future.

The Oso Landslide traveled much faster and farther than one might have expected based on the angle of the slope. Scientists found that groundwater levels were very high on the floodplain below the slope. When the slump hit the floodplain and river below, it became saturated with water, then turned into a flow with sufficient energy to move a long distance over nearly flat topography. This was not a simple or typical slump in that it turned into a fast-moving debris flow, but it is an example of a complex mass-wasting event.

6.3.6 Debris Flows

The city of La Conchita, California, is known for its slumps that turn into debris flows. The small town is located in Ventura County about halfway between the cities of Ventura and Santa Barbara. Here, steep 590 ft (180 m) bluffs descend to a coastal plain below. The bluffs are composed of loosely cemented sands, silts, and clays. The combination of steep slopes, loose material in the bluffs, and rainfall have led to two debris flows within the last three decades. Following a moderate amount of rain in early 1995, the slope above the town became unstable. On 4 March, a slump triggered by saturation of the slope turned quickly into a debris flow, which continued for several hours. The piece of the slope that broke loose was 1,150 ft (350 m) long by 400 ft (120 m) wide. Approximately 46 million ft^3 (1.3 million m^3) of material moved downhill. The toe of the debris flow intruded about two blocks into the town of La Conchita at the base of the bluff. Nearly 400 ft (120 m) of a road that descends the bluff above La Conchita was also destroyed. Nine homes were buried or otherwise destroyed, but thankfully there was no loss of life.

When it comes to landslides, history often repeats itself. In late December of 2004 and early January of 2005, the area around La Conchita received record rainfall. On 10 January 2005, the 1995 slump remobilized and again transitioned into a flow, sending a 7.1 million ft^3 (200,000 m^3) debris flow into the town (Fig. 6.17). The toe of the flow extended farther into La Conchita than it did in 1995. This time, 13 homes were destroyed and 23 more were so severely damaged they could not be reoccupied. Ten people lost their lives, including a mother and her three children.

A close examination of the slope above La Conchita reveals that the entire bluff was once the site of a massive landslide. Smaller slumps

Figure 6.17 La Conchita, California, was the site of debris flows in both 1995 and 2005. This aerial photo was taken in 2005. Ten people lost their lives when the flow destroyed and buried homes. Photo courtesy of U.S. Geological Survey. From Arbogast, 2017.

are clearly visible above the town to the north of the 1995 and 2005 slide area. The slope is unstable in many areas, poised to send additional debris flows crashing down onto the homes below. Residents have learned to be both aware and wary when rain falls. Though geologists have advised the town's residents that the risk remains and that future debris flows are inevitable, many people choose to live with the risk and continue to make their home in La Conchita.

6.3.7 Flows

Perhaps the most deadly and dangerous of all types of mass wasting, historically speaking, is the *flow*. A flow is the chaotic movement of fluid and solid material downhill under the influence of gravity. The chapter opening case study highlighted the Caraballeda debris flow in Venezuela in 1999. A debris flow is usually composed of a mixture of materials including soil, regolith, rocks, and other materials that are picked up and carried along with the flow. The Caraballeda debris flow was triggered by incredibly heavy rainfall. However, flows can be triggered by other events as well, such as earthquakes.

Peru, 1970

On 31 May 1970, a magnitude 7.9 earthquake occurred near the coastal city of Chimbote, Peru. Approximately 80 km (50 mi) inland was the Santa Valley, named for the Rio Santa that flowed within. The valley held fertile farms, and many of the farm workers lived in villages on the edges of the valley. The Cordillera Blanca, which forms the western edge of the valley, is an impressive mountain range, rising more than 12,000 ft (6,667 m) above the valley below. The two highest peaks are the Huascaran Norte and the Huascaran Sud.

The earthquake happened on a Sunday afternoon when many families were out enjoying the weekend. When the earthquake hit, a portion of Huascaran Norte about 3,000 ft (1 km) wide broke free and fell downward. It rode on a cushion of air, allowing it to reach a speed of 155 mi/hr (250 km/hr) down the steepest face of the mountain. This fall included not only rock, but also ice and snow. The ice and snow melted, turning the fall into a huge debris flow. Before the flow reached the base of the mountain, it struck a high hill and split into two lobes. One lobe covered the town of Yungay, burying 18,000 people within the city beneath debris (Fig. 6.18). The other lobe buried the town of Ranrahirca. About 450 of its 500 former residents had been killed in a debris flow following an earthquake in 1962, so the population in 1970 was quite small. The 1970 flow killed most of the remaining residents in Ranrahirca. The debris flow reached the floor of the valley, and momentum carried it approximately 250 ft (75 m) up the mountains on the opposite side of the valley. Mud, boulders, trees, and other material were swept down the Rio Santa at an estimated 20 mi/hr (35 km/hr). The debris flow continued to take lives as it roared down the Santa Valley. It destroyed a rail line, overtaking a train and killing all aboard. In all, about 70,000 people died as a result of that earthquake, and most of the deaths were from debris flows.

Other Flows

As stated previously, flows can be composed of many types of material. We have so far discussed debris flows composed of soil, regolith, larger rocks, and water. Another type of flow is an avalanche, which consists of snow mixed with compressed air (Fig. 6.19). In yet another form of flow, water mixes with soil after heavy rainfall and becomes a mudflow. In still another, water mixes with volcanic ash and forms a lahar, or volcanic mudflow (Fig. 6.20). (Lahars are discussed as a volcanic hazard in Chapter 3.)

6.3.8 Subsidence

Another form of Earth movement that occurs, and is hard to characterize as any of the above forms, is *subsidence*. Subsidence occurs when

(a)

(b)

Figure 6.18 Aftermath of the debris flow in 1970 that killed 18,000 people in this town of Yungay, Peru. (a) High altitude photograph of the former city of Yungay. Debris covers nearly the entire city. The X in the center of the image is the former location of Plaza de Armas, which was at the center of the city. Image S3820 from the Steinbrugge Collection, courtesy of NISEE/Berkeley. (b) A ground view of Yungay after the debris flow in 1970. The cluster of four palm trees near the center of the photo is the location of Plaza de Armas. The peak in the distance is Huascaran Norte, which was the source of the debris in the flow. Photo by Lloyd Cluff. Image S3810 from the Steinbrugge Collection, courtesy of NISEE/Berkeley. (c) The remains of a bus and truck covered in debris from the flow that roared down the Rio Santa Valley near the Plaza de Armas in Yungay, Peru, following the magnitude 7.8 Chimbote, Peru, earthquake. Photo by Lloyd S. Cluff. From NISEE-Berkeley; https://nisee.berkeley.edu/elibrary/Image/S3814. Image S3814 from the Karl V. Steinbrugge Collection.

(c)

Figure 6.18 (Continued)

Figure 6.19 A snow avalanche is composed of snow and air. Photo by Golden Rowell/Mountain Light/Alamy. From Fletcher, 2014.

ground level is lowered. This happens most often because of groundwater activity. The first of the two types of subsidence we will examine is a sinkhole (Fig. 6.21), a catastrophic and relatively rapid subsidence that resembles a rock fall. When groundwater flows through limestone, which is composed of calcium carbonate, that calcium carbonate dissolves. This leaves holes in the rock, known as *karst*. Karst can be of a variety of sizes, from tiny holes in limestone called pores to large cave systems (Fig. 6.22).

In some instances, caves may form very near the surface of the Earth. If the roof of the cave is left unsupported and becomes weakened, the roof of the cave free-falls to the bottom of the cave, leaving a circular hole in the ground called a sinkhole. Sinkholes form in limestone terrains throughout the world. One area in which sinkholes are common is south Florida. In this region, sinkholes can instantly be recognized. When a sinkhole forms in this environment, it often fills with water, leaving a landscape dotted with circular lakes. Every now and then, a new sinkhole forms. In May of 1981, a new sinkhole opened in Winter Park,

Figure 6.20 A lahar is a mixture of volcanic ash and water. The lahar that created this deposit, which has covered the landscape nearly to the height of the top of the first floor of the building in the photo, originated at Mount Pinatubo in the Philippines during its June 1991 eruption. © Roger Ressmeyer/CORBIS. From Fletcher, 2014.

Figure 6.21 A string of sinkholes near Roswell, New Mexico. Sinkholes such as these are formed when carbonic acid in groundwater dissolves limestone. Photo courtesy of University of Washington Libraries, Special Collections, John Shelton Collection, KC6480. From Strahler, 2013.

Figure 6.22 Karst is a name for the features created both at and beneath the surface when carbonic acid in groundwater dissolves limestone. Karst can produce unique landforms including caverns, sinkholes, and eventually valleys formed by collapse of a series of adjacent sinkholes. This landscape is located in Guangxi Province, China. Towers of limestone (a) are left behind after rock around them has dissolved. The limestone that used to exist in the flooded valleys below (b) completely dissolved. Dissolved minerals were carried away by streams. From Strahler, 2013.

Florida. The media documented its formation and enlargement over the course of several days (https://www.usgs.gov/media/images/winter-park-florida-sinkhole-1981). Sinkholes can also form under other circumstances. They can happen when buried pipes burst, eroding soil from around the pipe and causing the overlying soil to collapse, creating a large hole. They can also happen when human-made cavities, such as old mine shafts or tunnels, collapse. Sinkholes can appear suddenly, or they may form over the course of a few days. They may be preceded by a small amount of subsidence or ground cracking. They are often, however, catastrophic occurrences that occur without much warning.

A second type of subsidence is one that happens when humans pump too much groundwater from the ground. Water is stored in pores between grains of sand or other particles. When a small amount of water is pumped from these pores to the surface, those pockets eventually refill with water after rain falls, replenishing the groundwater supply. However, if too much groundwater is taken from the ground, the pores that once held water can collapse. This causes two things to happen. One is that the cavity is smaller than it once was, meaning less groundwater can now be stored in that area. The second is that the ground surface subsides. Subsidence due to groundwater withdrawal is a relatively slow process, taking years to recognize. However, it can be quite profound.

Our case study of this phenomenon is the San Joaquin Valley of California. This has long been a heavily agricultural area. While the soil is rich and the climate is perfect for growing a wide variety of fruits, vegetables, and nuts, rainfall is insufficient for these crops. As a result, the valley is irrigated. Farmers pump water from the ground in order to irrigate their crops. The result is that parts of the San Joaquin Valley have subsided more than 20 ft since the onset of agricultural irrigation in the 1920s. While the lowering of the ground surface may not seem rapid or dangerous enough to be addressed in a book on natural disasters, it is nevertheless a form of geologic hazard and a cause for serious concern. Groundwater in the San Joaquin Valley is being mined. That is, it is being taken at a rate that far exceeds nature's ability to replace it. The over-pumping has also caused changes in the structure of the rocks beneath the valley, causing them to hold less water. Eventually, water will become more scarce in the valley because groundwater will be unavailable as an irrigation source. Farmers will have to turn to other sources for water to irrigate their crops, or stop growing crops in this fertile area due to the unavailability of water.

6.4 Mass-Wasting Triggers

Mass wasting is not a type of natural hazard that scientists are able to forecast with any certainty like we can with hurricanes or other types of severe weather. In many cases we do know the general areas where it is likely to occur, and we know the conditions that can trigger it. Slopes that have seen past mass-wasting activity are likely to experience it again in the future. However, timing of these events is rarely predictable. There must be a trigger, or something that provides the last bit of force needed to make the mass-wasting event happen. Common triggers for mass-wasting events include earthquakes, heavy rainfall, melting of snow and ice, and volcanic eruptions. A trigger might be as small as the growth of an ice crystal within a crack. The trigger could also be altered groundwater flow from within a previous landslide deposit that allows water to accumulate unseen. It is also true that with some mass-wasting events, no trigger is ever identified. In all cases, there is a trigger but at times the exact triggering mechanism is unidentifiable.

6.5 Mitigation of Hazards

The first step in mitigation of hazards is to understand the hazards faced within a community. The United States Geological Survey and many state level geological surveys maintain data on where mass wasting has occurred in the past, and thus where it is likely to occur in the future. Figure 6.23 is a composite map of the continental United States indicating where mass-wasting hazards are greatest. The map may surprise you. Nearly every state on the map is at risk of some form of mass wasting. It may take the form of slumping soil in the rolling hills of Iowa or rock falls and debris flows throughout the steep canyons within the mountains of western Colorado.

The easiest and safest way to mitigate hazards is to avoid them. This means employing smart land-use planning with the aim of avoiding all areas in which mass wasting has occurred in the past. Looking at the map in Figure 6.23, you would be right to be surprised that this would preclude people from living in western North Carolina, eastern Tennessee, and nearly all of West Virginia, as well as all of western Colorado and the entire west coast of the United States. As you have correctly surmised, this is not possible. Many millions of people live in these areas.

The next best solution is to enact local zoning laws that prevent people from building in areas that are unsafe. Sometimes this means leaving some areas free from development, using those spaces instead for wilderness areas, wildlife refuges, recreation, farming, and livestock grazing. In other areas, it may include requirements that in order to obtain a building permit on a specific plot of land, a geological

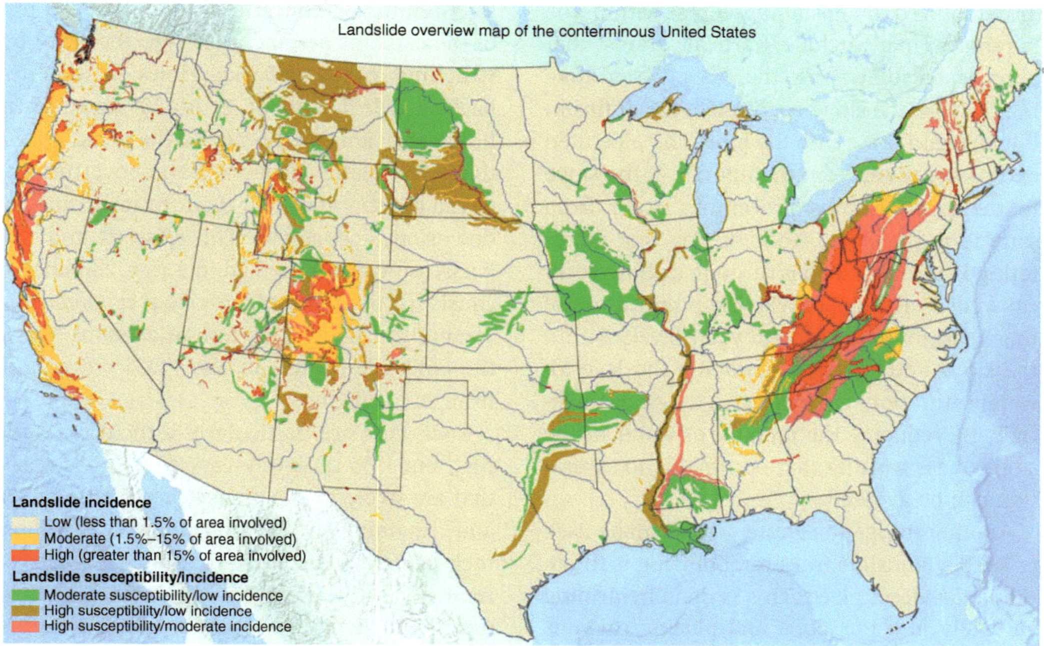

Figure 6.23 Map of landslide susceptibility in the contiguous United States. Courtesy U.S. Geological Survey. From Fletcher, 2014.

investigation, called a geotechnical survey, be undertaken. This type of scientific survey allows people to know before purchasing or building on a plot of land whether that land is likely to be involved in a landslide in the future. This information is also shared with insurance companies, and in most cases new structures built on landslide deposits become uninsurable without major mitigation measures (which can be expensive), taking away the possibility for most people of ever building on or occupying the land. Many communities with known landslide hazards have these laws in place to protect people, and to keep them from investing in land they cannot safely build on.

Failing avoidance, there are a number of mitigation strategies that can be employed to try to keep people safe in areas prone to mass wasting. Japan is one country that has invested heavily in early warning systems that raise alarms if rainfall exceeds levels known in the past to lead to mass landslides and debris flows. These thresholds have been determined by gathering data about past mass-wasting events and weather surrounding those events. This information is used to construct computer models that allow officials to warn people living in mountainous areas of an increased mass-wasting danger during heavy rainfall events. A number of other countries have followed suit, including Nepal, Korea, and Italy.

One way we can mitigate the hazards on slopes where mass wasting has already occurred is the process of buttressing. The process typically involves removal of the lower portion of the landslide and replacing it with compacted Earth fill behind a sturdy structure. Sometimes the entire landslide is replaced with compacted fill.

In slopes that are at risk of failing from saturation with water, some communities utilize a process of removing water from slopes, called dewatering. One way this is done is to drill into a slope horizontal boreholes, which are angled slightly downslope. These boreholes are stabilized by installing pipes within the boreholes that include vertical drains that allow water to flow into the pipes. Water flowing through the slope encounters the drains in the pipe and flows into the pipe. The pipe then carries the water from the interior of the slope to the exterior, thus preventing water from saturating the slope. Another way that dewatering is accomplished is through drilling vertical wells in the slope. Water is pumped out of the wells, lowering groundwater levels within the slope and therefore reducing the level of water saturation in the pores in the slope materials. By reducing the amount of water in the slope, mass wasting that occurs due to saturation can be avoided.

Another mitigation strategy that can be used to keep a landslide from remobilizing is to pin the landslide in place. This is done by drilling vertically into the slope and placing caissons (concrete cylinders), roughly 3 ft (1 m) in diameter, into the landslide deposit. They are extended well beyond the sliding surface at the base of the landslide (also called the failure plane). The caissons are then filled with steel-reinforced concrete. In addition, the top of the caissons are anchored into the slope by structures called tiebacks. Holes for the tiebacks are drilled about 20° to 30° down from horizontal and anchored to the rock below the failure plane. Tiebacks prevent overturning of the caisson and help them to resist further downslope movement. The caissons and tiebacks are typically drilled in a row, sometimes consisting of dozens of individual caissons, traversing the slope. This line of caissons keeps the landslide deposit in place and prevents it from remobilizing.

Rock falls are particularly difficult to forecast, but it is relatively easy to identify areas that are likely to experience them. Wire mesh and retaining walls are often used when rock-fall areas are located next to highways. In many hilly and mountainous areas, wire mesh, which resembles chain-link fencing, is draped over steep rock slopes (Fig. 6.24). The hope is that the mesh will slow down any rock falls and make the rocks less dangerous as they tumble toward the road. This may give passing drivers time to see the rock fall in progress and

(a) (b)

Figure 6.24 (a) Wire mesh is draped over a slope to protect the roadway below from falling rock. At the base of the slope is a cement barrier, placed there to catch smaller rocks that fall down the slope before they can reach the roadway. Photo © Marli Miller. (b) In this case, a slope is covered with wire mesh and a strong wire mesh catchment fence is located at the base of the slope to catch rocks before they enter the roadway. Photo © Marli Miller.

avoid hitting the debris. At the base of the slope, there may be a sturdy retaining wall (Fig. 6.24a), or a substantial mesh catchment fence (Fig. 6.24b). In other areas, sprayed cement called Shotcrete is also used to coat the slope, acting much like the wire mesh and holding soil and rocks in place so they have no chance to move downhill (Fig. 6.25).

Retaining walls as mitigation structures may take many forms. The type most of us think of is a wall of concrete, stone, or wood that is backfilled with earth, used to stabilize the base of a slope and prevent material from moving downhill. These retaining walls are commonly used in developed sites, including homes, businesses, and roadways. A structure sometimes used in conjunction with wire mesh is a short free-standing retaining wall like those often used in highway construction sites (see Fig. 6.24a). These are also known as K-rails, K-walls, or Jersey barriers. These are put in place on the sides of roads at the bases of slopes to prevent smaller rocks from reaching the road surface.

Yet another type of retaining wall is a structure called a gabion. A gabion is a wire cage that holds rocks. A gabion can act much like a concrete retaining wall and stabilize a slope. Gabions can also be used on the banks of streams to stabilize the slopes and prevent slumping (Fig. 6.26). They are strong enough in many cases to steer flow of a stream or even a small mudflow or debris flow.

In areas where rock slides are the most likely hazard, a typical scenario includes layers of rock tilted toward a roadway. Engineers have developed a mitigation strategy involving large bolts, called rock bolts, which are several feet long. They are drilled into the slope, anchoring the top layer of rock to a layer beneath and secured with a small metal plate. Engineers take into account the thickness and strength of the rock in the spacing of the rock bolts during installation. While this may not prevent slides, it may make them smaller and less frequent.

In the case of areas where karst is a distinct possibility, scientists can test for the existence

Figure 6.25 Sprayed cement, called Shotcrete, is used to stabilize slopes and limit mass wasting. © Marli Bryant Miller. From Fletcher, 2014.

Figure 6.26 Gabion used to stabilize a cut bank in a stream channel, to prevent erosion. Photo courtesy of Maccaferri USA.

of caves beneath the ground prior to the purchase of a piece of land or building on that land. Since cave systems can be completely closed and not visible from the surface, it is critical that scientists with the proper instruments be called upon to conduct a geotechnical survey to determine whether an area is underlain by a cave system before building structures in an area known to have karst. Consistently adhering to this practice ensures that new homes and businesses are not lost due to sinkhole formation.

Snow avalanches provide a separate and distinct hazard. Avalanches are to be expected anywhere with deep snow over a steep slope. Only the smallest trigger is needed to start an avalanche if enough snow is piled on a steep slope. Something as simple as a person stepping foot on the slope may be enough to dislodge thousands of cubic feet of snow. Skiers and snowmobilers are advised to use caution in backcountry areas because they can trigger avalanches. In heavy-use areas such as ski resorts, staff members can trigger controlled avalanches preemptively to prevent them from happening while people are on the slopes. They use an air cannon, artillery, or explosives to deliberately trigger an avalanche while the area is cleared of people. After the avalanche has occurred, the slope is safe for recreation for the day.

Another form of mitigation for avalanche hazard is a snow tunnel or snow shed. These structures are built in known avalanche corridors and are designed to provide a safe space during an avalanche. When an avalanche roars overhead, the avalanche tunnel is strong enough to survive the flow and provide shelter for anyone inside. These are used most often on highways traversing areas known to produce avalanches.

6.6 Conclusion

Mass wasting is the downhill movement of material under the influence of gravity. It takes on many forms, including both fast and slow movements. The major categories of mass wasting include creep, falls, slides, slumps, and flows. Mass wasting can involve any type of material, but is most common when water is involved. Saturation of a slope with water is one major cause of mass wasting.

There are a number of mitigation strategies that can be employed to protect populations in areas in which mass wasting is expected. Retaining walls, alteration and dewatering of slopes, rock bolts, wire mesh, geotechnical investigations, and preemptive triggering are all strategies that can be used to keep people safe.

6.7 Chapter in Review

Earth movements are also collectively known as mass wasting. Mass wasting is the downslope movement of material under the influence of gravity. Mass wasting occurs in many forms. The six major types are creep, falls, slides, slumps, flows, and subsidence. Mass wasting shapes our landscapes in a myriad of ways, including eroding of slopes and widening of valleys.

The slowest of all mass wasting is creep. Creep is the downslope movement of soil and regolith under the influence of gravity. It occurs because of shrinking and swelling in clays within the soil and freezing and thawing of water in the soil. Though creep is not a rapid, catastrophic form of mass wasting, it can be quite damaging and expensive. It affects even very gentle slopes of 10° to 15°, and causes displacement of fences, sidewalks, retaining walls, telephone poles, water lines, sewer lines, and foundations of buildings. Since these structures require repairs because of creep, this is an expensive type of mass wasting.

A second form of mass wasting is a fall. In this type of event, rock detaches from a steep slope and falls downhill, often with little or no ground contact. Pieces of rock often form a talus slope, which looks like an apron or rocks around the base of the larger slope. Rock falls are most common in areas with steep, rocky slopes.

A third form of mass wasting is a slide. A slide is characterized by material moving downhill above an inclined planar surface. Almost any type of material may be involved in a slide. One of the twentieth century's most memorable mass-wasting event was a slide near the newly constructed Vaiont dam in Italy in 1963. The dam was constructed in a steep, narrow valley in which rocks on one side were tilted toward the valley floor. When the reservoir behind the dam filled with water, clay layers within the rock sequence were saturated with water. The clays swelled and became slippery, creating a sliding surface. The rocks gave way and crashed into the reservoir, creating a large wave that inundated a nearby community.

A fourth type of mass wasting is a slump. Slumps are common mass-wasting events that occur when material moves downhill along a curved surface. Another name for a slump is a rotational slide. Slumps are most common in soft materials such as soil, silt, clay, or sand. They occur most commonly after a major rainfall event. They help shape our landscape as the primary cause of widening stream valleys. A major slump occurred in 2014 in the state of Washington. The Oso Landslide was actually a large slump that turned into a flow. It lowered the elevation of the slope by a few hundred feet and added to the elevation at the base of the slope. Later geotechnical investigation found that this was one small part of a larger landslide complex in the area. Slumps above the town of La Conchita, California, in 1995 and 2005 turned into debris flows, which caused destruction of homes and roadways, and in the 2005 debris flow, loss of life.

A fifth type of mass wasting is a flow. Flows are often the most deadly of mass-wasting events because they are often very rapid events. Heavy rainfall and saturation of a steep slope is one major trigger for a flow. Another common flow trigger is earthquake. The destruction of much of the city of Caraballeda, Venezuela, in December 1999 was from a massive debris flow caused by heavy rainfall in the mountains above the city. Much of the Santa Valley in Peru was affected by debris flows triggered by the 31 May 1970 earthquake centered near the city of Chimbote. Approximately 70,000 people died in that event, most from debris flows that affected the Santa Valley and mountains on either side.

A sixth type of mass wasting is subsidence. Subsidence can occur rapidly, such as during sinkhole formation. Sinkholes can be formed in multiple ways. They may form when thin roofs of caves in karst terrain collapse. They may also form due to the collapse of mine shafts or other human-made tunnels. They may also be created when water from a broken underground pipe erodes soil and creates a hole in the Earth. Subsidence can also occur slowly, over a period of decades. One way this happens is when too much groundwater is pumped from an area, compacting the material that once held the water. Over time, this lowers the elevation of the ground and eliminates groundwater storage. The San Joaquin Valley in California is one example of a location where this is happening.

Mitigation of mass-wasting hazards can take on many forms. The most effective way to prevent loss of life and property from mass wasting is not to build in areas where mass wasting is likely to occur. Failing this, there are a number of measures that can be taken to keep people and property safe. Some countries have embraced early warning systems that warn officials when rainfall amounts and rates exceed thresholds that caused mass wasting in the past. People may buttress slopes to stabilize them. They may use dewatering strategies to reduce water saturation within the slopes. Wire mesh can be draped onto steep slopes where rock falls are likely to occur. This slows rocks as they move downhill, hopefully giving people below a chance to see the rock fall in progress and avoid the area. Short retaining walls that are called K-rails, K-walls, or Jersey barriers can be placed next to roads to prevent small rock slides from reaching lanes of traffic. A different type of retaining wall meant to hold back soil and other loose debris can be used to prevent slumps and slides. A third type of retaining wall, called a gabion, is composed of a wire cage containing rocks. These can be used to stabilize slopes or steer flows into areas where they will do the least amount of damage. For slopes in which rock layers slant toward a populated area or a roadway, rock bolts can be drilled into the slope, stabilizing the rock face and preventing large slabs of rock from moving quickly downhill. Geotechnical surveys can also be used in many cases to detect past landslide activity or unseen karst.

Part II

Weather and Climate as Hazards

7

Atmosphere and Weather

Bethany D. Hinga

University of Nebraska at Kearney, USA

7.1 Introduction

Weather is defined as the state of the atmosphere at a specific time in a given location. Weather measurements we routinely make include temperature, relative humidity, atmospheric pressure, wind speed and direction, and precipitation type and amount. This set of measurements allows atmospheric scientists to not only describe the state of the atmosphere as it exists at the present time, but also to forecast what weather conditions will exist in a particular location in the near future. In this chapter, we will investigate weather in its many forms, and prepare you for the remainder of this section of the book, which covers the hazards that exist as a result of weather phenomena and climate change.

7.2 The Earth's Atmosphere

The Earth is enveloped by a life-giving, temperature-regulating *atmosphere* (Fig. 7.1). The atmosphere is a relatively thin layer of air that surrounds the Earth. It is composed of about 78% nitrogen, 21% oxygen, 0.9% argon, 0.04% carbon dioxide, and trace amounts of other gases.

The atmosphere is divided into four layers (Fig. 7.2). Closest to the Earth is the *troposphere*. The troposphere contains about 99% of all air molecules within the atmosphere. The troposphere is where we live, and where the vast majority of the Earth's weather takes place. The temperature of the troposphere drops dramatically with increasing elevation above the Earth. You may have seen evidence of this while flying in a plane. If you have watched the Outside Conditions screen in the in-flight entertainment, you have probably noticed the extreme cold temperatures outside the airplane. It is not unusual to see air temperatures of −50° F (−46° C) at the altitudes at which jets fly, around 7 mi (37,000 ft or 11 km) above the Earth. The top of the troposphere is called the *tropopause*. It is found approximately 10 mi (16 km) above the surface of the Earth.

Above the tropopause is a layer called the *stratosphere*. The stratosphere extends from the top of the troposphere (10 mi or 16 km above the Earth) to a height of about 31 mi (50 km). Within the stratosphere the temperature rises with increasing altitude. At the top of the stratosphere, the temperature is not much lower than it is at sea level! The stratosphere is home to the Earth's *ozone layer*, a protective layer of ozone that shields the surface of the Earth from excess ultraviolet radiation. The top boundary of the stratosphere is called the *stratopause*.

Above the stratopause is the *mesosphere*. The mesosphere extends from the top of the

Earth's Natural Hazards and Disasters, Advanced Textbook 7, First Edition. Bethany D. Hinga.

Figure 7.1 Photograph of the Earth from space. The atmosphere is a thin layer of gases above the surface of the planet. Image from NASA Johnson Space Center. From Arbogast, 2017.

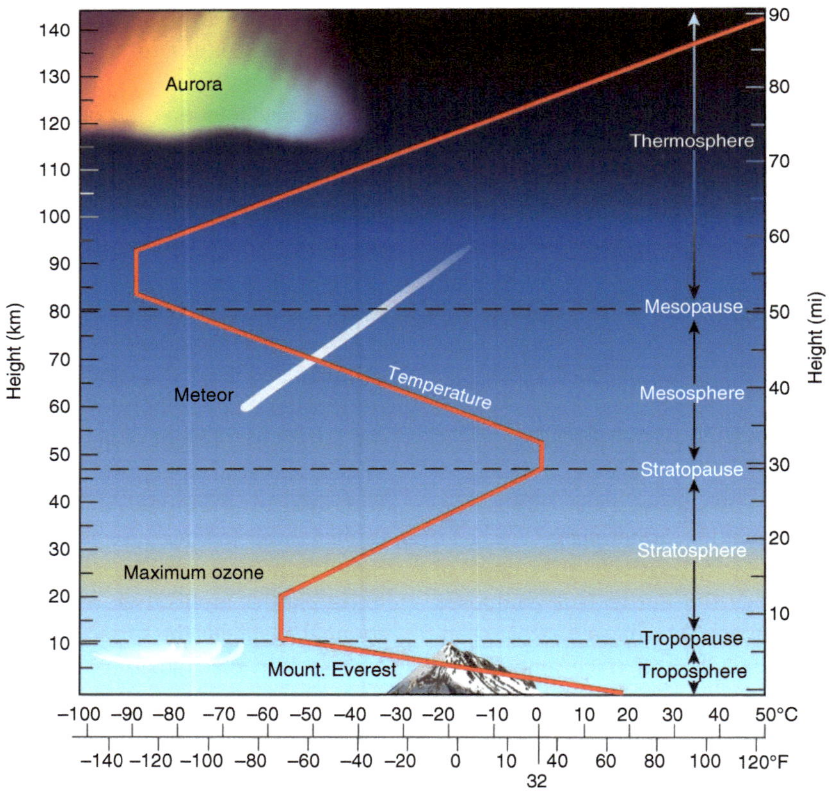

Figure 7.2 Vertical structure of the atmosphere, including temperatures. From Fletcher, 2014.

stratosphere at 31 mi (50 km) to about 50 mi (80 km) above the Earth's surface. As it does in the troposphere, the temperature of the atmosphere cools with increasing elevation within the mesosphere. At the top of the mesosphere is the *mesopause*. The temperature of the mesopause is, on average, the lowest temperature within the Earth's atmosphere at −139° F or −95° C. Above the mesopause is the *thermosphere*. Here, air temperature trends again are reversed, and the temperature increases with increasing altitude within the thermosphere. The thermosphere extends from the mesopause to an altitude of between 311 and 621 mi (500 and 1,000 km). The altitude of the top of the thermosphere is greater during times of increased sunspot activity and less during times of decreased sunspot activity.

Gravity pulls the majority of air molecules close to the surface of the Earth. As a result, about 99% of the atmosphere's air molecules are found in the lowermost layer of the atmosphere, the troposphere. Air is densest at the lowest elevations on the Earth, and becomes less dense with increasing altitude. The weight of a column of air from the surface of the Earth to the top of the atmosphere is called *atmospheric pressure*. Atmospheric pressure is greatest at or below sea level and decreases exponentially with elevation.

Many people are sensitive to changing altitudes as they drive or hike up tall mountains. In extreme cases, people may experience altitude sickness in which a lack of oxygen causes health problems. Many people experience severe headache, dizziness, shortness of breath, nausea, and vomiting. The only remedy for altitude sickness is to descend to a lower elevation. Mountaineers who spend time climbing mountains usually give themselves time for their bodies to acclimate to increasing elevations by ascending slowly, sometimes over a period of days. Those who climb the highest peaks in the world, such as Mount Everest and K2, not only take weeks to acclimate to the altitude, but in most cases also use supplemental oxygen to help them reach the summits of these mountains. Without it, they risk disorientation, severe altitude sickness, and death.

7.3 Solar Radiation

The Earth receives its warmth and light from the Sun in the form of *solar radiation*. Solar radiation travels to us as waves, and we can classify those waves on the *electromagnetic spectrum* (Fig. 7.3) based on their wavelengths. Our eyes have evolved to see wavelengths within the visible light spectrum. Radiation with wavelengths longer than those within the visible light spectrum are termed "infrared." *Infrared* radiation is not visible, but we can feel it in the form of heat. Scientists have developed sensors capable of detecting and measuring heat via infrared radiation. Moving to longer wavelengths, we encounter microwaves. Microwaves are used in microwave ovens to heat food and beverages. They are also used in radar. Radio waves, which have longer wavelengths than microwaves, are most familiar to us in the transmission of our favorite AM or FM radio stations.

Radiation with wavelengths that are shorter than those on the visible light spectrum are *ultraviolet*. Ultraviolet radiation is firmly within the human consciousness because we know that ultraviolet waves cause sunburn and skin damage. Ultraviolet radiation can be helpful, however, as it is excellent at killing germs. It is used in food preparation and water purification. Those in the aquarium trade often use ultraviolet sterilizers in their tanks. The sterilizers kill algae and harmful microorganisms and help keep fish and invertebrate populations healthy. X-rays are used in medical and dental imaging. Gamma rays are used to treat some forms of cancer. Overexposure to any of these shorter wavelengths of radiation can cause cell damage and even death.

When solar radiation enters the Earth's atmosphere, it can be absorbed or reflected back out into space. *Albedo* is the term for reflectance of solar radiation. Some materials

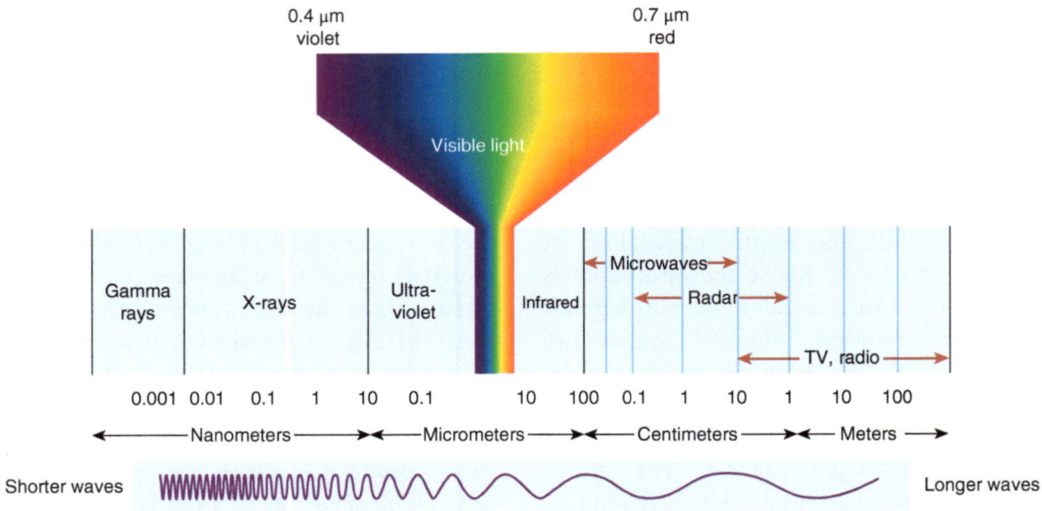

Figure 7.3 The electromagnetic spectrum. Solar radiation bathes the Earth in electromagnetic energy. We see visible light, but energy also comes to us in the form of ultraviolet (UV) and infrared energy, radio waves, X-rays, and gamma rays. From Arbogast, 2017.

have high albedo, meaning a high percentage of solar radiation that strikes these materials is reflected (Fig. 7.4). Light-colored objects typically have higher albedos than dark-colored objects. Snow and ice have very high albedos, up to 90% in smooth snow and ice fields. Asphalt has one of the lowest albedos of any material, as low as 5% to 10%. Most other materials are somewhere in between. Forests may have an albedo of 10% to 20% depending on type and density of foliage. Bodies of water have albedo ranges from 10% to 60%, depending on Sun angle. During sunrise and sunset when the Sun is low on the horizon, albedo is high. As the Sun rises toward midday, the albedo is reduced. Clouds can also reflect a great deal of energy away from the Earth's surface. Thick clouds have a high albedo while thin, wispy clouds have low albedo.

Solar radiation can also be scattered within the Earth's atmosphere. When radiation is scattered, it reflects off tiny particles, ice crystals, water droplets, and gas molecules in the atmosphere and is sent in many different directions. Clouds are responsible for scattering much radiation, as are dust particles and aerosols. *Aerosols* are tiny droplets of various compounds such as sulfur dioxide within the atmosphere. Remember from Chapter 3 that energetic volcanic eruptions can propel sulfur dioxide aerosols high into the atmosphere. The scattering that takes place as a result of a large volume of sulfur dioxide aerosols can cause absorption of solar radiation to be severely reduced, creating temporarily cooler conditions on the Earth's surface.

If solar radiation is not reflected or scattered, it is absorbed either by the atmosphere or the Earth's surface. Roughly 30% of solar radiation is reflected back into space, and the atmosphere absorbs about 23% of incoming solar radiation. The remaining 47% of solar radiation is absorbed by the Earth's surface.

Absorption of solar radiation occurs at different levels within the Earth's atmosphere. In the stratosphere, most of the Sun's ultraviolet radiation (such as ultraviolet B, or UVB radiation) is absorbed by the Earth's protective ozone layer. It is because of this layer within the Earth's stratosphere that life is able to exist on the surface of the planet. Without the ozone layer, too much ultraviolet radiation would

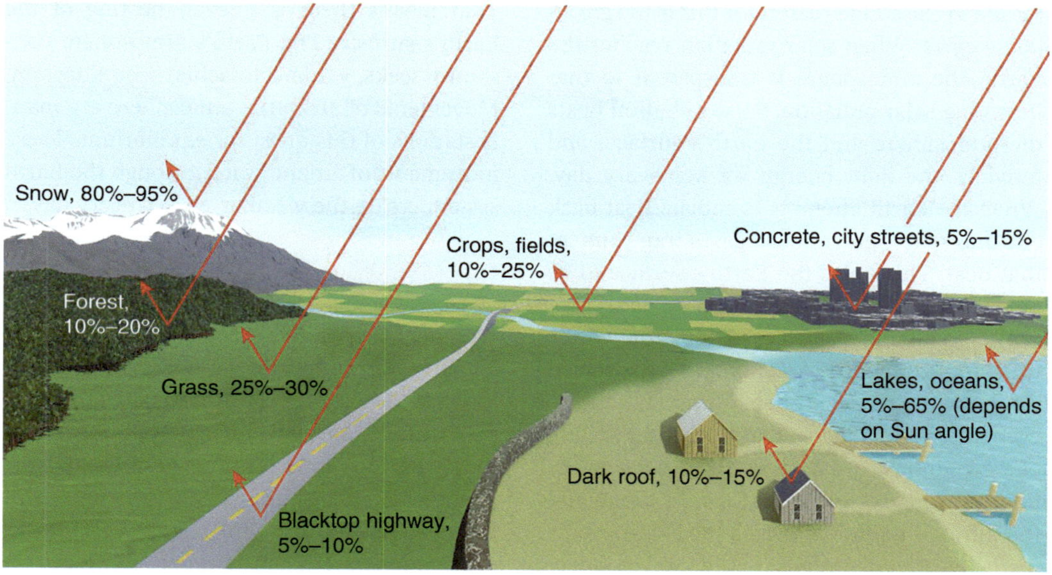

Figure 7.4 Albedo varies across the surface of the Earth. In general, darker areas absorb more energy and have low albedo. Brighter areas like snow and ice reflect a high percentage of energy, and have high albedo. (Length of arrow is not proportional to magnitude of albedo.) From Arbogast, 2017.

reach the Earth's surface, and cells within all life on Earth would be damaged. However, *ozone*, is not universally good. Ozone in the stratosphere is necessary for life on Earth. Ozone in the troposphere is a source of pollution and a major component of smog. It reduces air quality and endangers those with weakened or damaged respiratory systems.

You may have read about a hole in the ozone layer above the Earth. This is the result of environmental damage from decades of chlorofluorocarbon emissions and other assaults on the ozone layer. During the summer, the ozone layer above the polar region becomes thinned. It is so thin that in many years it disappears completely near the pole. Remember that ozone filters out the short-wavelength ultraviolet B (UVB) radiation from the Sun. The lack of ozone in these areas allows UVB radiation to reach the surface of the Earth. This is damaging to cells in living organisms. In areas with diminished ozone above, there is a higher incidence rate of nonmelanoma skin cancer. In addition, UVB plays a major role in the development of malignant melanoma. While the ozone hole is a serious environmental problem, it is not related to climate change.

Solar radiation that is absorbed by the Earth is converted to heat energy. Therefore, the Earth is constantly warmed by the Sun. However, a certain amount of energy leaves the Earth system every day. Solar radiation strikes only the part of the Earth that faces the Sun. For that reason, temperatures tend to drop at night as heat is emitted but there is no incoming solar radiation. However, temperatures do not drop as severely as they would if the Earth did not have an atmosphere. The Earth's Moon, for example, does not have an atmosphere. When the Sun hits the face of the Moon, the temperature can reach 260° F (127° C). During nighttime on the Moon, the temperature plunges to −280° F (−173° C).

Near the surface of the Earth, absorption of solar radiation within the atmosphere results in conversion of the radiation to heat energy. For this reason, the Earth's surface is habitable, which means it has the right temperature range for life to exist. One reason for this is the *greenhouse effect*. When solar radiation reaches the Earth, the atmosphere is transparent to that incoming solar radiation. Solar radiation heats the atmosphere and the Earth's surface, and provides the light energy we see every day. When the Earth attempts to radiate heat back out into space, greenhouse gases trap some of that heat. This helps the Earth's surface to be held within the habitable range.

The greenhouse effect was named after the warming seen within greenhouses constructed for growing plants under warm conditions, even in cold climates. The walls and ceiling of a greenhouse are transparent to solar radiation, so we can see into and out of them, and the infrared radiation warms the space inside. As the Earth radiates heat in the form of longer wavelength (infrared) radiation, that heat is trapped inside the glass walls of the greenhouse creating warm conditions within.

Greenhouse gases are increasing in volume within our atmosphere. As the volumes of these gases increase, they are causing the Earth to become warmer, much as the inside of a greenhouse warms. The climate change we are experiencing right now is a result of an excess of human-produced greenhouse gases. Greenhouse gases do occur naturally, and they keep the Earth comfortable for us. However, we are seeing the balance of those greenhouse gases shift dramatically. We now have more carbon dioxide, a greenhouse gas, in our atmosphere than has existed for almost 5 million years. This is causing our climate to change. Climate change and its relationship to our atmosphere (in particular the greenhouse gases in our atmosphere) will be discussed in depth in Chapter 11.

7.4 Weather

Solar radiation is what fuels the weather we see every day. Some areas of the planet receive more solar radiation and others receive less. That means there is uneven heating of the Earth's surface. The Earth's atmosphere constantly seeks, yet fails to achieve, equilibrium. Movements of air in the atmosphere are manifestations of this quest for equilibrium. These movements of air and water through the Earth system create the weather we see every day.

7.4.1 Weather Measurements

Weather observations are the basis for all studies of weather phenomena and climate. They indicate the status of the atmosphere at a snapshot in time, and give forecasters information they then use to determine how the atmosphere will change in specific locations over the next several days. In today's world, most observations are taken remotely using instruments that are in direct digital communication with National Weather Service (NWS) offices and other forecasting locations. A standard set of measurements is taken at the Earth's surface at a minimum of four times per day, at set 6 hr intervals dictated by international agreement. The agreement states that official measurements are made at 0000, 0600, 1200, and 1800 UTC. The numbers 0000, 0600, and so on, are 24 hr times. UTC specifies the time zone as Universal Coordinated Time, previously known as Greenwich Mean Time. The Eastern Time Zone in the United States is five hours behind UTC, or UTC-5. (The Eastern Time Zone is UTC-4 during daylight savings time). That means that observations in the Eastern Time Zone would actually be made at 0100 (1 a.m.), 0700 (7 a.m.), 1300 (1 p.m.) and 1900 (7 p.m.) local time. The Pacific Time Zone on the West Coast of the United States is 8 hr behind UTC (UTC-7 during daylight savings time). Observations would be made at 0400 (4 a.m.), 1000 (10 a.m.), 1600 (4 p.m.), and 2200 (10 p.m.) local time. In reality, many of our weather observation systems are automated now. Most readings are taken much more frequently than once every 6 hr.

Standard weather parameters recorded at each weather station include temperature, relative humidity, dew point, cloud cover, visibility, precipitation type and amount, atmospheric pressure, and wind speed and direction. Observations can be made manually by human observers, by automated instruments, or by a combination of human observers and automatic instruments. In the United States, it is common for surface weather stations to be almost entirely automated. In the 1990s, Automated Surface Observation Systems (ASOS) were introduced (Fig. 7.5). These automated weather stations are located at many of the nation's airports and National Weather Service offices. Though official observations are made four times per day, these ASOS stations allowed weather data to stream continuously. This was a major advance for the National Weather Service in that it freed scientists from making their observations, and allowed them to view continuous data, which is invaluable on unusually active weather days.

Scientists also use additional data sources to monitor the weather. This includes weather balloons, which are launched and carried high into the troposphere. They gather data about weather well above the Earth's surface. Weather satellites are in constant use across the globe, monitoring many aspects of the Earth's weather and the status of the atmosphere. Research aircraft also take direct measurements of the atmosphere. Data collected by all of these sources are used to populate computer models that help forecast future weather conditions.

Temperature is measured by a digital thermometer and is reported in degrees Fahrenheit or degrees Celsius. Daily maximum and minimum temperatures are reported routinely. *Relative humidity* is measured by an instrument called a psychrometer. It is reported as a percentage of humidity the atmosphere is carrying versus the maximum it could carry at a particular temperature. For example, 40% relative humidity means the atmosphere is carrying 40% of the moisture it could carry at that specific temperature. Warmer air can carry more moisture than colder air, so relative humidity is dependent on temperature. For example,

Figure 7.5 Weather station at Furnace Creek in Death Valley National Monument. Forecasters in the nearby National Weather Service office are able to report typical weather parameters such as temperature, relative humidity, dew point, atmospheric pressure, wind speed and direction, sky cover, and precipitation type and amount. Photo credit: GIPhotoStock/Photo Researchers, Inc. From Strahler, 2013.

100% relative humidity at a cold temperature will have a great deal less moisture in the atmosphere than 100% relative humidity at a warmer temperature. For that reason, dew point is considered to be a much better measurement of how much moisture is in the air.

Dew point is the temperature to which, for the amount of moisture currently in the air and for a constant air pressure, the air must be cooled to reach 100% relative humidity. At this temperature, dew or frost will begin forming on surfaces because the air cannot hold any more water vapor. Dew point directly measures the water vapor content of the air. The higher the dew point, the greater the amount of water vapor in the air.

Cloud cover is measured either by a sky-facing camera or by a human observer who estimates the percentage of the sky that is covered in clouds. Visibility is measured as a distance in the horizontal direction. It can range from unlimited visibility when the sky is clear to near-zero visibility in heavy fog or a severe blizzard. While automated systems can accurately assess the percent cloud cover, a trained observer has to determine the type of clouds present.

Clouds come in four basic varieties (Fig. 7.6). Cirrus clouds are found high in the atmosphere. The prefix *alto* refers to clouds in the midlevels of the atmosphere. Stratus clouds are blanketing clouds. Cumulus clouds are white, puffy clouds. Clouds with the designation *nimbus* carry precipitation. Often, cloud type and altitude are conveyed by combining terms. For example, a cirrostratus cloud is a blanketing cloud high in the atmosphere. An altostratus cloud is a blanketing cloud in the midlevels of the atmosphere. Puffy clouds in the midatmosphere seen on pleasant summer days are called altocumulus. Thunderstorm-producing clouds are called cumulonimbus clouds.

Precipitation amount is measured by a rain gauge and reported in inches or centimeters. There are three main types of rain gauges. Cylinder gauges are commonly used by the public and consist of a cylinder that is placed outside and allowed to collect precipitation. Weighing bucket gauges, as you probably surmised, weigh the precipitation to determine the amount of rain that has fallen in a particular location. Most commonly used at weather stations are tipping-bucket rain gauges. The tipping-bucket rain gauge allows automatic measurements of precipitation amount to be collected nearly continuously. The bucket has a known volume. Once the bucket is full, it tips and empties the water leaving it ready to record the amount of additional precipitation. Because of its design, it is possible to record rainfall amounts that are even greater than the volume of the measuring bucket.

Wind speed is measured by an instrument called an anemometer and is reported in miles per hour or kilometers per hour. Anemometer design has advanced in recent years. For decades, an anemometer was a series of cups that caught the wind and made the instrument spin. The rate of spin was converted to wind speed. It required an additional instrument, a wind vane, to determine wind direction. Today, there are ultrasonic anemometers that rely on pressure differences on either side of the instrument to detect not only wind speed but also direction.

Atmospheric pressure is measured using an instrument called a barometer. Historically, atmospheric pressure has been reported in inches of mercury. This is from an antiquated barometer design in which a tube of glass with a closed end was inverted and submerged in a container of mercury. When atmospheric pressure rose, it pushed down on the mercury and forced mercury far up into the glass tube. When atmospheric pressure fell, the level of mercury forced into the tube fell as well. Today, most barometers are digital. Atmospheric scientists use millibars as the standard unit of atmospheric pressure. One bar, or 1,000 millibars, is the average atmospheric pressure at sea level. Millibars can be converted to inches of mercury, so on television weather reports in the United States, inches of mercury is often still

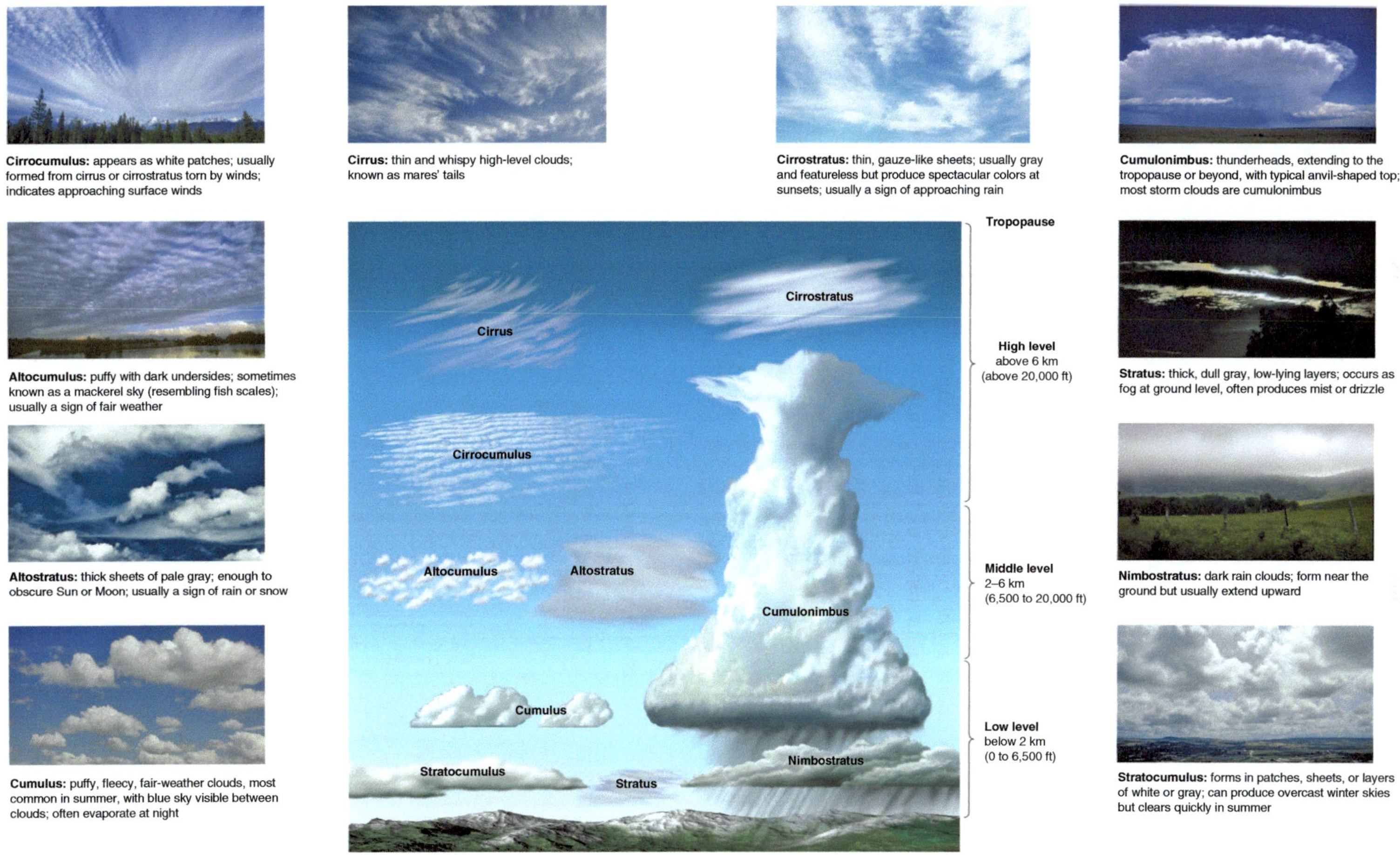

Figure 7.6 Common cloud types and their positions in the atmosphere. Arbogast, 2017.

used to describe atmospheric pressure. A barometer measures the weight of a column of air from the surface of the instrument up to the top of the atmosphere. In general, lower atmospheric pressures indicate bad weather such as storms and rainfall, whereas higher atmospheric pressure indicates fair weather. Take care in assuming that fair weather from higher atmospheric pressure means the weather is beautiful and mild. High pressure can accompany bitter cold or a deadly heat wave.

Multiple weather parameters can be considered together to give us information about specific conditions we can expect. For example, on hot summer days, you may hear television meteorologists talk about a *heat index.* Heat index is calculated using measurements of temperature and relative humidity (Fig. 7.7). For a given temperature, increases in humidity make the day feel hotter than it actually is. This apparent temperature (or "feels like" temperature) is the heat index. On days with high heat indexes, people are advised to moderate outdoor activity and drink plenty of water. On cold days, wind speed becomes a concern. Temperature and wind speed are used to calculate *wind chill* (Fig. 7.8). For a given temperature, increased wind speed makes the day feel colder than it actually is. This apparent temperature is wind chill. Low wind chill can hasten the onset of frostbite.

Major National Weather Service offices also are equipped with *Doppler weather radar.* Microwaves are emitted from an instrument located on a tower, above the ground. The waves travel outward in all directions until they encounter moisture in the atmosphere. The waves are then reflected back to the tower. The reflections are recorded and images are compiled using computers. Meteorologists use the weather radar data to augment their surface observations. This allows them to track weather systems in real time and issue up-to-the-minute forecasts.

The Doppler designation of the weather radar system is named for the Doppler effect, which is a situation that arises when an object is in motion. If an object is moving toward the

Heat index (Apparent temperature)

Air temperature, °F \ Relative humidity, percent	40	45	50	55	60	65	70	75	80	85	90	95	100
110	136												
108	130	137											
106	124	130	137										
104	119	124	131	137									
102	114	119	124	130	137								
100	109	114	118	124	129	136							
98	105	109	113	117	123	128	134						
96	101	104	108	112	116	121	126	132					
94	97	100	103	106	110	114	119	124	129	135			
92	94	96	99	101	105	108	112	116	121	126	131		
90	91	93	95	97	100	103	106	109	113	117	122	127	132
88	88	89	91	93	95	98	100	103	106	110	113	117	121
86	85	87	88	89	91	93	95	97	100	102	105	108	112
84	83	84	85	86	88	89	90	92	94	96	98	100	103
82	81	82	83	84	84	85	86	88	89	90	91	93	95
80	80	80	81	81	82	82	83	84	84	85	86	86	87

With prolonged exposure and/or physical activity

Extreme danger
Heat stroke or sunstroke highly likely

Danger
Sunstroke, muscle cramps, and/or heat exhaustion likely

Extreme caution
Sunstroke, muscle cramps, and/or heat exhaustion possible

Caution
Fatigue possible

To convert the heat index from °F to °C, use the following equation: $°C = (°F - 32) \times 5/9$

Figure 7.7 Heat index chart. To find the apparent temperature, plot the actual temperature and relative humidity. Where those two meet, the heat index, or apparent temperature, is found. For example, for a temperature of 92° F and a relative humidity of 60%, the heat index is 105° F. From Strahler, 2013.

Wind speed, mph \ Temperature, °F	40	35	30	25	20	15	10	5	0	−5	−10	−15	−20	−25	−30	−35	−40	−45
5	36	31	25	19	13	7	1	−5	−11	−16	−22	−28	−34	−40	−46	−52	−57	−63
10	34	27	21	15	9	3	−4	−10	−16	−22	−28	−35	−41	−47	−53	−59	−66	−72
15	32	25	19	13	6	0	−7	−13	−19	−26	−32	−39	−45	−51	−58	−64	−71	−77
20	30	24	17	11	4	−2	−9	−15	−22	−29	−35	−42	−48	−55	−61	−68	−74	−81
25	29	23	16	9	3	−4	−11	−17	−24	−31	−37	−44	−51	−58	−64	−71	−78	−84
30	28	22	15	8	1	−5	−12	−19	−26	−33	−39	−46	−53	−60	−67	−73	−80	−87
35	28	21	14	7	0	−7	−14	−21	−27	−34	−41	−48	−55	−62	−69	−76	−82	−89
40	27	20	13	6	−1	−8	−15	−22	−29	−36	−43	−50	−57	−64	−71	−78	−84	−91
45	26	19	12	5	−2	−9	−16	−23	−30	−37	−44	−51	−58	−65	−72	−79	−86	−93
50	26	19	12	4	−3	−10	−17	−24	−31	−38	−45	−52	−60	−67	−74	−81	−88	−95
55	25	18	11	4	−3	−11	−18	−25	−32	−39	−46	−54	−61	−68	−75	−82	−89	−97
60	25	17	10	3	−4	−11	−19	−26	−33	−40	−48	−55	−62	−69	−76	−84	−91	−98

Frostbite times: 30 minutes, 10 minutes, 5 minutes

To convert the wind chill index from °F to °C, use the following equation: °C = (°F − 32) × 5/9

Figure 7.8 Wind chill chart. To determine wind chill, or the temperature it feels like outside with the wind blowing, plot the temperature and wind speed. Where they meet, the number given is wind chill in degrees Fahrenheit. For example, an actual temperature of 10° F and a wind speed of 20 mph yields a wind chill of −9° F. From Strahler, 2013. https://www.weather.gov/bou/windchill

source of the waves, the wavelengths are shortened. If that same object is moving away from the source of the waves, the wavelengths are lengthened. The same is true if the object is stationary and the source of the waves is moving. This phenomenon can occur with any type of waves. One place in which the Doppler effect can be recognized is at a train crossing. As you sit in your car at a train crossing, the approaching train blows its whistle. The pitch of the whistle rises as the train approaches. The reason the pitch of the train whistle rises is because the wavelengths of the sound waves are being shortened as the train approaches. Shorter wavelength means the sound will have a higher pitch. As the train passes, the pitch of the whistle immediately lowers. This is because the train is moving away from you, and the wavelengths are getting longer. Longer wavelength means the sound will have a lower pitch.

In Doppler weather radar, the microwaves do not just record when they bounce into moisture in the sky (such as in a cloud). They also record whether the object is moving toward the radar station or away from it. Using this additional information, meteorologists can determine the wind direction within a storm and identify rotation. Rotation within a storm can indicate dangerous conditions that are precursors to tornado formation. This is very valuable information to have in real time. The use of Doppler weather radar has resulted in increased warning time before tornadoes strike because rotation in the clouds can be detected before it is seen on the ground. This has resulted in increased lead time in tornado warnings, allowing more people to take cover before a tornado strikes. Doppler weather radar has saved lives.

7.4.2 Planetary Air Circulation Patterns and the Coriolis Effect

As mentioned previously, solar radiation affects the Earth unevenly. The areas near the poles receive less sunlight than the equator, and so temperatures are markedly different near the poles than they are at the equator. The Earth's atmosphere serves as a giant conveyor belt, moving air from one part of the globe to another, seeking to even out the temperatures across the planet. Air flows northward and southward away from the equator toward the poles, warming the polar areas. Air also flows in the other direction, from the poles toward the equator, cooling equatorial regions.

Large-scale circulation within our atmosphere consists of three cells: Hadley cells, Ferrel cells, and Polar cells (Fig. 7.9). Nearest the equator is the *Hadley cell*. Hadley cells exist both north and south of the equator. Here, warm air is pulled upward into the atmosphere. In regions where it is filled with moisture, clouds form above the equatorial regions and drop precipitation. As the air moves away from the equator, it becomes drier and slightly cooler. It sinks near about 30 degrees north and south (30° N and S) of the equator, creating a high-pressure zone. Where the two Hadley cells meet at the equator is called the *Intertropical Convergence Zone* (ITCZ). The ITCZ is not positioned exactly at the equator. Its location changes with the seasons. It wanders northward during the Northern Hemisphere's summer, and southward during the Southern Hemisphere's summer.

Further from the equator, the second set of circulation cells forms. These are called *Ferrel cells*. Ferrel cells pull air downward from near 30° N and S of the equator and move it poleward to about 60° where it rises next to the Polar cells to their north and south. A high-pressure zone found at the boundary between the Hadley cells and Ferrel cells inhibits moisture from falling in many areas of the world, creating an arid or semiarid zone over many continents between about 20° and 40° N and S (Fig. 7.10). Most of the world's largest deserts, including the Mongolian and Sahara deserts, exist within this latitude range. At the northern and southern ends of the Ferrel cells form the

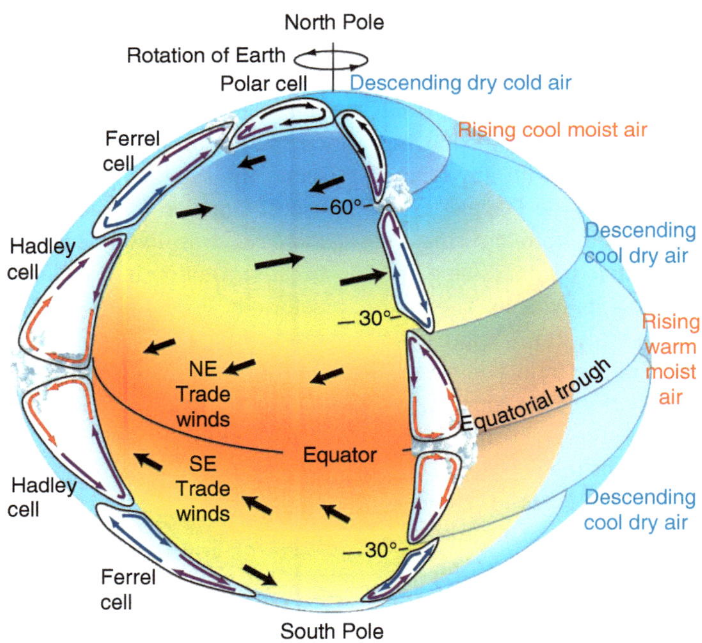

Figure 7.9 Global circulation patterns within the troposphere. Circulation cells are roughly defined by latitude. Polar cells are found between 60° N and the North Pole, and between 60° S and the South Pole. Between 30° and 60° N and S are Ferrel cells. Between 30° N and S and the equator lie the Hadley cells. From Fletcher, 2014.

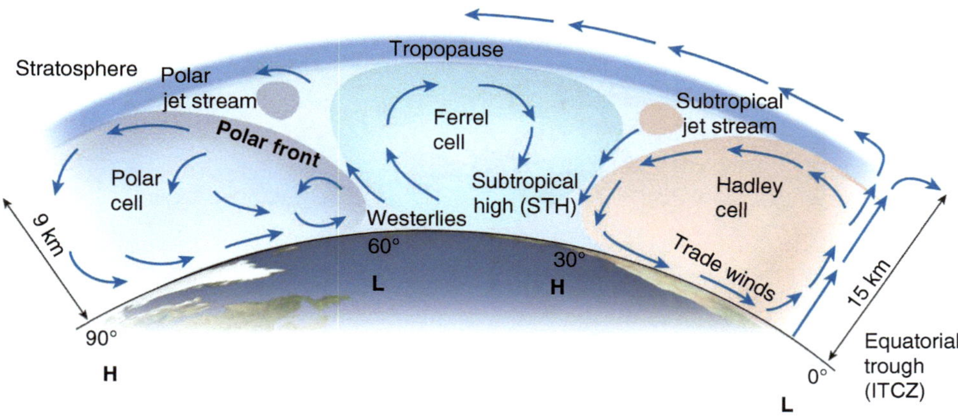

Figure 7.10 Large-scale atmospheric circulation. Warm, moist air near the equator rises. There it cools and condenses into clouds. Rain falls over the tropics. Cool, dry air descends around 30° north and south of the equator, creating conditions that are right, in some parts of the world, for deserts to form. The polar jet stream is located near the boundary between Polar and Ferrel cells. The subtropical jet stream is located between the Ferrel and Hadley cells. From Arbogast, 2017.

third set of circulation cells. These are the *Polar cells*. They extend from 60° N and S to the poles. Polar cells pull cold air from the poles toward the equator, bringing cool temperatures and moisture with them. Precipitation is more common in areas where winds near the surface are converging and subsequently rising. This occurs near the Intertropical Convergence Zone (ITCZ) and near 60° N and S latitude.

High in the troposphere, positioned between the Hadley cells and Ferrel cells, is a feature called the *subtropical jet stream*. A jet stream is a fast-moving stream of air that moves from west to east across the globe (Fig. 7.11). The subtropical jet separates the warmer air of the equatorial regions from more temperate air in the midlatitudes. The *polar jet stream* is found where the Ferrel cells and Polar cells meet. It separates the cold air of the polar regions from the more temperate air of the midlatitudes. The location of each jet stream changes frequently, both seasonally and within seasons.

On a nonrotating globe, air might move in an orderly fashion directly north and south from equator to pole and back again. However, the Earth does rotate, and that rotation causes deflection of air currents as they move above the planet. The *Coriolis effect* is a phenomenon in which motion of air within the atmosphere is deflected to the right in the Northern Hemisphere and to the left in the Southern Hemisphere (Fig. 7.12). In reality, atmospheric circulation within the Northern Hemisphere's Hadley cell is not directly north away from the equator, but is deflected to the right. That means that between the equator and approximately 30° N, upper-level winds blow approximately from the southwest to the northeast. Even that is not universal because surface level air masses have properties that can cause drag on the atmosphere above and change the direction of winds aloft. Seasons also change circulation patterns at the boundary between the troposphere and the stratosphere, causing variation in weather behavior patterns throughout the year. Whether an area of the Earth consists of land or ocean has an influence on air circulation as well. The Earth's atmosphere is a three-dimensional puzzle of global-level circulation, prevailing surface winds, and air masses with strong identifying characteristics like high or low pressure, high or low humidity, and high or low temperature. The result may seem

(a)

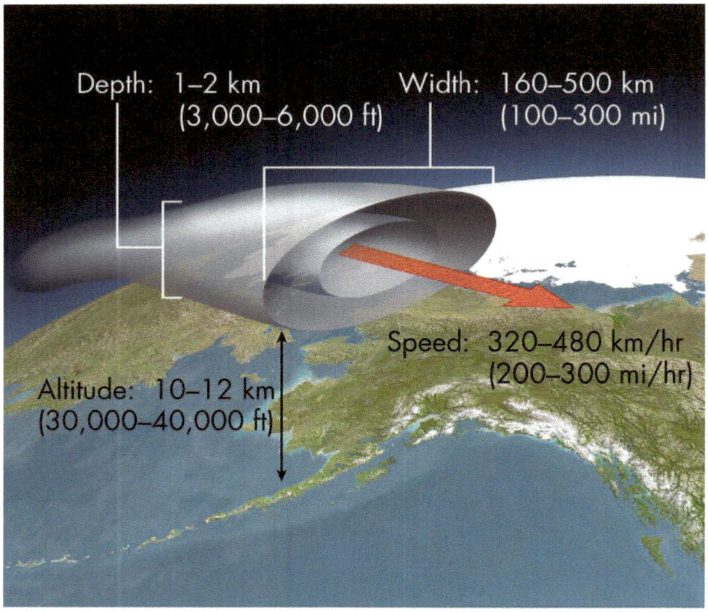

(b)

Figure 7.11 (a) A jet stream is a fast-moving stream of air that flows at high speeds relative to the air around it. (b) The jet stream in this photograph is identified by the band of clouds. Science Photo Library/NASA/Photo Researchers, Inc. From Arbogast, 2017.

Figure 7.12 The Coriolis effect is caused by the rotation of the Earth. It causes air to be deflected to the right in the Northern Hemisphere and to the left in the Southern Hemisphere. The Coriolis effect is responsible for rotation around high and low pressure systems, and for planetary scale air circulation. From Arbogast, 2017.

chaotic, but the reality is that each part of the Earth has predictable weather patterns governed by these atmospheric characteristics.

7.4.3 Surface Circulation of the Atmosphere

Figure 7.13 depicts the surface level circulation of the atmosphere across the globe. Near the equator is the Intertropical Convergence Zone (ITCZ) where the trade winds north and south of the equator converge. The trade winds within the tropics blow from the east to the west. In the midlatitudes are subtropical *anticyclones*. Poleward of the subtropical anticyclones are the westerlies. Near the poles are the polar easterlies. Where the westerlies and the polar easterlies come together, the warmer, less-dense westerlies override the cooler, more-dense polar easterlies. This boundary where the two air masses meet is known as the polar front. The polar front does not exist

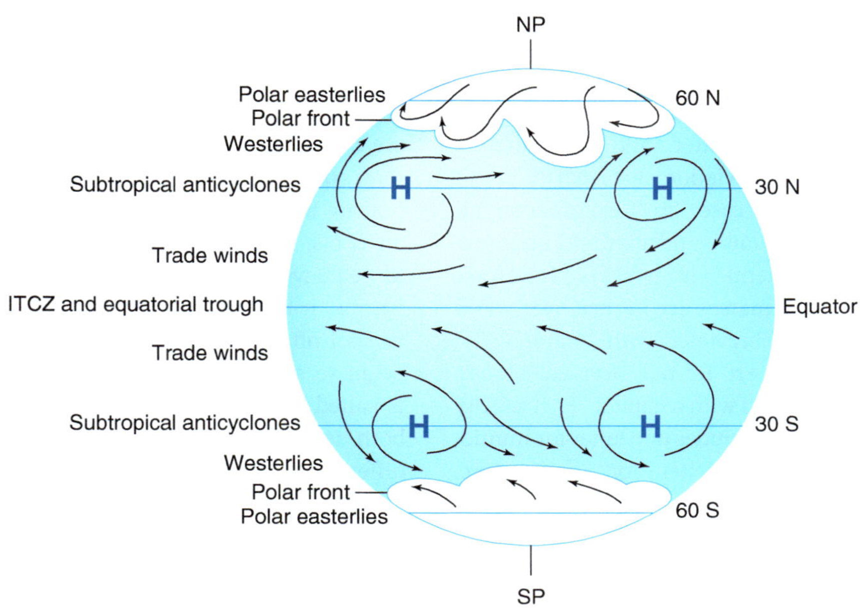

Figure 7.13 Idealized planetary-scale surface circulation of the atmosphere. H stands for high pressure. From Moran, 2012. Used with permission.

everywhere, and is not always terribly well defined even when it does exist. When the temperature difference across the boundary is great, the front is well defined. When the temperature difference across the boundary is small, the front may not be present at all. In areas where the front is well developed, this becomes an excellent place to form *extratropical cyclones*. Extratropical cyclones are cyclonic weather systems, which means they rotate counterclockwise (in the Northern Hemisphere) around a low pressure center. They are called extratropical because they form in the midlatitudes. Extratropical cyclones are typically a few hundred miles across. *Tropical cyclones* are large cyclonic storms that rotate around a low pressure center, but they form at low latitudes in the tropics.

7.5 Air Masses and Fronts

All over our planet, air masses exist. An *air mass* is a body of air that has similar temperature, humidity, and atmospheric pressure throughout. Air masses may be cold or warm and moist (humid) or dry. They are named for the areas over which they form (Fig. 7.14). Maritime air masses form over water, and are therefore humid. The abbreviation for maritime air masses is a lowercase "m." Continental air masses form over land, and are dry. The abbreviation for continental air masses is a lowercase "c." Tropical air masses are formed over warm areas, and are abbreviated with a capital "T." Polar air masses are formed over cold areas, and are abbreviated with a capital "P." Description of air masses includes one letter for humidity and a second letter for temperature. Continental polar air masses, for example, are abbreviated cP and contain cold, dry air. Maritime tropical air masses are abbreviated mT and contain warm, moist air. Continental tropical air masses are warm and dry, and are abbreviated cT.

Air masses are, by definition, different from the air around them. As an air mass moves into an area, the sharp contrast between the air mass and surrounding air creates a distinct boundary that is often the location of weather events. The boundary between air masses of different properties is called a *front*. The front is named for the type of air mass that is moving into an area. For example, if a cold air mass moves into an area occupied by warmer air, the front is called a cold front. A warm air mass that intrudes on an area occupied by a cooler air mass is a warm front. A cold front that descends from Canada onto the Great Plains during the month of February might bring temperatures that are significantly colder. In the summertime, the arrival of a cold front may not, despite its name, bring truly cold air. What it may do instead is bring a shift in wind direction and slightly stronger breezes. Temperatures may not be more than a few degrees different before and after passage of a cold front in summertime.

Warm air and cold air behave differently. Cold air is more dense than warm air, so cold air tends to sink, or to hug the ground as it advances into an area. Cold air that sinks has higher atmospheric pressure than warm air. Warm air is less dense than cold air, so it tends to rise above cooler air as it advances into an area. Warm air has lower atmospheric pressure than cold air. Sinking of cold air and rising of hot air together form *convection* cells. When air cools, it sinks. As it sinks and gets closer to sea level, it begins to warm. As the air warms, it rises. When that air rises, it begins to cool off. At some point the cooler air will begin to sink. This vertical circulation of air is called convection. Convection is related to both temperature and atmospheric pressure differences.

Cold fronts often bring stormy weather (Fig. 7.15). A strong cold front is one that has a high contrast in the temperatures on either side of the front. The stronger the cold front, the more likely it is to bring bad weather. Remember that cold air tends to sink or hug the ground as it advances into an area. When a cold air mass intrudes on a warm, moist air

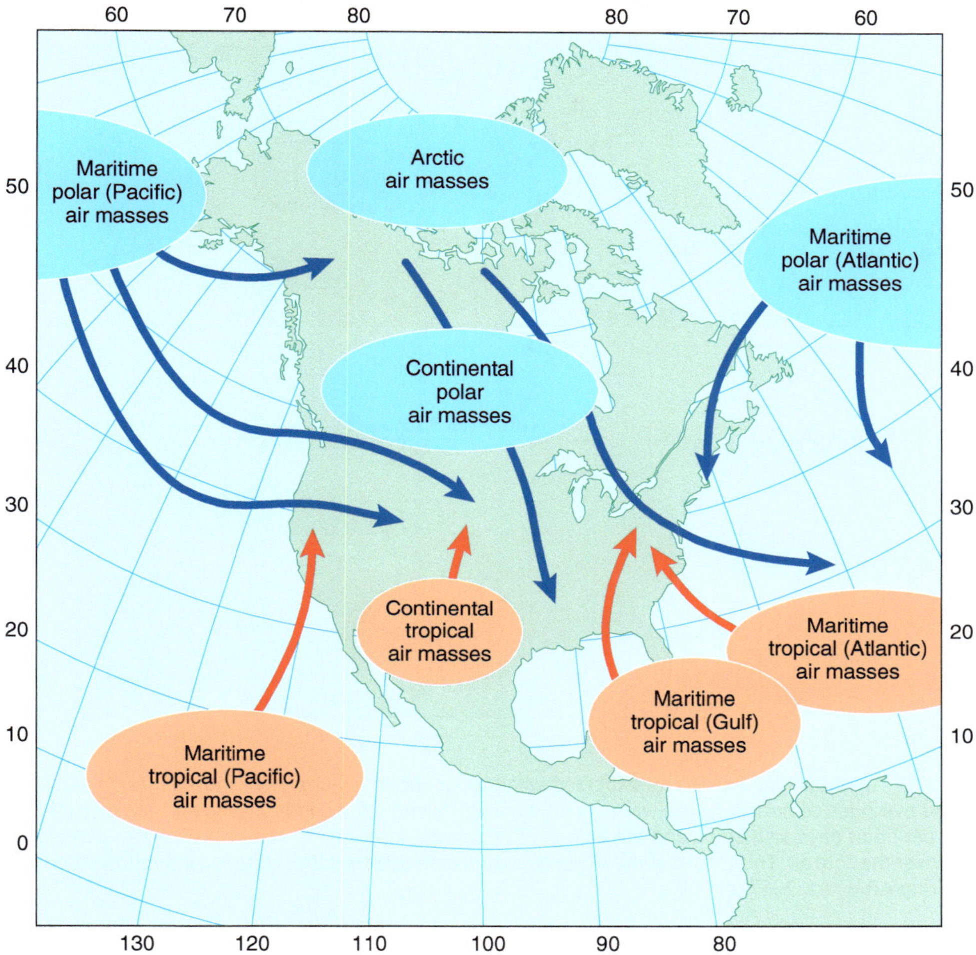

Figure 7.14 Commonly formed air masses over North America. The prefix "m" means maritime. Maritime air masses contain a great deal of moisture, so they are humid. The prefix "c" means continental. Continental air masses form over land, so they are dry. P refers to polar, so these air masses are cold. T refers to tropical, so these masses are warm. Therefore, an mP air mass is maritime Polar, or cold and moist. From Moran, 2012. Used with permission.

mass the front is a sharp, nearly vertical boundary where the warmer moist air is abruptly forced up and over the colder air. The moisture in the previously warm air begins to cool. It will condense and form clouds that can bring snow storms in the winter, and rain and thunderstorms in the spring, summer, and fall.

When a warm front moves into an area, the result is typically much less dramatic. The warm air moves up and over the existing cold air. The boundary between the two air masses has a shallow angle, so the warm air rises gradually over the cold air. The result is rain showers or the appearance of stratus clouds for some length of time. If the front moves very slowly, observers may notice cirrus clouds first, then approaching altostratus clouds, then stratus clouds and rain showers. This process may take several hours or even a day or two.

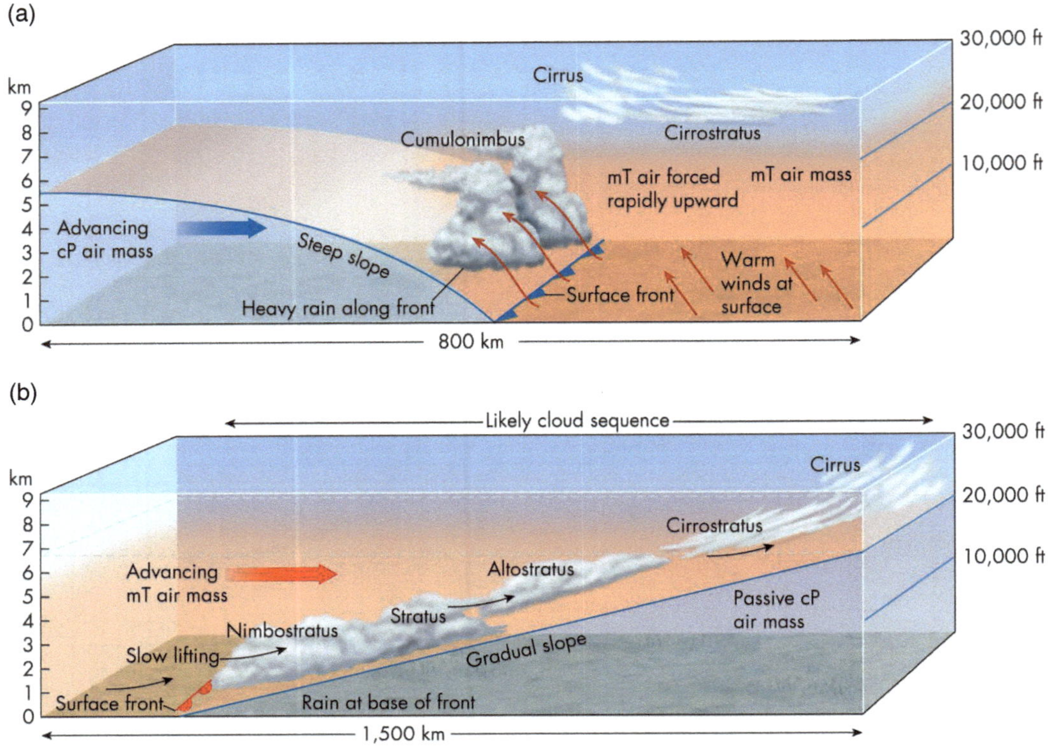

Figure 7.15 Approaching fronts and expected weather. (a) When a warm front moves into an area, the warm air overrides cooler air. As the warm air rises, moisture condenses and rain showers are often the result. Over a period of days, rain stops and the clouds lift. (b) When a cold front approaches, warm air is forced up and over the cold air. This results in rapid condensation of moisture in the warm air, and thunderstorms often form. From Arbogast, 2017.

7.6 High and Low Pressure Systems

In addition to air masses moving across the surface of the Earth, high and low pressure systems are also on the move (Fig. 7.16). As we learned earlier in this chapter, high pressure systems often bring fair weather and low pressure systems may bring precipitation. In addition to these features, we also see that wind patterns are fairly predictable around high and low pressure systems. Wind blows from areas of high pressure to areas of low pressure. The greater the change in atmospheric pressure, the stronger the wind will blow. So if a strong low pressure system moves into an area, the wind will blow very hard.

Atmospheric pressure changes not only because of fronts and air masses moving through regions, but also because of seasons and due to wind and temperature changes. Some areas of the world experience a monsoon climate, in which part of the year is dry and the other part of the year is rainy. In India, for example, people depend heavily on monsoon rains for agriculture as well as water for drinking and household use. More than 80% of the year's precipitation typically falls during the monsoon rains. Monsoons occur because of seasonal changes in the heating of the land and the ocean surface. In the spring and summer, relatively cool air over the ocean and warm air over the land create a horizontal air pressure gradient in which the air pressure is

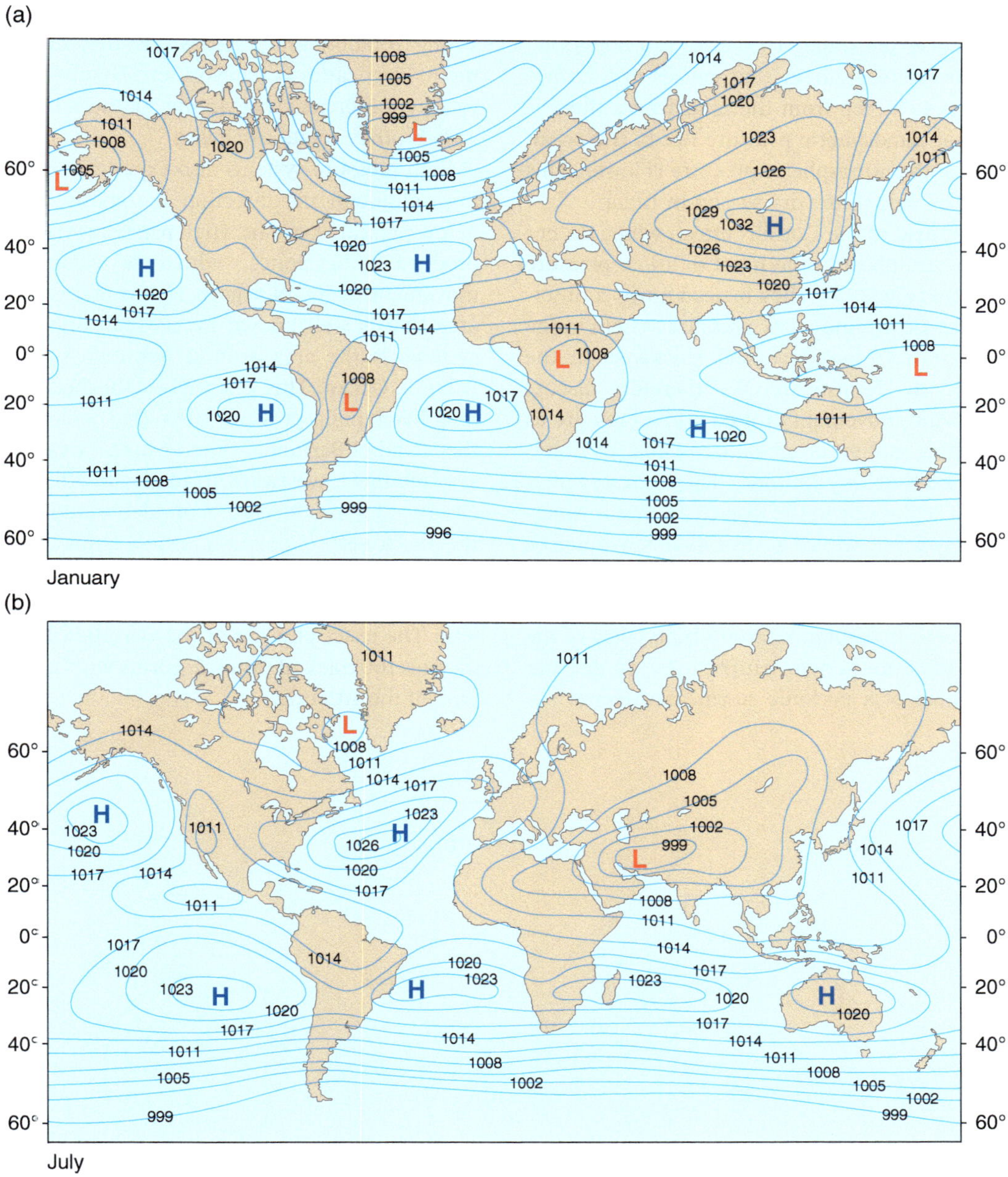

Figure 7.16 Typical global high and low pressure systems in (a) January and (b) July. Numbers on the maps are millibars of atmospheric pressure. From Moran, 2012. Used with permission.

higher over the sea and lower over the land. This causes warm, humid air from the high pressure area to move from the ocean to the land where there is lower atmospheric pressure. This humid air rises, cools and condenses into clouds, and causes rain to fall. Therefore, the spring and summer are the rainy season in India. In the fall and winter, the land becomes cooler than the ocean. Cooler air over the land condenses and descends, creating higher pressure over the land and lower pressure over the water. The result is a horizontal air pressure gradient that causes dry surface winds to sweep from the land where there is higher atmospheric pressure out to the ocean where the atmospheric pressure is lower. This corresponds to the dry season in India.

Because of the Coriolis effect, winds do not blow straight into the heart of a low pressure system (Fig. 7.17). They instead move counterclockwise around a low pressure system in the Northern Hemisphere and clockwise in the Southern Hemisphere. We therefore discuss the wind movement around a low pressure as cyclonic motion or *cyclonic rotation*. Cyclones can exist at any latitude. They can occur in polar regions or in the tropics. In North America, cyclonic winter storms that move across the continent in the winter months often have a comma shape. The shape of the storm system is due to that rotation around the heart of the low pressure system. Hurricanes are another example of a cyclonic storm. Hurricanes are low pressure systems, and therefore have very strong rotation and high winds that define them. Hurricanes in the Northern Hemisphere rotate counterclockwise. In the Southern Hemisphere, they rotate clockwise.

High pressure systems also experience rotation, although the direction is opposite that of low pressure systems. High pressure systems produce anticyclones, which is clockwise movement in the Northern Hemisphere. One area that is well known for its high pressure systems is the area around Bermuda in the Atlantic Ocean. This is a well-established and long-lived high pressure area that sits near the island of Bermuda. When tropical weather systems move westward through the Atlantic Ocean, they are steered around Bermuda by the clockwise circulation around this high pressure system. Tropical storms typically travel just to the south of this high pressure system. The exact position of the Bermuda High dictates the track of tropical storms within this area of the Atlantic.

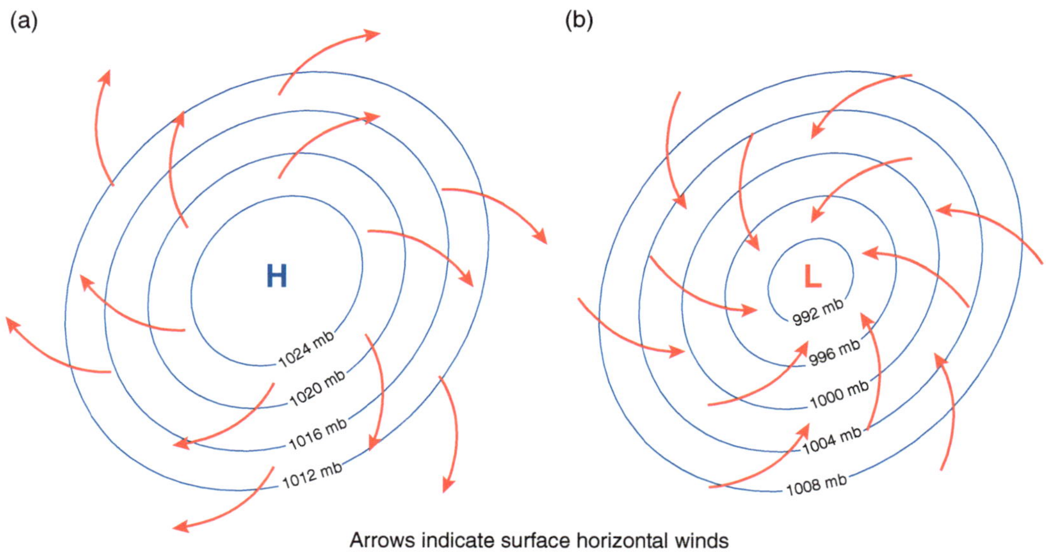

Figure 7.17 Rotation of air around high and low pressure cells. Winds blow clockwise around a high pressure cell, which is an anticyclone (a). Winds blow counterclockwise around a low pressure cell, which is a cyclone (b). From Moran, 2012. Used with permission.

7.7 Conclusion

In this introductory chapter to the weather section of this book, we have examined the many characteristics, measurements, and movements of the atmosphere that create the weather we experience every day. While each location on Earth may experience different weather on any given day, weather patterns are predictable from one season to the next and are forecastable. We can use weather measurements taken by scientists, automated instruments, weather balloons, and satellites to build a model of the atmosphere. Multiple measurements allow the model to recognize trends in the data. This model is then used to forecast what weather conditions can be expected in a few minutes, hours, or days. The ability to forecast weather, as we will see in subsequent chapters, allows people to prepare for adverse weather well in advance of these weather events.

This chapter serves as the foundation for discussing many types of severe weather, including heat waves, winter storms, thunderstorms, tornadoes, floods, and hurricanes. Convection, horizontal pressure gradients, heating and cooling, movement of air masses, and the Earth's rotation combine to form weather systems. In each of the next three chapters you will discover the unique conditions that bring about extreme weather events, and learn how to keep yourself and your family safe when severe weather strikes. Following those discussions is a chapter dedicated to climate change, which is causing changes in our global weather patterns. Any discussion of severe weather is incomplete without spending some time discussing how our changing climate is bringing about more extreme weather events.

7.8 Chapter in Review

The Earth's atmosphere is a relatively thin layer of air that surrounds the planet. The composition of the atmosphere is about 78% nitrogen, 21% oxygen, 0.9% argon, 0.04% carbon dioxide, and trace amounts of other gases. The Earth's atmosphere is composed of four layers with transition zones between them. The lowermost layer is the troposphere. The troposphere contains about 99% of all air molecules within the atmosphere and is where the vast majority of our weather occurs. The temperature of the atmosphere drops with increasing altitude in the troposphere. The top of the troposphere is called the tropopause. The next layer of the atmosphere is the stratosphere. In the stratosphere, the temperature increases with increasing altitude so the top of the stratosphere is almost as warm as the surface of the Earth at sea level. The stratosphere is also the location of the Earth's ozone layer. The ozone layer absorbs most of the Sun's ultraviolet radiation, making the surface of the Earth a more hospitable place to live since levels of ultraviolet radiation at the surface of the planet are fairly low. The top of the stratosphere is called the stratopause. The layer above the stratosphere is called the mesosphere. Temperatures drop again within the mesosphere such that the top of the mesosphere is the coldest part of the Earth's atmosphere. The outermost layer of the Earth's atmosphere is the thermosphere. In the thermosphere, the temperature trend is again reversed, and the thermosphere becomes warmer with increasing altitude.

Our weather is caused by differences in solar heating around the globe. The atmosphere seeks equilibrium, which means cool air moves from the poles toward the equator, and warm air moves from the equator toward the poles. This movement of air causes the weather we see at the surface.

The Earth is warmed by solar radiation. Solar radiation comes in a broad spectrum of wavelengths. This is called the electromagnetic spectrum. We see light from the Sun, which is a form of electromagnetic radiation located within the visible light spectrum. Some light is reflected back into space. Reflectance of a material is called its albedo. Snow and ice can have an albedo of 90% while dark objects like asphalt have albedos of less than 10%. Warmth comes primarily from wavelengths longer than those of visible light, termed infrared radiation. Ultraviolet radiation comes from wavelengths shorter than those of visible light. The Earth has a moderated climate because of the greenhouse effect, which is a phenomenon in which the Earth's atmosphere retains heat radiated by the Earth. Without the greenhouse effect of our atmosphere, temperature extremes would make it impossible for life to exist on the planet. However, increases in carbon dioxide and other greenhouse gases from human activity are causing the greenhouse effect to be intensified, which means higher temperatures worldwide. Higher temperatures are causing changes in global weather patterns and climate.

Weather is the set of atmospheric conditions that exist in a specific location at a given time. Scientists make a number of surface measurements of the Earth's atmosphere a minimum of four times each day. These include temperature, relative humidity, dew point, percent cloud cover, visibility, precipitation type and amount, wind speed and direction, and atmospheric pressure. In addition, satellites constantly monitor the Earth, as do radar installations. Weather balloons also provide data taken directly from the atmosphere above the Earth's surface. These measurements define the state of the atmosphere at that location and time of day. Scientists can use this information to forecast weather conditions. On hot days, temperature and humidity are used to calculate a heat index. Heat index is an apparent temperature and reflects the fact that higher humidity causes it to feel warmer on a hot day. Wind chill is also an apparent temperature, but is calculated using temperature and wind speed. On cold days, higher winds causes the air to feel colder than the actual temperature.

Scientists track movement of storms and other atmospheric phenomena using Doppler weather radar. Doppler weather radar allows scientists to detect not only precipitation in the atmosphere, but also the direction the precipitation is moving. When scientists see this rotation within clouds, they can issue tornado warnings before tornadoes touch down, saving lives.

The atmosphere is always in motion. Globally, there are three cells per hemisphere that move moisture and heat from one part of the atmosphere to another. Closest to the equator are Hadley cells, which transport heat and moisture poleward. Between about 30° and 60° north and south latitude are Ferrel cells. These cells bring cooler, drier air from higher latitudes toward the lower latitudes. Polar cells bring warmer air toward the poles. Between the Hadley and Ferrel cells a subtropical jet stream forms. This is a fast-moving stream of air in the upper troposphere that travels from west to east across the Earth. Between the Ferrel cells and Polar cells is the polar jet. It is another fast-moving stream of air in the upper troposphere, which also travels from west to east. The polar jet moves poleward in the summer and toward the equator in the winter. It generally separates areas of cold air from areas of warmer air.

The Coriolis effect is a result of a rotating globe. It is a phenomenon in which air is diverted to the right in the Northern Hemisphere and to the left in the Southern Hemisphere. Air moves from areas of high pressure to areas of low pressure. The greater the difference in pressure from one location to another, the harder the wind blows. The air does not move in a straight line. Instead, it is deflected and causes weather systems to rotate. Air moves counterclockwise around a low pressure system and clockwise around a high pressure system in the Northern Hemisphere. Rotation directions are opposite in the Southern Hemisphere. Hurricanes are large low pressure systems, so the winds around hurricanes rotate counterclockwise in the Northern Hemisphere. Rotation around a low pressure system is called cyclonic rotation, and anticyclonic rotation is characterized by rotation around a high pressure system.

Circulation of air in the atmosphere is called convection. Cool air tends to sink. As cool air sinks, it is warmed at lower elevations and rises. When the warm air rises and cools at higher elevations, it begins to sink. This is convection. Convection is also related to air pressure differences. Cool air has higher atmospheric pressure, and warm air has lower atmospheric pressure. When there is an atmospheric pressure gradient (a difference in air pressure between two locations), air moves from areas of high atmospheric pressure to areas of low atmospheric pressure. Seasonal horizontal pressure gradients cause monsoon climates.

Bodies of air exist that are distinct from the air around them. Air masses may be warm or cold, and moist (humid) or dry. These air masses move across the planet, and interactions with the surrounding air at their boundaries often brings weather events. When a warm, moist air mass moves into an area of cooler air it can bring rain showers. When a cold air mass moves into an area of warm, moist air it can bring severe thunderstorms. Boundaries between air masses are called fronts.

8

Severe Weather

Bethany D. Hinga

University of Nebraska at Kearney, USA

8.1 Blizzard on the Plains

In the post-Civil War era, the United States government sought to expand settlement into the area west of the Mississippi River. It offered free land to homesteaders who paid a modest filing fee and farmed the land for 5 years. The program brought pioneers westward into the Great Plains. Immigrants from as far away as Germany and Ukraine sold their belongings and headed to America on ships to take advantage of this offer.

Life on the prairie was hard. Much harder, in fact, than they had been led to believe it would be by those who enticed them to take advantage of this free land. Each family typically bought a pair of oxen to help them break the sod and plow the ground so they could plant crops. They also brought chickens and cows to help provide milk and eggs, and horses to haul wagons. The soil was rich, so hopes were high for a bountiful harvest in the fall. Some years brought that bountiful harvest. In other years, summer brought drought and fast-moving fires. In still other years, late in the summer when crops were nearing maturity, huge clouds of grasshoppers descended on the fields and ate everything green they could find. In unfortunate years, all of the hard work of the spring and summer seasons spent in the fields yielded little more at harvest time than the amount of grain the pioneers had used to seed their fields. When winter came, there was little food. Fuel for heating their homes came from the hay they cut during the summer. They twisted it into small bundles that they fed into the stoves in their sod huts. Humans and animals alike suffered from hunger. And then there was the cold.

Winters on the prairie in the 1800s were brutal. On the worst of days, the temperature slipped to −10° F to −20° F (−23° C to −29° C), and snow blanketed the landscape. Every few weeks a winter storm would come through the area carrying with it a new course of snow and strong winds that created large drifts. After each snow storm, the skies would clear and the real cold would settle in. Weather observers across Montana, the Dakota Territory, and Nebraska routinely measured temperatures of −40° F during cold snaps.

The winter of 1887–1888 was typical of many. The cycle of winter storms and extreme cold started early that year, well before Thanksgiving. Debilitating cold held a firm grip on the region. On 11 January 1888, another few inches of snow fell in the eastern half of the Dakota Territory, Nebraska, Minnesota, and Iowa. When people woke up the next morning, however, they noticed that the temperature had risen considerably from the day before. On 11 January, the temperature across much of the region was approximately −20° F (−29° C).

Earth's Natural Hazards and Disasters, Advanced Textbook 7, First Edition. Bethany D. Hinga.

On 12 January the temperature had risen to 10° above zero (−12° C), increasing throughout the morning to 20° F (−7° C) or even higher. One resident of Yankton in eastern Dakota Territory noted that the Sun was warm enough it melted some snow and ice from the windows around 1 p.m. A warm south breeze brought with it the tantalizing hint of a thaw.

Such a large increase in temperature was met with great relief and happiness by those who lived in the region. Families took the opportunity to let livestock out of their barns to get some exercise and fresh air. The people also took pleasure in getting outside. Some tended to outside chores that they could not do with temperatures far below freezing. Others headed into town to restock their supplies. Many children headed to school for the first time in weeks, since it was warm enough they could walk to school comfortably. The weather was so warm (compared to what they had experienced in the weeks before) many people left behind heavy coats in favor of lighter jackets. Most people were buoyed by this wonderful turn in the weather. They knew the unseasonably warm weather would not last, but they welcomed the respite from the cold. Others were skeptical of this turn in the weather. The sky had a strange color to it, and others described a strange texture to the air, which disquieted some who had grown so wary of the unpredictable nature of the weather in their new homes on the plains.

What the people in the Dakota Territory, Nebraska, Minnesota, and Iowa did not know is that a large and fast-moving mass of bitterly cold air was making its way east-southeast from Montana. When weather observers in Montana and Wyoming telegraphed their instrument readings to forecast offices to the east early on the morning of 12 January, the forecasters did not understand how quickly the cold air was moving. It hit the western parts of the Dakota Territory and Nebraska early in the morning while those farther east were enjoying the dawning of a warm day. Forecasts issued in St. Paul, Minnesota, were of a nice balmy day for most of the residents of Nebraska, the Dakota Territory, and Minnesota with cold air returning that night. In reality, the cold air mass was moving incredibly fast, around 45 mi/hr (72 km/hr). It would reach central Nebraska and the central Dakota Territory about 2 p.m. and eastern Minnesota and Iowa by 6 p.m.

The accompanying blizzard was a dark, roiling monster. All who witnessed the arrival of the storm described the onset of the blizzard as sudden and violent. One weather observer stated that he was standing outside enjoying the warmth of the day, when he noticed the warm breeze stop. The air became eerily still. Just a short few moments later, the wind shifted so it was blowing strongly out of the northwest and it got cold, bitterly cold. With the wind came a thundering noise and a wall of fine snow. The snow was the texture of fine sand. It immediately clogged people's noses and stuck to their faces creating masks of ice. The wind and snow hit simultaneously and with great force. Within just a few minutes, the snow was so thick and was blowing so rapidly that visibility was reduced to zero. The sound of the wind was deafening. Temperatures fell as quickly as they had risen the night before, significantly more than 10° F per hour. In addition, many reported eerie electrical effects from the storm. Telegraph wires glowed and crackled with electricity. Telegraph operators who attempted to use their machines were severely shocked. In people's homes, pokers they used to stoke their fires glowed with electricity. When they attempted to touch the pokers or open their stoves to add more fuel, they too were shocked. There is even one report of a family freezing to death because they were unable to touch the stove to add more twisted hay to it to keep the fire burning. Meteorologists later attributed this strange electrical activity to the very rapid movement of warm humid air past cold dry air. This created an immense buildup of static electricity.

The timing of the blizzard's arrival was particularly tragic for the people of the central Dakota Territory and Nebraska. To the west in Wyoming and western Nebraska, the blizzard arrived before many people left home so they were safely inside when the worst arrived. To the east in Iowa and Minnesota, the blizzard arrived after they were home for the day, so they too were safe. In much of Nebraska and the Dakota Territory, however, the storm hit midday while many people were outside the home. Farmers were working with livestock or performing necessary repairs to barns, fences, and outbuildings. Families were on the road between town and home. Children were at school.

Near the town of Ord, in central Nebraska, teacher Minnie Mae Freeman (Fig. 8.1) was holding classes in her one-room sod school house (Fig. 8.2). When the storm hit around noon, the wind tore the door off its leather hinges, blowing it into the school room. Some of the students helped her put the door back in place, but the wind just blew it in again. She asked the students to nail the door shut. She took stock of her supplies and knew she had

Figure 8.1 Nebraska teacher Minnie Freeman. Ms. Freeman saved many children in her care during the Children's Blizzard of 1888. Photo from the Nebraska State Historical Society.

Figure 8.2 Ms. Freeman and several of her students gather in front of the sod school house they fled the day of the Children's Blizzard in 1888. The roof blew off of the building the day of the blizzard. This photo was taken after the roof had been replaced. Photo from the Nebraska State Historical Society.

enough coal to last the night. She decided she would keep the children in the school house with her until the storm relented. However, the storm had other plans. Shortly after the door was secured, the wind blew part of the roof off the school. Snow billowed inside and she knew that there was no way she could keep them there. She decided to take the students to the home of the family with whom she boarded, about half a mile away. Legend says that she found some rope and tied all of the children together, though at least one student disputed this account. The 17 of them trooped across the prairie and made it to the house safely, if quite tired and cold. She was heralded for her bravery. She received 80 marriage proposals and the tune "Song of the Great Blizzard 1888, Thirteen Were Saved or Nebraska's Fearless Maid" was written about her exploits during the storm.

In another school house in Groton, Dakota Territory, a somewhat larger four-room school house welcomed children that morning. An 8-year-old named Walter Allen gleefully arrived at school, looking forward to performing his duties as row monitor. He got out his slate, wiping rag, and his prized perfume bottle full of water. He sprayed his slate with water and wiped it down, getting ready for the lessons that day. When the storm struck near 10:30 a.m., the students abruptly stopped their lessons and flew to the windows to look at the sky. In no time at all, the wind plastered the windows with fine snow. Teachers left their classrooms to confer with one another. They decided to dismiss school for the day and send the children home. Young Walter and the other row monitors distributed jackets and capes to the children in their rows. By the time all of the children were dressed in what meager winter clothes they had brought with them that day, the teachers realized they could not send the children out in that weather. About that time, five drays, flatbed wagons drawn by horses, arrived at the school. The drivers told the teachers they were there to take the children home. The teachers counted heads, loaded the children onto the drays, and sent them on their way. Walter realized as the drays pulled away that he'd left his precious perfume bottle in his desk. He could not bear the thought of it freezing in the cold school room and cracking so he hopped down off the back of the dray and ran into the school to get it. In the short minute it took him to retrieve the bottle, the drays had vanished into the blizzard. Visibility was so poor he could see nothing in front of him. Snow stung his eyes, clogged his nostrils, and blew down the neck of his coat. He decided to walk home. He didn't make it very far before he was incapacitated by the blowing snow and the bitter cold. A few hours later after his brothers made it home safely, they realized he was missing. His older brother set out to find him. He was found near the school still clutching his perfume bottle. Miraculously he was still alive. He eventually recovered and lived in the area the rest of his life.

Many that day were not as fortunate as young Walter Allen or the students in Miss Minnie Freeman's class. Farmers died when they became disoriented and could not find their way home after making sure livestock were safe in their barns. Many children died when the blizzard struck as they were on their way home from school. Parents died going out into the storm to search for their children. One teacher decided to lead his students to the house in which he boarded. Along the way, five of the male students, all of them immigrants from eastern Europe including a set of brothers, became separated from the group and got lost in the zero visibility conditions. They were found dead many days later, still huddled together.

Etta Shattuck was a teacher in Holt County, Nebraska. She decided after experiencing dwindling attendance throughout the winter that she would end the school term early and return home to her family. She closed her school and set out the morning of 12 January to get her orders signed so she could get paid. Then she would be on her way home to Seward,

Nebraska. As soon as she left the Murphy family's house where she had been boarding, the blizzard hit. She was not far from the house when she realized the danger she faced. Because of the blowing snow, she could no longer see the house to return to its safety and warmth. She was, however, next to the property's pasture fence. She reasoned that if she followed the fence, she would eventually get back to the house. She kept a hand on the fence and followed it for what seemed like forever. She began to doubt her reasoning. In her panic and confusion, she crawled under the fence and eventually wandered upon a haystack. She dug in to the haystack and crawled inside. She was found alive 3 days later by Mr. Murphy and a hired hand. They dug her slowly out of the haystack and carried her into the house. She was suffering from severe frostbite. Her father received word that she had been found and he spared no expense to bring doctors to tend to her. Unfortunately, she lost both legs to frostbite and contracted an infection after the surgery. She died in early February.

The blizzard was dubbed "The Children's Blizzard" or "The Schoolchildren's Blizzard" because many of the estimated 250 to 500 fatalities that day were children trying to make it home from school. An excellent account of the storm was written by author David Laskin in 2005. His book is entitled *The Children's Blizzard.*

8.2 Severe Weather Events

Major weather events can happen anywhere on the globe. Over the next several pages we will cover a number of common weather occurrences that people experience in various climates. As you will find, not all weather events need to be of the violent type. Even fair weather days can be deadly.

8.2.1 Heat Waves

One of the deadliest weather events of the last 100 years occurred in western Europe in 2003. In late July 2003, temperatures soared to more than 100° F (38° C), which is 18° F or 10° C higher than normal (Fig. 8.3). Since much of the population lives in un-air-conditioned houses and apartments, the temperature took a terrible toll on the health of those who were unable to find relief from the unrelenting heat. Over 35,000 people are known to have died as a direct result of the heat. France was hardest hit, with over 14,800 deaths. Germany had the second highest death toll, with 7,000 deaths. About 2,000 people died in both Italy and Spain, and over 1,000 died in the United Kingdom. The United Kingdom recorded its first 100° F (38° C) day ever on 10 August 2003. The populations with the most deaths included the very old, the very young, and those with already compromised health. A second heat wave hit Europe in 2006, causing hundreds of deaths. Germany and Poland were at the heart of the 2006 heat wave, though several countries experienced temperatures that were well above average.

The summer of 2018–2019 was a brutal one in Australia. Australia is known for its hot, dry summers and massive bush fires that erupt during the hot season. The continent is setting new records for heat almost every year. In January 2019, the temperature in Adelaide reached 116° F (46.6° C) on many days, the hottest temperature on record for any major city in Australia. By mid-January, northwest Queensland had recorded 40 consecutive days over 104° F (40° C). Despite the brutally hot temperatures, only a few hundred human deaths were reported. The wildlife of the area, however, was severely distressed by the heat. In November of 2018, approximately 23,000 flying foxes died from the heat over the course of just 2 days. They simply dropped out of the trees. More continued dying as the summer's heat intensified. In mid-January hundreds of wild horses were discovered dead as well.

In addition to the high temperatures, the country is also suffering from a crippling multiyear drought. The drought contributed to an intensely active wildfire season across

Figure 8.3 Temperature anomaly map of western Europe from 20 July to 20 August 2003. Warmer colors indicate higher than normal temperatures. Much of France and Germany (areas in deep red) experienced temperatures of 18° F (10° C) higher than usual. Image by Reto Stockli and Robert Simmon of the NASA Earth Observatory, based on data from the MODIS land team. From https://earthobservatory.nasa.gov/images/3714/european-heat-wave

Australia and Tasmania. As bad as the summer of 2018–2019 was, the summer of 2020–2021 was worse. Fires raged across eastern Australia, and killed an estimated 3 billion animals. While about 2.9 billion of these animals were reptiles, over 61,000 of the animals were koalas. Koalas were already in sharp decline in eastern Australia, and this bushfire was devastating to the population in that area.

One cause of heat waves is a phenomenon called atmospheric blocking. Atmospheric blocking occurs when a center of high or low pressure sets up over a region in a way that prevents other weather systems from moving through. A blocking high, or an area where a high pressure system sits stationary over an area, often for several days, can cause a heat wave. A major heat wave in the Pacific Northwest of the United States and southwestern Canada occurred in June of 2021 as a result of a blocking high. During this event, the highest temperature ever measured in Canada occurred at 121.3° F, or 49.6° C. The heat wave caused multiple wildland fires, forced businesses to close, damaged roads and railways, disrupted cultural events, and melted snowcaps, which resulted in flooding. The estimated economic impact of the heat wave was $8.9 billion. More than 1,400 people died, with approximately 800 deaths occurring in western Canada and the rest in the United States.

In many areas of the world, it is not common to have air-conditioning in buildings. One way to give people relief during heat waves is to set up cooling sites, which are buildings that can accommodate people who need access to an air-conditioned space because of age, infirmity, or other condition. Cooling sites are often equipped with water and other drinks so people can stay hydrated, plus medics who can diagnose and begin treatment for heat-related illnesses. Heat illnesses can include heat cramps, heat exhaustion, and heat stroke. Heat cramps are often caused by heavy sweating and dehydration. Heat exhaustion is characterized by heavy sweating, cold, pale, clammy skin, a fast, weak pulse, nausea or vomiting, muscle cramps, weakness, dizziness, headache, and fainting. Victims of heat exhaustion should be moved to a cool place and given water. Cool, wet cloths on the body or a cool bath can also help. Heat stroke is a life-threatening condition characterized by high body temperature (103° F / 39.4° C or higher), hot, red, dry, or damp skin, a fast, strong pulse, headache, dizziness, nausea, confusion, and fainting. Since this is a medical emergency, one should call 911 and move the person to a cool place. Do not give the person anything to drink, but help lower the individual's temperature with cool, wet cloths or a cool bath.

8.2.2 Drought

A lack of appreciable precipitation in an area over a period of time is a *drought*. One of the best documented droughts in history was that of the 1930s in the United States. Terrible soil erosion and thick billowing clouds of dust led to the decade of the 1930s being called the Dust Bowl years. There were many factors that contributed to the Dust Bowl. First, the Homestead Act of 1862 promised 160 acres of free farm land to families moving west into the Great Plains. This brought hundreds of thousands of inexperienced farmers west in the decades that followed. There was a mythology about the Plains that led people to believe "the rain follows the plow." People were told that if they plowed the land and converted the land from prairie to farmland, the climate would be altered and rain would become more plentiful. This was nonsense, but people plowed up and planted land that was never well suited to agriculture because of the semiarid climate. Rising grain prices during World War I further incentivized the plowing under of native prairie grasses in order to plant wheat, corn, and other row crops. When the Great Depression started in 1929, the price of wheat crashed and families responded by plowing up even more land to plant larger crops in hopes of breaking even during a bumper harvest.

Severe drought hit the Great Plains in 1930, and the dry conditions lasted until 1939. Farmers plowed the earth and planted crops in the spring, but the rains that would have allowed the crops to grow never came. Crops failed for years on end, and families became destitute. As a result, some 2.5 million people decided to move their families westward to California. This was the largest migration in United States history. Over 440,000 people from Oklahoma alone moved west, and up to one third of those families settled in the San Joaquin Valley. The derogatory term "Okie" was coined to describe these poor migrants that moved to California in search of land to farm, or farming jobs on large corporate farms. Most did not find the wealth they hoped for in California. They lived in shacks and endured terrible working conditions for very low wages.

The poor farming practices in the Plains states severely damaged the soil and farming potential of those areas. Bare soil from plowed fields where failed crops had been planted had no established root systems to keep it in place. Millions of tons of topsoil blew away from the states of Kansas, Colorado, Nebraska, Oklahoma, Texas, and New Mexico. Terrible thick clouds of dust rolled across the landscape. The dust turned daytime to dusk and coated everything in sight. It formed drifts so people had to shovel the dust away from doors of homes and barns (Fig. 8.4). Dust blew thousands of miles eastward, reaching New York and Washington, D.C., as well as ships in the Atlantic Ocean. Many people also contracted dust pneumonia in which the dust invaded the lungs and sinuses and made it hard to breathe. The number of deaths that occurred from inhalation of dust is unknown, but the number is thought to be somewhere between several hundred and several thousand. The

Figure 8.4 Dune of dust outside a home near Liberal, Kansas, in March of 1936. Photo by Arthur Rothstein. Library of Congress. https://www.loc.gov/pictures/item/2012647036/

worst single dust storm of the Dust Bowl era occurred on 14 April 1935. It started in Oklahoma. The press dubbed it "Black Sunday" because the dust was so thick at times it was difficult to see. Upwards of 3 million tons of topsoil fueled the storm that day.

President Franklin D. Roosevelt formed new government agricultural agencies to help people learn better farming practices and to plant wind breaks to prevent topsoil erosion. The Soil Erosion Service, forerunner of the National Resource Conservation Service (NRCS), and the Prairie States Forestry Project were both established in 1935. They helped those who stayed behind on the plains to learn more about their environment. The Dust Bowl finally came to an end in 1939 when rains returned to the region.

Another drought that garnered a great deal of attention was one that hit Ethiopia in the mid-1980s. It became famous when musicians from across the United States and the United Kingdom came together in 1985 to record the song "We Are the World" for famine relief. The crisis began in 1983 when the area was plunged into drought. Vegetation failed to grow. This included both grasses on grazing land used for cattle as well as crops people planted in order to make a living and feed their families. Problems were exacerbated by border region conflicts and inability or unwillingness of government officials to feed people within regions known to be antigovernment in their sentiments. International relief organizations donated food and money to the Ethiopian government. The government took control of the donations and decided to distribute them only to certain areas. Hundreds of thousands of people were forced to relocate in hopes of receiving food aid. They were left without land, money, or other resources they needed to rebuild their lives once the drought was over. The terrible convergence of drought, failed harvests, armed conflict, and government policies led to food insecurity for more than 8 million people, and a desperate famine in which a million people are thought to have died.

Today, multiple government agencies monitor the world for drought conditions. When precipitation falls below that expected based on climate data for the region, the area is noted to be entering a drought.

Two major drought indexes are used in the United States to convey information to the public about the severity of droughts within the borders of the country. The most popular is the Palmer Drought Index, which is compiled every week by the National Weather Service and the U.S. Department of Agriculture. It categorizes each of the 344 climatic divisions of the United States into one of seven moisture categories. Three of those seven moisture categories are drought categories: (1) moderate drought, (2) severe drought, and (3) extreme drought. The second drought index used in the United States is the Standardized Precipitation-Evapotranspiration Index (SPEI). The U.S. Drought Monitor (Fig. 8.5) uses the SPEI. This report is produced weekly by the National Drought Mitigation Center at the University of Nebraska at Lincoln. It focuses only on areas that are in various stages of drought, from abnormally dry on the mild end to exceptional drought on the severe end. The maps are updated weekly and can be accessed at https://droughtmonitor.unl.edu/. People living in areas of drought must be aware of their water usage. Adhering to city and state water

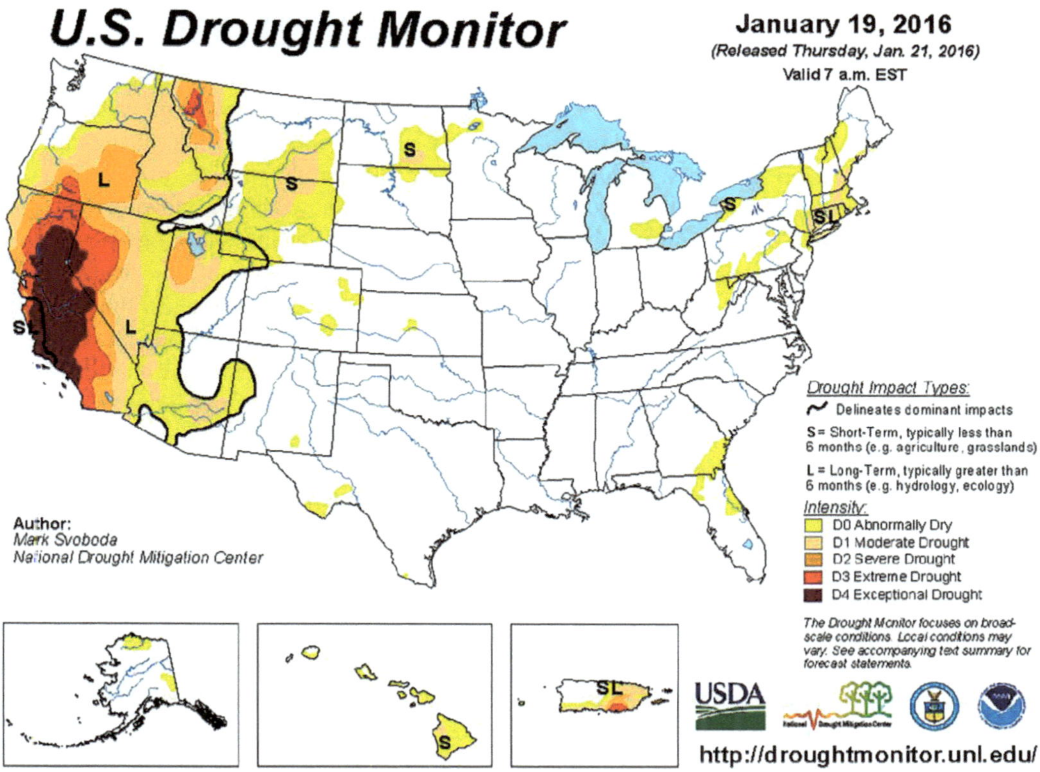

Figure 8.5 U.S. drought map issued by the United States Drought Monitor on 19 January 2016. Warmer colors indicate more severe drought. The letter S stands for short-term impacts, and L stands for long-term impacts. From Arbogast, 2017.

conservation policies during times of drought can save water for critical usage, such as for drinking, bathing, washing dishes and clothes, and providing water for livestock and crops.

8.2.3 Winter Storms

Winter storms often accompany an extratropical cyclone. Mild winter storms include freezing or frozen precipitation, without excess wind or bitterly cold temperatures. Blizzards are severe storms characterized by high wind and reduced visibility from falling or blowing snow. In order to be classified as a blizzard, a severe storm must have sustained winds of at least 35 mi/hr (56 km/hr) for at least three consecutive hours, and sufficient snow to reduce visibility to a quarter mi (400 m) or less. The fiercest of winter storms is the severe blizzard. This is a blizzard in which the temperature drops significantly, snowfall is extreme, and winds are severe. The Children's Blizzard from the chapter opening is an example of a severe blizzard.

A severe blizzard in January of 1978 is remembered by many as the worst storm to hit the Upper Midwest in its history. Originally forecast as only a winter storm, the blizzard grew out of the unlikely circumstances in which two extratropical lows combined over the state of Ohio to form one very large severe storm. On 25 January 1978, one low pressure system existed over North Dakota and a second low pressure system was present just north of the Gulf of Mexico in Louisiana. The southern low was carrying a great deal of humid air, and the northern low was accompanied by bitterly cold air. During the day, meteorologists watched these two lows and realized they were on a collision course. What developed was a storm of epic proportions. Temperatures dropped to 0° F (−18° C) and wind speeds increased across some parts of the Ohio River Valley to 50 to 70 mi/hr (80 to 113 km/hr), at times gusting to 100 mi/hr (161 km/hr). Wind chills hovered around −40° F to −60° F (−40° C to −51° C). Heavy snow fell across the region. The winds downed tree limbs and power lines, and overturned trailers. Many people were left without power and others were stranded in their homes or schools because roads were blocked by deep snowdrifts. Many people had to abandon their cars on highways because the roads became impassable (Fig. 8.6). Volunteers with four-wheel-drive vehicles helped rescue stranded motorists. Almost all roads throughout the region were closed for a period of 2 days. Emergency personnel used snowmobiles and other unorthodox methods of travel to make rescues and to transport doctors to hospitals so they could work (Fig. 8.7). In all, approximately 70 people died in the blizzard.

Severe winter storms bring a multitude of hazards. Bitterly cold temperatures pose a risk to anyone who spends time outside. With sufficiently cold temperatures, hypothermia can set in within just a few minutes. If a person is unable to find warm shelter in short order, death is a distinct possibility. Heavy snowfall is another hazard of winter storms. Rapidly falling and blowing snow can considerably reduce visibility. As illustrated in the chapter opening case study, people may lose their way in rapid snowfall and blowing snow, so it is advisable to stay inside during heavy snow. Drivers are urged to stay off the roads during low visibility conditions because the inability to see obstacles or to make out the lanes or sides of a road makes vehicular crashes likely. Many fatalities occur each year due to car accidents in winter weather conditions.

In addition to cold temperatures, heavy snowfall, and high winds, ice is another major hazard from some winter storms. Ice storms occur when a cold air mass moves into an area occupied by warmer air. Precipitation in the form of liquid rain forms in the warmer air mass. As it falls through the cloud, it remains liquid until it reaches the ground. Ground-level air, however, is below freezing. The rain freezes on whatever surface it touches. Ice coats roads, bridges, overpasses, trees, power lines, and so on (Fig. 8.8). With a layer of ice coating them, power lines are stressed and break, leaving

Figure 8.6 Cars buried to their roofs in Vandalia, Ohio, during the Blizzard of 1978. Photo from Wright State University Special Collections and Archives.

Figure 8.7 A snowmobile with stretcher in tow near Vandalia, Ohio, in 1978. This was the necessary method by which patients were transported to hospitals by first responders. Roads were completely blocked by snow, so the only vehicles capable of reliably navigating the roads were snowmobiles. Photo from Wright State University Special Collections and Archives.

Figure 8.8 Icicles hanging from a tree. The tree's needles and branches are also coated with ice. From Arbogast, 2017.

thousands without power until linemen can get out to repair the damage. Tree limbs break, blocking roads and damaging roofs, cars, and other objects beneath them. Tree limbs touching electrical lines can cause the power lines to arc, again causing people to lose power. Ice forms first on bridges and overpasses, then on roadways. Drivers may encounter unexpected ice and lose control of their vehicles, causing crashes.

8.2.4 Polar Vortex

In the winter months, it is possible for the polar jet stream to bend southward in an event called a Rossby wave (Fig. 8.9). A Rossby wave occurs when the polar jet stream weakens and shifts southward. This is popularly known as a *polar vortex*. A polar vortex (in the Northern Hemisphere) brings bitterly cold temperatures, often accompanied by high winds, into areas far south of where the normal polar jet stream is typically located. Actual temperatures dip severely, often into the range of −20° F to −30° F (−29° C to −34° C), and wind chills may reach −60° F (−51° C) or colder.

During a January 2019 polar vortex, the city of Cotton, Minnesota, saw an actual temperature of −56° F (−49° C). The lowest calculated wind chill in the same polar vortex event was −77° F (−61° C) in Thief River Falls, Minnesota. Temperatures in Chicago were so cold that the city's commuter rail company used "fire snakes" or rags soaked in diesel fuel that they laid next to the train tracks and lit on fire to prevent the tracks from contracting. The extreme contraction that can take place during a polar vortex can cause rails to separate at their joints, which can lead to a risk of derailment and expensive repair bills.

People who went outside for any length of time in areas under the effect of the polar vortex risked frostbite and death. A student at the University of Iowa succumbed to hypothermia outside a campus building, and a FedEx worker died outside a delivery hub in Illinois. These were just two of dozens of casualties from this weather event.

One action cities can take to help people survive cold temperatures is to set up warming shelters. This can be a lifesaving action for anyone who is homeless or does not have access to heated spaces due to power outages in their homes. Warming shelters are effective in preventing hypothermia and death. Early stages of hypothermia are characterized by shivering, fatigue, loss of coordination, confusion, and disorientation. Sufferers of late stage hypothermia experience no shivering, but blue skin, dilated pupils, slowed pulse and breathing, and eventually loss of consciousness. If you find a person experiencing hypothermia, move the victim to a warm room, remove any damp or wet clothing, and wrap the person in a warm blanket. Warm the center of the body first (chest, neck, head, and groin) using an electric

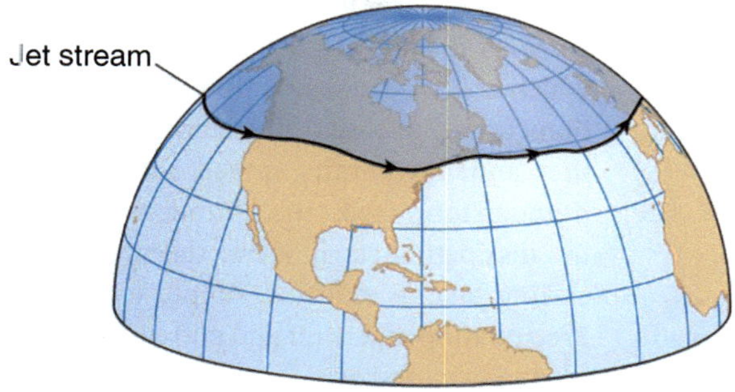

(a) Polar jet stream with small undulations

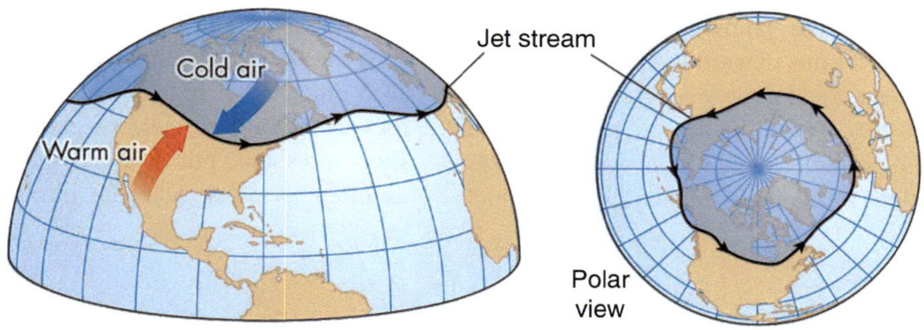

(b) Rossby waves develop, causing north–south motion of large masses of warm and cold air

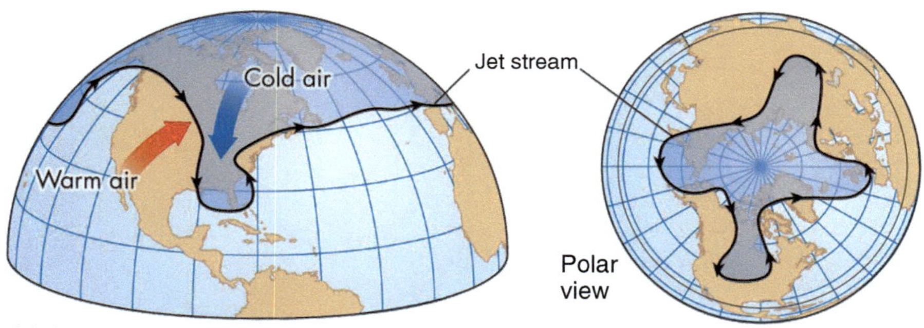

(c) Cells of cold air break off from the larger air mass, forming isolated cyclones of cold air

Figure 8.9 A polar vortex is the result of a southward bend in the polar jet stream, called a Rossby wave. This occurs during the winter months and brings bitterly cold temperatures to the Midwest. Actual temperatures may dip into the range of −20° F to −30° F (−29° C to −34° C), and wind chills may reach −60° F (−51° C). (a) Typical conditions during winter in theNorthern Hemisphere, in which the polar jet stream is located near (or north of) 45°N latitude. Cold air is separated from warm air by the strong polar jet stream. (b) When the jet stream is weakened, possibly by a warming Arctic, cold air moves southward and warm air moves northward, forming Rossby waves. (c) A fully developed polar vortex, in which Rossby waves bring cold air to much lower latitudes. From Arbogast, 2017.

blanket if available. Otherwise, skin-to-skin contact under loose, dry layers of blankets, clothing, towels, or sheets may be used. Provide the person with warm beverages if the person is conscious. Seek medical attention.

8.2.5 Thunderstorms

Thunderstorms are rain storms accompanied by lightning and thunder. Thunderstorms are typically formed when warm moist air is forced upward by a cold front entering the area or by topography (the landscape). As the warm, moist air streams upward in updrafts, it begins to condense and form towering cumulus clouds (Fig. 8.10). If conditions are favorable and updrafts continue, clouds build even taller. Precipitation forms and begins to fall, which means the storm has reached its mature stage. During the mature stage, downdrafts occur, in which cool air sinks and flows out in all directions from the base of the storm. As precipitation and downdrafts spread throughout the thunderstorm cell, the thunderstorm enters the dissipating stage. Here, the updrafts stop, cutting off the moisture supply to the storm. Precipitation tapers off and ends, and the cloud gradually vaporizes.

Each individual thunderstorm cell is relatively small (just a few miles across) and has a typical lifespan of no more than half an hour or so. However, multiple cells can form near each other either in a line or a cluster. When they do, individual cells keep their relatively short life spans, but new cells pop up nearly as often as older cells die out, so it looks

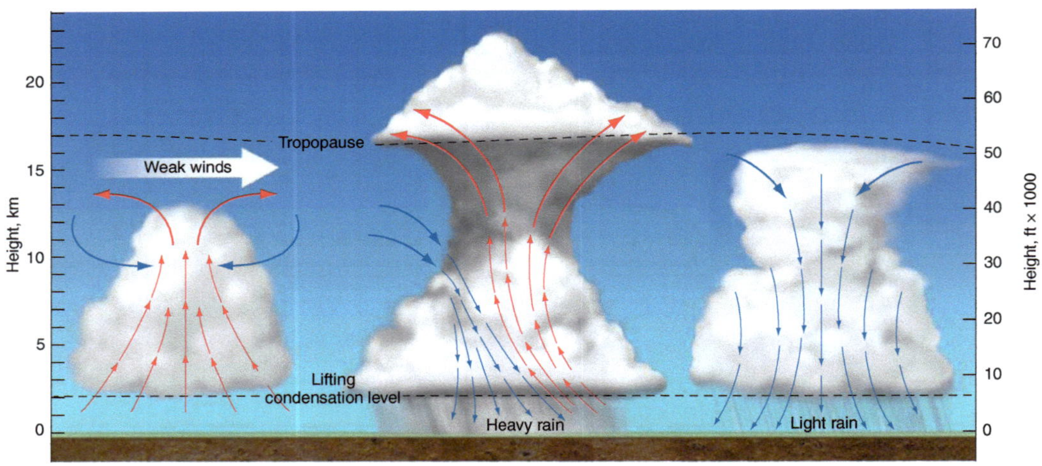

▲ CUMULUS STAGE
Vertical motions are limited by mixing with cool, dry environmental air aloft. But the mixing adds water droplets that evaporate and cool the surrounding air, leading to instability.

▲ MATURE STAGE
In this stage there are well-organized updrafts and downdrafts. Updrafts, which can reach as high as the tropopause, spread out to form an anvil cloud. Downdrafts are created by falling precipitation and entrainment of cooler, drier environmental air.

▲ DISSIPATING STAGE
Dissipation occurs when cool, dry environmental air mixes into the cloud, inhibiting convection and latent heat energy release.

Figure 8.10 The three stages of thunderstorm formation. The first stage is building of towering cumulus clouds with updrafts and condensation of water into clouds. Second is the mature stage in which the thunderstorm is sustained by continued cloud development and latent heat of condensation, but also produces downdrafts. The third and final stage is the dissipating stage in which the storm loses energy and dies. It is characterized by downdrafts. From Strahler, 2013.

like a continuous line or cluster of thunderstorms that exists for much longer than any individual cell.

An elongated line of interacting thunderstorm cells is called a *squall line* (Fig. 8.11). It is often seen developing on the edge of the warm quadrant of a mature extratropical cyclone ahead of and parallel to the cold front. In this case, the circulation around the larger low pressure system adds strength to the updraft and allows thunderstorm cells to form quite efficiently. Squall lines, and in fact all thunderstorms, are most common in the spring, summer, and fall of each year. However, strong extratropical cyclones have been known to produce squall lines in the winter, even producing thundersnow (Fig. 8.12). Cold air associated with an extratropical cyclone destabilizes the warmer air of the atmosphere in front of it and encourages rapid convection. Rapid convection causes thunderstorms to form. Though severe thunderstorms (thunderstorms accompanied by hail, high wind, and/or tornadoes) can occur anywhere in the country, conditions are favorable within the midcontinent more commonly than in other places in the country.

A multicell cluster of thunderstorms is called a *mesoscale convective complex*. This is a nearly circular cluster of interacting thunderstorm cells, which can measure up to a few hundred miles across. They typically do not develop when a front moves through an area. Instead, they tend to occur when a low pressure system moves into the area and warmer, moist air moves into the low pressure zone. Mesoscale convective complexes can move rather slowly (9–18 mi/hr or 15–30 km/hr) but in many cases they move very fast (greater than 40 mi/hr or 64 km/hr). They can carry with them considerable amounts of rainfall. They are responsible for about 80% of rainfall during the growing season in the Midwest and Great Plains. These types of storm clusters can have a life span from 6 to 24 hr.

The most energetic and dangerous of thunderstorm types is the *supercell thunderstorm* (Fig. 8.13). Supercell thunderstorms are

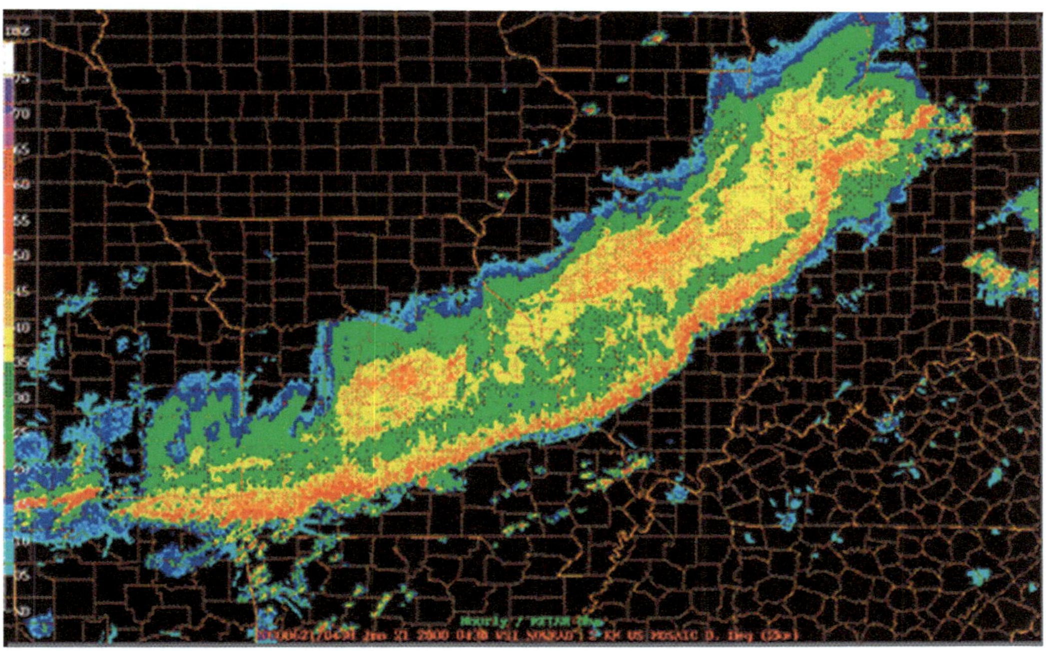

Figure 8.11 Squall line, indicated by bright red and yellow colors, that extends from southwest Michigan to southeastern Kansas. A squall line is a group of storms arranged into a line. From NOAA. From Arbogast, 2017.

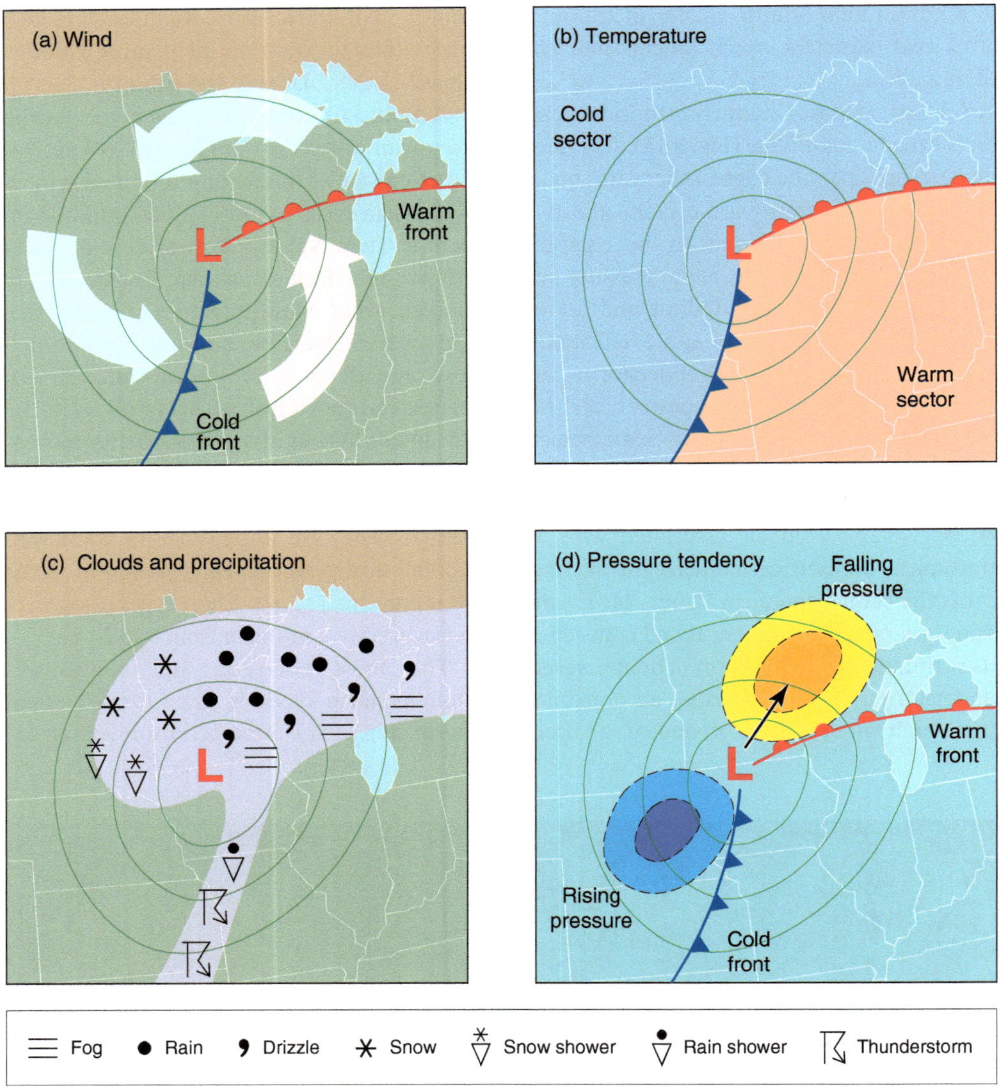

Figure 8.12 Weather conditions expected around a low pressure center. (a) The wind blows counterclockwise around a low pressure zone. A warm front is depicted as a red line with half-circles. A cold front is depicted by a blue line with triangles. Both are moving counterclockwise around the center of low pressure. (b) The warm sector is the area between a warm front and a following cold front. The warm front has already brought warm air into the area, and the cold front has not yet reached the area. Weather in this quadrant tends to be dry, with no precipitation. (c) A warm front advances warm air into an area of cooler air. This causes minor precipitation in the form of drizzle and rain to form. As the warm air mass encounters colder and colder temperatures, snow may form. At the same time, cold air is moving into an area of warmer air. This creates a line of thunderstorms, some of which may be severe. (d) Rising warm air in advance of the warm front causes an area of falling atmospheric pressure. Sinking cold air in the wake of a cold front creates an area of rising atmospheric pressure. From Moran, 2012. Used with permission.

(a)

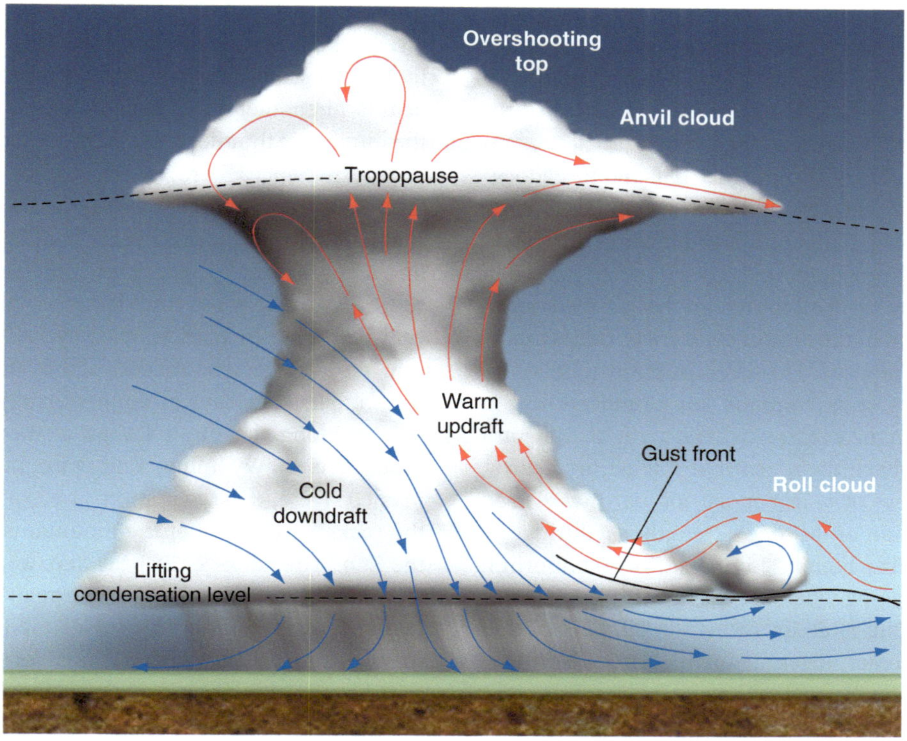

(b)

Figure 8.13 (a) An idealized severe thunderstorm. The strongest of severe thunderstorms are supercell thunderstorms, characterized by towering cumulonimbus clouds, an anvil shape, and an overshooting top. They can also produce rotation. If conditions are right, a wall cloud can form and tornadoes can develop and drop from the base of the wall cloud. From Strahler, 2013. (b) Photo of a supercell thunderstorm from near Imperial, Nebraska, in May of 2019. Photo by Mike Coniglio, National Severe Storms Laboratory.

long-lived intense, persistently rotating, large thunderstorm systems. They form when conditions allow exceptionally strong updrafts. Updrafts in supercells may reach speeds of 150 to 175 mi/hr (240 to 280 km/hr). These storms are accompanied by high winds and heavy rainfall, and may also produce hail and tornadoes.

Solar heating and air density differences caused by temperature or moisture differences cause atmospheric convection, so it would make sense if all thunderstorms formed during the heat of the day. In most areas, this is exactly what happens. However, in many parts of the country it is common for thunderstorms to form in the evening or even at night. Meteorologists have theorized that this occurs because the flow of warm moist air northward from the Gulf of Mexico strengthens during the nighttime hours. Thunderstorms then form in the evening when the northward flow of warm moist air is at its strongest. This air collides with cooler, drier air inland and causes thunderstorm formation.

Worldwide, continental interiors within the tropics have the most thunderstorms in any given year. The Amazon Basin in South America and the Congo Basin in Africa are examples of this phenomenon. Scientists keep track of the number of thunderstorm days per year each state experiences here in the United States (Fig. 8.14). A thunderstorm day is defined as a day in which thunder is heard. Thunderstorm frequency generally increases from north to south, with the most thunderstorm days per year occurring in Florida, with over 100 each year on average. The second highest number of thunderstorm days at approximately 70/yr is seen along the Front Range of the Rocky Mountains in Wyoming, Colorado, and New Mexico. Here, solar radiation is absorbed by the rocks on the side of the mountains facing the sun. The rocks emit heat and warm the air adjacent to the rocks. The warm air begins to rise and is replaced by cooler air flowing westward from the Great Plains located to the east. In this manner, convection begins.

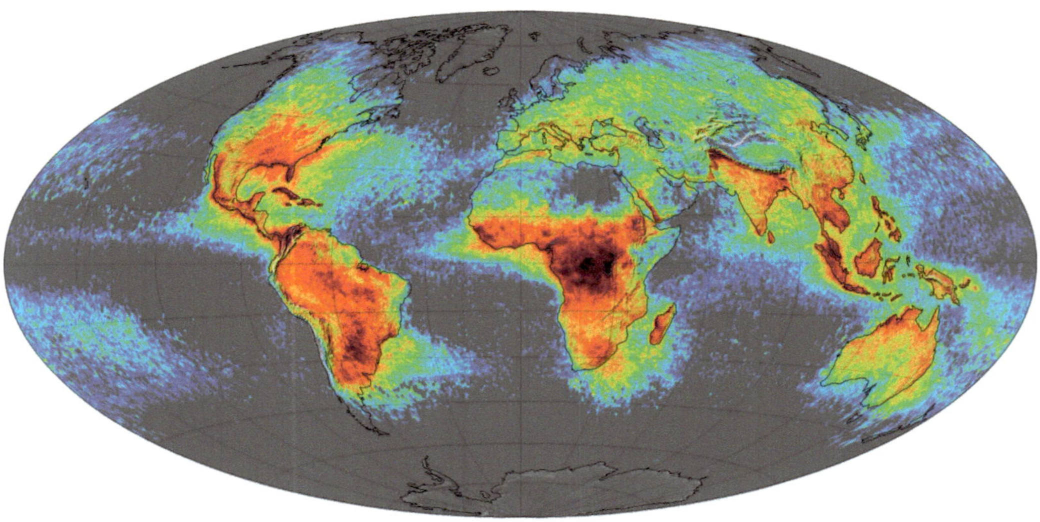

Figure 8.14 Map of lightning flashes per square kilometer per year. The cooler colors (blues and greens) indicate fewer lightning occurrences per year, and the warmer colors (orange, red, and black) indicate greater numbers of lightning occurrences per year. From NASA's Earth Observatory. https://eoimages.gsfc.nasa.gov/images/imagerecords/6000/6679/global_lightning_lrg.jpg

Convection fuels the thunderstorms. In this case, daytime heating plays a major role in thunderstorm development. It is common in the summertime to see afternoon thunderstorms form over the mountains.

Some parts of the United States rarely experience thunderstorms. Alaska sees very few thunderstorms because the warm moist air needed to fuel them is hard to come by. Thunderstorms are also rare in Hawaii where warmer dry air above the trade winds limits vertical development of clouds. Thunderstorms also do not tend to form very often in coastal areas downwind of cold ocean, such as along the coast of California. San Francisco experiences only about two thunderstorm days per year, and San Diego only five.

In the United States, the National Weather Service is responsible for issuing weather warnings and watches. A watch means that conditions are right for a specific weather event to occur. Warnings mean that the weather event is underway. For example, a severe thunderstorm watch means atmospheric conditions are right for severe thunderstorms to form, but they have not yet begun to form. A severe thunderstorm warning means that severe thunderstorms have formed, and the warning is issued for locations in the path of the storm. The same is true of tornado watches and warnings. If a tornado watch is issued, it means that atmospheric conditions are right for tornadoes to form. A tornado warning means that tornadoes have been confirmed either on weather radar or by spotters on the ground, and the warning is issued for people in the path of the tornadoes. If you are under a severe thunderstorm or tornado watch, simply remain aware throughout the watch interval of the weather conditions around you. If you are under a severe thunderstorm or tornado warning, take shelter and listen to NOAA weather radio or your local media for additional instructions.

8.2.6 Thunderstorm Hazards

Lightning

Lightning and thunder always accompany thunderstorms. Lightning is a brilliant flash of light caused by electrical discharge (Fig. 8.15). Lightning can occur within a single cloud, cloud to cloud, cloud to air, or between a cloud and the

Figure 8.15 Lightning storm. Photo credit: Lyle Leduc/Getty Images, Inc. From Arbogast, 2017.

Earth's surface. Thunder is a sound wave produced by rapid intense heating of the atmosphere from a lightning bolt. The rapid heating of the atmosphere causes a sharp increase in air pressure, which generates a shock wave. The shock wave creates sound waves, which we hear as thunder.

Lightning within a single cloud, or between two clouds, is the most common type that occurs, constituting about 80% of all lightning. The most dangerous type of lightning strike is between the cloud and the ground. Thankfully, cloud-to-ground lightning represents only 20% of lightning strikes. What causes lightning to strike the ground? On a clear day, the Earth's surface has a negative charge. When a cumulonimbus cloud forms, charges within the cloud separate so the upper part of the cloud becomes positively charged and the base of the cloud is negatively charged (Fig. 8.16). The charges separate because of rapid convection within the cloud. Warm air moves upward in updrafts. Moisture within the cloud condenses and freezes near the top of the cloud causing ice particles to form. The ice moves downward because of gravity and is caught in updrafts

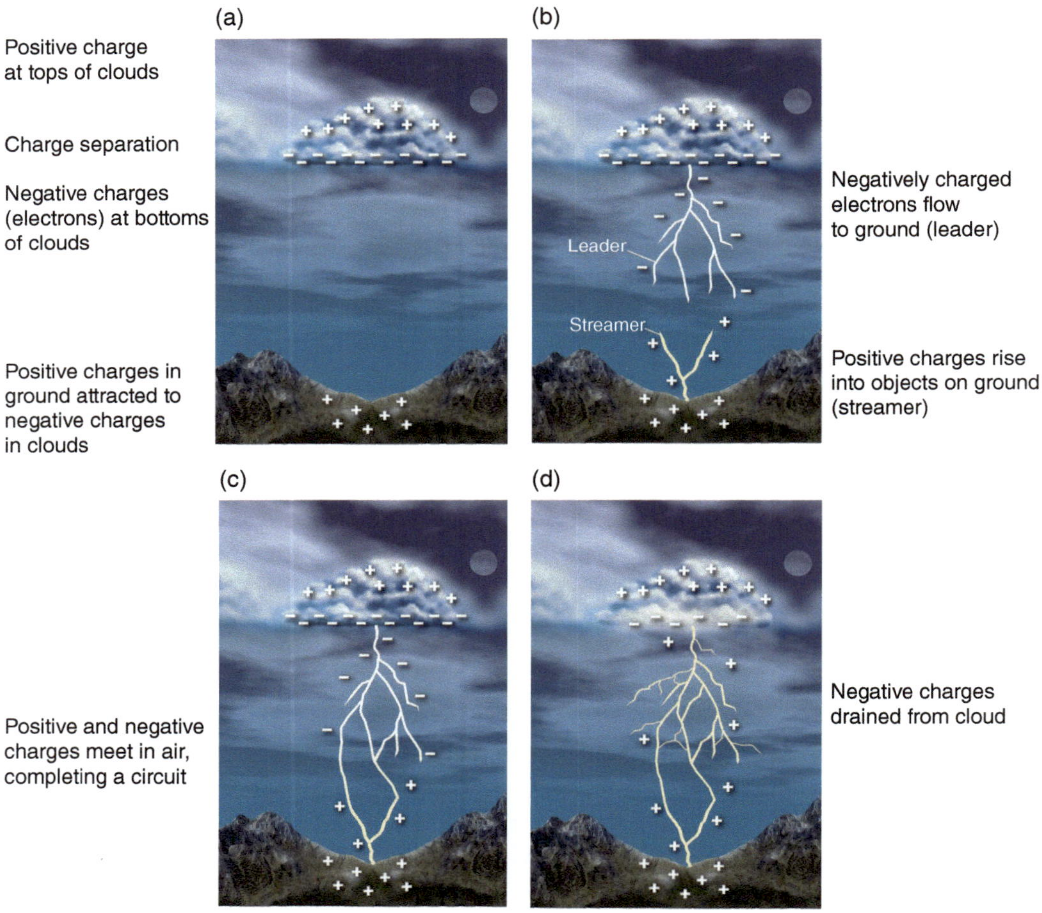

Figure 8.16 (a) During a thunderstorm, the bottom of a cloud is negatively charged, and the ground is positively charged. (b) Negative charges reach toward the ground in a leader, and positive charges reach upward in a streamer. (c) The leader and streamer connect, creating a flash of lightning. (d) The negative charges have now drained from the cloud, and the cycle is set to repeat. From Arbogast, 2017.

so it also moves upward in the cloud. Electrons are sheared off of the ascending ice particles, and these electrons accumulate on descending particles. This creates an imbalance in electrical charges, with an excess of negative charges on the bottom of the cloud and a larger number of positive charges at the top of the cloud. This induces a positive charge on the ground.

To create cloud-to-ground lightning, a stepped (zigzagging) leader of negatively charged particles moves downward through the cloud toward the ground. This is called a negative leader. As the negative leader nears the bottom of the cloud, it attracts a streamer of positive charges from the ground. The two connect and a large electrical current flows from the cloud to the ground. At the same time, a wave of positive charges zips up to the cloud, creating the light flash we see, which is called the return stroke. The return stroke travels approximately 31,000 mi/s (50,000 km/s) and can reach a temperature of 50,000° F (27,760° C).

Of all thunderstorm hazards, lightning kills the most people on an annual basis. The number of lightning deaths per country each year varies considerably around the world. In the United States, about 55 deaths each year are attributed to lightning. However, in some countries deaths number in the thousands. In India, for example, it is not uncommon to see 2,000 deaths a year from lightning strikes. Death toll per capita is somewhat higher in developing countries than in industrialized nations for two primary reasons. In developing countries, much farming is still done by people who labor by hand. That leaves people out in the elements when thunderstorms come, and unfortunately many people in agricultural areas are struck by lightning while going about their daily farm work. In addition, in many developing countries, homes are simply constructed and do not have a large network of wires for electricity or extensive indoor plumbing. Wires and plumbing help divert electricity into the ground if a house is struck by lightning. If a building lacks these features, lightning that strikes a roof can jump to the people inside, killing them where they seek shelter. This simply does not happen in the United States and other developed countries. Deaths within the United States happen primarily during leisure-time activities.

What can you do to protect yourself when a thunderstorm arrives and lightning threatens? If you can hear thunder, lightning is close enough to strike you. It does not have to be raining in order for lightning to strike. Quit whatever outdoor activity you are doing and seek shelter in a house or other building, avoiding contact with conductors of electricity. Do not talk on a land line telephone or sit down to work at your plugged-in computer. These objects are plugged into the wall, and connected to wires that readily carry electricity. Do not shower or bathe, or wash dishes. Plumbing also conducts electricity quite well. Cars are generally quite safe because a lightning strike to the roof of a car sends electricity through the car frame, protecting people inside as long as they do not touch any metal elements connected to the frame. If you are caught in the open, find shelter under a cliff, in a cave, or in a low area. Stay clear of isolated tall structures such as trees, telephone poles, and flagpoles since lightning is more likely to strike tall objects. Stay away from metallic objects that conduct electricity such as wire fencing, wire clotheslines, bicycles, and golf clubs. Steer clear of high elevation areas like mountain or hill tops and roofs. Stay away from bodies of water. Do not continue mowing a lawn.

If you are out in the open without shelter and feel your hair standing on end and your skin tingling, or if you hear a buzzing in metal objects near you, such as the metal frame inside a frame back pack, a lightning strike might be imminent. Drop low to the ground, leaving as little of your body as possible touching the ground. Electricity that strikes the ground can travel for over 100 ft (30 m), and travel through your body if you're lying flat on the ground. Crouch on the balls of your feet with your hands on your knees. Make yourself as small a target as possible.

Downbursts

Thunderstorms can produce strong downdrafts called *downbursts*. When a downburst hits the Earth's surface, it moves horizontally as potentially damaging wind. Downbursts can cause winds at the Earth's surface in excess of 125 mi/hr (200km/hr). They can do just as much damage as a tornado, uprooting trees, overturning mobile homes, travel trailers, and semitrailers, flattening crops, and destroying buildings. Downbursts are classified as either *microbursts* or *macrobursts* based on the size of the area affected. Macrobursts are more than 2.5 mi (4 km) across, can last up to 30 min, and have surface wind speeds of up to 130 mi/hr (210 km/hr). Microbursts are smaller, less than 2.5 mi across. They exist for shorter periods of time (typically less than 10 min) but can carry winds up to 170 mi/hr (270 km/hr). Microbursts are especially dangerous for aircraft (Fig. 8.17). Scientists proved that two plane crashes in 1975 were the result of pilots flying into microbursts they could not see or detect with the instrumentation they had at the time. Now, commercial airplanes are required to have microburst detection systems on board so pilots have enough advanced warning of microbursts and can avoid them.

Derechos

Derechos are straight-line winds produced by squall lines or mesoscale convective complexes. They are at least 240 mi (400 km) long and have gusts of at least 58 mi/hr (93 km/hr). They usually last for several hours, and can move across the entire width of the United States. Derechos are associated with the bowed shape of rapidly moving bands of showers and thunderstorms. Approximately 20 of them occur every year. Like downbursts, derechos can be as destructive as tornadoes. Between 1986 and 2003, there were 377 derechos that caused 155 fatalities.

On the afternoon of 10 July 2011, a line of thunderstorms formed over western Nebraska. An associated system of derechos formed as

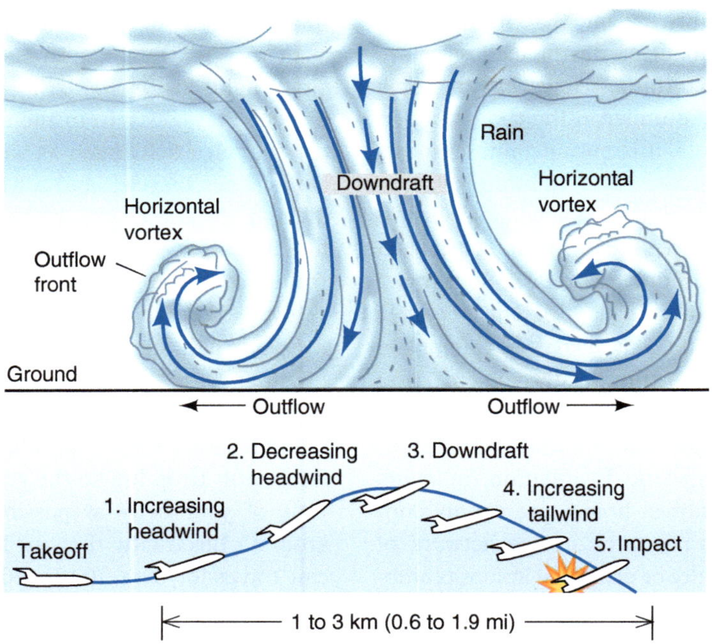

Figure 8.17 Microbursts are particularly dangerous for aircraft. A microburst is a type of downdraft, which is a strong downward-directed gust of wind that accompanies many thunderstorms. From Strahler, 2013.

well. Wind gusts were between 70 and 80 mi/hr (113 and 129 km/hr). The system swept eastward, hitting Chicago about 8 a.m. the next morning. Derechos caused extensive damage to sturdily built structures, downed power lines, and uprooted large trees. More than 800,000 people were left without power. The derechos continued eastward throughout the day and only dissipated once they reached Maryland near sunset on 11 July.

In August of 2020, a derecho swept across the Midwest. It traveled from South Dakota to Ohio in 14 hr. The derecho flattened crops, downed trees, and destroyed homes, businesses, and vehicles. Officials estimated damages at $7.5 billion, making it one of the most expensive weather disasters of 2020.

Flash Floods

During thunderstorms, rainfall can be considerable. When a thunderstorm system becomes stationary or moves very slowly across an area, it can drop significant amounts of rainfall. If the ground is unable to absorb the rainfall, it will run off and collect in low spots and creeks. This causes flash flooding. Though flash flood watches and warnings are often issued around the same time as a thunderstorm watch or warning, these floods can catch people by surprise and lead to people venturing into areas that are unsafe. If a roadway is covered with water, do not drive across it. It is likely that you cannot see the bottom of the water, which means you will be unaware if the road is washed out, is deeper than you expect, or has large potholes that could trap your car in flood waters. Never attempt to cross a stream on foot if the water is above your knees. Know your environment and be aware of rising water in the event of heavy rainfall in your location or nearby.

Hail

Hail is frozen precipitation that falls from storm clouds during intense thunderstorms. It consists of balls or lumps of ice that are more than 0.2 in. (5 mm) in diameter. Hailstone size varies considerably and is usually reported as a comparison to the size of a common object such as a pea, a dime, a tennis ball, or a grapefruit (Table 8.1). The largest recorded hailstone in the United States fell on Vivian, South Dakota in 2010. It was 8 in. (20.3 cm) in diameter and weighed 1.94 lb (879 g).

Hail forms within cumulonimbus clouds that have strong updrafts and a great deal of moisture in the form of supercooled water droplets. Within the cloud, an ice pellet is transported vertically through the cloud via updrafts. As it travels through the cloud, supercooled water adheres to the ice pellet, coating it with a thin layer of ice. Then it may follow a number of different paths. It may fall slowly through the cloud, becoming coated with additional ice before it falls to Earth. Or it may descend through the cloud, then be caught by another

Table 8.1 Updraft speed required to form hailstones of various sizes.

	Measurement		Updraft speed	
Hailstone size	**in.**	**cm.**	**mph**	**km/h**
BB	< 1/4	< 0.64	< 24	< 39
Pea	1/4	0.64	24	39
Marble	1/2	1.3	35	56
Dime	7/10	1.8	38	61
Penny	3/4	1.9	40	64
Nickel	7/8	2.2	46	74
Quarter	1	2.5	49	79
Half dollar	1 1/4	3.2	54	87
Walnut	1 1/2	3.8	60	97
Golf ball	1 3/4	4.4	64	103
Hen egg	2	5.1	69	111
Tennis ball	2 1/2	6.4	77	124
Baseball	2 3/4	7.0	81	130
Tea cup	3	7.6	84	135
Grapefruit	4	10.1	98	158
Softball	4 1/2	11.4	103	166

Source: From the National Weather Service; https://www.weather.gov/jetstream/hail

updraft and cycled upward again to be coated with multiple layers of ice. An ice pellet can make numerous passes through the cloud. Scientists have counted more than 20 layers of ice in a single hailstone. In general, the stronger the updraft the larger the ice pellets can become. As an example, in order to form a penny-sized hailstone, an updraft of 40 mi/hr (64 km/hr) is required. Grapefruit sized hail requires an updraft of nearly 100 mi/hr (160 km/hr). Hail falls from clouds in a column called a hailshaft. The narrow zone on the ground where hail accumulates is called a hailstreak. A single thunderstorm can produce multiple hailstreaks, and a typical hailstreak is about 1.2 mi (2 km) wide and 6.2 mi (10 km) long.

Hail can be incredibly damaging to crops, livestock, and property (Fig. 8.18). Thunderstorms that are likely to produce large hail often happen within the growing season. Hail can decimate a crop within just a few minutes. Crop losses due to hail total nearly $1 billion every year. Large hail stones can dent cars, break windows, leave dents in roofs, and injure people and livestock. In rare cases, hail can also accumulate to great depths on the ground. In August of 1980, drifts of hail in Orient, Iowa, were reported to be 6 ft (1.8 m) deep. In September of 1988, pea-sized hail formed 18 in. (46 cm) drifts on the north side of Milwaukee, Wisconsin. Snowplows had to be called out to clear the hail from the roads. In July of 2019, a freak hail storm dropped 3 ft (1 m) of hail on the city of Guadalajara, Mexico.

Tornadoes

Tornadoes are the most violent of weather phenomena (Fig. 8.19). They are violently rotating columns of air that come into contact with the ground. Their wind speeds can reach more than 200 mi/hr (322 km/hr), and they can produce incredible types and amounts of damage. They are most commonly produced from supercell thunderstorms, the most severe of thunderstorm types. Most tornadoes are small and strike sparsely populated areas, but even major cities are not immune to tornadoes. Unlike most weather phenomena, conditions favorable to the formation of tornadoes are not found everywhere in the world. North America has the highest frequency of tornado occurrences in the world.

Tornadoes have been reported in all 50 states and the provinces of southern Canada (Fig. 8.20), but the majority of tornadoes strike the midcontinent area that includes the

Figure 8.18 Hail can be incredibly damaging to crops, livestock, homes, and property. Photo by Luca, Adobe Stock, Asset ID 629040227.

Figure 8.19 A tornado is a violently rotating column of air that touches the ground. It descends from the wall cloud, which is found at the base of a supercell thunderstorm. From Strahler, 2013.

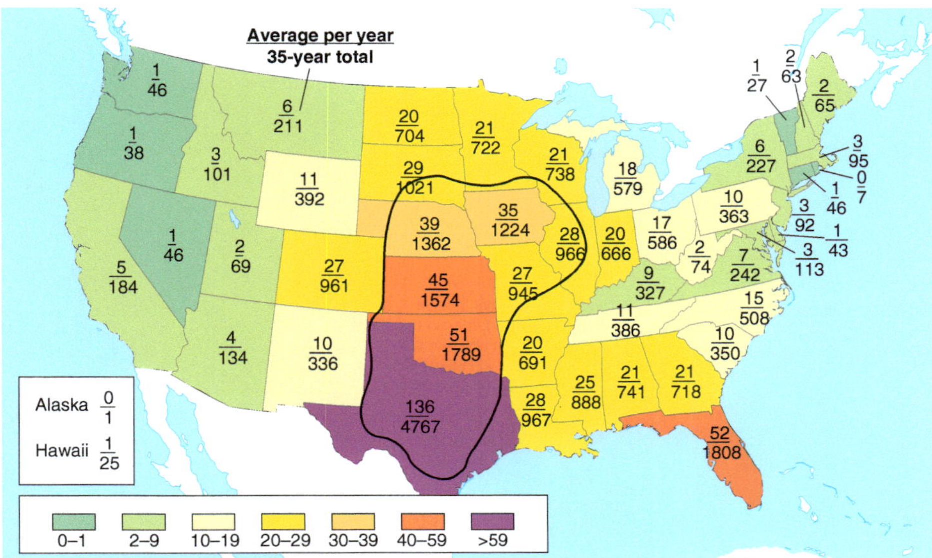

Figure 8.20 Map of tornado occurrences in the United States over a 35 year period. The top number in each state is the average number of tornadoes in that state, and the bottom number is the total number of tornadoes to strike that state during a 35 year period. The approximate boundary of Tornado Alley is outlined in black. From Arbogast, 2017.

panhandle of Texas, Oklahoma, Kansas, Nebraska, and into the southeastern area of South Dakota, the area known as tornado alley. It is within tornado alley that most significant (EF-2 or greater) tornadoes occur. However, the entirety of the Midwest and the Southeast also see tornadoes with some frequency.

The United States experiences an average of 1,300 tornadoes in a typical year. About half of tornadoes occur during the warmest part of the day, between 10 a.m. and 6 p.m. Almost 75% of tornadoes occur between the months of March and July, and peak tornado activity occurs during May and June. Extratropical cyclones often form when there is a large difference in temperature between the northern part of the country versus the southern part of the country. This occurs in the late spring. During these months, the atmosphere is relatively unstable as the Earth makes the transition from spring to summer. Increasing solar radiation warms the ground, and the ground warms the lower troposphere. However, the upper troposphere is still fairly cool. The temperature and moisture differences between the upper and lower troposphere as well as between the northern and southern United States provides a setting conducive to forming extratropical cyclones. This means conditions are also favorable for strong thunderstorm formation. Severe thunderstorms often form in the warm quadrant of an extratropical cyclone.

Tornadoes form most commonly in supercell thunderstorms. Supercell thunderstorms have vigorous updrafts, sometimes more than 150 mi/hr (240 km/hr). These types of storms are very energetic and well organized, and can last for hours. They can also produce more than one tornado. The formation of a tornado within a supercell occurs when these strong updrafts interact with horizontal winds within the storm system. Horizontal wind shear is a situation in which the wind strengthens and turns clockwise with altitude. This creates horizontal rotation within the cloud. When this horizontal rotation interacts with the updrafts, it tilts the axis of the rotating air from horizontal to vertical (Fig. 8.21). This allows the rotating shaft of

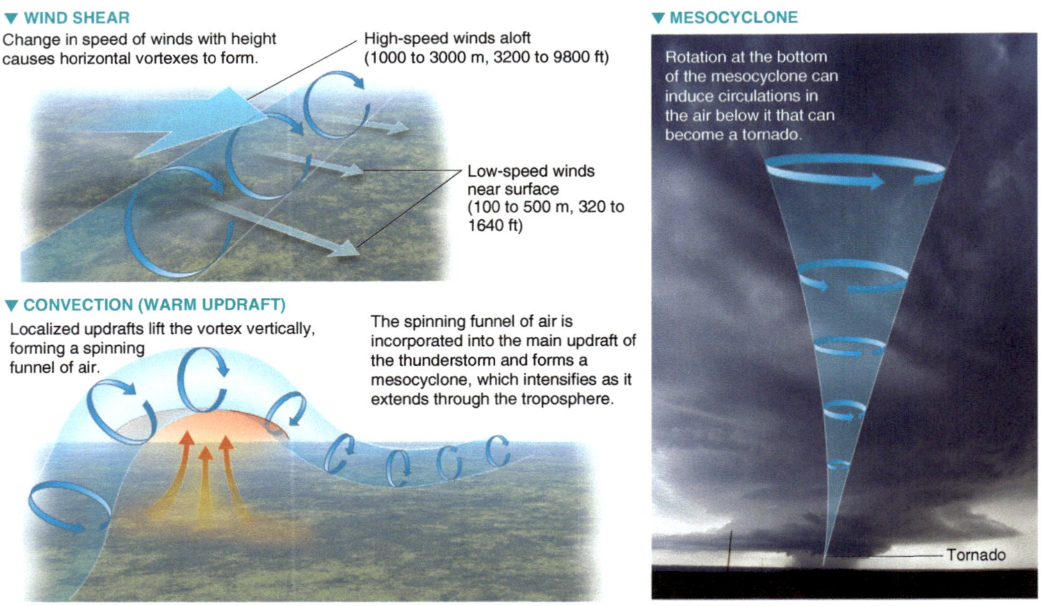

Figure 8.21 A tornado forms within a supercell thunderstorm when conditions are correct. Wind shear creates horizontal vortexes. Updrafts tilt the horizontal vortexes so they are vertical, producing a mesocyclone. A tornado forms if the base of the mesocyclone descends to the Earth's surface. From Strahler, 2013.

air to reach down toward the ground, and it becomes a tornado. Tornadoes usually drop out of a low hanging part of the supercell cloud, called a wall cloud. The wall cloud may occur in a part of the cloud where precipitation is not falling.

Atmospheric scientists are better able to detect tornadoes today than at any point in the past thanks to advances in atmospheric monitoring capabilities such as advanced weather radar. Using weather radar, rotation within a storm is easy to detect. On Doppler weather radar, a storm that has strong rotation and may be producing a tornado is recognized by a hook-shaped echo on the radar image and rapidly shifting wind direction in the same location (Fig. 8.22). A tornado is likely to form at the tightest curve of the hook. When a hook echo forms, meteorologists can issue a tornado warning. The warning may come several minutes before a tornado actually forms and hits the ground. This is a vast improvement over conventional weather radar, which could not detect wind direction like Doppler weather radar can.

Tornadoes are classified using the Enhanced Fujita (EF) Scale (Table 8.2, Fig. 8.23). The EF scale classifies tornadoes based on damage done and estimated wind speeds within the tornado. The smallest of tornadoes cause only light damage and have wind speeds of approximately 65–85 mi/hr (105–137 km/hr). These are classified as EF-0 tornadoes. The most severe tornadoes that cause incredible amounts of damage have wind speeds in excess of 200 mi/hr (322 km/hr) and are classified as EF-5 tornadoes. Thankfully most tornadoes that occur are on the low end of this scale, EF-0 and EF-1. In an average year, only one tornado may be an EF-5, the most destructive type of tornado (Fig. 8.24).

Tornadoes may occur as single events, or as outbreaks where multiple tornadoes form in the same general area and last for several hours. The most deadly tornado event in United States history occurred on 18 March

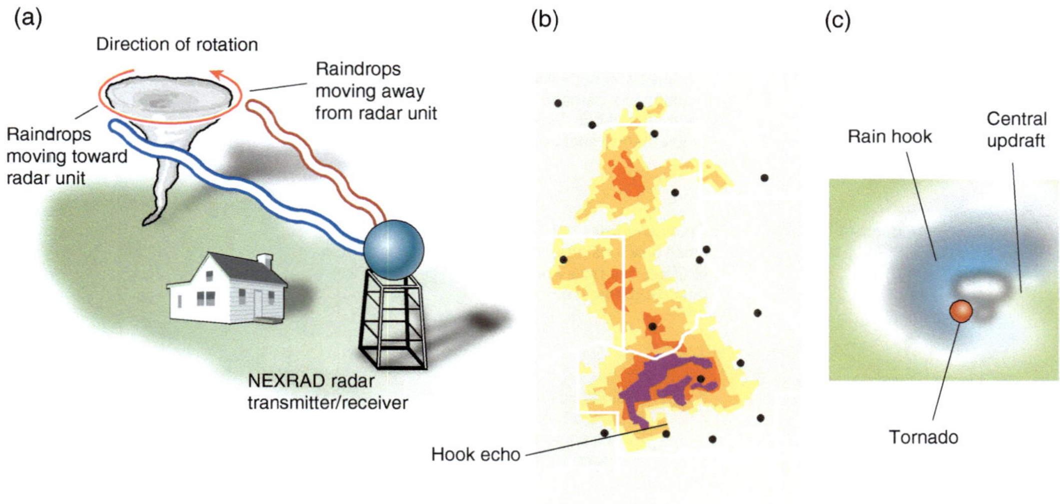

Figure 8.22 (a) Doppler radar provides valuable information about tornado formation. This type of radar is able to detect moisture moving both toward and away from the radar station. When moisture is moving both toward and away from the radar station within a relatively small geographic area, this is an indicator of rotation within the cloud. (b) Direction of moisture movement is displayed on a map, with red and orange areas indicating moisture moving away from the radar station, and blue and purple indicating moisture moving toward the radar station. In this case, a hook echo pattern is present. (c) The hook echo indicates aggressive cloud rotation, which usually means a tornado is present at the tightest curve of the hook. From Arbogast, 2017.

Table 8.2 The Enhanced Fujita (EF) Scale for tornado classification.

		Wind speed		
EFscale	**Class**	**mph**	**km/h**	**Description**
EF0	Weak	65–85	105–137	Gale
EF1	Weak	86–110	138–177	Moderate
EF2	Strong	111–135	178–217	Significant
EF3	Strong	136–165	218–266	Severe
EF4	Violent	166–200	267–322	Devastating
EF5	Violent	> 200	> 322	Incredible

Figure 8.23 Examples of damage caused by tornadoes at each level on the Enhanced Fujita Scale. From Strahler, 2013.

(a)

(b)

(c)

Figure 8.24 Damage from an EF-5 tornado that touched down in Greensburg, Kansas, in May of 2007. (a) Ground level view of the damage. Trees are stripped of their branches, and many houses are leveled. Photo by Greg Henshall/FEMA. (b) Aerial photo of Greensburg before the tornado. Kansas Division of Emergency Management. (c) Aerial photo of Greensburg after the tornado. In much of the city, damage is total. Kansas Division of Emergency Management. All images in Arbogast, 2017.

1925 and is known as the Tri-State tornado. The tornado (or more likely, tornadoes) formed in southeastern Missouri, traveled through southern Illinois, and finally ended in southwest Indiana. The tornadoes moved incredibly quickly, at a maximum forward speed of 73 mi/hr (118 km/hr), and were on the ground for three and a half hours. This tornado outbreak caused 695 deaths and left more than 11,000 people homeless.

A record-breaking tornado outbreak occurred in late April of 2011. Over the 3 day period of 25–28 April, 342 tornadoes were confirmed in 21 states from Texas to New York. The most violent day of the outbreak was 27 April. On that day, a strong upper level low pressure system was moving through the Midwest accompanied by a cold front. Ahead of the cold front, warm moist air was streaming northward from the Gulf of Mexico. As we might predict, massive supercell thunderstorms formed ahead of the cold front where the warm moist air was forced over the advancing cold air mass. A total of 200 tornadoes occurred that day. Four of them were EF-5s and 11 were EF-4s. An EF-5 tornado ripped through the city of Smithville, Mississippi, and did incredible damage. Newly constructed brick homes were swept clean of their foundations. Bark was removed from trees. Asphalt was removed from sections of roadways. Another EF-5 tornado hit Hackleburg, Alabama. It was the first EF-5 tornado to strike Alabama in more than 10 years. About 75% of the city was destroyed. This tornado had a track length of over 130 mi (210 km), making it the longest track length of any tornado in the outbreak. This single tornado killed 72 people, the largest death toll from a single tornado in Alabama history. A second deadly tornado also struck Alabama that day, killing 40 people in Tuscaloosa. The EF-4 tornado continued from Tuscaloosa into the suburbs of Birmingham where it destroyed entire neighborhoods. This outbreak was the fourth deadliest in history, killing more than 300 people, 235 of which were in Alabama. More than 2.000 people were injured as well.

On 3–4 April 1974, a tornado outbreak occurred in the east-central United States. It affected 13 states from Illinois and Indiana southward to Alabama and Mississippi. A total of 148 tornadoes touched down during the outbreak. Ninety-five of those tornadoes were EF-2 or greater, and 30 were either an EF-4 or EF-5. The death toll from the outbreak was 335, and more than 6,000 people were injured. The hardest hit areas were in the states of Ohio, Indiana, and Kentucky. The single tornado that gained the most attention was an EF-5 that struck Xenia, Ohio, on the afternoon of 3 April. The tornado destroyed the downtown area of Xenia, and left homes in splinters. The hospital narrowly missed destruction. It was the strongest tornado of the outbreak, leaving 34 people dead and causing $250 million damage. This makes it one of the most expensive tornado disasters in history.

The year 2023 was an active severe storm year with 1,423 tornadoes reported (Fig. 8.25). A number of tornado outbreaks in the late winter and early spring of 2023 killed 60 people, more than double the number of tornado victims in all of 2022. Most of the storms, as we might predict, occurred in the springtime, in March through June. However, more than 100 tornadoes occurred in January of 2023, and nearly 50 tornadoes occurred in December of 2023. This is an anomaly from the usual weather pattern in which very few tornadoes occur during the winter months.

Tornado Safety

If a tornado threatens, seek shelter. The safest place to be is in an underground storm cellar or a basement. If it is not possible to get underground, find a sturdy building and take shelter in an interior room or hallway on the lowest floor of the building, away from windows. If possible, cover yourself with something to protect yourself from flying debris. Avoid large structures such as auditoriums, gymnasiums, and supermarkets that have wide free-span roofs. If you are caught in a vehicle, do not try to outrun the tornado. Stop and seek shelter

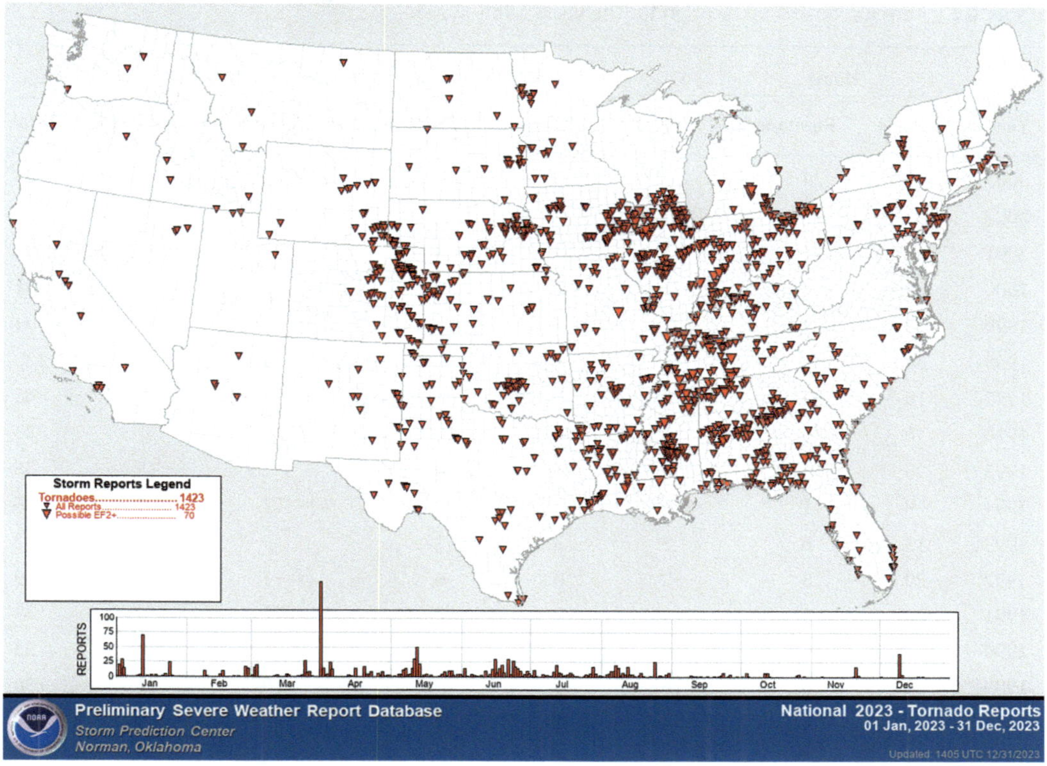

Figure 8.25 Map showing tornadoes reported in 2023 accompanied by a graph of tornado activity by date for 2023. An unusually large number of tornadoes occurred in January and December of 2023. From the NOAA Storm Prediction Center.

inside a building. If that is not possible, get out of the car and lie flat in a ditch, ravine, or dry creek bed. Statistically speaking, the most dangerous place to be in a tornado is in a mobile home (Table 8.3), so if you live in a mobile home find shelter elsewhere. Find shelter in a permanent structure nearby, or lie in a low spot such as a ditch.

8.3 Conclusion

While the specific hazards are unique to each location, no place on the planet is free from severe weather, because every location on the planet experiences weather. Heat waves, often caused by atmospheric blocking, can affect any location on the planet, and usually result in a string of record high temperatures. Thousands of people can die in these events from heat-related illnesses. Winter weather can turn severe very quickly and result in fast-moving blizzards with blinding snow and high winds, ice storms that coat power lines, trees, and roads resulting in loss of electricity from falling power lines and trees, as well as vehicular crashes on icy roads. Polar vortexes result in extremely low temperatures and wind chills that can cause damage to infrastructure and death for individuals who find themselves outside for any length of time. Thunderstorms have numerous hazards including extreme rainfall that can cause flash flooding, lightning, and tornadoes.

We have greater capacity today to forecast weather events than at any time in our past.

Table 8.3 Tornado deaths by location for the years 1985–2003.

Year	Home		Vehicle	Business	School/Church	Outdoors	Others	Total
	Mobile	Permanent						
2003	25	24	0	1	0	3	1	54
2002	32	15	4	0	1	3	0	55
2001	17	15	3	3	0	2	0	40
2000	28	7	4	0	0	2	0	41
1999	36	39	6	3	0	9	1	94
1998	64	46	16	1	0	3	0	130
1997	16	38	3	3	0	8	0	68
1996	16	6	3	0	0	0	1	26
1995	8	10	4	0	0	1	15	38
1994	9	11	1	0	20	2	3	46
1993	13	6	7	3	1	3	0	33
1992	20	18	0	0	0	1	0	39
1991	20	3	4	0	0	12	0	39
1990	7	11	14	15	5	1	0	53
1989	12	8	16	4	9	0	1	50
1988	21	6	3	2	0	0	1	33
1987	24	7	3	0	22	3	0	59
1986	7	3	3	0	0	0	2	15
1985	28	40	4	0	0	0	22	94
Total	403	313	98	35	58	53	47	1007

Source: Data from the National Weather Service; https://www.spc.noaa.gov/climo/torn/locations.html

This is making it easier for people to keep themselves safe in extreme weather events. However, the risk to loss of life and property is still severe even when we can forecast weather events. Heat waves bring unbearably high temperatures to regions. If air-conditioning is not common in these areas, many people will suffer as a result of their inability to stay cool. Winter storms can strike quickly and cause ice storms and near whiteout conditions. Severe thunderstorms can generate lightning, hail, heavy rain that causes flash flooding, and tornadoes. Each of these hazards is forecastable, and therefore survivable. With the information in this chapter, readers have basic understanding about weather hazards and how to stay safe when severe weather strikes. Pay attention to weather forecasts and severe weather alerts and prepare ahead of time to mitigate the effects of extreme weather events.

8.4 Chapter in Review

No place on Earth is immune from the effects of severe weather. Heat waves, winter storms, droughts, severe thunderstorms, and tornadoes can be devastating experiences for people as well as for the environment. Heat waves occur when an area experiences much higher temperatures than are typical for that location and season. One cause of heat waves is a phenomenon called atmospheric blocking. Atmospheric blocking occurs when a center of high or low pressure sets up over a region in a way that prevents other weather systems from moving through. A blocking high, or an area where a high pressure system sits stationary over an area, often for several days, can cause a heat wave. A heat wave in Europe in 2003 killed an estimated 35,000 people from heat-related illnesses. A major heat wave in the Pacific Northwest of the United States and southwestern Canada occurred in June of 2021 as a result of a blocking high. The estimated economic impact of the heat wave was $8.9 billion, and more than 1,400 people died.

Drought is a condition that exists when there is a lack of precipitation for a period of time. The worst drought in American history was the Dust Bowl, a period of 9 years in the 1930s in which little rain fell across the Plains states, causing crops to fail. Soil was eroded at record rates. Millions of farmers were left destitute from a lack of income as a result of failed crops. Many moved west toward more fertile ground in California.

Winter storms carry with them cold temperatures, high winds, and large amounts of snow and possibly ice. They are often part of an extratropical cyclone, which develops around a low pressure system in the midlatitudes. A polar vortex is a phenomenon in which the polar jet dips southward (in the Northern Hemisphere) bringing brutally cold temperatures.

Thunderstorms are often produced when a cold air mass moves into an area causing warm moist air to rise. The rising moisture condenses into towering cumulonimbus clouds, which give us thunderstorms. In the United States, Florida has the most thunderstorm days each year, but the Rockies also have large numbers of thunderstorm days each year. Thunderstorms are formed when convection (vertical circulation of air that brings warm, moist air upward and cool dry air downward) causes towering, moisture rich clouds to form.

All thunderstorms produce lightning and potentially heavy rainfall. A person can be protected from lightning by seeking shelter inside or crouching on toes and fingertips to become a small target with very little contact with the ground. Severe thunderstorms carry high winds, heavy rain, and possible hail. Supercell thunderstorms are more likely to produce hail, torrential rainfall, damaging winds, and tornadoes. Hazards of severe thunderstorms include lightning, downbursts, derechos, flash floods, hail, and tornadoes. Tornadoes are classified based on the Enhanced Fujita Scale. EF-0 tornadoes are the least damaging, and EF-5 tornadoes are the most damaging. A person can be protected from tornadoes by seeking shelter in the interior of the lowest floor of a building away from windows. Avoid areas with large roof spans such as supermarkets, gymnasiums, and big box stores. If caught outside or in a vehicle, lay flat in a ditch or other low spot. Never stay in a vehicle or mobile home during a tornado.

9

Floods

Bethany D. Hinga
University of Nebraska at Kearney, USA

9.1 Introduction

Extremes in precipitation, both too much and too little, can have catastrophic consequences for populations that live in affected areas. In an average year, floods kill more people in the United States than tornadoes, hurricanes, or lightning. Floods can come on suddenly and unexpectedly, leaving people and animals stranded in areas of rising waters with very few escape routes. They can damage infrastructure and change the course of streams. In this chapter, we will examine several types of floods that occur and their root causes. We will also examine mitigation strategies for homeowners and communities.

9.2 Central China Flood

The Yangtze River (Chang Jiang) is the longest river in Asia. Its headwaters are in the Tibetan Plateau. It flows eastward across southern China, eventually emptying into the Yellow Sea near Shanghai. The 10 provinces surrounding the Yangtze River in central China are home to nearly 500 million people. In 1931, the river valley was heavily populated, both by people who lived at a subsistence level, and by those who grew food for the major cities of the region. Both populations lived near the river where small floods often replenished silt in the rich soil, making the region perfect for growing rice and other crops.

The winter of 1930–1931 was particularly harsh. Abundant snowfall throughout the river basin led to an excess of water when the spring melt began. Small-scale flooding often happens in this area as snow and ice begin to melt. River waters are often held back in the spring by ice jams in the river, which contributes to small-scale flooding. In the spring of 1931, an extremely powerful East Asian Monsoon brought unusually heavy rainfall. This, together with melting snowfall, swelled streams to near flood stage and saturated the ground, raising the water table to nearly ground level. Exacerbating the situation was heavy rainfall in the summer of 1931. In an average year, central China typically has two cyclonic storms. In 1931, the month of July saw seven cyclonic storms, producing 1.5 times the annual volume of precipitation in that single month.

Extreme rainfall in July led to serious flooding. As the waters rose, hundreds of thousands of people were driven from their homes. Flooding became even more severe throughout the first half of August as rain continued to fall. By mid-August, approximately 34,000 mi^2 (88,000 km^2) of land were completely inundated and 8,000 mi^2 (21,000 km^2)) of land were partially flooded. Famed aviator Charles Lindbergh and his wife Anne visited China in the

Earth's Natural Hazards and Disasters, Advanced Textbook 7, First Edition. Bethany D. Hinga.

(a)

(b)

Figure 9.1 (a) Charles and Anne Morrow Lindbergh at the city wall of Nanking, China (now Nanjing), in September of 1931. The Lindberghs stopped in China on their world tour of 1931. They happened to be in China during the catastrophic floods on the Yangtze River. (b) An aerial photo of flooding on the Yangtze River, likely taken by Charles Lindbergh. Photos courtesy of the Yale University Library Manuscripts and Archives. Box 11, Folder 11 of the Lindbergh Picture Collection (MS 325B).

late summer of 1931 (Fig. 9.1). Lindbergh flew around the area and took a number of aerial photographs that allowed people to view the extent of the floods.

Millions were displaced from their homes by flood waters. Many people fled the flooded countryside for the cities, but the cities were flooded as well. Conditions deteriorated within the region throughout the summer and early fall of that year as more and more land slipped beneath the water. The crops on which farmers and city dwellers alike depended were flooded and ultimately failed. Grain storage facilities were also overcome by flood waters, causing the grain to rot. There was little food left for the people of the region. As a result, famine set in. During the flooding, over 150,000 people drowned in floodwaters. Tens of millions of people, which represented nearly 60% of the population in some provinces, were left homeless. This sent many people into refugee camps. In the months that followed, disease ravaged the refugee population. The wet landscape provided an excellent home for mosquitos, and mosquito-borne diseases such as malaria killed up to 300,000 people. A spike in the number of freshwater snails caused a dramatic increase in schistosomiasis, an illness caused by a parasitic worm that uses freshwater snails as hosts. When the parasite leaves the snail and enters the water, people who touch the water have a very real risk of being infected with schistosomiasis. Typhoid, cholera, and dysentery were also spread through contaminated water. Measles and smallpox were transmitted in the overcrowded refugee camps. The actual death toll of this flood event is unknown but estimates place it between 850,000 and 4 million, with many scholars accepting 2 million as a reasonable figure. The death toll includes not only flood deaths from drowning, but also deaths in the aftermath of the flood from disease and starvation.

Like most natural disasters, it was a combination of natural occurrences and human activity that turned this excess rainfall into a deadly event. The Yangtze River has flooded more than 1,500 times since records began approximately 200 BCE. Floods are therefore common events. However, this flood was much larger than most. Certainly, the excess rainfall throughout central and south China caused much of the misery. However, there were a number of additional factors that caused this natural event to turn into a disaster.

Millions of people lived within the areas adjacent to the Yangtze River, known as the *floodplain*. In an ideal world, people would not live in floodplains. However, the river provided much-needed water for large cities and small farms alike. In addition, the soil on these floodplains was fertile, giving people incentive to live on and farm the low-lying land near the river. The river carries a large volume of silt. Small floods carry silt onto the floodplain, contributing to the soil's fertility. This silt can also fill the stream channel, forcing the water over its banks. China had a long history of managing the silt volume in this river through regular dredging prior to the 1931 floods. Civil and international wars, however, had interrupted removal of silt from the river channel in the previous few years, making the flood worse than it might have otherwise been. *Levees* and dams that struggled to control the river were overtopped or destroyed. People who relied on those features to give them some measure of protection were surprised when they failed, allowing flood waters to swamp areas that normally were not covered. In addition to the river expanding its banks and causing death by drowning, disease and famine were ruthless killers. The nation was unable to prevent or control the rapid spread of disease during and immediately after the floods, which led to immense loss of life. The ongoing war between nationalists and the Communist Party also prevented some aid agencies from distributing food and medicine throughout the flood zone.

Many people did survive by eating fish, aquatic plants, and waterfowl. Some people earned money by renting parts of their undamaged homes. Others used their family boats as water taxis. Wealthy families, religious organizations, and benevolent societies supplied food and shelter to the best of their ability. Other nations contributed food, money, and scientific expertise. Ultimately, the government took stock of the nation's flood control infrastructure and built thousands of miles of new and improved levees and dikes.

9.3 The Nature of Streams

The term *stream* is used to describe a body of flowing water confined within a channel, regardless of the size of the channel or volume of flow. The term *river* is often used to denote a larger stream. Chances are you live near a river or stream and have a firm picture in your mind of what a stream looks like. Streams are quite variable in their appearance. Some are wide and curvy. Others are narrow and straight. Some flow over steep terrain and actively work to cut downward into the rocks below. Some of the largest rivers flow great distances over nearly flat ground. Still others only flow during heavy rainstorms. The shape of a stream has a great deal to do with the terrain over which it flows and the climate of the area it drains.

The area that contributes water to a particular stream is called its *drainage basin* (Fig. 9.2). An observer first identifies which particular stream he or she is interested in investigating, then the drainage basin for that stream can be defined. A drainage basin, also called a *watershed*, is outlined by high points in the area, called *divides*. Divides are high elevations in the landscape, such as the tops of ridges in mountainous terrain. Rain falling on a divide will either flow down one slope or the other, in this way dividing drainage basins from each other. Within the drainage basin for a particular stream, all water within its boundaries

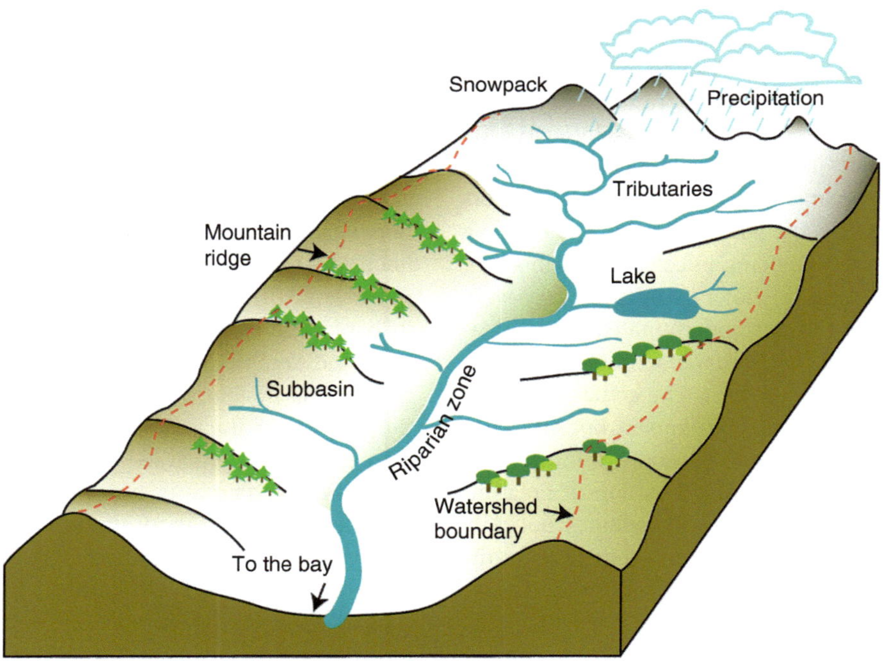

Figure 9.2 A hypothetical watershed that gathers water from precipitation and melted snow into tributaries that lead to a major stream. The watershed boundary is outlined in red. From The Watershed Project. http://thewatershedproject.org/wp-content/uploads/2017/02/watershed_schematic.jpg

drain into that stream. For a small stream in the mountains, a drainage basin may be tiny, encompassing only a single canyon or valley. However, that one small stream likely empties into a larger stream. In that case, the small stream is called a *tributary* to the larger stream. A drainage basin for a larger stream includes the drainage basins for all of the smaller tributaries that contribute to it. The Mississippi River (Fig. 9.3), as an example, has a huge watershed that covers almost half of the contiguous United States. It includes the watersheds for the Missouri River, Arkansas River, Red River, and Ohio River as well as all of those rivers' tributaries.

Each river or stream has a beginning and an end (Fig. 9.4). The *headwaters* of a stream is the location where the stream begins. It is found at the highest elevation along the stream channel. The *mouth* of a stream is the location where the stream empties into a larger body of water, and is at the lowest elevation along the stream channel. This may be a lake, another river, or the ocean. In many cases, a delta forms at the mouth of a stream, especially if the body of water into which the stream flows is a lake or the ocean. A *delta* is a fan-shaped deposit that forms when sediment is dropped out of the river as it enters another body of water and slows down (Fig. 9.5).

9.3.1 Stream Parameters

Scientists monitor stream channels for fluctuations in the stream's character. They use a variety of parameters to describe the state of a stream, often from several locations along the waterway.

The slope of the stream channel, called the *gradient*, varies with distance along the stream. In general, the gradient of a stream is greatest near the headwaters and least near the mouth. Gradient is most often expressed in feet per mile (or meters per kilometer). To calculate

NASA

Figure 9.3 Mississippi River watershed is the largest drainage basin in the United States. It includes 3.2 million km^2, and the smaller drainage basins of the Ohio, Missouri, Arkansas, and Red Rivers. From Fletcher, 2014.

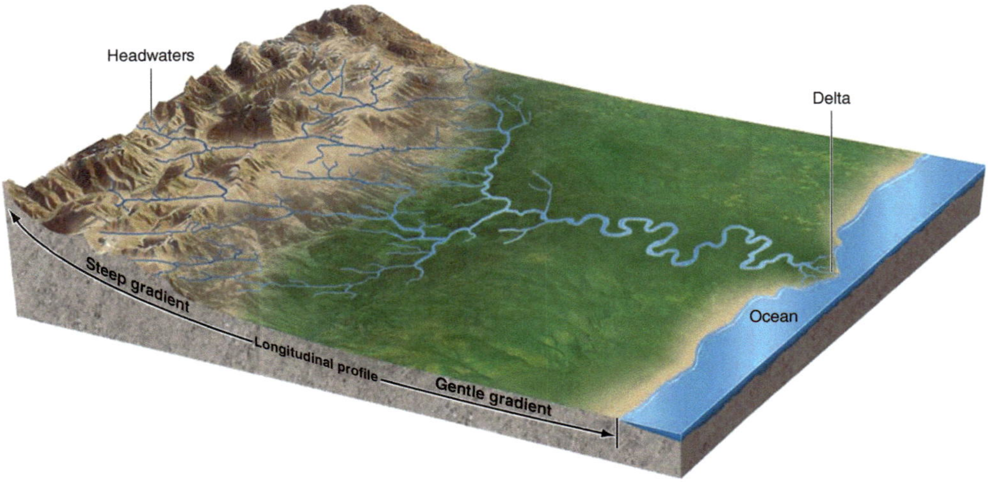

Figure 9.4 Profile of an ideal stream from headwaters to mouth. The gradient of the stream is steepest near the headwaters, and gentlest near the mouth. The mouth of this stream is found at the ocean. There, a deposit of sediment called a delta forms where water and sediment pour into the ocean. From Fletcher, 2014.

Figure 9.5 Satellite image of Egypt and the Sinai Peninsula. The green fan-shaped feature at the top of the image is the Nile River Delta, formed where the Nile River empties into the Mediterranean Sea. Courtesy of Visible Earth/NASA. From Murck & Skinner, 2016.

the gradient of a specific segment of the stream, find the elevation at the beginning and the end of that stretch of stream. Measure the length of the stream segment, following all bends and twists in the river. Gradient is equal to the elevation change (in feet) divided by the distance traveled by the stream (in miles). The result is a value expressed in feet per mile, or meters per kilometer.

Another stream characteristic that scientists routinely calculate is the stream's *discharge*. Discharge is defined as the volume of water moving past a specific point in the stream in a given period of time (typically each second). Two measurements are needed to calculate a stream's discharge. The first is the cross-sectional area of the stream at the specific location where discharge is being measured. Second, the stream's velocity must be measured at that location. When you multiply the stream's cross-sectional area (square feet or square meters) and the velocity (feet per second or meters per second), the result is a volume (cubic feet per second or cubic meters per second). This is done for thousands of streams both large and small through the use of permanent stream gauges, installed and monitored by the United States Geological Survey. When the stream gauge is installed, scientists measure the cross-sectional area of the stream channel at that location. It therefore becomes a known parameter. Instruments in the stream automatically measure the velocity on a regular basis, allowing for automatic calculations of discharge. These data are invaluable to the people who monitor streams for flooding.

Discharge is plotted on diagrams called *hydrographs* (Fig. 9.6). Hydrographs are records of a stream's discharge at a specific point on that stream, plotted over time. When monitored for a period of a few years, it is possible to identify the typical discharge for a specific stream, called *base flow*. Base flow is the sustained low flow in a river during dry or fair weather conditions. It usually represents the amount of water supplied to a stream by groundwater, or water that is stored and migrates below the surface of the Earth. Rainfall events and periods of snowmelt contribute additional flow to the stream. These events are easy to see on a hydrograph because they appear as peaks above the typical base flow.

When discharge increases at any given point in the stream, the volume of water increases because the height of water within the stream channel increases, and the water moves faster. The height of water within a stream channel, above a defined location such as the bottom of that stream channel, is called its *stage*. In the United States, stage is usually reported in feet. During flood events, there is often discussion about how far above flood stage the water will climb. Flood stage is the height of water in the river at which it is considered to be in flood. Flood thresholds can also be expressed in terms of discharge. For example, a river may have a base flow of 2,000 ft^3/s (cubic feet per second) (56.63 m^3/s (cubic meters per second)), and a flood stage discharge of 10,000 ft^3/s (283 m^3/s). Any discharge above 10,000 ft^3/s describes the river in flood. The magnitude (size) of the flood can therefore be determined by its discharge.

Yet another calculation scientists make of stream parameters is that of *recurrence interval*. A recurrence interval, or return period, is the average length of time between events of a specific size. Since these are averages, it is very likely that they will occur earlier or later than the stated recurrence interval. You have likely heard sizes of floods referred to as the "100 year flood" or the "500 year flood." These are theoretical *return periods* for floods of a certain size at the same location within a stream. These recurrence intervals are calculated using discharge records for the stream.

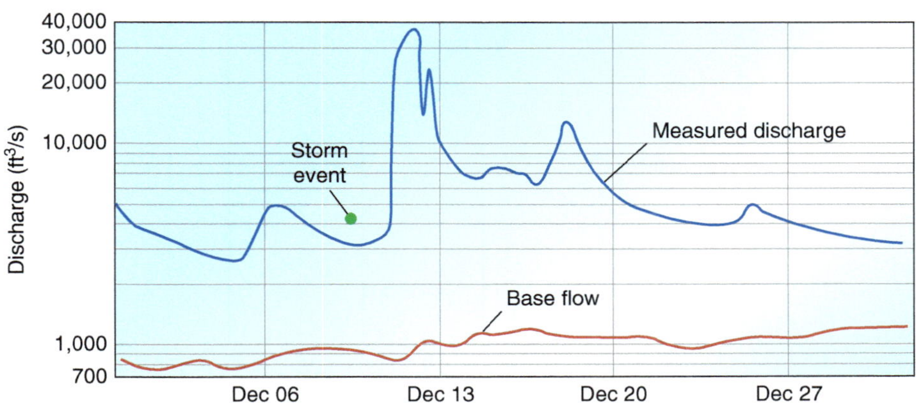

Figure 9.6 Hydrograph of the Rappahannock River near Fredericksburg, Virginia. Base flow for this stream averages around 1,000 ft^3/s, but after a heavy rainfall, discharge at this location reached nearly 40,000 ft^3/s. Data from U.S. Geological Survey. From Arbogast, 2017.

9.3.2 Procedure for Calculating Recurrence Interval and Flood Probability

To calculate recurrence interval and flood probability, first list peak discharge for each year within your records. Next, each year's discharge is ranked in order from largest to smallest. The largest peak discharge is assigned the rank of 1. Second largest peak discharge is assigned the rank of 2. Continue this until you have assigned a rank to each peak discharge. Recurrence interval (R.I.) for each peak discharge in the list is calculated by

$$\text{Recurrence interval} = (N + 1)/M$$

where N = the number of records you have in years and M = the rank of that event. For example, if you have 50 years of records for a stream, and the particular event you're looking at is ranked 6 in the list, the recurrence interval is equal to

$$\text{R.I.} = (50\ \text{years} + 1)/6 = 8.5\ \text{years}.$$

In reality, the flood of that size will probably not recur in 8.5 years. It is very likely to occur earlier or later. A flood of that size may even occur in two successive years. The recurrence interval is an expression of how often a flood of that size will occur, on average. However, given the fact that nature rarely adheres to such regular patterns, a better way to express the recurrence interval is to calculate flood probability using the formula

$$\text{Probability of a flood of this size occurring in any given year} = (1/\text{Recurrence Interval}) \times 100\%.$$

For a recurrence interval of 100 years, the probability of having a flood of that size in any given year is (1/100) × 100%, or 1%. In our example above, the odds of having a flood with a recurrence interval of 8.5 years is 1/8.5, or 0.12, which equals a 12% chance of having a flood of that magnitude in any given year.

For each stream with sufficient data, it is possible to calculate *flood frequency curves* (Fig. 9.7).

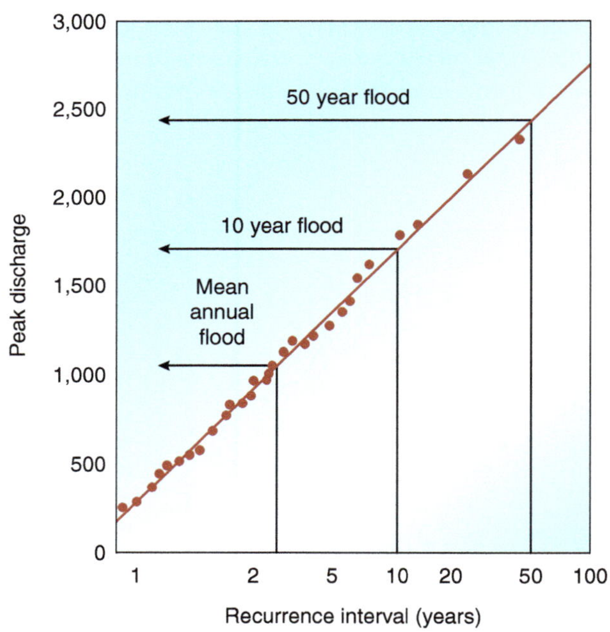

Figure 9.7 Flood frequency data for the Skykomish River at Goldbar, Washington. Flood frequency curves are used to determine the discharge that can be expected for a given flood's recurrence interval (bottom scale). In this example, the 10 year flood would have a peak discharge of approximately 1,750 ft^3/s. From Arbogast, 2017.

These allow us to determine what the discharge will be for a given recurrence interval. Recurrence interval in years is plotted along the X axis, usually on a logarithmic scale. The logarithmic scale allows us to compress the recurrence interval scale into a smaller space. Equal distances along the X axis are between powers of 10. Discharge is plotted along the Y axis. Peak discharges are plotted as points on the graph. A best-fit line is drawn between the points. Once the line is drawn, peak discharge for any recurrence interval can be read off the chart. In the example given in Figure 9.7, the 10 year flood has a peak discharge of about 1,750 ft^3/s. Another way to think about this is that in any given year, there is a 10% chance that a flood with a discharge of 1,750 ft^3/s will occur. The 100 year flood has a peak discharge of approximately 2,750 ft^3/s. Therefore, there is a 1% chance in any given year that a flood with a discharge of 2,750 ft^3/s will occur.

9.3.3 Stream Shapes

In many ways, the gradient of a stream determines the shape of the stream. Steeper gradients found near the headwaters of a stream are more likely to have straight stream segments. These young streams actively erode downward into the rocks over which they flow. Streams with steep gradients have great erosive power, and have the capability to carry not only a great volume of sediment, but large particles such as boulders.

When a stream makes the transition from a steep mountain stream to the shallow gradient at the base of the mountains, the stream velocity slows considerably. Sediment that could easily be carried by faster-moving water on the steeper segments of the stream become too heavy for a slower stream on less steep ground to carry. It is here at the change in gradient from steep to shallower that we see the stream change character and shape. Sediment drops out of the water and is deposited at the base of a slope. Here, a fan-shaped deposit called an alluvial fan is likely to form (Fig. 9.8). If the stream continues onward with sufficient flow, the stream may be *braided* (Fig. 9.9). The stream channel becomes choked with sediment that continues to drop out of the stream, and the water in the stream has to make its way around these

Figure 9.8 Straight stream segment with an alluvial fan. Sediment that was carried by the steep stream in the mountains is dropped at the base of the slope, where the landscape becomes less steep. The result is this fan-shaped deposit. Photograph © Dr. Marli Miller, used with permission.

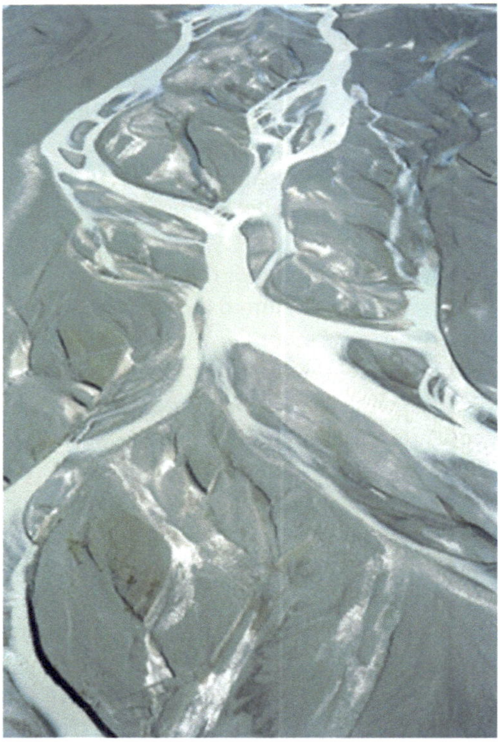

Figure 9.9 Braided stream. A stream is braided when the river's multiple channels intermingle. From Murck & Skinner, 2016.

deposits of sand and gravel. One area of the United States that is known for braided streams is central Alaska.

On nearly flat plains, a more common stream shape is that of a *meandering stream*, or one that has many bends and turns (Fig. 9.10). Meandering streams flow in sinuous patterns across a nearly flat landscape. Bends in the stream are called *meanders*. Meanders are formed when water scours the outer portion of a bend in the stream, creating an ever-changing landscape of bends in the stream. The outer part of a meander is called a *cut bank* (Fig. 9.11). Erosion of the stream bank happens at cut banks because the velocity of the stream is greatest there. The faster water has the ability to erode material from the edge of the stream. On the inside of a meander, the water velocity is slower, and because slower water cannot carry as much sediment as faster water, deposition of sediment occurs where the water slows down. The depositional features formed here are called *point bars*. Point bars are generally composed of sand or gravel.

Meandering streams cut a broad swath through the landscape, creating wide plains.

Figure 9.10 Meandering stream segment. Note the pronounced bends in the stream, and nearly flat, low-lying floodplain adjacent to the stream. From Arbogast, 2017.

Figure 9.11 Aerial photo of a single meander in a meandering stream. The outside of the meander is the location of a cut bank (left), where erosion occurs because velocity is highest within the stream channel. Where velocity is lowest, on the inside of the meander, sediment is deposited and forms a point bar (light colored area on right). Photograph © Dr. Marli Miller, used with permission.

These low-lying areas next to the stream are called floodplains. The floodplain is inundated with water when a stream floods. Both meandering and braided streams tend to have wide floodplains. When floods hit these areas, the water expands laterally away from the stream, making the stream much wider than usual.

Still another type of stream is the *intermittent stream*. This is a channel that only has water flowing within it during part of the year. It may flow, for example, when rain falls or if it is fed by snowmelt during the warm season. Intermittent streams can flow within a number of different land features. *Wadi*, *arroyo*, *wash*, and *gulch* are terms that are used to refer to stream valleys that are dry much of the year. They are common in semiarid and arid climates, and are prone to rapid flooding once rain begins falling or the spring snowmelt begins.

9.4 Types of Floods

The National Oceanic and Atmospheric Administration (NOAA) is the United States' official agency responsible for monitoring and studying weather. They define five different types of floods: flash floods, river floods, storm surge, coastal floods, and inland flooding.

9.4.1 Flash Floods

Flash floods most commonly occur when heavy rainfall occurs over an area, causing water to rise quickly (Fig. 9.12). They can also happen when water is released suddenly from an ice jam, something that requires no rainfall. (An ice jam occurs when blocks of ice hold back water in a river, almost like a small dam.) A third scenario in which flash floods occur is when a dam or levee fails and releases great quantities of water. Flash floods due to excess rainfall are more likely to occur if the ground is already saturated, or if the ground is not terribly porous, such as in areas that have little or no soil cover over bedrock. Flash floods can happen on any type of terrain. They can happen on relatively flat ground, and can be particularly deadly in steep-walled canyons.

The United States' deadliest flash flood was one that occurred in Pennsylvania in 1889, referred to as the Johnstown Flood (Fig. 9.13). Heavy rainfall on 30–31 May caused minor flooding in several communities within the Little Conemaugh River valley, including the city of Johnstown, home to about 30,000 people. The 10 in. of rain that had fallen on the river valley had also affected Lake Conemaugh, the centerpiece of the South Fork Fishing and Hunting Club, which hosted the likes of

Figure 9.12 Flash flood damage to a road. The dry intermittent stream channel to the left became flooded following a heavy rainfall. Photograph © Dr. Marli Miller, used with permission.

Figure 9.13 Residents carrying a deceased person to the makeshift morgue in Johnstown, Pennsylvania, in May of 1889. The wreckage of the city, which was flooded when the South Fork Dam on the Little Conemaugh River failed, is in the background. Photo from The Johnstown Flood of 1889 Collection at Pennsylvania Highlands Community College; http://digitalcollections.powerlibrary.org/cdm/singleitem/collection/acacc-jtf/id/4783/rec/7

Andrew Carnegie. The lake was approximately 2 mi (3.2 km) long, 1 mi (1.6 km) wide, and 60 ft (18 m) deep at the dam. The South Fork Dam was an earthen dam, and it had not been particularly well maintained. With the heavy rainfall that day, the dam showed obvious signs of instability. People made heroic efforts to save the dam. They attempted to add height to the dam itself, and dug a secondary spillway to relieve pressure from the existing dam. Word was passed to a train engineer that there were mudslides across the tracks in a few places within the valley. He also received word from a freight agent at South Fork that the dam was likely to give way soon. He thanked them for the warnings and promised to pass word on to those in the valley that a dam break was likely. He did as promised, warning people along the way that the dam was failing and they should "fly to the hills." Few people in the valley were concerned. They had heard about imminent dam breaks before and they had never materialized. Unfortunately, shortly after 3 p.m. on 31 May, the dam gave way.

People who witnessed what came next described a roiling 40 ft (12 m) high, half-mile (0.8 km) wide wall of water and debris created as the lake emptied in 45 min. The communities of South Fork, Mineral Point, Woodvale, and East Conemaugh were hit hard by the flood. Flood waters destroyed, then carried the remains of homes and barns. Animals and people, both alive and dead, were caught up in the flood as well. The city of Johnstown was located 14 mi (22.5 km) downriver from the dam. When the flood reached Johnstown, it struck the Pennsylvania Railroad Company's Stone Bridge and the train on it. The flood destroyed the train, leaving mangled wreckage in its place. It then deposited debris 40 ft (12 m) high over a 30 ac (12 ha) expanse. Soon after, a kerosene tanker tangled in the debris caught fire and burned for 3 days. The flood destroyed many homes and businesses in the center of Johnstown and caused a great deal of damage at the Cambria Iron Works and other large employers in town. The water kept moving, scooping up more debris, more animals, and more people. It flowed through Nineveh, Bolivar, and other communities downstream. Amazingly, a few people did survive being caught up in the flood waters. Those were the lucky ones who landed on floating debris. A 6 year old girl was rescued as she hung on to a floating mattress.

News of the disaster spread quickly. Over 100 newspapers and magazines sent journalists to cover news of the flood. Hospitals sent medicine and bandages. Doctors from other communities rushed to Johnstown to tend to the wounded. People sent money, clothing, food, and materials to rebuild homes and businesses. Makeshift morgues were set up in Johnstown so families could identify and claim the 2,209 dead. Even Clara Barton and her Red Cross assisted in building hotels for people and warehouses for supplies. They were an integral part of the disaster response. Just over a month later, some businesses on Main Street reopened. The Cambria Iron Company was open on 6 June. People took advantage of the resources they were given to recover from the disaster. However, floods continued to plague the town for many years after the flash flood that claimed so many lives. Finally, after another major flood in 1936, real efforts were made to control flooding within the community. Flooding has been reduced in the city of Johnstown, but another major flood occurred in 1977.

Another major flash flood occurred in 1976 in the Big Thompson Canyon near the city of Loveland, Colorado. Circumstances were markedly different from that in Johnstown, Pennsylvania in 1889. The Big Thompson River usually flows through the rocky canyon at a depth of about 18 in. (46 cm). However, on 31 July 1976, a freak thunderstorm dropped 12–14 in. (30–36 cm) of rain at the top of the canyon in just 4 hr. Waters moved downhill swiftly into the canyon, swelling the river to a 20 ft (6 m) wall of water. Highway 34, which runs through the canyon just above the normal height of the river, was washed out in multiple places. About 4,000 people were in the canyon that day. Many heard of the flood by word of mouth since power lines and electric lines were brought down by the flood waters. Many tried to escape via Highway 34, and several of them were caught in the flood waters in their car and drowned. Several who escaped did so by climbing trees or trekking up the wall of the canyon. The flood caused 143 deaths and injured 150 people. Some bodies were never found. Others were transported as many as 25 mi (40 km) away. The flood caused $35 million in damage. It also destroyed 418 homes, 52 businesses, and 438 vehicles as well as several bridges and the highway. When the highway was rebuilt, the State of Colorado erected signs throughout the canyon advising people to climb their way to safety instead of driving. These signs no doubt saved lives when another flash flood damaged the highway through the canyon in 2013.

The Big Thompson Canyon flood was an excellent example of what can happen when an unusually heavy rainfall event affects a steep-walled canyon with only the smallest of floodplains. Flash flooding also happens on much less dramatic terrain, though it is no less dangerous when it does. Many desert cities have intermittent streams in their midst. Since these intermittent streams have water in them only when rain falls, people can get complacent and forget that these areas turn into raging rivers. The city of Las Vegas, Nevada, has experienced a number of flash floods. The latest (as of this writing) was in July of 2018 when 2–3 ft (60–90 cm) of water covered portions of the major highways through Las Vegas and nearby Henderson.

Loss of life in flash floods often happens when people attempt to drive through flood waters and realize too late that the water is deep and moving incredibly quickly. Just 6 in. (15 cm) of fast-moving water can carry a car off the road. A national public safety marketing campaign has been underway since 2003, urging people to "Turn Around, Don't Drown." If water is covering a road, turn around rather than trying to drive through the water. It is impossible to see what lies beneath the water's surface. The water could be hiding the fact that the road is washed out, or that a sinkhole has formed in the street.

9.4.2 River Floods

River floods are different from flash floods in that they develop over a longer period, typically weeks or months. They are usually the product of heavy rainfall over a prolonged period. They can also happen because of melting of heavy snowfall, or because of ice jams on rivers in the springtime.

One of the United States' most severe river floods occurred in 1993 on the Missouri and Mississippi rivers, as well as their tributaries (Fig. 9.14). This flood was about a year in the making. The summer of 1992 and winter of 1992–1993 were both wetter than usual, creating higher than normal soil saturation levels. Between April and August of 1993, much of

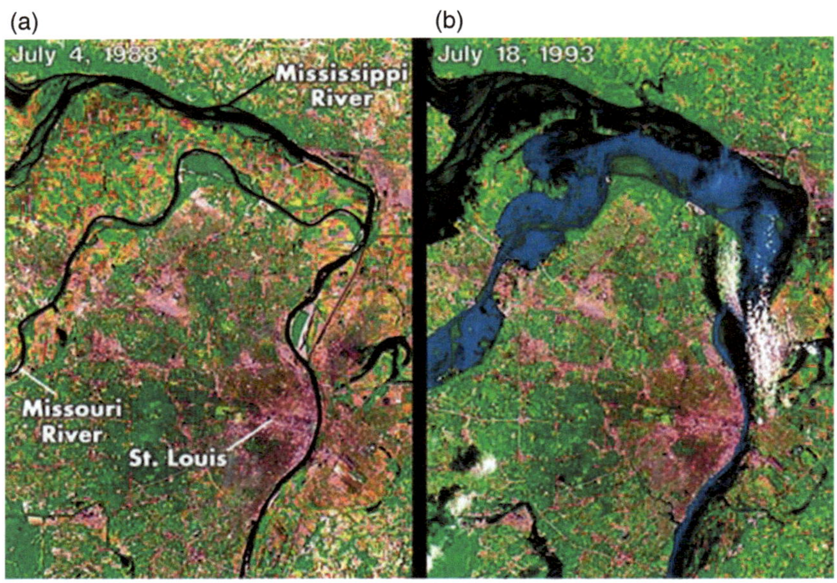

Figure 9.14 Satellite images of the confluence of the Mississippi, Missouri, and Illinois rivers near St. Louis, Missouri. (a) The first image is from a "normal" year (taken 4 July 1988), and (b) the second from 19 August 1993 during the height of the flooding. Image courtesy of NASA/Science Photo Library/Photo Researchers. From Arbogast, 2017.

the upper Midwest received more than twice the normal amount of rainfall. In parts of east-central Iowa, rainfall was two times to three and a half times normal rainfall for that time period, and 12 in. (30 cm) more than the normal annual rainfall in the area. In July, many parts of the flood area experienced rainfall on at least 20 days, which is significantly more than the average of 8–9 rainy days in that month. In some areas of the upper Midwest, there was measurable rain every day between late June and late July. Some areas received up to 4 ft (1.2 m) of rainfall. For nearly 2 months, bridges were inaccessible on the Mississippi River from Davenport, Iowa, to St. Louis, Missouri (Fig. 9.15). The Mississippi River at St. Louis reached a record height of 47 ft (14.3 m) on 20 July. Much of the city of Des Moines, Iowa, was under water beginning 9 July from record flooding on the Skunk, Iowa, and Des Moines rivers. It took weeks for the waters to recede so people could return to the flooded areas. Many communities along the affected rivers experienced breached levees, which means people who thought their homes were protected by these flood control structures lost those homes quickly once a breach appeared in the nearby levee. The rain fell on already saturated ground, meaning that the only place the water could go was to flow across the land and end up in rivers and streams. Flooding occurred in North Dakota, South Dakota, Nebraska, Kansas, Minnesota, Iowa, Missouri, Wisconsin, and Illinois. Nearly 150 streams were affected. This was the most severe flooding to happen in this part of the country in the approximately 150 years that records have been kept. The death toll of this major nine-state flood event was 50 people. Damages approached $15 billion.

In 1997, the Red River of the North, which flows northward through North Dakota and Minnesota into Manitoba, experienced severe flooding. The winter of 1996–1997 saw record snowfall. In eastern North Dakota where the normal annual snowfall was about 38 in. (0.97 m), they experienced snowfall of about 110 in. (2.8 m). This left the area with a great deal of extra meltwater to contend with once the spring thaw got underway. They unfortunately experienced all of the factors that lead to severe spring floods. There was heavy snowfall in the winter. The ground was frozen, preventing infiltration of meltwater into the

Figure 9.15 Aerial photograph of flooding along the Mississippi River in 1993. Photo courtesy of the U.S. Geological Survey. From Arbogast, 2017.

ground once melting began. They also had a late spring thaw and above-average snowfall during the spring thaw, including a blizzard in early April. Since the Red River of the North flows northward, the upper reaches of the watershed in the south melt sooner than the lower part of the river in the north. When this occurs, ice jams are created. An ice jam is a temporary natural dam that blocks river flow downstream. Water therefore backs up and floods the landscape upstream of the ice jam. All of these factors came together to cause some of the most severe flooding the region had seen in 100 years. Artificial levees built in Grand Forks, North Dakota, and East Grand Forks, Minnesota, were overtopped or breached in mid-April of 1997. The river crested at 53.99 ft (16.46 m). Nearly 4 ft (1.2 m) of water covered 75% of the city of Grand Forks, North Dakota. Virtually all of the city of East Grand Forks, Minnesota, was flooded. A large fire in flooded downtown Grand Forks destroyed 11 buildings and 60 apartment units.

The year 2008 was another year of record-setting flooding. Areas in Iowa and Illinois that were largely spared from the worst of the 1993 floods found themselves in the midst of record-setting flooding following excess precipitation in the spring and early summer. In July of 2008, the Cedar River in Iowa crested at 32 ft (9.8 m), which was 12 ft (3.7 m) higher than the previous record set in 1929 (Fig. 9.16). Approximately 20,000 people had to evacuate their homes, which is a significant portion of its population of 120,000. An estimated 439 city blocks were under water. Levee breaks along the Mississippi River caused many people in rural areas and small towns to evacuate. The University of Iowa experienced flooding that damaged several buildings on campus. Record flooding occurred in the city of Des Moines, Iowa, on the Raccoon and Des Moines rivers. Damages across the affected area were estimated to be nearly $10 billion.

Spring of 2019 brought devastating flooding to the Midwest, including Nebraska, Iowa, and Missouri. The region had experienced a harsh winter with heavy snowfall. In mid-March, the temperature warmed considerably and the snow melted quickly. Immediately following an episode of rapid snowmelt, a spring blizzard dumped several additional inches of

Figure 9.16 Extreme flooding in Cedar Rapids, Iowa, along the Cedar River, 2008. Photo courtesy of the Iowa Air National Guard.

snow on the area. It was again followed by a warm spell, which contributed extra water to area rivers. Shortly thereafter, a period of heavy rainfall began. Rivers and streams were overwhelmed, causing widespread severe flooding. In Iowa, flooding occurred on many major rivers throughout the state. The Iowa National Guard was deployed to help shore up levees with sandbags in an effort to prevent water from rising into residential neighborhoods and business districts. Unfortunately, the rivers rose and covered floodplains immediately next to the major rivers.

In eastern Nebraska and western Iowa, rural communities were hit hard by flooding. Dams broke, levees were breached and overtopped, and rivers rose well out of their banks. Many bridges were washed out, leaving communities isolated. Grain bins that stored the previous autumn's harvest were flooded and destroyed, leaving farmers without income from those grains. The floods struck during calving season. Tens of thousands (perhaps hundreds of thousands) of head of livestock, including newborn calves, were carried away by the floodwaters. Other cattle were stranded on isolated patches of land by floodwaters. The loss of mature cattle and the year's calves was a severe emotional and financial blow to farmers in the region. Farmers who count on those livestock for their income were left with the prospect of severely reduced income for the year. The Nebraska National Guard enacted Operation Hay Drop in which they airlifted large bales of hay to livestock.

People expected that the floodwaters would recede in due course. However, an unusually rainy spring meant that many fields were too wet for farmers to plant. In mid-May, which is usually the end of planting season, only about 67% of farms were planted. Farmers were concerned about their livelihoods. Tariffs enacted in 2018 and 2019 dried up the international grain market, leaving many farmers with soybeans and corn they could not sell to their usual customers overseas and depressed domestic prices. The additional combination of losing stored grain and livestock, and reduced crop volumes in the upcoming year spelled disaster for thousands of farmers. Many wondered how long they could afford to hold onto their land and make a living from farming.

During the floods of 2019, infrastructure was severely disrupted. Bridges were destroyed, roads were damaged, and many roads were covered with water not for days, but for months. Interstate 680 that connects Iowa and Nebraska in north Omaha, Nebraska, was closed for 3 months. Interstate 29, which runs along the western edge of Iowa and Missouri was closed even longer as floodwaters from the Missouri River and its tributaries remained high throughout the entire spring. Even as late as December 2019, floodwaters could be seen standing in fields. The U.S. Army Corps of Engineers was responsible for assessing damage to levees as a result of the floods. At the end of June 2019, the estimated cost to repair levees was over $900 million.

9.4.3 Storm Surge

Storm surge happens when a tropical storm comes ashore and floods the coastal area with seawater. This is a natural part of landfall for hurricanes. The high winds of the hurricane causes an increase in water height as the storm approaches shore (Fig. 9.17). In general, the stronger these winds are, the higher the storm surge is. (The height of a storm surge is also affected by the size and severity of the storm.) However, the shape of the ocean floor offshore causes variation in storm surge heights. Areas that have a wide and shallow continental shelf, such as that seen in the Gulf of Mexico, can produce storm surges of 20 ft (6 m) in height. Areas with narrow continental shelves, such as south Florida, may only see 8 or 9 ft (2.4 to 2.7 m) storm surges for the same strength hurricane.

Storm surge is one of the greatest threats to life and property in a hurricane. A storm surge causes destruction of structures in the direct

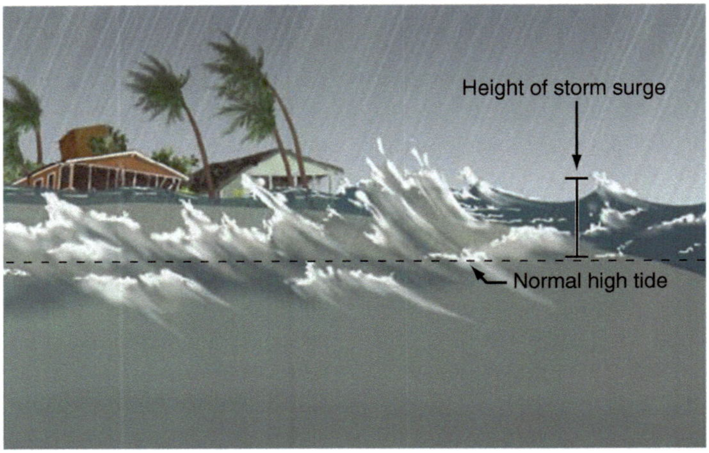

Figure 9.17 Storm surge rushes onshore when a hurricane makes landfall. Storm surge can be several feet above normal high tide level, causing severe and sudden flooding in coastal communities. A storm surge combined with a high tide can lift the sea level so that structures are subjected to continuous pounding of heavy surf. From Strahler, 2013.

path of the surge, and can lead to severe flooding as water rushes over the landscape. Water has great capacity to erode sand, not only from beaches but also from base material beneath homes, roads, and other structures within the surge's reach. In estuaries or other coastal habitats, storm surge can increase the salt content of the water, forcing mobile animals to flee inland, and killing fragile plants that had adapted to a certain level of salinity.

The greatest loss of life in a hurricane in the United States can be attributed to storm surge. In 1900, the community of Galveston, Texas, was hit by a major hurricane. Landfall occurred at high tide, which turned the storm surge into a more dangerous, higher storm tide of 8–15 ft (2.4–4.6 m). Nearly the entire city was inundated with water and at least 8,000 people died as a result.

In 2008, Hurricane Ike wandered across the warm waters north of the Lesser Antilles and into the Caribbean. It hit the Turks and Caicos, then moved on to the Bahamas and Cuba as a Category 4 storm. It then moved into the Gulf of Mexico and weakened to a Category 2 storm as it made landfall on the northern part of Galveston Island. The Bolivar Peninsula of Texas, slightly to the northeast of where the eye made landfall, experienced storm surges of 15–20 ft (4.6–6 m). Galveston Bay saw storm surge of up to 10 ft (3 m). On the Bolivar Peninsula, homes were wiped clean of their foundations and entire neighborhoods disappeared (Fig. 9.18).

9.4.4 Coastal Floods

The fourth type of flood is a *coastal flood*. Coastal flooding occurs when extremely high tide carries water farther inland than usual, causing water to cover low-lying areas with elevations just above the average high tide. This tends to happen most often at new and full Moon phases, when the Moon and Sun are pulling together on the Earth's waters, causing more extreme tides than usual. The problem of coastal flooding is exacerbated when high winds or storms accompany high tide. As sea levels continue to rise due to climate change, coastal flooding is expected to only get worse. Some areas such as the Florida Keys, the Maldives in the Indian Ocean, and the Marshall Islands in the Pacific are experiencing coastal flooding on a daily basis. In the

Figure 9.18 Storm surge damage in a neighborhood on the Bolivar Peninsula of Texas from Hurricane Ike, 2008. The house in midimage was the only structure left standing in Gilchrist, Texas, following the storm surge from Hurricane Ike. From National Weather Service HGX and Galveston County OEM.

Maldives, they have shored up dikes and seawalls in many heavily populated areas to try to stem the encroachment of the ocean, but this is not likely to control coastal flooding for much longer. Homeowners in areas that have begun to experience coastal flooding are concerned about their homes flooding as well as falling property values. Some are concerned, with good reason, about their property becoming uninhabitable as coastal flooding becomes more frequent and waters encroach on their homes.

In some areas of the world, not only is sea level rising, but the land is sinking. The city of Venice, Italy, is located on the shore of the Adriatic Sea. It is famous for its canals. Unfortunately, the city is also becoming increasingly famous for its coastal flooding (Fig. 9.19). Subsidence, or sinking, of this region of Italy has led to more frequent flooding of the city.

9.4.5 Inland Flooding

Inland flooding is most commonly associated with tropical or extratropical storms that make landfall and continue to drop large volumes of precipitation as they move inland. All hurricanes that move onto land produce inland rainfall, and it is quite common for that rainfall to cause floods, even hundreds of miles from the coast.

Occasionally, hurricanes stall over land and dump incredible volumes of rainfall on the area. This was the case in the late summer of 2017. Hurricane Harvey originated in the Caribbean as a tropical depression in mid-August. It crossed the Yucatan Peninsula of Mexico and entered the Gulf of Mexico on August 22, where it developed quickly into a tropical storm, then a rapidly intensifying hurricane. Harvey was a Category 4 hurricane with 130 mile per hour (209 kilometer per hour) sustained winds when it made landfall on August 25. It came ashore about 30 mi northeast of Corpus Christi, Texas. Significant damage occurred throughout the area. The hurricane weakened to a tropical storm, then moved northeast, paralleling the Texas coast near Houston. Here, it stalled for five long days. The warm Gulf of Mexico supplied a great deal of moisture to the storm for the duration of

Figure 9.19 A sinking coastline has left Venice vulnerable to coastal flooding. The Adriatic Sea is encroaching on the city, and often causes flooding of more than Venice's famous canals. Here, high water has inundated the Piazza San Marco and an outdoor café. Photo © M. Smith/Sipa Press. From Strahler, 2013.

its stay on the Texas coast. Torrential rain and flash flooding occurred in the Houston area, and the rain kept falling. Ultimately, Harvey dumped a record-setting 52 in. (132 centimeters) of rain on some parts of the Houston metro area. For comparison, average annual rainfall for the Houston area is 50 in. (126 centimeters). All low-lying areas were flooded, and water crept upward into areas that had never experienced flooding before. Entire neighborhoods were submerged. First responders were out in force to rescue people in flooded areas. On August 29, the storm moved back onshore near the Texas-Louisiana border, then continued northeastward through Louisiana, Mississippi, and Tennessee.

Hurricane Harvey claimed 80 lives and displaced thousands of families (Fig. 9.20). One year after the hurricane, nearly 10 % of people displaced by Harvey had not returned home. Fifteen percent of the hundreds of thousands of homes damaged in the storm were still uninhabitable. Thirty percent of people in the 24 county area affected most by the storm say their lives were still at least somewhat disrupted. Damages from Hurricane Harvey were estimated to be $125 billion, making it one of the most expensive natural disasters in U.S. history.

9.5 Flood Mitigation

As with other natural hazards, there are a number of strategies that can be utilized to reduce risk from floods. The most obvious and reliable way to protect populations against floods is to prevent homes and businesses from being built in floodplains.

When a person decides to purchase a home and applies for a mortgage, part of the process of purchasing that home is a flood survey. Floodplain maps of the area are checked to be sure the property is not in a known

Figure 9.20 Members of the Kentucky Air National Guard's 123rd Special Tactics Squadron fly on board a U.S. Coast Guard helicopter near Houston, 29 August 2017. Airmen are patrolling the area in motorboats, looking for people trapped in homes or on rooftops because of the massive flooding. Once the residents are safely in the boats, Airmen provide any necessary medical care and transport them to the nearest shelter. U.S. Air National Guard photo, https://www.af.mil/News/Photos/igphoto/2001802560/

floodplain. If the property is at risk, the homeowner is required to purchase flood insurance in order to protect the mortgage company's investment in the home.

The federal government is the source of flood insurance in the National Flood Insurance Program. The federal government pays claims when an area floods. The federal flood insurance program is fraught with controversy. In many ways, it encourages people to purchase land and homes in floodplains because the insurance premiums are subsidized, making it more affordable and enticing for homeowners to live in these areas. This puts both life and property at risk. The program is also in dire financial straits. Premiums paid by homeowners are lower, in fact, than the cost of paying claims every year. Premiums do not accurately reflect the risk of flooding, as evidenced by the fact that they are lower than the cost of paying claims every year. This is especially true when a part of the United States experiences a catastrophic storm such as a Hurricane Sandy, Harvey, or Maria. While low premiums are helpful for the homeowner's budget, it means that the National Flood Insurance Program is chronically underfunded and even homeowners who pay their flood insurance premiums regularly cannot always count on the federal government to pay their claims when disaster strikes. Another problem with the National Flood Insurance Program is that the agency has spent billions of dollars repairing homes that are rebuilt in the same floodplain, and flood over and over again. In this scenario, it is not uncommon for flood insurance poured into the home after multiple repairs to equal more than the home is worth. A better choice may be for the federal government to purchase the person's home and require the homeowner to relocate to another parcel of land outside the floodplain. The offer for buy-out and relocation out of flood-prone areas is sometimes extended to homeowners after floods, but it is not mandated. Congress has recognized that changes need to be made to the National Flood Insurance Program, but thus far any changes have been small.

Many countries, including the United States, have spent billions of dollars to construct levees, dikes, dams, and canals in order to control the flow of rivers and protect cities from flooding. The terms *levee* and *dike* are sometimes used interchangeably. Both refer to structures that run parallel to the river and help prevent water from flooding adjacent land when the river rises. Levees may be natural or human-made. Dikes are human-made structures. Some people also draw the distinction that dikes protect land that would normally be underwater from flooding. The Netherlands has created excellent artificial flood control structures that effectively protect coastal communities from floods for decades. Levees, dikes, and dams are integral parts of their flood control system.

The Mississippi River Delta region has a complex network of canals, levees, and dams to keep the river in its banks. Hundreds of thousands of people rely on this system to keep their homes safe from floods. Unfortunately, Hurricane Katrina tested many of the levees in the New Orleans area to their limits. Katrina's storm surge did a great deal of damage in coastal communities when the storm made landfall. Storm surge also was a major reason the levees failed in New Orleans. Flooding in New Orleans and surrounding communities occurred because the levees on the river and multiple canals failed (Fig. 9.21).

A mock disaster exercise was conducted in New Orleans in 2004. Named "Hurricane Pam," the exercise brought together hundreds of local, state, and federal emergency management professionals in a simulation meant to expose underlying weaknesses in the planned emergency response to a major hurricane making landfall in the New Orleans area so they could be fixed before an actual hurricane occurred. It simulated a Category 3 storm hitting southeast Louisiana. It became clear during the exercise that there were multiple major weaknesses in the hurricane disaster plan for New Orleans. Recommendations were made for shoring up these weaknesses, which included repairs and enhancements to levees in the New Orleans area. Unfortunately, those repairs were not completed by the time Hurricane Katrina, a nearly exact analog for the fictional Hurricane Pam, hit in 2005. The result was failure of the levee system in areas predicted by emergency management officials in the Hurricane Pam exercise. Billions of dollars have since been poured into the flood control structures in Louisiana, but they have not yet been tested against a major river flood, direct strike from a hurricane, or inland flooding from a storm that made landfall to the west.

Figure 9.21 Flooding in New Orleans, Louisiana, was largely the result of failure of flood control structures. Levees were overtopped or broken, canal walls failed, and pumping stations, which keep water out of the city every day, stopped working. Photograph ©Yann Arthus-Bertrand/Altitude. From Strahler, 2013.

Figure 9.22 Channelized bayou in Houston, Texas. Photograph © Dr. Marli Miller, used with permission.

Some cities have sought to control flooding with the use of concrete channels (Fig. 9.22). Cities throughout the desert Southwest employ this technique to channelize floodwaters. This can be an incredibly effective technique when the channels are carefully and intentionally designed to contain the maximum flood the area is expected to experience. While the channels are immensely effective, problems come to downstream communities when the channels are ended abruptly. Water that was contained within the channel moves very quickly thanks to the lack of resistance in the streambed from rocks, uneven terrain, and vegetation. When the stream encounters the end of a concrete channel, the water has enormous erosive power and can scour the landscape. Floodwaters spread out throughout the natural floodplain and leave people and communities downstream vulnerable to both flash flooding and erosion of infrastructure.

The aforementioned mitigation strategies are largely a matter of city, county, state, or federal action. The most important thing individual homeowners can do to protect themselves from flooding is to not buy a home within a floodplain. However, in the event of unprecedented rainfall rates, or the case of a historic flood, nearly any home can experience flooding. A homeowner or business owner can use sandbags to protect a building from intruding water. In coastal areas where storm surge and coastal flooding are possibilities, homeowners should consider elevating their homes to a height above the maximum expected surge. In some cities devastated by Hurricane Sandy, for example, homeowners were told they needed to elevate their homes to comply with new floodplain standards drawn up by county governments after the storm (Fig. 9.23). Some funds were made available through the government and nonprofits to do this work, but certainly not enough assistance was provided to help all homeowners. The process of elevating homes in the area continues as of this writing (2023).

Scientists are working to develop a standard flood index, which is a scale much like the Enhanced Fujita Scale used to categorize tornadoes, or the Modified Mercalli Intensity Scale used to describe damage and shaking due to an earthquake. A number on this scale would

Figure 9.23 Elevated home in Keansburg, New Jersey. The home was elevated to comply with new floodplain standards instituted following Hurricane Sandy, which struck the area in 2012. Photo by Liz Roll/FEMA. From https://www.fema.gov/media-library/assets/images/69333

be determined by the height of the water expected at a specific location given particular rainfall, tide, ice jam, or snowmelt conditions. Flood forecasting using a standard scale such as this would alert homeowners in more concrete terms as to what level of flooding they are likely to experience in their location. In turn, the hope is that people would take advantage of these forecasts to make informed decisions about when and by what route they need to evacuate. Forecasting accurate flood indices would require a great deal of real-time monitoring and computer modeling. The technology exists to do this. A standard flood index also requires agreement among scientists of a worldwide standard scale. At the current time, there is no standardized flood index, but hydrologists and atmospheric scientists are working toward this goal. A number of suggested flood scales with 5 and 10 categories have been proposed, but scientists worldwide have not adapted a standard scale or a standard methodology for computing each number on that scale. Expect this to happen in the near future, as more scientists in more countries around the world acquire the technological capability to implement such a model.

9.6 Chapter in Review

A body of flowing water confined within a channel is a stream. Many streams are referred to as rivers, and there is no official criteria for naming a stream as a river. The component of water in a stream that comes from groundwater is called base flow. Base flow can be measured after a period of fair weather in which no precipitation has fallen, contributing surface water to the stream.

The land area that contributes surface water to a specific stream is called a drainage basin or watershed. Divides are high points on the landscape that separate the land into drainage basins. A smaller stream often empties into a larger stream. The smaller stream is a tributary to the larger stream. The place where a stream begins is called the stream's headwaters. Headwaters are found at the highest elevation along the stream's path. The place where a stream empties into a larger body of water is called the mouth of the stream. The mouth of a stream is found at the stream's lowest elevation.

A recurrence interval, or return period, is the average length of time between the same sized events on the same area. Floods are often named for their theoretical recurrence interval. An example is the 100 year flood. In reality, a 100 year flood rarely happens every 100 years. A flood of this size has a 1/100, or 1%, chance of occurring in any given year. The volume of water flowing past a certain point in a stream is the stream's discharge. Flood frequency diagrams depict the relationship between recurrence interval and discharge.

Streams take on a variety of shapes. Streams in mountainous landscapes are often fairly straight, and they have extremely narrow floodplains. These rivers flood quickly. A river at the base of a mountain may become choked with sediment it can no longer carry once the velocity decreases. It becomes a braided stream. Streams flowing over relatively flat terrain are often meandering, with many bends and turns called meanders. Meandering streams have wide floodplains. Intermittent streams carry water only after a rainfall event. They are prone to flash floods.

There are five types of floods. The first type is a flash flood. It occurs relatively quickly after rainfall, release of water after breach of an ice jam, or after dam or levee failure. A second type is a river flood. This type occurs most commonly when an area receives prolonged rainfall, and can be exacerbated by melting of heavy snowfall or other circumstances that saturate the ground with water, making it impossible for rainfall to soak into the ground. River floods occur over weeks or months. A third type is a storm surge, which occurs when a tropical storm or hurricane comes ashore. A storm surge may be up to 20 ft (6 m) or more high. It floods coastal communities, adding to the danger of a hurricane's landfall. A fourth type is coastal flooding, which occurs when low-lying coastal land is flooded by daily tides, or the extreme tides that occur during new Moon and full Moon phases. As climate change continues to cause sea levels to rise, coastal flooding will become a more common problem. The fifth type of flood is inland flooding. This is flooding that occurs inland from coasts as a result of tropical or extratropical storms that originate in the oceans.

Homeowners can protect themselves against flooding by not building or buying homes in floodplains. For homeowners who do purchase homes in known floodplains, they are required to purchase federal flood insurance. This flood insurance, while helpful to those whose homes are likely to flood, is also controversial because its existence encourages people to build or buy in known floodplains. The program is chronically underfunded and runs at a deficit. Some areas are looking at requiring people who live in floodplains to relocate once their homes have been flooded, but this is not yet the norm. Some communities use levees, dikes, canals, and dams to control streams so floodwaters are diverted away from communities. However, these structures can fail and produce major catastrophic flooding, such as that which occurred during Hurricane Katrina in 2005. Other communities build concrete channels in which rivers flow. While this works relatively well in the communities in which they exist, downstream communities without the channels can be flooded. When the channel no longer exists, the water is poured out onto the natural landscape, causing flash flooding to occur during even relatively minor rainfall events.

In some areas that have been affected by flooding, municipalities and insurance companies are requiring homeowners to elevate their homes. This is especially true in coastal neighborhoods that have been affected by storm surge. Scientists are working toward developing a standardized flood index that would provide more concrete information to those in the path of floods so they have greater ability to make informed decisions about evacuation.

10

Hurricanes

Bethany D. Hinga

University of Nebraska at Kearney, USA

10.1 Introduction

Hurricanes are the largest and most powerful weather systems on the planet. They likely take their name from Haracan (alternate spellings Huricán, Hurakán, and Hurakā), storm god of the Taino people who lived on Caribbean islands prior to and during the time of Spanish exploration. According to the Online Etymology Dictionary (https://www.etymonline.com/word/hurricane), Spanish explorers, as early as the late 1540s, used the word *huracan* to describe hurricanes. Some confusion in Spanish between an initial *-h* and an initial *-f* led to no fewer than 39 different spellings of the word used in the late 1500s, including *forcane, herrycano, harrycain,* and *hurlecane.* The Portuguese began calling these storms *furacão.* The first English usage of the word is in Richard Eden's *Decades of the Newe Worlde or West India*, published in 1555. Eden uses the word *furacanes.* Shakespeare used the term *hurricano* in his plays *King Lear* and *Troilus and Cressida* but appears to have been referring to waterspouts rather than hurricanes.

A hurricane is a type of *tropical cyclone.* Tropical cyclones are large low-pressure systems that form over warm seas. They include tropical depressions, tropical storms, and hurricanes. A hurricane is a large and intense cyclone that has sustained winds of at least 74 m/hr (119 km/hr). Tropical cyclones have different names, depending on their locations. In the northern Pacific, tropical cyclones are often named typhoons instead of hurricanes. In the Indian and southern Pacific oceans, they are simply called tropical cyclones.

10.2 Hurricane Maria, 2017

Hurricane Maria can trace its origins to the northern Atlantic Ocean during the week of 10 September 2017. On 16 September, it was located about 700 mi (1,127 km) east-southeast of the Lesser Antilles, which mark the eastern extent of the Caribbean. On that day, a loose cluster of thunderstorms became strong enough and sufficiently organized to be described as a tropical depression. Just 3 hr later, it had sustained winds of 50 mi/hr (80 km/hr). It was upgraded to a tropical storm and given the name Maria. On 17 September, when its maximum sustained winds reached 74 mi/hr (119 km/hr), it became Hurricane Maria.

Maria intensified rapidly within the next 18 hr from a weak Category 1 hurricane into a fierce Category 5 storm. It first made landfall on the island of Dominica in the Lesser Antilles at 9:15 p.m. on 18 September as a Category 5 hurricane (Fig. 10.1). Roofs were ripped from many buildings and several wooden

Earth's Natural Hazards and Disasters, Advanced Textbook 7, First Edition. Bethany D. Hinga.

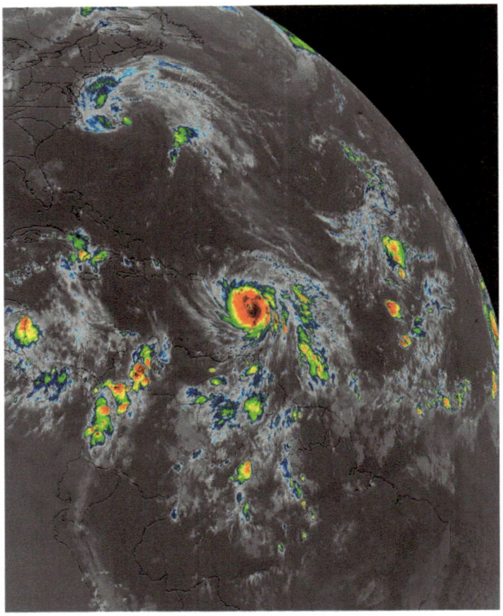

Figure 10.1 Enhanced satellite image of Hurricane Maria as it made landfall on Dominica on 19 September 2017. Hurricane Maria was a Category 5 storm when it struck Dominica. Damage was widespread and catastrophic. The storm to the northwest is Tropical Storm Jose. Image courtesy of the National Oceanic and Atmospheric Administration.

homes were completely destroyed (Fig. 10.2). Electricity was knocked out to the entire island. Most of the island's lush vegetation was defoliated. Large trees were uprooted, and branches were down everywhere. Debris from destroyed buildings as well as rocks and soil from minor landslides were strewn across the landscape, making it difficult to navigate the island's roadways (Fig. 10.3). The island had been struck by a glancing blow from Hurricane Irma just two weeks before and residents had little time to make repairs before Maria struck. Weakened and damaged structures were more vulnerable than before to the high winds from Hurricane Maria, and damage was severe.

Losing very little strength after its landfall on Dominica, Maria continued to the northwest over the warm open waters of the Caribbean. Its next stop was the U.S. Virgin Islands and the U.S. territory of Puerto Rico in the northeast Caribbean. As Maria approached the islands, it weakened very slightly from a Category 5 storm to a strong Category 4 storm (Fig. 10.4). Its maximum sustained winds were 155 mi/hr (240 km/hr) when it hit Puerto Rico at 6:15 a.m. on 20 September (Fig. 10.5),

Figure 10.2 Damage on Dominica the morning after Hurricane Maria made landfall. This photo was taken on the route between Castle Comfort and Roseau, and shows debris strewn across the road, power lines down, and people starting to work to clear the road. https://commons.wikimedia.org/wiki/File:Morning_after_Hurricane_Maria_(37372721465).jpg

Figure 10.3 Image taken in Dominica the morning after Hurricane Maria, a Category 5 storm, made landfall. People are surveying the damage, which includes defoliation of the area, roofs blown from buildings, and damage to well-built structures. https://commons.wikimedia.org/wiki/File:Morning_after_Hurricane_Maria_(36975223580).jpg

Figure 10.4 GOES satellite image of Hurricane Maria, taken on the morning of 18 September 2017. U.S. Navy, accessed from Wikimedia Commons; https://commons.wikimedia.org/wiki/File:A_GOES_satellite_image_of_Hurricane_Maria_in_the_Caribbean_Sea._(36905931020).jpg

Figure 10.5 Enhanced satellite image of Hurricane Maria making landfall in Puerto Rico on 20 September 2017. The remnants of Hurricane Jose, then Tropical Storm Jose, can still be seen off the east coast of the United States. Image courtesy of the National Oceanic and Atmospheric Administration.

making landfall at the city of Yabucoa. Many structures were heavily damaged or destroyed, especially those constructed of wood. As on the island of Dominica, vegetation was stripped of leaves and branches, leaving the usually verdant landscape looking barren and brown. Roads were rendered impassable by fallen trees, washed out bridges, and debris. The island's fragile electrical grid was destroyed, leaving the entire island without power. Water distribution systems on the island failed due to the lack of electricity, leaving most of the island's residents without access to clean water. Cell phone towers were destroyed preventing people from communicating with first responders as well as friends and loved ones. With the cell towers and electricity went Internet access. In addition, the island suffered torrential rainfall. Some areas in the center of the island received more than 35 in. (89 cm) of rain. The unprecedented rainfall caused record flooding, particularly in the mountainous interior of the island. With flooding came landslides. A storm of this magnitude had not hit Puerto Rico since Hurricane San Felipe in 1928.

In the first few days after the storm, Puerto Rico's officials made it clear that damage was catastrophic and the island needed a great deal of help to recover. The meager relief supplies the territorial and federal governments had prepositioned in Puerto Rico had been sent to the U.S. Virgin Islands just 2 weeks prior because they too had been struck by Hurricane Irma. There were no waiting relief supplies such as emergency meals, clean drinking water, fuel, batteries, tarps, and generators. Many people had to rely on water from creeks, which was often contaminated with leptospirosis, a microorganism that causes a severe intestinal illness. Others were able to find wells with hand-operated pumps to provide drinking water. Food in refrigerators and freezers rotted. It was difficult to get around the island because of the number of downed trees blocking roads. Hospitals were damaged, and many closed for days after the hurricane. Before they could reopen, they had to find enough generators to allow them to provide necessary services and run vital medical equipment. People did the best they could to help each other get what they needed. Help from the outside, however, was slow in coming.

The U.S. government activated emergency response teams. Cargo planes flew generators, water, prepackaged meals, and medical supplies to the island. However, distribution of relief supplies was slow. Ships full of supplies set out for Puerto Rico, but the ports were not immediately ready to receive them since they too had received damage. Though supplies made it to the island, there was no plan in place for distribution of those supplies to areas that needed them. There were few truck drivers available to move supplies to where they were needed. To complicate matters, additional storm systems hit Puerto Rico in the weeks following Maria. As a result, ships carrying supplies often had to wait offshore for storms to pass in order to safely enter ports on the island.

Once it became apparent that electricity to the island would not be restored anytime soon and therefore most hospitals would not have the resources they needed to treat critically ill patients, the U.S. Navy hospital ship *USS Comfort* was deployed to Puerto Rico. Its arrival did not seem to be coordinated well with agencies on the ground, so for several days it stayed offshore and patients were ferried to it. It was underutilized until it was positioned closer to shore, making it more accessible.

There were many such inefficiencies noted by residents of the island in the weeks that followed. The island's leaders became increasingly frustrated and expressed that frustration through both official channels and the news media. San Juan mayor Carmen Yulín Cruz became an outspoken critic of the federal government by asserting that it had failed to recognize the severity of Puerto Rico's situation and was not offering as much help as it could. An estimated $90 billion in damage occurred on the island due to Hurricane Maria. The federal government committed more than $60 billion in funds to help the island recover. That included $30.3 billion in public assistance, $200 million in hazard mitigation, $20 billion in Housing and Urban Development (HUD) Community Development Block Grant Disaster Recovery Program funds, $9.5 billion in power reconstruction and permanent works, $1.5 billion in FEMA (Federal Emergency Management Agency) individual assistance.

According to one seasoned emergency management official, those who worked on recovery efforts described it as "the perfect storm" (Joyce Flinn, personal communication). The island had not been hit by a major hurricane since Hurricane Georges in 1998. As a result, the island's government had become complacent and had failed to prepare adequately for a major hurricane. Government had not prepositioned enough food, water, generators, fuel, or medical supplies to be able to reach all areas of the island in the event of a disaster. The island had no secondary communication system, such as a network of trained shortwave radio operators, that could fill in the gaps for first responders and relief workers until landline telephone and cell phone service was restored. It also did not have disaster plans in place that would have allowed it to mobilize people quickly to start the recovery process. The closest thing government had to a hurricane disaster response plan was a tsunami response plan drawn up years before. Tsunamis and hurricanes are fundamentally different disasters that affect the island in very different ways. Tsunamis affect only coastal areas, whereas the effects of hurricanes are felt across the entire island. As a result, the tsunami emergency plan was inadequate for hurricane recovery.

The weather patterns that kept ships from docking at the island's ports for several days after the hurricane delayed vital supplies from reaching the island in a timely fashion. FEMA was terribly short staffed partially because of a number of retirements the previous year and slow replacement of those individuals. Thus, there was a shortage of trained and ready staff to address the situation. Some retired emergency management officials were brought in to fill gaps in the response. As is typical in large-scale disasters, emergency management officials from state agencies across the United States were also pulled in to assist. State emergency management personnel, acting alongside FEMA representatives, were vital to the response effort.

Residents were frustrated with the lack of help from the government. There were shortages of food and water, and rural areas had to deal with lack of electricity for many months. A strange scandal played out regarding repair of the electric grid. A tiny company in Montana was awarded a contract worth over $300 million to restore electricity to the island. The contract was awarded without accepting bids from other larger and more capable electric companies, which raised eyebrows. After realizing that the company was unable to provide the necessary services, and after many accusations of cronyism and corruption, the contract was canceled. Larger power companies ultimately ended up traveling to the island and restoring power (Fig. 10.6).

Figure 10.6 U.S. Army Corps of Engineers working to help restore power in San Lorenzo, Puerto Rico, 7 months after Hurricane Maria. https://commons.wikimedia.org/wiki/File:USACE_working_in_San_Lorenzo,_Puerto_Rico,_7_months_after_Hurricane_Maria_(2017).jpg

Norms of society on the island also made the recovery more difficult. U.S. law states that when emergency disaster relief funds are used, infrastructure is to be restored to predisaster levels. In the instance that an upgrade must be made to bring the infrastructure up to the most recent safety codes, the state (or in this case the territory) must provide funds to pay for that upgrade. The electrical grid in Puerto Rico was in very poor shape before the hurricane, and could not be restored to that level after the hurricane because doing so would not comply with U.S. electrical code. Therefore, the island's government had to find money to complete the upgrades. The island was in the midst of severe financial difficulties that had persisted for years, and did not have funds readily available to make these upgrades. Power companies did the best they could to get power restored to the island, but this law made the process more complicated. Electricity was restored first to the larger cities, which were more accessible and where the electrical grid was in relatively good shape before the hurricane. Then electric companies worked toward rural areas in the interior of the island. Power was not restored to the entire island until a year after the hurricane.

Families on the island became incredibly frustrated with lack of approval to pay for repairs to their property or, in many cases, allocation of funds to help them rebuild their homes. U.S. law states that disaster funds can pay only to replace homes for which families hold the title. On the island, many homes had been built on family plots without proper legal titles ever being issued. Therefore, many families were ineligible to secure federal aid to repair or rebuild their homes.

The death toll, initially reported at just 17, was later revised to 2,975 after a study conducted by George Washington University and the University of Puerto Rico. Many deaths initially went unreported. Others were not reported correctly by doctors, undertakers, and other professionals who had never dealt with a mass casualty situation. Additional people who died after the hurricane from secondary causes were included in the updated total. This includes people who were unable to undergo dialysis treatments or take supplemental oxygen due to the island-wide power failure and hospital closures, people who died of intestinal parasites from drinking contaminated water, and people who died later of injuries sustained in the storm.

Hurricane Maria was a tragic disaster. Puerto Rico will be recovering for many years to come. No matter how many outside factors may have contributed, this will likely be remembered as an epic failure of government to help those in most dire need in a timely manner.

10.3 Tropical Cyclones

The tropics is the area of the Earth located between the Tropic of Cancer (23.5° N) and the Tropic of Capricorn (23.5° S). In this region of the planet, there is very little variation in temperature with changing seasons. Air over tropical seas is uniformly warm and humid. Because of this, weather systems within the tropics are not driven by fronts as they are outside the tropics. They are instead driven by convection and winds. It is not uncommon to see daytime heating and strong convection lead to afternoon thunderstorms within the tropics, especially in the summer. The location of the storms is somewhat variable with seasons, likely due to migration of the *Intertropical Convergence Zone* (ITCZ). It moves northward in the Northern Hemisphere's summer and southward in the winter. These summer thunderstorms are typically short-lived events, though occasionally longer-lasting, tropical squall clusters can develop. These look similar to squall lines seen near storm systems in the midlatitudes, and can carry strong thunderstorms with heavy rain and high winds.

10.3.1 Requirements for Tropical Cyclone Formation

Tropical cyclones form at fairly low latitudes. They may either stay in tropical waters or veer into higher latitudes during their life cycles. Tropical cyclones need four conditions in order to form. The first condition necessary for the formation of a tropical cyclone is a large expanse of warm ocean. In order to form a tropical cyclone, sea surface temperatures typically need to be above 80° F (26.5° C) to a depth of 150 ft (45 m). This typically occurs between the months of June and November in the Northern Hemisphere. The peak of hurricane season occurs in September in the North Atlantic basin. The peak of hurricane season in the eastern Pacific is in August, and in the Southern Hemisphere, hurricane season is at its peak between January and April.

Warm ocean water is the underlying energy supply for tropical cyclones. The warmth of the water allows for large-scale evaporation. Water evaporates from the ocean, and the resulting water vapor is buoyed by its warmth. Through convection, it is carried upward and begins to condense into clouds. Latent heat energy is released into the environment when the water vapor condenses into clouds. The storm gains energy and strength through these processes of evaporation and condensation. When a storm drifts over land, it is cut off from its energy supply and it begins to weaken. This occurs quite rapidly, such that as soon as a hurricane makes landfall it begins to lose strength. Hurricanes simply do not survive long as hurricane-strength storms as they move inland. However, they can carry great amounts of moisture inland, causing immense amounts of rainfall.

The second condition necessary for the formation of a tropical cyclone is weak vertical wind shear (change in wind with height). Wind shear enables drier air to be imported into the tropical cyclone. Dry air inhibits thunderstorm formation, and is therefore detrimental to storm formation and maintenance of conditions necessary for continued latent heat release that powers a tropical cyclone. It may seem counterintuitive that weak wind shear allows thunderstorm formation. However, if upper level wind shear is strong, it has the ability to weaken thunderstorms, eventually causing them to dissipate.

The third condition that must be present to form a tropical cyclone is a triggering mechanism. In extratropical storms in the mid-latitudes, a common trigger for thunderstorm formation is collision of a warm, humid air mass with a cold air mass. However, in the tropics where the air is almost uniformly warm and humid, other types of air masses do not move throughout the region. Instead, the triggering mechanism in the Atlantic is often an *easterly wave* off the coast of Africa. An easterly wave is a disruption in the area's trade winds featuring a weak low pressure system that moves from east to west (Fig. 10.7). The disturbance of trade winds often results in air moving around this low pressure system. When the air currents converge on the east side of the wave, these converging winds can cause and organize convection.

The fourth requirement for forming a tropical cyclone is a strong *Coriolis effect* that allows the storm to rotate. The Coriolis effect is a phenomenon in which the Earth's air is diverted to the right in the Northern Hemisphere and to the left in the Southern Hemisphere due to the rotation of the planet. At the equator, the Coriolis effect is zero. Therefore, in order to develop into a rotating tropical storm and eventually a hurricane, it is necessary for the storm to form at least 5° north or south of the equator. There have been a few rare exceptions to this rule, including Typhoon Vamei, which formed at 1.5° N near Singapore in December 2001. Its circulation actually extended into both hemispheres, which is exceedingly rare for any type of storm.

Tropical cyclones form in many areas around the world where all of the necessary conditions are present to form tropical storms (Fig. 10.8). The most intense tropical cyclones are called by different names in different parts of the world. In the Atlantic, Caribbean, and eastern Pacific, they go by the name hurricane. In the southwest Pacific and Indian Oceans, they are called tropical cyclones. Off the east coast of Asia, they are called typhoons.

Most hurricanes that strike the United States originate in the Atlantic Ocean, Caribbean Sea,

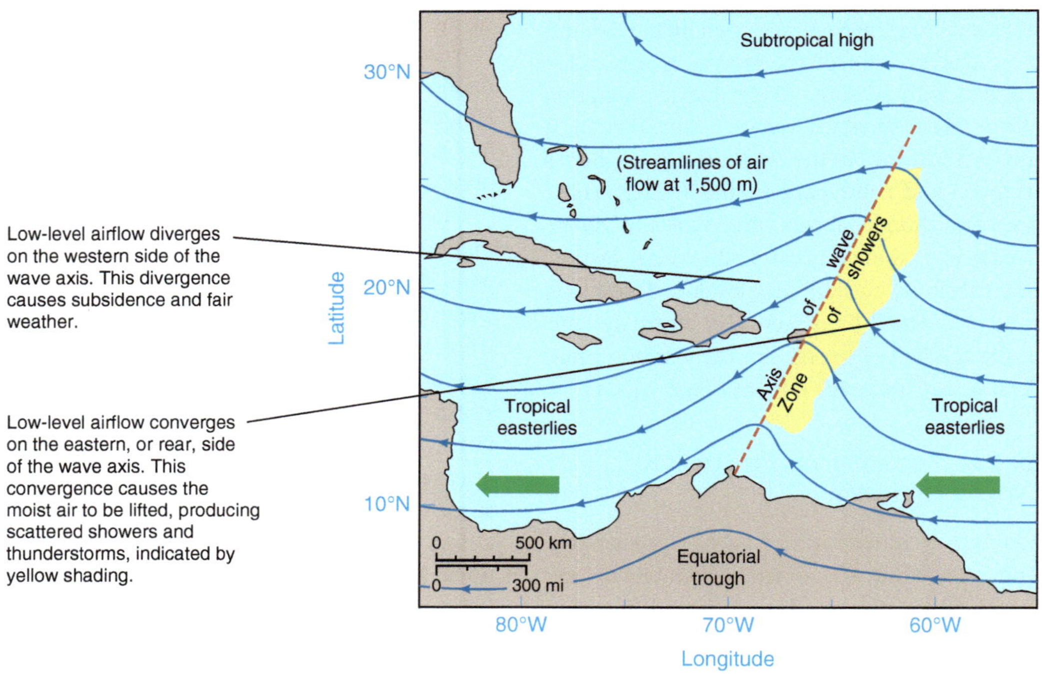

Figure 10.7 Diagram of an easterly wave in the tropics near Puerto Rico. From Strahler, 2013.

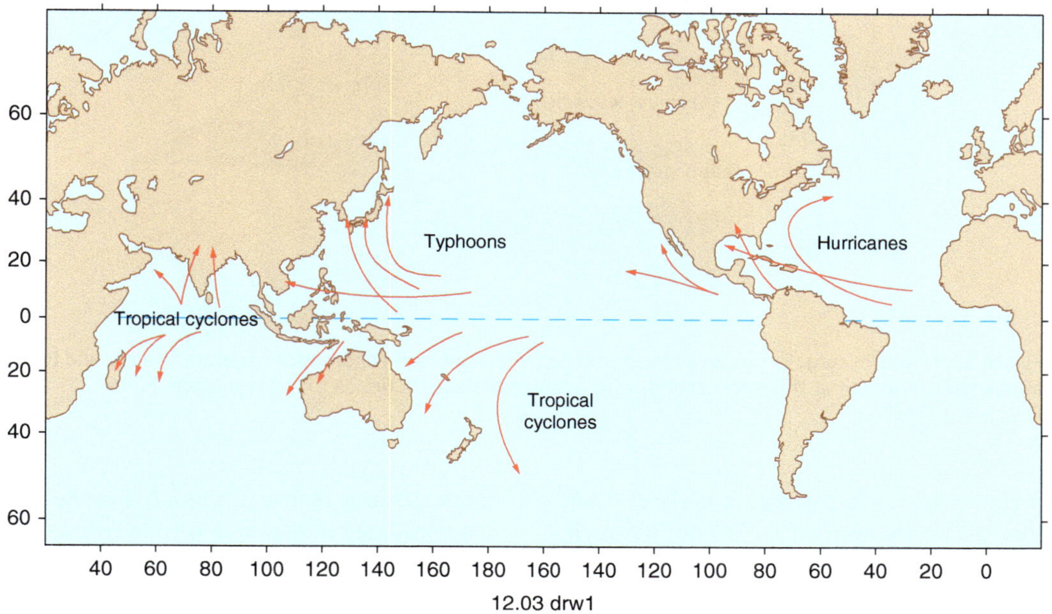

Figure 10.8 Areas of the world in which tropical cyclones form. From Moran, 2012. Used with permission.

or Gulf of Mexico. Tropical storms may also originate in the eastern Pacific Ocean off the west coast of Mexico and Central America. In the vast majority of cases, storms that originate off the coast of Mexico do not impact the United States. This is because the ocean off the coast of California is rather cold. In rare cases, however, storms have brushed past the California coastline bringing high surf, high winds, and heavy rainfall.

Hurricanes occasionally make it to the central Pacific where they can impact the Hawaiian Islands. In September of 1992, Hurricane Iniki struck the island of Kauai as a Category 4 storm with 140 mi/hr (225 km/hr) maximum sustained winds and gusts to 170 mi/hr (274 km/hr). Winds severely damaged hotels, stranding many tourists in their resorts while the hurricane raged around them. Thousands of terrified guests were forced to take refuge in their hotels instead of enjoying the tropical vacation they had imagined. Countless homes and businesses were damaged. Trees were uprooted. Waves of more than 30 ft crashed on the shoreline. Telephone service and electricity were knocked out island wide. By the time Iniki left the island, it had destroyed or severely damaged 90% of the island's buildings and done over $2 billion worth of damage. Seven people lost their lives. Thankfully, hurricanes in Hawaii are relatively rare occurrences. However, Hurricane Lane passed just west of the island of Hawaii in 2018. It dumped up to 52 in. (1.3 m) of rain on portions of eastern Hawaii, causing widespread flooding and minor landslides.

In the western Pacific, tropical storms form east of the Philippines and east of Australia. One of the worst typhoons in recent memory hit the Philippines in 2013. Super typhoon Haiyan (referred to locally as Yolanda) made landfall just south of Tacloban City on 7 November with sustained winds of 190 mi/hr (306 km/hr. It caused widespread devastation and killed an estimated 7,360 people. *Storm surge* was to blame for the large death toll (Fig. 10.9). Storm surge is a dome of water that rushes ashore when the hurricane makes landfall. It was 24.6 ft (7.5 m) tall when it struck land. Most of the land within a few miles of the ocean is

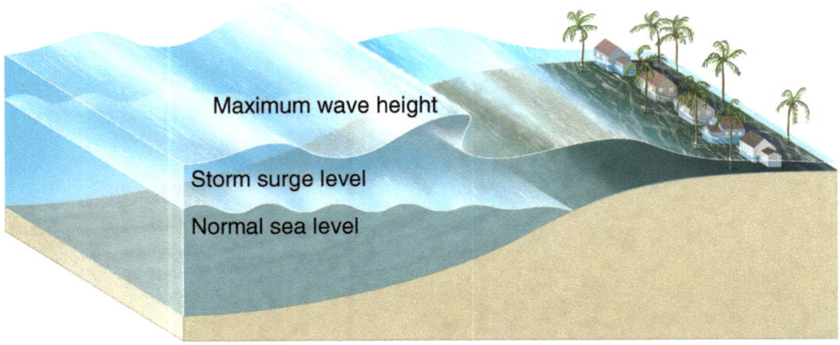

Figure 10.9 Storm surge is caused by the combined effects of high wind, which generates waves, and lower atmospheric pressure at the storm center, which causes the water surface to bulge upward. From Fletcher, 2014.

no more than 16 ft (5 m) above sea level. Therefore the waves reached far inland, destroying homes and taking lives as it did so. The storm was the most powerful to strike this stretch of coastline since records have been kept. Many of those who died lived in poorly constructed homes on the coastal plain. A reported 1.14 million homes were destroyed in the typhoon, leaving millions of people homeless.

Tropical cyclones also form in the Indian Ocean on both sides of the Indian subcontinent and in both the Northern and Southern hemispheres. The deadliest hurricane in all of recorded history was the Bhola Cyclone, which struck Bangladesh (then known as East Pakistan) in the early morning hours of 12 November 1970. Maximum sustained winds were 140 mi/hr (225 km/hr) and the storm brought with it a 20 ft (6 m) storm surge. Cyclones that hit this area of the world are particularly deadly because the land in coastal Bangladesh has a very low elevation, and is nearly flat. Therefore, the storm surge rushes inland with nothing to stop it. People had nowhere to run for refuge. The cyclone claimed at least 300,000 lives, and perhaps more than 500,000 lives. The coastal district of Thana saw incredible losses. Approximately 167,000 people died in that district alone. It is estimated that nearly 46,000 fishermen lost their lives in the cyclone. Many who were not killed were injured and unable to work. The storm destroyed 65% of the coastal region's fishing industry. About 80% of the protein consumed in that area is from fish so this was a major blow to the nutritional capacity of the region. In all, losses included 85% of homes and businesses, and totaled over $490 million. Though meteorologists knew the storm was coming, it was nearly impossible to warn people in the path of the cyclone because communication systems were so poor at that time. Thus, the storm hit when people had very little warning, and had not adequately prepared for landfall. At that time there were very few storm shelters as well, leaving people with no safe space during the storm. Since that time, many cities in Bangladesh have built vertical escape structures. These are sturdy buildings several stories tall that allow people to move to higher elevations to avoid flooding from cyclones and seasonal rains.

10.3.2 Life Cycle of a Tropical Cyclone

When an organized cluster of thunderstorms forms over the tropical ocean, this is called a *tropical disturbance*. If conditions are right, a tropical disturbance intensifies. Air pressure falls, and a low pressure center forms. The storms begin rotating counterclockwise around the low and winds intensify. When maximum sustained winds reach 23 mi/hr (37 km/hr) the system is called a *tropical depression*.

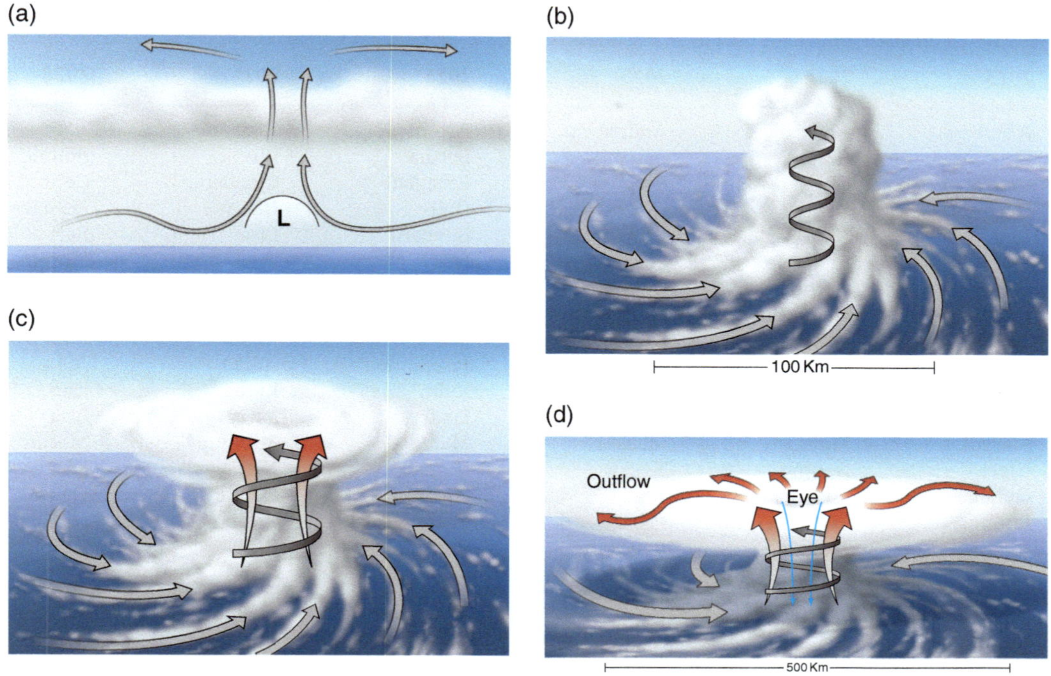

Figure 10.10 Development of a tropical cyclone. From Strahler, 2013. (a) A low pressure center forms over the ocean. Warm air and moisture rise in the vicinity of the low pressure center. (b) Thunderstorms form around the low pressure center, and rotation of the thunderstorm cluster begins. (c) The thunderstorms organize into a single tropical storm, which strengthens with continued exposure to large expanses of warm water. (d) The tropical storm intensifies into a hurricane that has rotates around an eye.

Tropical depressions are given a number (Tropical Depression 14, for example). Numbers are assigned chronologically based on date of classification (Fig. 10.10).

As convection and rotation intensify and the air pressure at the center of the storm drops, the thunderstorms become part of a larger, more organized weather system with broad circulation. When maximum sustained winds reach 39 mi/hr (63 km/hr), the tropical depression is given a name from a predetermined list (Table 10.1; Table 10.2) and is classified as a *tropical storm*. With further development, a tropical storm can intensify into a hurricane. A storm is classified as a *hurricane* when maximum sustained winds reach 74 mi/hr (119 km/hr).

Hurricanes feature strong cyclonic rotation around an intense low. Because of the Coriolis Force, this cyclonic rotation is counter-clockwise in the Northern Hemisphere and clockwise in the Southern Hemisphere. Maximum sustained wind speeds may reach more than 150 mi/hr (241 km/hr) (Table 10.3). Hurricanes are classified into *categories* based on maximum sustained winds. The scale used for this classification is the *Saffir-Simpson Hurricane Wind Scale*. The weakest storms, with maximum sustained winds just over the threshold for designation as a hurricane, as Category 1 storms. The most intense storms with maximum sustained winds over 157 mi/hr (253 km/hr) are Category 5 storms (Fig. 10.11).

In general, more intense hurricanes tend to have lower atmospheric pressures at their centers. Typhoon Tip formed in the western Pacific Ocean near the Philippines in 1979. It had the lowest atmospheric pressure ever measured in a hurricane at 870 millibars, and its wind speeds were a maximum 190 mi/hr (306 km/hr), making it an extremely strong Category 5 hurricane.

Table 10.1 Names for Atlantic tropical cyclones for the years 2023–2028.

2023	2024	2025	2026	2027	2028
Arlene	Alberto	Andrea	Arthur	Ana	Alex
Bret	Beryl	Barry	Bertha	Bill	Bonnie
Cindy	Chris	Chantal	Cristobal	Claudette	Colin
Don	Debby	Dexter	Dolly	Danny	Danielle
Emily	Ernesto	Erin	Edouard	Elsa	Earl
Franklin	Francine	Fernand	Fay	Fred	Farrah
Gert	Gordon	Gabrielle	Gonzalo	Grace	Gaston
Harold	Helene	Humberto	Hanna	Henri	Hermine
Idalia	Isaac	Imelda	Isaias	Imani	Idris
Jose	Joyce	Jerry	Josephine	Julian	Julia
Katia	Kirk	Karen	Kyle	Kate	Karl
Lee	Leslie	Lorenzo	Leah	Larry	Lisa
Margot	Milton	Melissa	Marco	Mindy	Martin
Nigel	Nadine	Nestor	Nana	Nicholas	Nicole
Ophelia	Oscar	Olga	Omar	Odette	Owen
Philippe	Patty	Pablo	Paulette	Peter	Paula
Rina	Rafael	Rebekah	Rene	Rose	Richard
Sean	Sara	Sebastien	Sally	Sam	Shary
Tammy	Tony	Tanya	Teddy	Teresa	Tobias
Vince	Valerie	Van	Vicky	Victor	Virginie
Whitney	William	Wendy	Wilfred	Wanda	Walter

Source: From the National Oceanic and Atmospheric Administration; https://www.nhc.noaa.gov/aboutnames.shtml

Table 10.2 Names for eastern north Pacific cyclones for the years 2023–2028.

2023	2024	2025	2026	2027	2028
Adrian	Aletta	Alvin	Amanda	Andres	2Agatha
Beatriz	Bud	Barbara	Boris	Blanca	Blas
Calvin	Carlotta	Cosme	Cristina	Carlos	Celia
Dora	Daniel	Dalila	Douglas	Dolores	Darby
Eugene	Emilia	Erick	Elida	Enrique	Estelle
Fernanda	Fabio	Flossie	Fausto	Felicia	Frank
Greg	Gilma	Gil	Genevieve	Guillermo	Georgette
Hilary	Hector	Henriette	Hernan	Hilda	Howard
Irwin	Ileana	Ivo	Iselle	Ignacio	Ivette
Jova	John	Juliette	Julio	Jimena	Javier
Kenneth	Kristy	Kiko	Karina	Kevin	Kay
Lidia	Lane	Lorena	Lowell	Linda	Lester
Max	Miriam	Mario	Marie	Marty	Madeline
Norma	Norman	Narda	Norbert	Nora	Newton
Otis	Olivia	Octave	Odalys	Olaf	Orlene
Pilar	Paul	Priscilla	Polo	Pamela	Paine
Ramon	Rosa	Raymond	Rachel	Rick	Roslyn
Selma	Sergio	Sonia	Simon	Sandra	Seymour
Todd	Tara	Tico	Trudy	Terry	Tina
Veronica	Vicente	Velma	Vance	Vivian	Virgil
Wiley	Willa	Wallis	Winnie	Waldo	Winifred
Xina	Xavier	Xina	Xavier	Xina	Xavier
York	Yolanda	York	Yolanda	York	Yolanda
Zelda	Zeke	Zelda	Zeke	Zelda	Zeke

Source: From the National Oceanic and Atmospheric Administration; https://www.nhc.noaa.gov/aboutnames.shtm

Table 10.3 Saffir-Simpson Hurricane Wind Scale.

Category	Wind speed (mph)	Damage at landfall	Storm surge (feet)
1	74–95	Minimal	4–5
2	96–110	Moderate	6–8
3	111–129	Extensive	9–12
4	130–156	Extreme	13–18
5	157 or higher	Catastrophic	19+

Source: From NASA; https://spaceplace.nasa.gov/hurricanes/en/

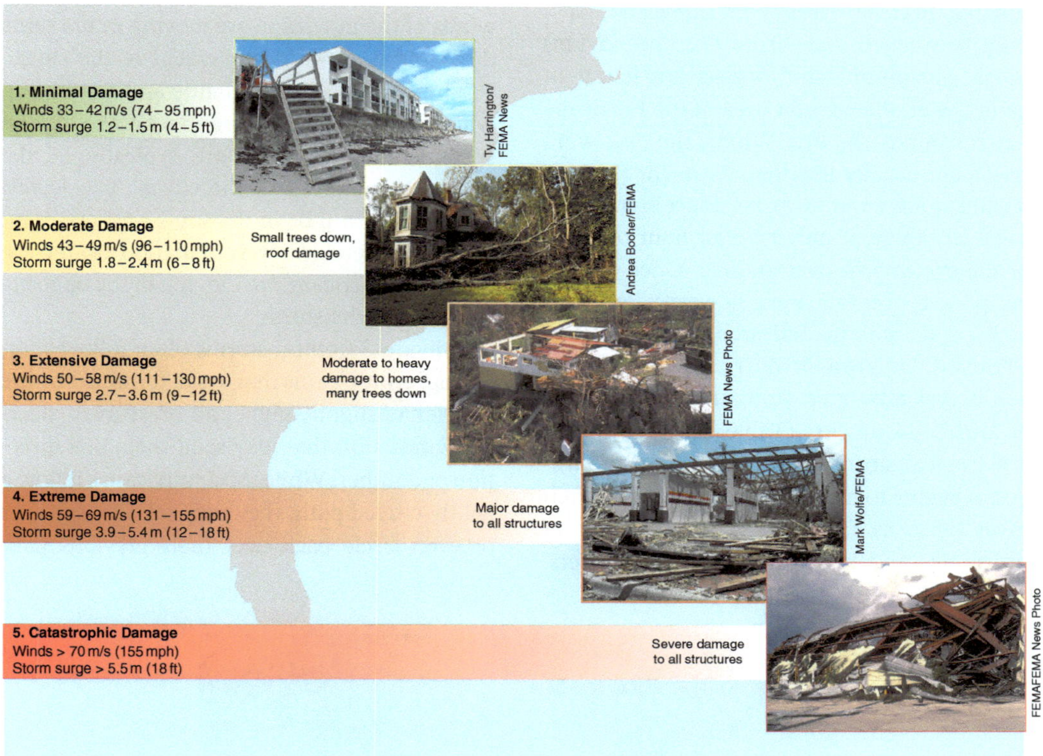

Figure 10.11 Typical damage inflicted by hurricanes, indicated by their Saffir-Simpson category. The most damaging hurricanes are Category 5 storms. From Strahler, 2013.

In fact, Tip still holds its spot as the strongest hurricane on record. The lowest air pressure measured in an Atlantic hurricane was 882 millibars in Hurricane Wilma in 2005. Wilma's maximum sustained wind speed was 185 mi/hr (298 km/hr), making it a very strong Category 5 storm like Typhoon Tip. The strongest winds recorded in an Atlantic hurricane came from Hurricane Allen in 1980. Its maximum sustained wind speed was 190 mi/hr (306 km/hr) but its atmospheric pressure was 899 millibars, slightly higher than that of Hurricane Wilma.

10.3.3 Wind Speed Variations Within the Storm

On the outer edges of a hurricane are spiral-shaped bands of clouds and thunderstorms (Fig. 10.12). Wind speeds in these outermost bands are fairly low, barely more than a breezy summer day. However, wind speeds intensify closer to the storm's center. The center of the storm is identified by the appearance of an *eye*, which is a relatively calm and clear zone at the center of rotation. On satellite images, such as in Figure 10.12, the eye is a circular feature at the center of the storm that is devoid of heavy cloud cover. The eye of a hurricane is typically between 6 and 40 mi (10 and 65 km) across. Maximum wind speeds are found in the eye wall, the circular area of the hurricane that surrounds the eye. Within the eye, wind speeds are usually less than 16 mi/hr (25 km/hr) and cloud cover is sparse. Since hurricanes travel so slowly, it may take an hour or more for the eye to pass over an area. Once the eye has passed over an area, however, residents are hit again with the full strength of the storm.

Forward progress of hurricanes in tropical seas is typically fairly slow, usually less than 15 mi/hr (24 km/hr) (Table 10.4). However, as a tropical storm moves away from the tropics, it begins to move faster. Hurricanes often travel twice as fast once they move out of the tropics and into more temperate waters. When this happens, a storm becomes an extra-tropical cyclone. Like hurricanes, extra-tropical cyclones can cause a great deal of damage because they still carry high sustained winds, and the high winds in these storms can extend more than 200 mi from the center of the storm. Hurricane Sandy caused a great deal of damage in New Jersey and New York as an extra-tropical cyclone in 2012.

Winds are strongest near the center of the storm in the eyewall. Even within the eyewall, there is variation in the strength of the winds in each quadrant of the storm. Winds are highest in the quadrant of the hurricane in which the winds of the hurricane are moving in the same direction as forward movement of the storm. For hurricanes in the Atlantic and Caribbean, which tend to move north and west, the highest sustained wind speeds are typically in the northeast quadrant of the storm. The lowest wind speeds will be in the southwest quadrant where counterclockwise circulation around the eye of the hurricane are moving in the opposite direction of the storm.

Individual tropical storms often follow wandering, erratic paths that are determined by the location of slightly higher air pressure or steering winds aloft that may send them one direction or another. While most hurricanes follow a gently curved path, they may take sharp turns or even circle back over their previous path

Figure 10.12 Hurricane Katrina at its peak intensity on 28 August 2005. From NOAA; http://www.aoml.noaa.gov/keynotes/keynotes_0815_katrina_anniversary.html

Table 10.4 Average forward speed of tropical cyclones by latitude.

Latitude	km/hr	Speed knots	mph	Number of cases
0°–5°N	25.9	14.0	16.1	186
5°–10°N	22.0	11.9	13.7	4,678
10°–15°N	19.2	10.4	11.9	7,620
15°–20°N	17.4	9.4	10.8	7,501
20°–25°N	17.5	9.4	10.8	8,602
25°–30°N	20.1	10.8	12.5	6,469
30°–35°N	27.1	14.6	16.9	3,397
35°–40°N	39.0	21.0	24.2	1,120
40°–45°N	49.3	26.6	30.6	264
45°–50°N	51.5	27.8	32.0	34
50°–55°N	51.4	27.8	32.0	15
55°–60°N	55.8	30.1	34.7	1

Source: From NOAA; http://www.aoml.noaa.gov/hrd/tcfaq/G16.html
Notes: In general, storms move faster at higher latitudes.

(Fig. 10.13). However, patterns in hurricane paths do exist (Fig. 10.14). In the northern Atlantic, storms are generally steered to the west and north around a high pressure system. The high, known as the Bermuda High, the Bermuda-Azores High, the Azores High, the North Atlantic Anticyclone, or the North Atlantic High, is a stable feature and has been influencing the paths of hurricanes in the North Atlantic at least as long as people have been paying attention to storm tracks.

Once a storm makes landfall, it begins to weaken. This does not mean it immediately ceases to be a dangerous storm. It simply means that its wind speeds decrease and it will be downgraded in reverse order from how it was designated as it grew. It is downgraded from a hurricane to a tropical storm, to a tropical depression, and finally ends up as a cluster of storms that may move over land for some time. Thunderstorms resulting from tropical storms often carry a great deal of moisture and high winds with them. They frequently cause flooding along their paths and can spawn supercell thunderstorms and tornadoes.

10.4 Hurricane Hazards

10.4.1 Heavy Rains and Inland Flooding

Inland flooding is flooding that occurs as a result of heavy rainfall. Rainfall amounts associated with the landfall of hurricanes are frequently in the range of 5–10 in. (13–25 cm) but can be much higher. The result is widespread severe flooding that can happen even very far inland from where a hurricane makes landfall.

Tropical Storm Allison struck Galveston Island, Texas, on 5 June 2001. As the storm moved onshore, it followed a meandering path and hit Houston twice with unprecedented rainfall. Rainfall at the Port of Houston for the period 4–9 June, which included both of Allison's paths over Houston, was 36.99 in. (94 cm). From Texas, it moved offshore into the Gulf of Mexico and strengthened into a subtropical cyclone (one that has characteristics of both tropical and extratropical cyclones). It dumped 29.86 in. (75.8 cm) of rain over Thibodaux, Louisiana, when it came ashore again. Allison then moved eastward through Mississippi and sat over Florida for a

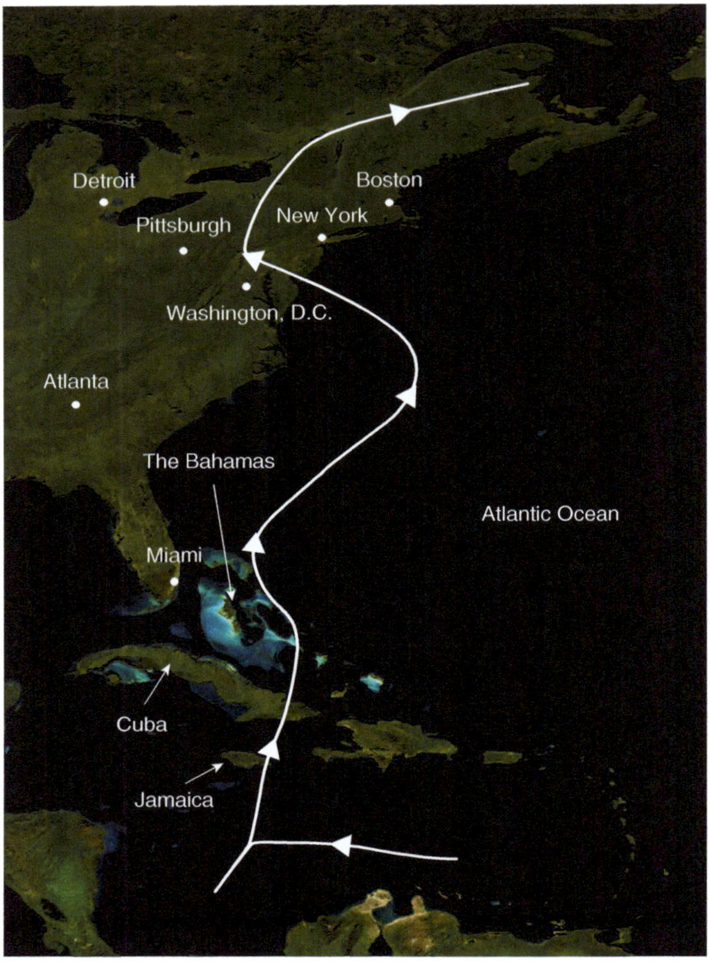

Figure 10.13 Path of Hurricane Sandy, 2012. It followed a wandering path from the tropics northward into New England and beyond. From Arbogast, 2017.

day. Tallahassee received 10.13 in. of rain between 11–12 June. Allison eventually meandered to the northeast and dissipated over the North Atlantic near Nova Scotia. Damage from Allison totaled at least $6.4 billion, and the storm claimed 41 lives. Most deaths were due to inland flooding from the storm's intense rainfall.

The Houston area became ground zero again for catastrophic flooding from Hurricane Harvey in August of 2017. The hurricane made landfall at Rockport, Texas, near the city of Corpus Christi on 26 August 2017. It was a Category 4 storm with maximum sustained winds of 130 mi/hr (209 km/hr). It was the strongest storm to hit the Texas coast since 1961. Harvey caused serious damage to most of the structures within the city of Rockport and both rain and the storm surge flooded the town (Fig. 10.15). The storm turned slightly northeast and headed very slowly up the coast to the Houston area. Wind speed dropped considerably, but the storm was carrying immense amounts of moisture. Over the next 5 days, Harvey dropped record amounts of rainfall over the Houston area. Much of the Houston metro area received more than 40 in. (101 cm) of rainfall. The record for the area during Hurricane Harvey was 51.88 in. (131.8 cm) at Cedar Bayou on the east side of Houston.

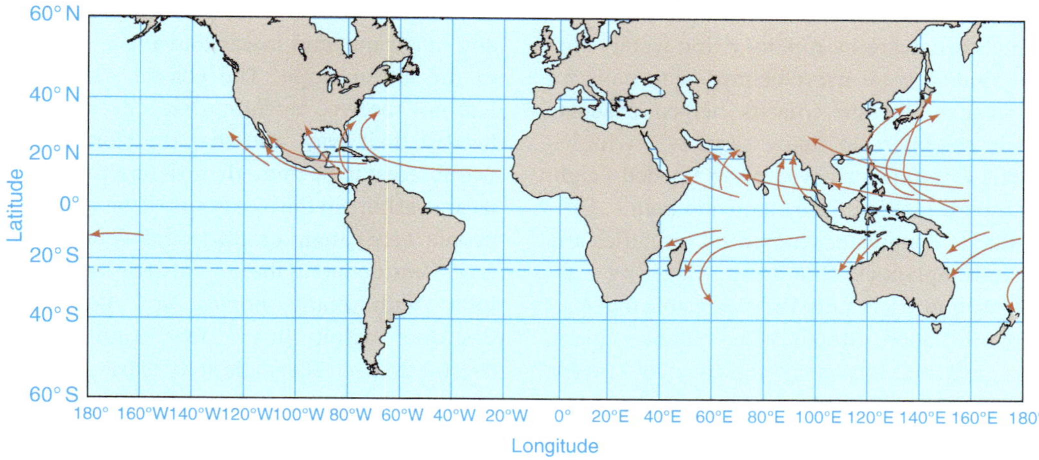

Figure 10.14 Typical tropical storm tracks in the world's oceans. From Strahler, 2013.

Figure 10.15 An aerial view shows a home damaged by Hurricane Harvey in Rockport, Texas, 28 August 2017. Army National Guard photo by Sgt. 1st Class Malcolm McClendon; https://www.defense.gov/observe/photo-gallery/igphoto/2001800459/

Public schools closed for more than a week and colleges and universities delayed the start of their semesters as far north as Bryan/College Station, Texas. Flights into and out of the major international airline hub of Houston were canceled. Many major highways were completely impassable as they were under several feet of water. Neighborhoods that had never seen flooding before were inundated with floodwaters. Thousands of people had to be rescued from their homes because they were caught unaware by rapidly rising water. First responders were overwhelmed with calls for assistance. Private citizens helped evacuate their neighbors in many cases when individuals with boats showed up in flooded neighborhoods looking for a chance to help.

Rising floodwaters caused oil refineries and chemical plants to shut down, triggering a rise in gasoline prices until they were able to restart once the flood subsided. One chemical plant, Arkema Inc., located in the town of Crosby, Texas, housed chemicals that required constant cooling. Without electricity during the inland flooding of Hurricane Harvey, the cooling process was interrupted for several days. Chemicals overheated and caught fire, leading to an explosion. Damages from Hurricane Harvey nearly equaled those of Hurricane Katrina at $125 billion, making it second in the list of costliest U.S. natural disasters. Sixty-two people lost their lives.

10.4.2 Storm Surge

Storm surge is a dome of water that builds as a hurricane approaches shore. It is formed due to low pressure over the ocean and high winds within the storm. As a storm's eyewall moves onshore, the storm surge rushes onshore as well. In general, the higher the storm category, the higher the storm surge is likely to be. For Category 1 hurricanes, the storm surge may be barely discernible among the wind-whipped waves. For Category 5 hurricanes, the storm surge can be more than 16.4 ft (5 m) tall. The height of a storm surge is somewhat dependent as well on the shape of the seafloor just offshore, the forward speed of the hurricane, and the direction of hurricane motion relative to the orientation of the shoreline.

Prior to 1971, storm surge was the number one killer in hurricanes. Many of the largest death tolls from hurricanes were the result of massive storm surges hitting unsuspecting populations without much warning or awareness of the hazard. The 1970 Bhola Cyclone in Bangladesh saw the highest death toll for any hurricane, and it was due almost entirely to the storm surge. The coastal population had not received warning that a cyclone was headed their way, and that the storm surge was forecast to be more than 20 ft (6 m). When it came ashore in the wee hours of the morning, people were caught unaware and were vulnerable. Destruction in the coastal area was nearly total, with many homes and businesses destroyed within just a few minutes. The coastal area of Bangladesh is extremely flat, so the storm surge was able to travel a great distance inland. The result was the great tragedy of the loss of perhaps half a million or more people. Unfortunately, the scene was repeated in the Chittagong area of Bangladesh in 1991. A cyclone with top wind speeds of 150 mi/hr (240 km/hr) and a storm surge of 15 ft (4.6 m) came ashore on 29 April. It killed approximately 135,000 to 145,000 people and left as many as 10 million people homeless. Finally, after the 1991 hurricane, the government of Bangladesh began to improve warning systems and build shelters for those who live in the flat delta area of southern Bangladesh.

The largest death toll for a hurricane in the United States was the result of a storm surge. A hurricane made landfall in early September of 1900 at Galveston, Texas (Fig. 10.16). The storm surge in that hurricane was 15 ft (4.6 m). The city of Galveston is built on a barrier island whose top elevation at the time was 8 ft (2.7 m) above sea level, so water washed over nearly the entire island. About 3,600 homes were destroyed and approximately 8,000 people were killed. In response to the disaster, the U.S. Army Corps of Engineers built a seawall along the shoreline within the city of Galveston. At the same time, they embarked on an ambitious project to raise the elevation of the city an average of 10 ft (3 m) above its prior elevation. The combination of the seawall and increased elevation undoubtedly saved lives when Hurricane Ike struck just north of Galveston in 2008 (Figs. 10.17, 10.18). The storm surge during that hurricane was 10–15 ft

Figure 10.16 Relief workers removing a body from the wreckage of a building in Galveston, Texas, in September of 1900. More than 8,000 people lost their lives when a 15 ft (4.6 m) storm surge flooded the island on 8 September 1900. Photo courtesy of the Library of Congress; https://www.loc.gov/pictures/resource/cph.3c23881/

(3–3.6 m). A great deal of structural damage was done to buildings on the waterfront, but flooding within the city was held to a minimum. The storm surge from Hurricane Ike was worst to the east of Galveston Island on the Bolivar Peninsula. The seawall does not extend to that part of the island, so the coastline is vulnerable to storm surge. There, entire neighborhoods disappeared as homes were swept off of their foundations and crumpled in the waves.

10.4.3 Hurricane Katrina

Storm surge was also largely to blame for the high death toll in Hurricane Katrina. Hurricane Katrina formed over the Caribbean. It made its first landfall in southern Florida as a Category 1 storm. The hurricane largely held together as it crossed the state and made its way into the Gulf of Mexico. There, the hurricane rapidly intensified and became a Category 5 storm in the central Gulf of Mexico. Models indicated that the most likely path would take the hurricane very close to New Orleans. On 26 August, areas in coastal Louisiana and Mississippi began making preparations for landfall of a major hurricane. Approximately 1.2 million people evacuated the greater New Orleans area, but about 100,000 stayed behind despite a mandatory evacuation order. Most of those people had no means to evacuate because they had no transportation, little money, or were infirm in some way. Nearly 10,000 people headed for the city's shelter of last resort, the Superdome (home to the New Orleans Saints football team).

On the morning of 29 August, Hurricane Katrina came ashore as a strong Category 3 hurricane at the border of Louisiana and Mississippi. Storm surges were tremendous, and occurred as far away as the Alabama shoreline. The storm surge was as high as 24–28 ft (7.3–8.5 m) in parts of Mississippi, and up to 12 ft (3.7 m) high in Alabama. In the New Orleans area, the storm surge was 19 ft (5.8 m) in the eastern part of the city. The storm surge destroyed many of the flood control features in southern Mississippi and in the New Orleans area. The city of New Orleans is especially at risk for flooding because it occupies a bowl between Lake Pontchartrain and the Mississippi River. Much of the city lies several feet below sea level. It relies on an elaborate system of dams and levees, as well as an extensive pump system to keep water out of the city. The storm surge put pressure on all of these systems. There were 50 breaches on levees throughout the city. Water rushed into adjacent areas, flooding much of the city (Figs. 10.19, 10.20, 10.21). In addition, the pump system failed, which led to additional flooding. Much of the city was flooded to a depth of more than 10 ft (3 m). Water depths over 15 ft (4.6 m) were not uncommon. Many of the 1,800 deaths within New Orleans were caused by drowning in the floodwaters that resulted from levee breaches. Levees were breached by the strong flow of water from the storm surge.

Figure 10.17 Aerial view of damage on Galveston Island from Hurricane Ike, 2008. U.S. Air Force photo by Staff Sergeant James L. Harper, Jr.; https://www.af.mil/News/Photos/igphoto/2000680065/

(a) (b)

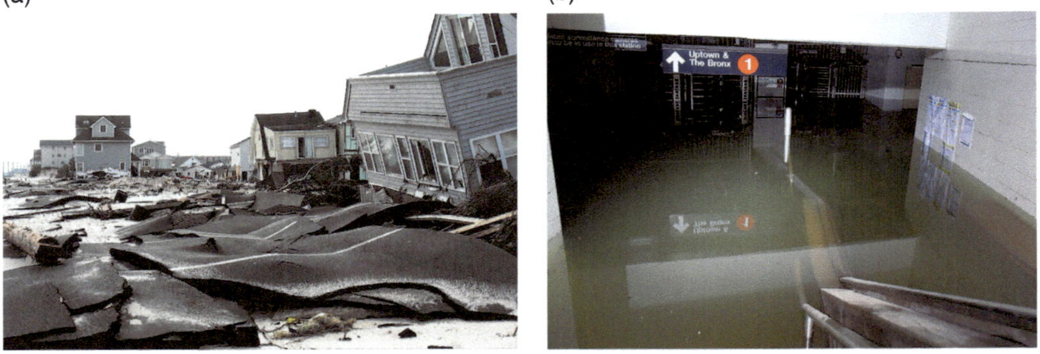

Figure 10.18 (a) Hurricane Sandy made landfall near New York City and did a tremendous amount of damage to structures in its path. The storm was responsible for over $68 billion in losses. (b) During Hurricane Sandy in 2012, storm surge flooded large parts of Lower Manhattan and completely submerged many subway tunnels in the city. From Arbogast, 2017.

At the Superdome, thousands of people arrived to the city's shelter of last resort to find that it had not been stocked or staffed adequately for the number of evacuees that arrived. There was not enough food or clean drinking water, and the bathroom facilities were inadequate to handle that many people for the duration of the storm. They huddled in the darkness as the storm raged throughout the day on 29 August. Part of the arena's roof blew off during the worst of the storm, leaving people exposed to the elements. They were hot, hungry, thirsty, and

Figure 10.19 Aerial view of flooding in New Orleans following Hurricane Katrina on 30 August 2005. Aerial view of a flooded neighborhood with streets and homes totally submerged. New Orleans was evacuated as a result of floods from hurricane Katrina. Thousands of people were rescued from the flood waters. They tried to escape the water by moving to their roofs and attics. Photo by Jocelyn Augustino/FEMA Photo.

Figure 10.20 Aerial photograph of flooded downtown New Orleans, with the Superdome in the center of the image. U.S. Navy photo by Photographer's Mate Airman Jeremy L. Grisham; https://en.wikipedia.org/wiki/List_of_Louisiana_hurricanes_(2000%E2%80%93present)#/media/File:Navy-FloodedNewOrleans.jpg

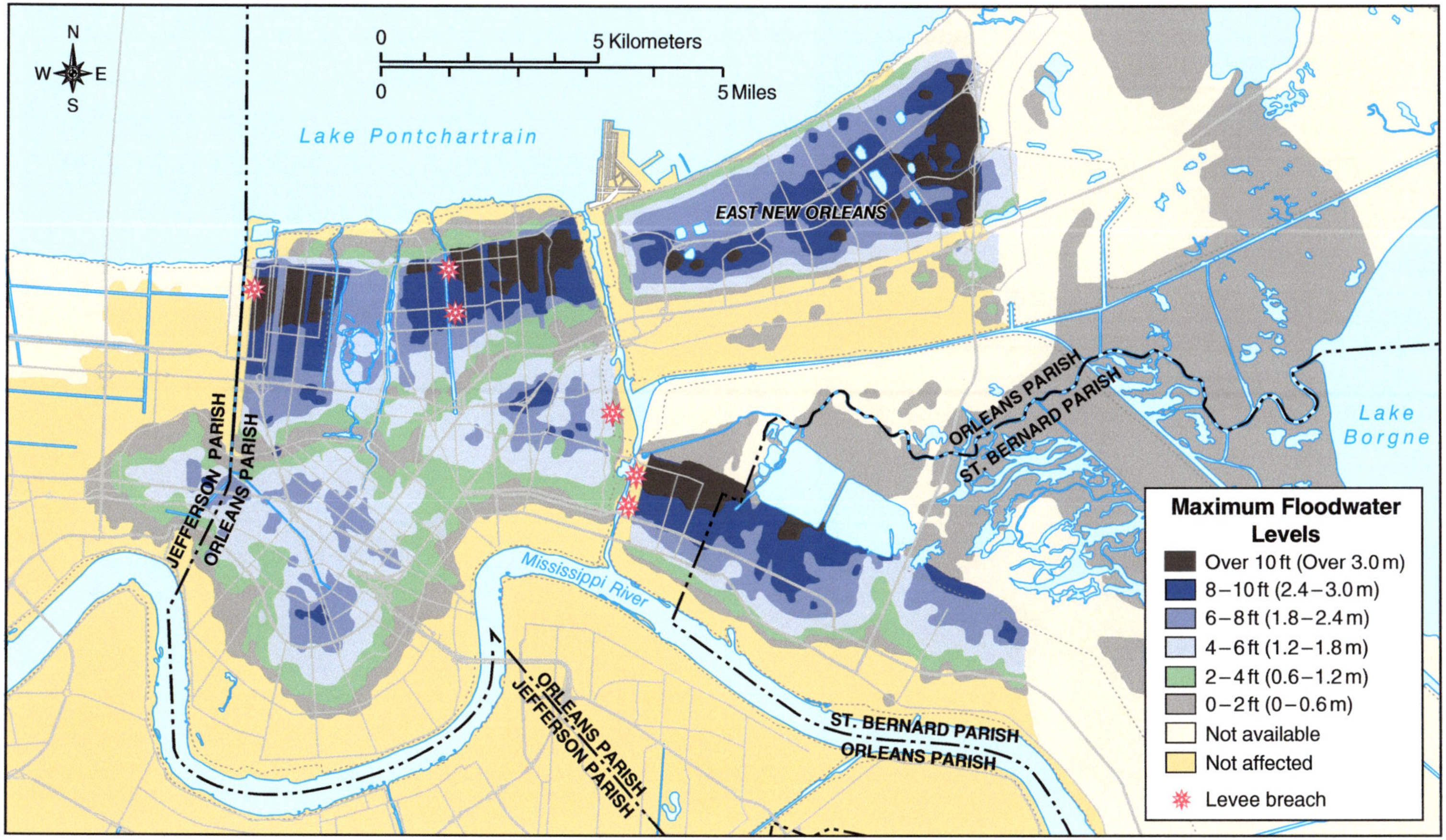

Figure 10.21 Maximum floodwater levels in the New Orleans area after Hurricane Katrina in 2005. Some of the most severe flooding occurred due to levee breaches. From Strahler, 2013.

angry. They had been told this place would provide shelter in the storm, but nobody prepared the shelter for them. Government officials did not seem to know they were there because help did not arrive in the day or two after the storm. Within the rest of the city, people were shocked by the magnitude and speed with which the floods happened. Many people caught in their homes had to climb to higher floors, or onto their rooftops to escape floodwaters. The Coast Guard rescued scores of stunned citizens who never imagined the city could flood so completely or so quickly. As much as 80% of the city was under water during the height of the crisis.

In the aftermath of the hurricane, the federal government made a number of missteps. The news media found the people at the Superdome and had to alert the government to their plight. People had to walk past bodies floating in the flooded streets to try to find shelter, food, and water. People were suffering all over New Orleans and the government did not seem to have a solid plan to address the city's problems. It took many days to get pumps restarted to start ridding the city of its flood waters. People needed places to stay. For those with no means of transportation, Federal Emergency Management Agency (FEMA) officials ordered hundreds of buses to take people to cities that had fully functioning relief services. Many people ended up in Dallas, Houston, Atlanta, Baton Rouge, and in states even more distant, often with no promise of transportation back to New Orleans once floodwaters receded and city services were restored. What was intended to be a temporary displacement ended up being permanent for many people. The name Katrina even today conjures up horrific scenes of destruction and human misery. The response to Katrina is considered by many to be one of the largest failures of domestic affairs in the George W. Bush presidency.

10.4.4 High Winds

High winds are one defining characteristic of hurricanes, so the fact that high wind is a primary hazard of hurricanes is not a surprise. The highest winds in a hurricane are found near the eyewall. This is where wind speeds are measured in order to determine the category of the hurricane on the Saffir-Simpson scale. Winds diminish with increasing distance from the eyewall of the hurricane. It is common for meteorologists to report the distance to which hurricane-force winds, those at least 74 mi/hr (119 km/hr), extend away from the eye of the storm. For the largest of hurricanes, hurricane-force winds may extend as much as 300 mi (483 km) away from the eye of a hurricane. High winds, storm surge, and rainfall are the three most damaging events in a hurricane, and at the point of landfall is where they make their greatest impact. The point of landfall is the location where the eye of the hurricane moves ashore.

10.4.5 Tornadoes

Another hazard associated with hurricanes is tornadoes. Once hurricanes move onshore, they often spawn tornadoes. After they make landfall, hurricanes typically take on characteristics of a cluster of strong thunderstorms. The atmosphere is unstable because of rapid convection. Wind shear is created by different wind speeds at various elevations. The combination of these factors leads to formation of supercell thunderstorms and tornadoes. In the North Atlantic, Gulf of Mexico, and Caribbean, tornadoes tend to form in and forward of the northeast quadrant of the storm.

In 2018, Hurricane Michael came ashore in the Florida panhandle as a Category 5 storm with maximum sustained wind speeds of 160 mi/hr (257 km/hr). The storm produced devastating wind damage, dumped copious amounts of rain over Florida and surrounding states, and generated 17 tornadoes over Florida, Georgia, South Carolina, and Virginia, primarily to the northeast of the storm where it made landfall. Four of the tornadoes were classified as EF-1s, and the remaining 13 were EF-0s. This is typical of the tornado output from many hurricanes.

10.5 Mitigation

Hurricanes are the most powerful storm systems on the planet. They bring with them strong winds, heavy rainfall, and storm surge. They can also spawn severe thunderstorms and tornadoes. Those who live in coastal areas or on islands in warm waters must be aware of developing and approaching storm systems, and take action to protect themselves and their property as a storm nears. Those in inland areas must be prepared for heavy rainfall, severe thunderstorms, and possible tornadoes. Most deaths in hurricanes occur due to inland flooding.

For those who live in coastal communities within the reach of storm surge, the best course of action is to evacuate inland and to higher ground. The water within a storm surge carries a great deal of energy, and that energy can demolish even well-built human-made structures. Until fairly recently, storm surge was the leading cause of death from hurricanes. It is important to respect the power of a hurricane. Therefore, it is best to vacate structures near the shoreline well before the storm makes landfall.

Wind is another major hazard of hurricanes. Hurricane-strength winds can extend a few hundred miles outward from the center of a large storm, causing damage far from where the hurricane makes landfall. Trees can be uprooted, power lines brought down, roofs blown from homes, and any material not tied down can be blown about and cause damage of its own. Homeowners and those responsible for the physical integrity of businesses often install plywood over windows in an effort to prevent flying debris from breaking glass.

Evacuation is also advisable for those who may live a few miles inland, but in low-lying areas. Inland flooding caused by heavy rainfall is currently the leading cause of death from hurricanes. In 2017, Hurricane Harvey made landfall in southern Texas, but wandered northeastward up the coast and dropped record amounts of rainfall on the Houston metro area. Inland flooding displaced thousands from their homes. Many people had to be rescued as the waters rose quickly around their neighborhoods and blocked escape routes. During Hurricane Katrina, communities were flooded almost instantly when levees on rivers and canals in the New Orleans metro area were breached, flooding inland areas to the roofs of homes. Hundreds of people drowned as a result, and thousands were rescued from rooftops and by boat, often after crawling onto their rooftops to escape floodwaters.

10.6 Conclusion

Hurricanes are the most powerful of all storm systems on the planet. They can be hundreds of miles wide, affecting people even far from the eye with high winds, torrential rainfall, thunderstorms, tornadoes, and flooding. In this chapter, we have reviewed not only how these low pressure systems form, but also how and why they move, how they are classified, and what hazards are to be expected when a hurricane approaches. Modern weather forecasting methods allow scientists to give populations days of warning prior to landfall. Hurricanes have immense potential to damage property, to change the landscape due to defoliation and landslides, and to take the lives of those who are unwilling or unable to escape the path of hurricanes and their resulting wind, storm surge, inland flooding, and tornadoes. Through the study of hurricanes, it is the author's sincere hope that if you live in an area where hurricanes or tropical storms are expected, that you will prepare for that eventuality by purchasing flood insurance, storing nonperishable food and clean water, and paying close attention to the weather as the summer approaches. The next time you and your family find yourselves in the path of a hurricane, you will recognize the risk in your area. Understand and follow the orders of your local officials. Take care of your family, check on your neighbors to see if they need help preparing for the storm and evacuating if necessary, and respect the power these storms bring.

10.7 Chapter in Review

A hurricane is the most powerful type of tropical cyclone. In the northern Atlantic Ocean, a hurricane often begins with an easterly wave. Disruption of trade winds around this weak low pressure center is sometimes enough to generate thunderstorms. If thunderstorms become organized, the cluster of thunderstorms can become a tropical disturbance and is given a number. If the thunderstorms in the tropical disturbance organize further and a low pressure zone develops within them, they become a tropical storm. A tropical storm is given a name from a predetermined list. If the atmospheric pressure at the center of the storm drops further, the storm becomes more organized, develops strong rotation, and forms an eye. Once maximum sustained winds reach 74 mi/hr (119 km/hr) near the eye wall, the storm becomes a hurricane.

Hurricanes need four conditions in order to form. (1) They typically form in wide expanses of warm water, where the sea surface temperature is at least 80° F (26.5° C) to a depth of 150 ft (45 m). (2) They must also have a triggering mechanism such as an easterly wave, which is a disruption of trade winds near a weak low. (3)Developing storms must have weak vertical wind shear to prevent the tops of the storms from shearing off, weakening the storm. (4) There must also be a strong Coriolis effect, which exists more than 5° north or south of the equator.

Tropical cyclones can form in any warm ocean. They are called by different names in various parts of the world. In some oceans, they are called tropical cyclones. In others they are typhoons, and near North America they are called hurricanes.

Hurricanes are classified on the Saffir-Simpson scale. This scale is based on maximum sustained wind speed within the storm, which occurs in the eyewall of the storm. The weakest hurricanes are Category 1 storms, and the strongest are Category 5 storms. Weak hurricanes can certainly cause a great deal of damage. Wind speed near the center of the storm must be at least 74 mi/hr (119 km/hr), which can bring down trees and power lines, cause damage to roofs, and cause loose objects to become projectiles. Hurricane-strength wind speeds can reach more than 100 miles from the center of the storm, which means wind damage can be significant that far from the point of landfall. Heavy rainfall, can cause inland flooding. This can be severe, damaging, and deadly. Some of the worst flooding experienced in the Houston area occurred after landfall of Tropical Storm Allison in 2001. Allison dumped more than 40 in (1.02 m) of rain on parts of Houston causing record flooding.

The most powerful hurricanes, Category 5 storms, have maximum sustained winds of more than 157 mi/hr(252 km/hr) near the center of the storm, and hurricane-strength winds (at least 74 mi/hr or 119 km/hr) that extend as far as 300 mi (483 km) or more from the eye. Storm surge can be higher than 20 ft (6 m), which along with the wind can destroy human-made structures at and near the coast. The inland flooding hazard remains severe with this type of storm as well.

Mitigation strategies include boarding up windows to prevent glass from breaking, securing loose objects such as lawn furniture, grills, and toys, and evacuating coastal and low-lying areas. People in coastal communities should keep a close eye on weather reports during hurricane season, which is in the summer and early autumn of the year. In the Northern Hemisphere, hurricane season runs from June 1 to November 30. Peak hurricane season is in the latest part of the summer and early autumn when the ocean is at its warmest. In the northern Atlantic, peak hurricane season is from August to October.

11

Climate Change

Jillian Gregg[1] *and John Helms*[2]

[1] *Oregon State University, USA*
[2] *Sterling College, USA*

11.1 Slushballs and Greenhouses: What the Geologic Record Says About Paleoclimate Compared to the Current Stable Climate That Has Supported Human Civilization Over the Past 10,000 Years

11.1.1 Slushballs and Greenhouses

The climate of the Earth has not always been as hospitable and temperate as that which we enjoy today. (Full disclosure: As of this writing, I am 'enjoying' a brutal Iowa winter!) The Earth's history began in extreme heat, and perhaps a completely molten surface. Since the Earth's crust solidified, and the atmosphere and oceans formed, the Earth's climate has oscillated between prolonged periods of hot and cold (Fig. 11.1). The most intense phases of these global cold snaps are referred to as *glacial maxima,* or *ice ages.* Some geoscientists have suggested that the Earth's surface has at times during its long history been nearly covered entirely with ice and snow: a scenario referred to as *Snowball Earth.* Others have suggested that the while the extent of ice and snow cover has been quite extreme, it is unlikely to have reached all the way to the equator, a scenario referred to as *Slushball Earth,* and the warmer periods between these ice ages are referred to as *interglacial* periods.

11.1.2 Current Stable Climate

Despite these dramatic changes between snowball and hot house phases of the Earth throughout the geological record, the Earth's current temperature, CO_2, and sea levels have all been relatively stable since the beginning of human civilization 10,000 years ago, which coincided with the end of the last ice age (Fig. 11.2, red box). These stable climate conditions have allowed humans to rely on dependable agricultural production to support large human populations living in established towns and cities, which enabled the development of stable agrarian lifestyles. Many of the world's largest cities have been constructed on the edge of the sea, which has not previously presented a problem because there was no indication that sea level could change. Despite these stable climate conditions upon which human civilizations have thrived for the past 10,000 years, human emissions of CO_2 and other greenhouse gases are now causing the temperatures and sea levels to rise (Figs. 11.2 and 11.3). Since there is currently no efficient method to scrub greenhouse gases from the atmosphere, the decisions we are making today are expected to impact the

Earth's Natural Hazards and Disasters, Advanced Textbook 7, First Edition. Bethany D. Hinga.

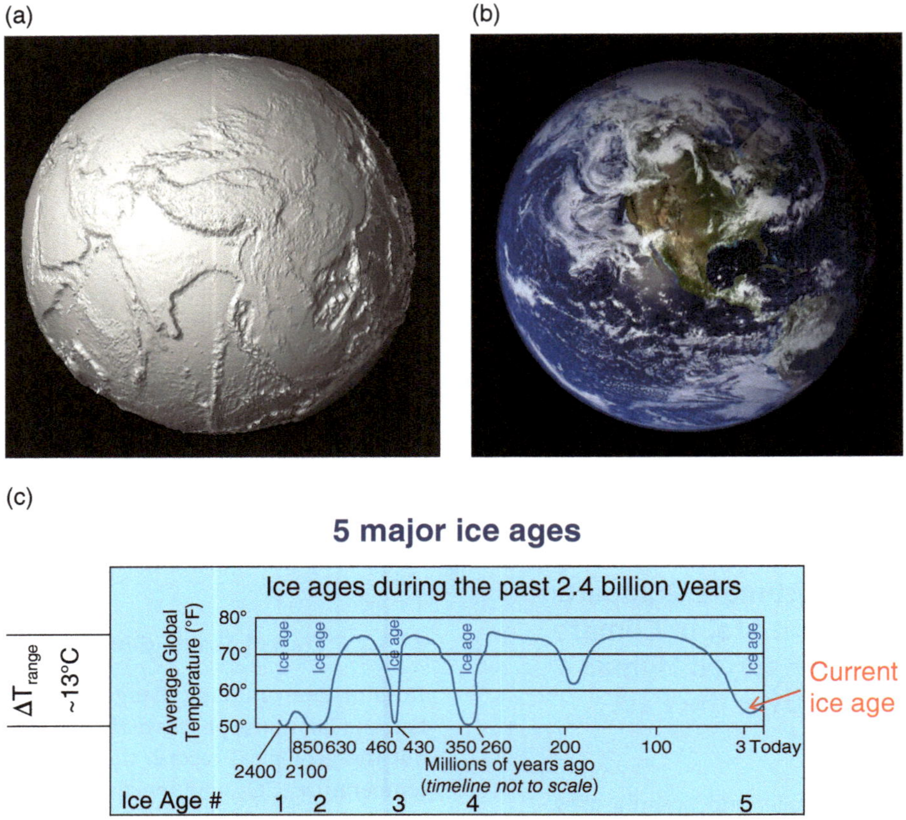

#	Ice Age	MYA
5	Current ice age (aka Quaternary glaciation)	~3 to now
4	Karoo ice age	~350 to 260
3	Andean-Saharan glaciation	~460 to 430
2	Sturtian/Marinoan glaciation	~850 to 630
1	Huronian glaciation	~2400 to 2100

http://geology.utah.gov/surveynotes/gladasked/gladice_ages.htm

Figure 11.1 (a) An artistic depiction of Snowball Earth where ice is thought to have fully covered the surface of the Earth. (b) Current image of the Earth from space by NASA, Blue Marble. (c) Simplified chart showing five major ice ages over the history of the Earth. Images courtesy (a) Imglee under CC BY 4.0, (b) NASA under the public domain, and (c) Adamo et al., 2021, under CC BY 4.0.

global climate for millennia (Fig 11.2, green and blue lines).

In this chapter, we will explore the mechanics of climate change and how climate change is likely to impact the frequency and severity of natural disasters. We will also examine our community resiliency and mitigation efforts in the face of these challenges.

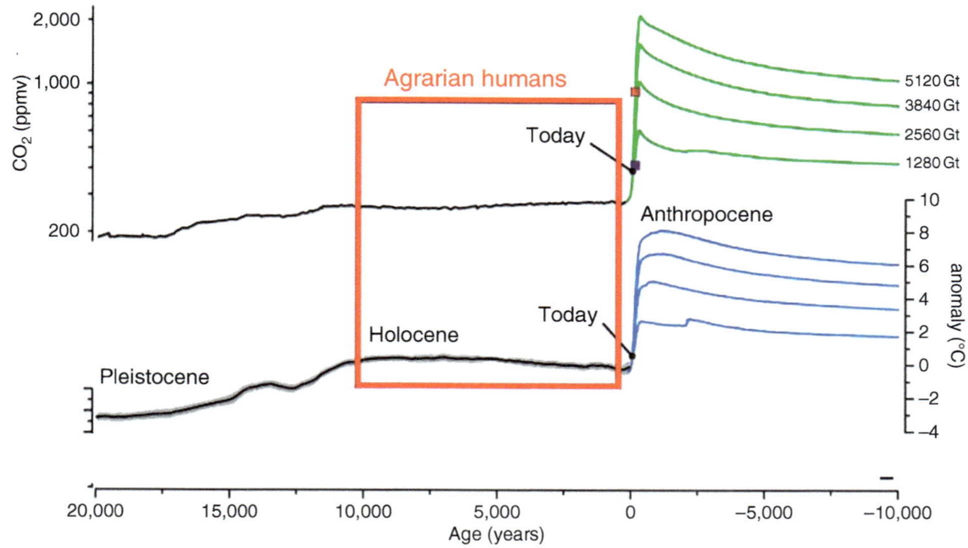

Figure 11.2 The average global temperature and CO_2 concentrations over the past 20,000 years and projected 10,000 years into the future. The stable conditions over the past 10,000 years enabled development of agrarian lifestyles and the development of civilization. Courtesy Peter Clark after Clark et al., 2016; edited by author.

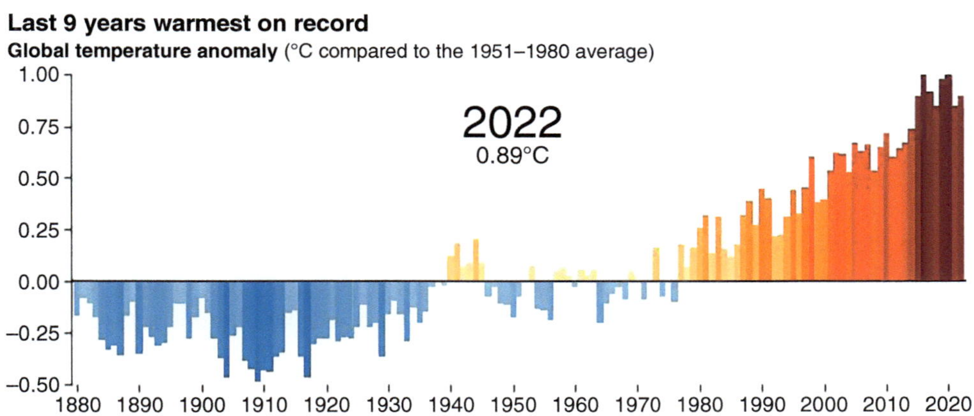

Figure 11.3 Global temperature anomalies from 1880 to 2022 compared with the twentieth century average. Plotting these data as the anomaly, or difference from the average, allows one to focus just on the change compared with the average. Here, red values show years that have been warmer than average, with the coolest years as the darkest red. While blue values show years that have been cooler than average, with the coolest years as the darkest blue. The nine warmest (darkest red) years on record have been the most recent 9 years. The average global temperatures for each year are obtained by averaging all the data for each of the Earth's 1° latitude by 1° longitude grid cells, and then averaging these together. This method ensures that results are not biased toward regions with more weather stations. Image courtesy NASA under the public domain.

11.2 Climate Change Basics

11.2.1 Climate Versus Weather

To begin our discussion, it is important to clarify the difference between climate and weather. *Climate* refers to the prevailing or typical weather patterns for a given region based upon long-term averages, typically longer than 30 years. Thus, climate is simply the average of the weather conditions over 30 years or more. If 30 years or longer of the global weather data is significantly different than a previous 30 year or longer period, we call this climate change.

By contrast, *weather* refers to a short-term or even instantaneous set of atmospheric conditions including temperature, precipitation, wind, cloud cover, and so on. It is what actually happens on a particular day (weather) rather than what you might generally expect based on 30 year averages for a given season and location on the planet (climate). Reports of colder than average temperatures for a given location on Earth in one particular year, therefore, are not evidence of a cooling climate but rather an interesting excursion of weather.

As the Earth's temperature changes between glacial and interglacial phases over millennia or exhibits a significant temperature difference between subsequent 30 year intervals within a century, it represent a change in the global climate that we refer to as climate change. It is important to note that planetary climate change is not random. In this chapter, we will discuss the many physical and human factors that can account for changes in the Earth's climate and assess which factors could account for the comings and goings of ice ages and which can account for the current rise in global temperatures since the Industrial Revolution.

11.2.2 The Earth Is Warming

To assess whether the Earth's climate has changed over the past century, scientists have combined many different sources of temperature data for locations across the Earth's surface, then averaged them together across both time and space to obtain the average global temperature for each year. There has been sufficient weather station coverage across the globe to calculate the global average temperatures since about 1880; we call this the Instrument Record. Scientists use a term called *temperature anomaly* to compare each year's globally averaged temperature to a long-term (~30 year) baseline average. The temperature anomaly is determined by subtracting the baseline from the yearly average. Thus, cooler than average years have negative temperature anomaly, while warmer than average years have positive temperature anomaly. Examination of the temperature record as summarized in this manner is shown in Figure 11.3, which shows that any 30 year period toward the end of this record has been substantially warmer compared with any 30 year period toward the beginning. In fact, as of 2022, the nine warmest years during the instrument record were the most recent 9 years.

Because temperatures have been significantly warmer for the most recent 30 years compared with any previous 30 year period in the mid- or early 1900s, we call this climate change. When the only variable we are discussing is temperature, we call this *global warming*. By contrast, the term *global climate change* refers to the combined impacts of global warming and the concomitant impacts on droughts, floods, extreme weather, fires, ocean acidification, and sea level rise that will all be discussed as we progress through the chapter.

Although average global temperatures have warmed, this does not mean that temperatures have risen equally throughout the globe. Figure 11.4 provides a visualization of how global warming varies by location. Temperatures might rise 5° C degrees in one region and drop 2° C in another location. This figure shows a greater increase in temperature over land surfaces than over the oceans, which have greater thermal inertia and where the warm

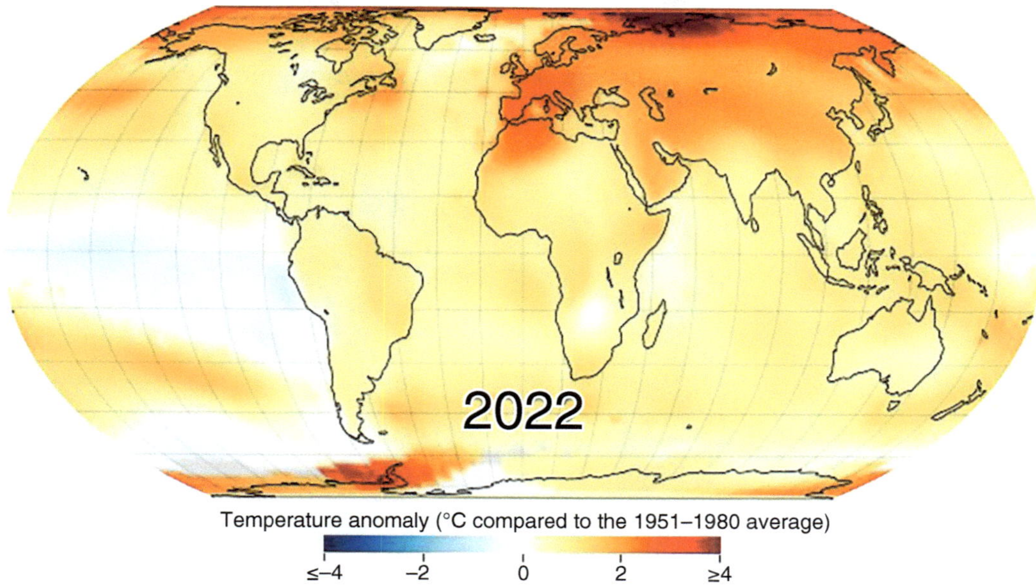

Figure 11.4 Spatial comparisons show greater warming over land than in the ocean and the greatest warming in the Arctic. This so-called polar amplification is intensified in the Arctic by the large-scale melting of sea ice. Open water is darker than sea ice, thus reducing the Earth's albedo (reflectivity). Image courtesy NASA under the public domain.

temperatures are able to mix to depth. It also shows the four times greater warming in the Arctic where melting sea ice is reducing the amount of reflected radiation contributing to a phenomenon called *polar amplification*, in which cooler high latitude regions are warming more rapidly.

Now that we have seen that global temperatures vary not only between years but also between regions of the Earth, and that even as most areas of the Earth have warmed, some locations have cooled. The question is whether the entire Earth system is warming. To address this question, we must calculate the planetary average through all years and not bias our analysis to a subset of data for less than 30 years or for one particular region without considering the average across the full globe.

In addition to the temperature record, many other physical and biological factors have confirmed the warming of the globe over the past century. The world's glaciers and polar ice sheets are melting, sea levels are rising, species are migrating to higher elevations and latitudes, and frost-free growing seasons are expanding (Fig. 11.5). Of the 10 factors where the Intergovernmental Panel on Climate Change (IPCC, see Box 11.1) collated data from all available instruments and countries, data from the variety of sensors as analyzed by a variety of scientists show the same pattern where all factors expected to rise in a warming world are rising (sea surface temperature, land temperature, troposphere temperature, absolute humidity of the atmosphere), and all those expected to decline with warming are declining (snow, glacier, Arctic, Greenland and Antarctic ice melt). For example, the instrument data sets from the U.S. NASA (shown in Fig. 11.3), NOAA, and the U.K.'s HADCET data centers all show a similar pattern of the rising global temperatures. All satellite and instrument data have been fully summarized, compared, and presented in the many IPCC reports that have summarized the scientific findings approximately twice per decade since 1995. A summary of the most recent IPCC findings for policymakers is provided in Box 11.1.

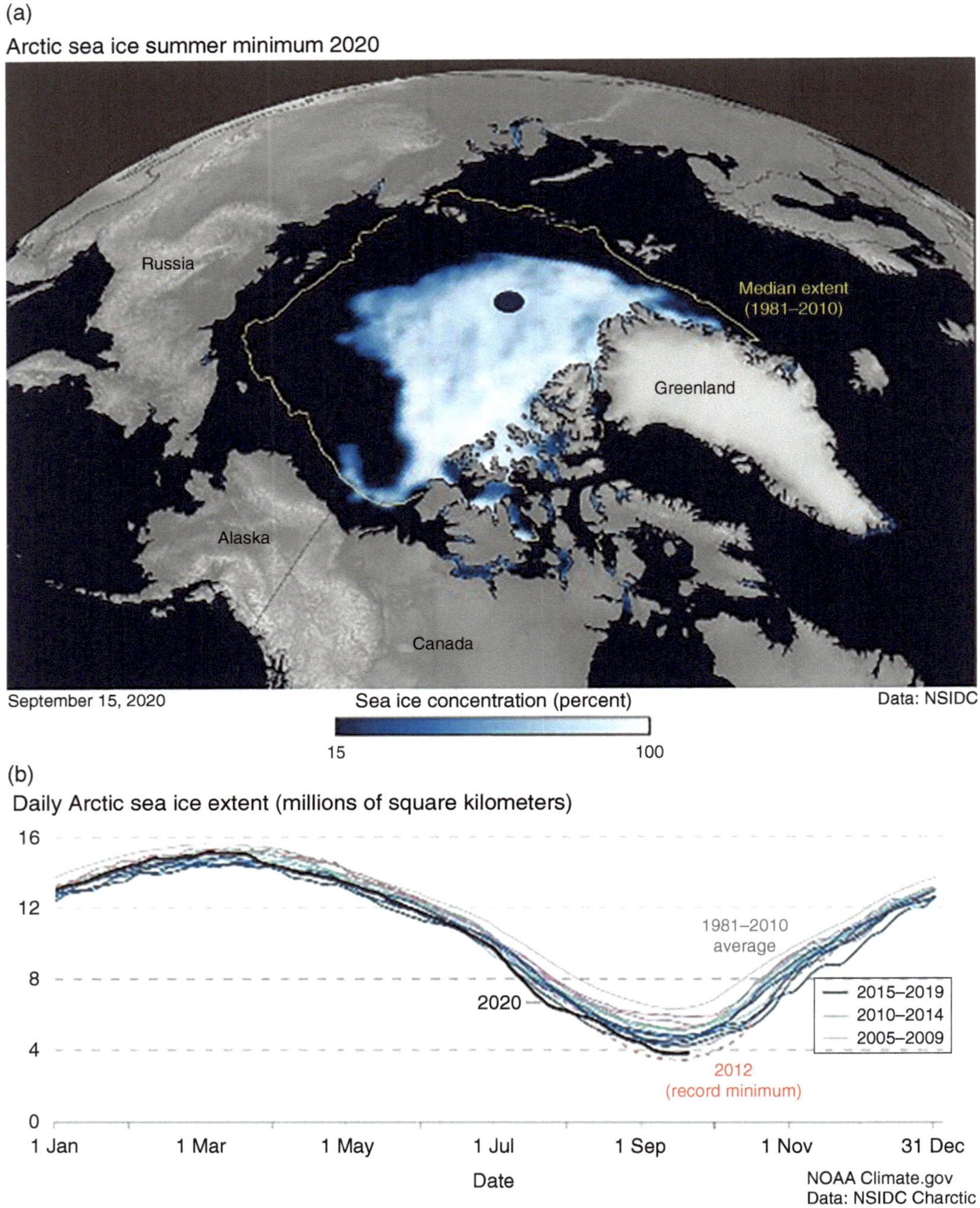

Figure 11.5 Warmer temperatures have caused an increase in Arctic sea ice melt. (a) Sea ice concentration (light blue to white) on 15 September 2020, the day of the summer minimum extent. The gold line is the median extent for 1981–2010: half of the years had smaller extents, half had larger. (b) A graph of daily ice extent since 2005. Years 2005–2009 are light purple, the record-low year 2012 is salmon, other years for 2010–2014 are light green, and years 2015–2019 are blue. The 2020 daily extent line is in black. Credit: NOAA Climate.gov, based on data from the National Snow and Ice Data Center. Image courtesy NASA under the public domain.

Box 11.1 Intergovernmental Panel on Climate Change.

In 1988, the United Nations (U.N.) General Assembly established the Intergovernmental Panel on Climate Change (IPCC) to prepare recommendations to the United Nations and its member nations based on the most up-to-date scientific information and available technology regarding climate change. The first IPCC Assessment Report was released in 1990 and pointed to the continued need for international cooperation to deal with the massive scope and dire consequences of global climate change culminating in the adoption of the U.N. Framework Convention on Climate Change (UNFCCC) in 1992. A second IPCC Assessment Report was released in 1995, which played an instrumental role in the adoption of the Kyoto Protocol in 1997. The Kyoto Protocol was the first treaty to set emissions reduction targets for greenhouse gases by the member states. The United States was a signatory to the treaty, but failed to ratify the treaty and did not participate substantively.

Since then, about twice each decade (every 5–7 years), the IPCC has released a new and updated assessment report outlining the best available predictions and providing guidance to the United Nations and its member states on how to avoid the worst impacts of climate change. In 2007, with the release of the Fourth Assessment Report, the IPCC was awarded the Nobel Peace Prize jointly with former U.S. Vice President Albert Gore, Jr., "for their efforts to build up and disseminate greater knowledge about man-made climate change, and to lay the foundations for the measures that are needed to counteract such change." The Fifth Assessment Report (2014) was used as an important touchstone in negotiating the Paris Climate Agreement intended to strengthen global efforts to limit climate change to 2°C or less and to enhance the ability of member states to survive the already inevitable impacts of climate change. The Sixth Assessment was released in 2022 and urges that there is a "rapidly closing window of opportunity to secure a livable and sustainable future for all." We present a subset of the IPCC's most recent findings from its Summary for Policy Makers as follows:

A) Human activities, principally through emissions of greenhouse gases, have unequivocally caused global warming, with global surface temperature reaching 1.1°C above 1850–1900 in 2011–2020. Global greenhouse gas emissions have continued to increase, with unequal historical and ongoing contributions arising from unsustainable energy use, land use, and land-use change, lifestyles and patterns of consumption and production across regions, between and within countries, and among individuals (high confidence).

B) Continued greenhouse gas emissions will lead to increasing global warming, with the best estimate of reaching 1.5°C in the near term in considered scenarios and modeled pathways. Every increment of global warming will intensify multiple and concurrent hazards (high confidence). Deep, rapid, and sustained reductions in greenhouse gas emissions would lead to a discernible slowdown in global warming within around 2 decades, and also to discernible changes in atmospheric composition within a few years (high confidence).

C) Climate change is a threat to human well-being and planetary health (very high confidence). There is a rapidly closing window of opportunity to secure a livable and sustainable future for all (very high confidence). Climate resilient development integrates adaptation and mitigation to advance sustainable development for all and is enabled by increased international cooperation including improved access to adequate financial resources, particularly for vulnerable regions, sectors, and groups, and inclusive governance and coordinated policies (high confidence). The choices and actions implemented in this decade will have impacts now and for thousands of years (high confidence).

11.2.3 Greenhouse Gases Are Increasing and They Trap Heat in the Atmosphere

At the same time that global temperatures have been rising, greenhouse gases have also been on the rise. Figure 11.6 shows the rise in the five most important greenhouse gases since 1979 as tracked by the National Oceanographic and Atmospheric Administration's (NOAA) Annual Greenhouse Gas Inventory (AGGI). While CO_2, which is released from the burning of fossil fuels, accounts for 60%–70% of all warming since 1850, the AGGI also monitors methane, nitrous oxide, and two chlorofluorocarbons (CFC-12 and CFC-13) that were banned by the Montreal Protocol because they damage Earth's protective ozone layer. Together, these five gases account for ~96% of global warming since 1850.

Table 11.1 compares various factors for the six most important greenhouse gases since the international ban on CFCs by the Montreal Protocol (more details follow). Note that methane emissions arise not only from agriculture, landfills, and permaculture (Fig 11.6) but also from natural gas leakage (Table 11.1) during both extraction and distribution.

Greenhouse gases are climate forcers that act to warm the globe. A list of the gases most responsible for absorbing and reradiating, thereby trapping radiative heat energy within the atmosphere are given in Table 11.1. While water is the strongest greenhouse gas, either as vapor or cloud droplets, and is responsible for 55%–60% of the Earth's *greenhouse effect*, it is not included in this list because humans have no control over atmospheric water vapor. When the air warms, more water evaporates into the air, especially from open water bodies like the oceans, which cover 70% or the Earth's surface. When the air cools, the water vapor condenses and rains out of the atmosphere. These processes take place on a timescale of days to weeks, rather than the decades to millennia lifetime of the greenhouse gases shown in Table 11.1.

In addition to the naturally occurring greenhouse gases (CO_2, CH_4, N_2O, and H_2O), humans have also developed many new chemicals and emitted them into the atmosphere. Many of these chemicals also act as greenhouse gases and have the potential to cause substantial warming. The most concerning set of these chemicals are the various forms of chlorofluorocarbons (CFCs), which came into production as early as the 1920s and have increased from zero to parts per trillion levels in the atmosphere as shown in Figure 11.6d. CFCs were used as refrigerants in air- conditioning, refrigeration, and heat pumps, and also as propellants in aerosol sprays, but were determined to have detrimental impacts on the stratospheric ozone layer, which shields the Earth's surface from harmful UV radiation. CFCs were banned by the Montreal Protocol due to their stratospheric ozone depleting properties. It turned out that they also have global warming potential 6,200–7,100 times more potent than CO_2. Although CFCs were replaced with hydrofluorocarbons (HFCs), which have no ozone depleting properties, their Global Warming Potential (GWP) is up to 12,000 times more potent compared with CO_2. Therefore, a crucial step that the Obama administration took in 2016 with regards to global warming, was to sign the Kigali Amendment to the Montreal Protocol, which called for a ban on HFCs and is thought to have prevented global warming of 0.2–0.4°C by 2100. HFCs are now being replaced by alternatives such as hydrofluoroolefins (HFOs) that have no Ozone Depletion Potential (ODP) and a Global Warming Potential of <1 compared to CO_2.

11.2.4 The Greenhouse Effect

Thus far, examination of the data has shown that average global temperatures and global greenhouse gas concentrations have risen over the same time frame. But you can't see temperature and you can't see CO_2, so it can be more difficult to understand how greenhouse gases cause global warming. The rise in greenhouse

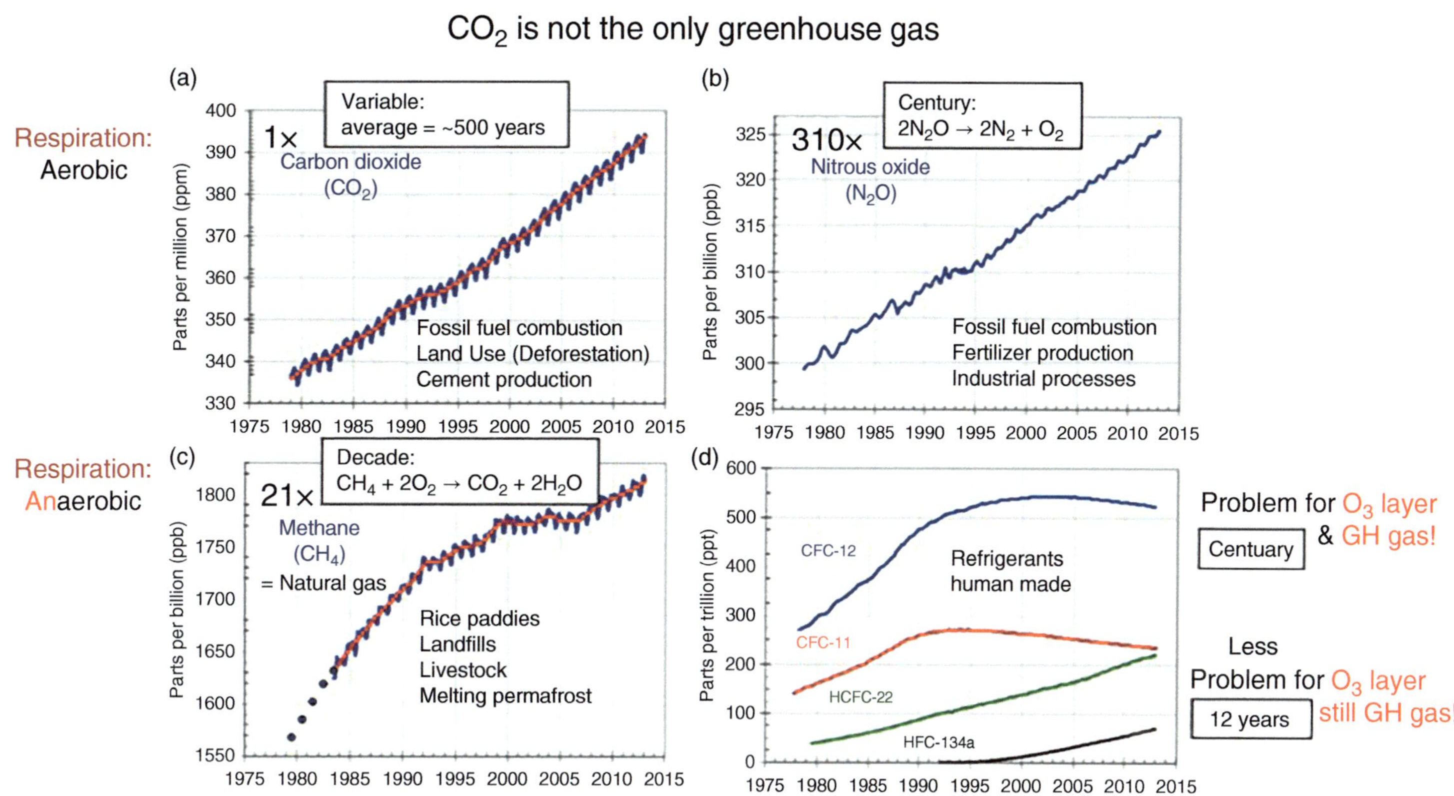

Figure 11.6 The rise in the five primary greenhouse gases over time, as tracked by the National Oceanographic and Atmospheric Administration's (NOAA) Annual Greenhouse Gas Inventory (AGGI). The AGGI monitors (a) carbon dioxide (primarily from aerobic respiration), (b) methane (primarily from anaerobic respiration), (c) nitrous oxide, and (d) two chlorofluorocarbons (CFC-12 and CFC-13). Together, these five gases account for ~96% of global warming since 1850. Note that CFCs were banned by the Montreal Protocol of 1987 because they damage Earth's protective ozone layer. CFCs were replaced with HCFC and HFC (also shown in (d)) which are not harmful to stratospheric ozone. However, CFCs, HCFCs, and HFCs are all strong greenhouse gases, so HCFCs and HFCs are now being replaced by HFOs, which are not destructive to the ozone layer and are only weak greenhouse gases with a global warming potential less than that of CO_2. The text in each panel shows the primary sources of each gas. Boxes show the atmospheric lifetimes. And top left of panels (a), (b), and (c) show the global warming potential compared with CO_2 over 100 years. Image courtesy NOAA under the public domain, edited by author; https://research.noaa.gov/article/) /587/ArticleID/2359/NOAA%E2%80%99s-greenhouse-gas-index-up-41-percent-since-1990

Table 11.1 The six most important greenhouse gases still emitted today.

Greenhouse gas (GHG)	Concentration (ppm) Preindustrial	Concentration (ppm) 2021	Global warming potential (GWP)	Atmospheric lifetime (in years)	Primary current sources
Carbon dioxide (CO_2)	278	414	1	50 to 200	Fossil fuel use, cement production, land use changes
Methane (CH_4)	0.722	1.895	21	12 ± 3	Fossil fuel use, agriculture
Nitrous oxide (N_2O)	0.275	0.334	310	120	Largely agriculture, approx. 1/3 are anthropogenic
Hydrofluorocarbons (HFCs)	0 (humanmade)	0.002–0.250	150 to 11,700	1.5 to 209	Replacing CFCs (CFCs deplete ozone)
Perfluorocarbons (PFCs)	0 (humanmade)	0.005–0.09	6,500 to 9,200	2,600 to 50,000	Aluminum & semiconductor manufacturing
Sulfur Hexafluoride (SF_6)	0 (humanmade)	0.0000112	23, 900	3,200	Electrical infrastructure and electronics manufacturing

Source: Data from NASA and EPA.
Notes: Note that the CFCs and HCFCs from Figure 11.5 have been banned by the Montreal Protocol and Kigali agreement, respectively, and are replaced here by two other important synthetic greenhouse gases: perflourocarbons (PFCs) and sulfur hexaflouride (SF_6). Preindustrial and 2021 concentrations, global warming potential (GWP, a.k.a. heat-trapping capacity), atmospheric lifetime, and primary sources are shown. Carbon dioxide is the most abundant and is assigned a GWP of 1.

gases is correlated with the rise in temperature, but in order for correlation to indicate causation one must invoke a mechanism.

In 1856, Eunice Foote was the first scientist to demonstrate that CO_2 absorbs heat. She filled glass cylinders with CO_2, air or hydrogen gas and found that the bottles filled with CO_2 warmed the fastest. She hypothesized that changing amounts of greenhouse gases in the atmosphere would alter the global climate. This experiment has been repeated many times including in several YouTube videos. You can try it yourself using soda bottles if you have access to a CO_2 source and thermometers. Results will work best in the summer by moving the bottles from an air-conditioned building to a shaded but warm outdoor area, so heat is being added from IR radiation.

This demonstration showing that CO_2 gas-filled cylinders/bottles warm faster than air-filled containers could be explained either by the fact that CO_2 molecules have a greater mass compared with the atmosphere or because CO_2 has vibrational modes (wobbly bonds) that absorb and then reradiate infrared (IR) radiation (the greenhouse effect). This puzzle was solved in 1861, when John Tyndall designed a ratio spectrophotometer, which documented that greenhouse gases absorb low energy infrared (IR) radiation (radiative heat energy) in a process that would ultimately be termed the *greenhouse effect*.

Gases comprising the Earth's atmosphere have varying capacities for absorbing different parts of the solar radiation spectrum. As discussed previously, those gases capable of absorbing the infrared radiation, often thought of as thermal radiation, are considered *greenhouse gases.*

Dry air in the Earth's lower atmosphere (currently) contains 78.08% nitrogen (N_2), 20.95% oxygen (O_2), 0.93% argon (Ar), and 0.04% carbon dioxide (CO_2). It also contains very small amounts of other gases such as volatile organic compounds (VOCs), which are typically human-made compounds that are emitted from solids or liquids at the surface, and other small hydrocarbons such as methane (CH_4), nitric oxides (NO_x), sulfur oxides (SO_x), hydrogen (H_2), helium (He), and so on. In addition, there may be water vapor that varies significantly spatially, seasonally, and with local weather patterns (e.g., there may be high, moderate, or low humidity). Water, either as vapor or cloud droplets, is also responsible for most (55%–60%) of the Earth's greenhouse effect, or heating of the atmosphere due to absorption of radiative heat energy by greenhouse gases (Fig. 11.7). A list of gases most responsible for absorbing and reradiating heat within the atmosphere is provided in Table 11.1.

Figure 11.8 shows how the greenhouse effect that works in the Earth's atmosphere is similar to what happens when you leave your car

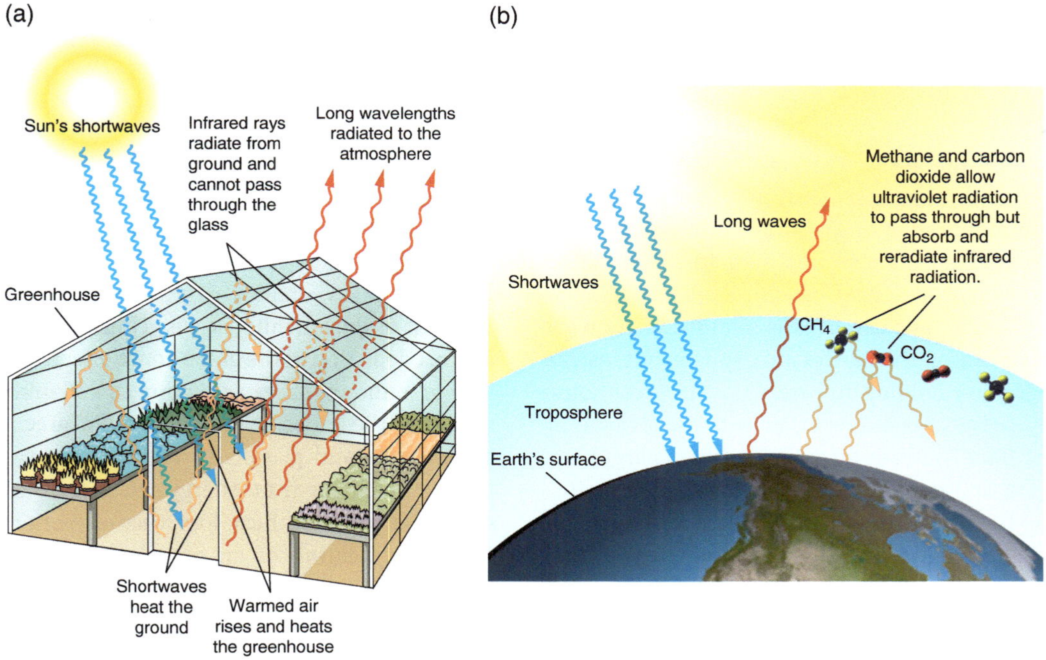

Figure 11.7 The greenhouse effect works by absorbing and reradiating heat close to the Earth's surface. (a) A greenhouse is a good analog for warming by greenhouse gases. In a greenhouse, incoming solar radiation (primarily light energy) enters through the glass. Some of the light energy is absorbed, and much is reflected (not shown). This reflected short wavelength energy is able to escape the greenhouse. However, the absorbed radiation changes form to become longer wavelength infrared radiation (radiative heat energy), which is trapped since it cannot travel through the glass. Through this process, the area inside a greenhouse is warmer than the air outside. (b) In much the same way, solar radiation enters the Earth's atmosphere. Some of that radiation is absorbed by the Earth's surface, and some is reflected back toward space. Reflected, high-energy, short-wavelength radiation is able to escape the Earth's atmosphere, but the absorbed solar radiation changes form to infrared radiation (radiative heat energy), which is absorbed by greenhouse gases such as carbon dioxide and methane, and is reradiated in all directions. About half of this heat is reradiated back toward the Earth's surface. From Murck & Skinner, 2016.

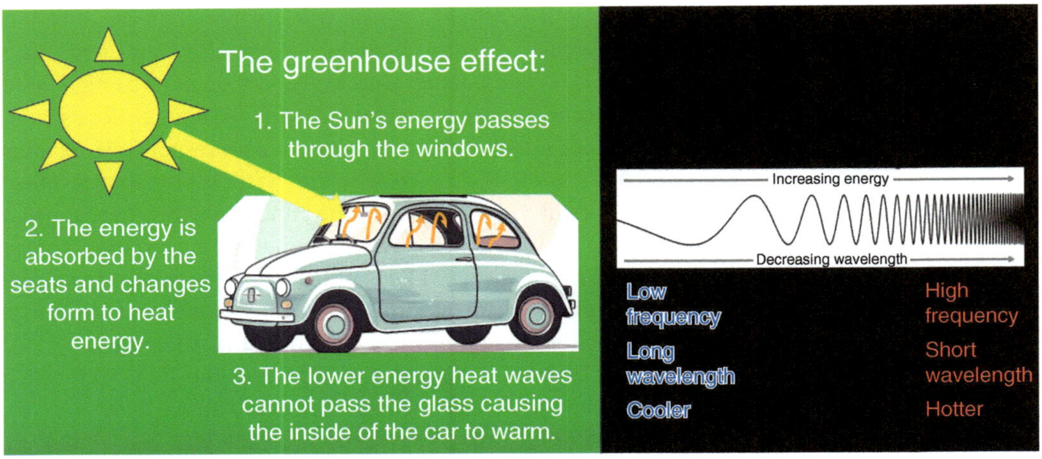

Figure 11.8 Depiction of the greenhouse effect in action as it warms up a car left sitting in the sunlight. The high energy short wavelengths of sunlight (light energy) can pass through the glass. But once that energy is absorbed by the seats and reradiated as infrared radiation (heat energy), the lower energy heat waves cannot escape, so heat builds up in your car. Image by Jillian Gregg, with car image from alexokov at freepic.com.

sitting in the Sun. The high energy sunlight can pass through the glass. But the low energy infrared radiation (radiative heat energy) cannot escape. These opposing forces result in the buildup of heat, causing the warm temperature in your car. This is almost exactly what happens in the Earth's atmosphere, but instead of glass holding the heat in, the greenhouse gases absorb and reradiate with the low energy infrared radiation (radiative heat energy) acting to warm the atmosphere.

It is easy to envision that glass could prevent the passage of radiative heat energy: you can't put your fist through glass! But it is much more difficult to envision that gases could also prevent the loss of heat. To illustrate that this is nonetheless true, we have created identical images taken with a standard versus an infrared camera. The standard camera shows the light energy that we see with our own eyes, while the infrared camera shows only the radiative heat energy depicted in various colors. The results show that light, as depicted in the standard camera images, can easily pass either through glass (Fig. 11.9a) or through the refrigerant gas held in an airtight bag (Fig. 11.9b). However, the warmer temperatures, depicted as red and yellow in the IR camera images, cannot pass through either the glass or the greenhouse gas. Thus, we have a direct depiction of the greenhouse effect, and why greenhouse gases are acting as a blanket to warm the globe. Note that while greenhouse gases and glass both prevent the passage of radiative heat energy, glass is solid, so it also prevents heat loss via convection. Thus, the heating of an actual greenhouse (or a car left in sunlight with closed windows) would be much faster than that of the atmosphere where greenhouse gases are acting amidst a convective atmosphere. The blanket on the globe is also a metaphor, as blankets also function more as a physical barrier to convective and conductive heat losses and less via absorption and reradiating of radiant heat energy, which is the primary mechanism by which greenhouse gases warm the Earth's atmosphere.

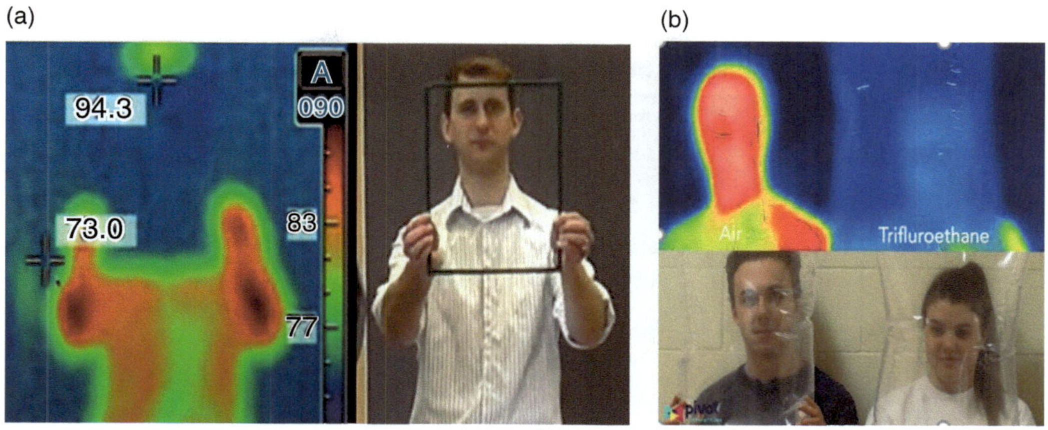

Figure 11.9 Identical photographs taken by infrared versus standard camera. Note that in the infrared images, the colors show the different temperatures (dark red = hottest, dark blue = coolest), so you can actually see the temperatures. Comparison of these photographs demonstrates heat loss being blocked by (a) glass, and (b) a refrigerant gas on the right compared with regular air on the left held in airtight bags. Image (a) courtesy Toby Alt/YouTube; (b) courtesy Vernier Scientific with screenshots by the author.

11.2.5 The Earth's Temperature Is Controlled by the Balance of Incoming Solar Radiation, Albedo, and the Greenhouse Effect

Planetary temperatures are controlled not only by the concentration of greenhouse gases in the atmosphere, but also by the balance of incoming versus reflected solar radiation in an overall planetary energy balance (Fig. 11.10). Here we will explain each of these processes beginning with the original source of radiative energy from the Sun, reflected solar radiation, called albedo, then revisit the concept of recycled radiative heat energy within the atmosphere via the greenhouse effect.

Examination of NASA's depiction of the Earth's energy balance in Figure 11.10 shows that 23% of the incoming solar radiation is reflected by clouds and 7% is reflected from the surface of the Earth, so approximately one third (~30%) of total incoming solar radiation is reflected. Another 23% of the incoming solar radiation is absorbed by clouds, so only ~48% of the total incoming solar radiation at the top of the atmosphere reaches the surface of the Earth. Once absorbed by the Earth's surface, the high energy, short wavelength, light energy that passed through the greenhouse gases with no problem is absorbed and reradiated at longer wavelengths. As such, it changes form to become lower energy infrared radiation (radiative heat energy), which cannot pass easily through the greenhouse gases as it leaves the Earth's surface, as described in section 11.2.3.

Keep in mind that the Earth's energy balance has nothing to do with money and the transition to renewable energy. Instead, it considers the balance of incoming solar radiation and outgoing heat energy to determine the temperature of the planet. If there is more outgoing than incoming energy, the Earth cools. But if there is more incoming than outgoing energy, the Earth warms. If there is no difference, the planetary temperature remains the same.

Recall that per the First Law of Thermodynamics, energy cannot be created or destroyed but can change form. Here, the incoming solar radiation is not created or destroyed but changes form into heat energy. You can experience this as the difference between wearing dark or light colors on a sunny day. Lighter

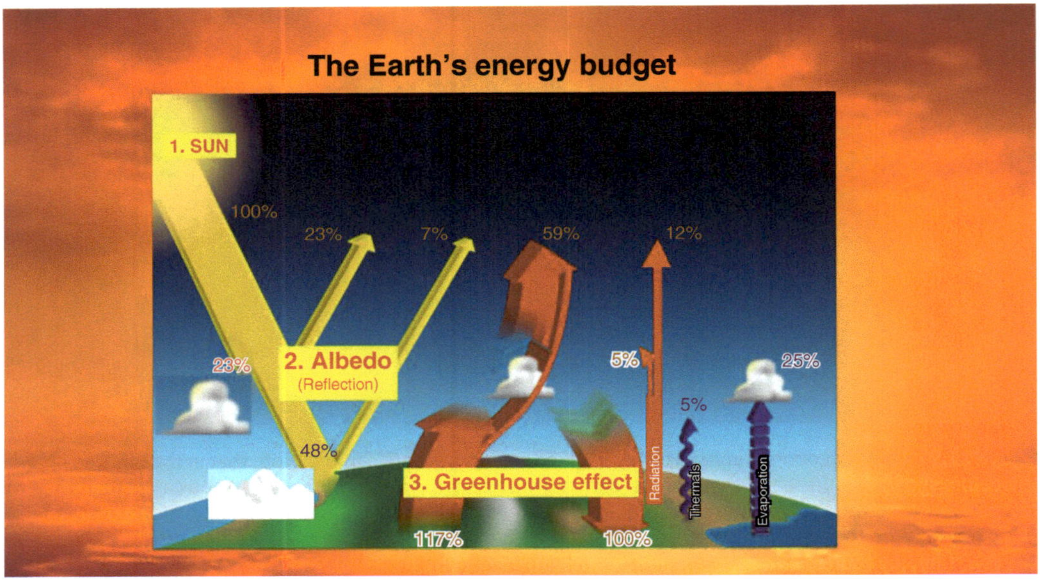

Figure 11.10 Depiction of the significant fluxes of energy in the Earth's energy balance. High energy, short-wavelength solar radiation is shown in yellow. Reflected solar radiation is labeled as Albedo. Low energy, long-wavelength infrared radiation is shown in red and shows the recycling of radiant heat energy due to the Greenhouse effect. Nonradiative heat losses are shown in purple: convective and conductive heat transfer are labeled as thermals, and the latent heat of vaporization is labeled as evaporation. Values show the global average percentages compared with the incident solar radiation at the top of the atmosphere. Note that only half of the solar radiation incident at the top of the atmosphere makes it to Earth's surface, and that the surface of the Earth receives twice as much heat back-radiated from the greenhouse effect as it does from the Sun. Image courtesy NASA Energy Budget Story Board under the public domain, modified by the author with percentages from NASA Earth Observatory: Climate and Earth's Energy Budget.

colored clothing reflects more visible sunlight, so you stay cooler, while darker colored clothing absorbs visible sunlight, so you get even hotter. That warmer heat energy is reemitted through "vibrational relaxation" as infrared (IR) radiation. Ultimately, light energy has changed into heat energy.

The Sun is the primary heat source warming the Earth, and this works in a way similar to that which we just explained. Once the high energy visible light hits the Earth's surface, it is absorbed and transformed into lower energy infrared radiation, also called heat (or heat energy), which is ultimately emitted back to space. The balance between incoming solar radiation (primarily light energy) and outgoing infrared radiation (radiant heat energy) determines the temperature of the planet. When the incoming and outgoing energy are balanced, the climate is stable. If there is more energy entering than leaving the system, the Earth warms. And if there is more energy leaving than entering the system, the Earth cools.

The Sun is very hot, so emits high energy radiation in visible spectrum (Fig. 11.11). By contrast, the Earth is much cooler, so even after being warmed by the Sun, the peak radiation emitted is in the cooler, nonvisible, infrared (IR) range (Fig. 11.11). The emission of lower energy IR radiation matches our observations that the Earth doesn't emit visible light. Instead, we see the Earth and other planets and the Moon because they reflect a portion of the incoming solar radiation, making them visible from space when illuminated.

The transformation of energy from incoming solar radiation (primarily light energy) to

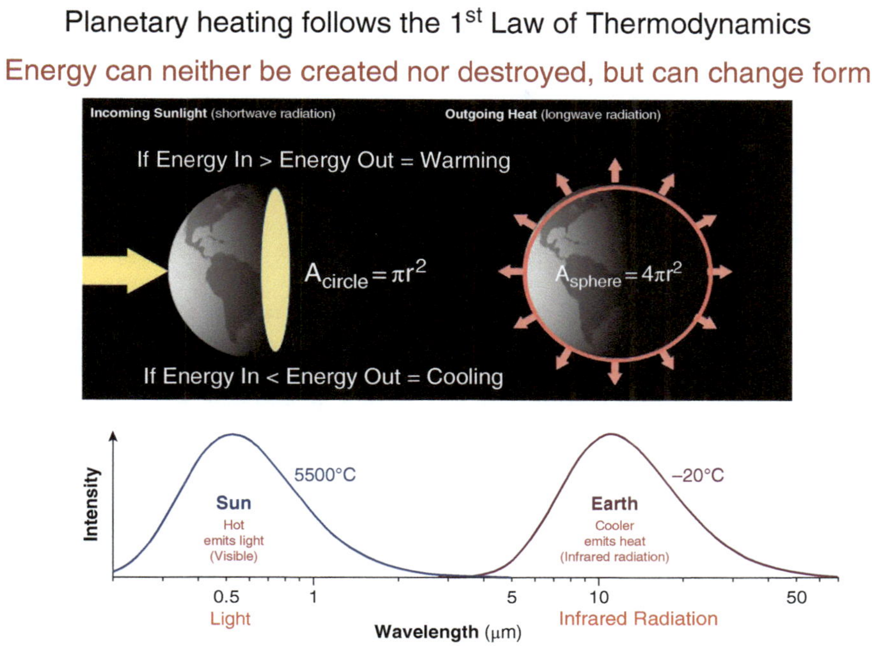

Figure 11.11 Transformation of incoming solar radiation (light energy) to outgoing infrared (IR) radiation (heat energy). Bottom panel shows the shorter wavelength of incoming solar radiation, compared with the longer wavelength of the outgoing IR radiation. Top panel shows the four times greater surface area for the emission of the lower energy IR radiation to keep up with the interception of higher energy incoming solar radiation. Images courtesy NASA, modified by author.

reradiated infrared radiation (often considered heat or heat energy) follows the First Law of Thermodynamics where the energy is not destroyed but has changed form. It initially seems inconceivable that reradiation of lower energy IR radiation could keep up with the input of higher energy solar radiation. However, the Earth only intercepts solar radiation on half of its surface at any one time, which is functionally only a disc of the incident solar radiation ($A = \pi r^2$; Fig. 11.11). By contrast, release of the transformed, lower energy, IR radiation is continuously emitted across the Earth's entire surface ($A = 4\pi r^2$; Fig. 11.11) 24 hr/day. This 4 times greater surface area for outgoing IR makes up for much of the lower energy released via IR radiation, so that an equilibrium temperature can be reached. Other factors that contribute to Earth's overall energy balance are conductive and convective heat loss from thermals, and the latent heat of vaporization, where absorbed heat fuels phase transformations of water from solid and liquid to vapor as depicted in Figure 11.10.

Albedo

The reflectiveness of any surface is determined by its color. As an example, we use pool balls in Figure 11.12. White surfaces reflect all wavelengths of light, so white is much cooler. Black surfaces instead absorb all wavelengths of light so black objects are warmer. Red objects absorb all wavelengths but reflects the red light, so are intermediate in temperature.

Climate scientists refer to the reflectivity of a planet as the planet's *albedo*. High reflectivity denotes high planetary albedo, while low reflectivity denotes low planetary albedo. Incoming solar radiation that strikes light colored surfaces like snow or ice is reflected back

(a)

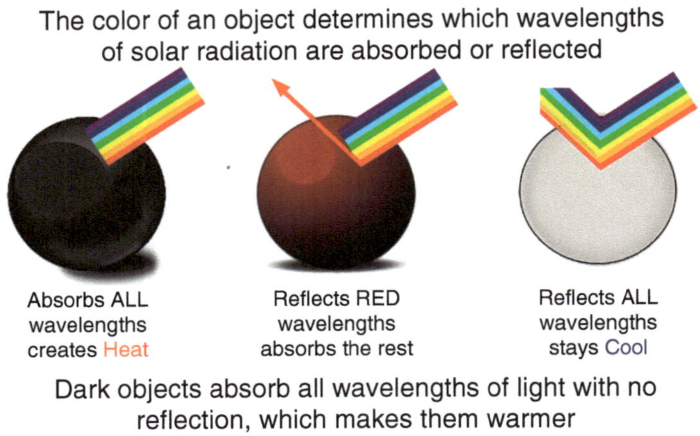

(b)

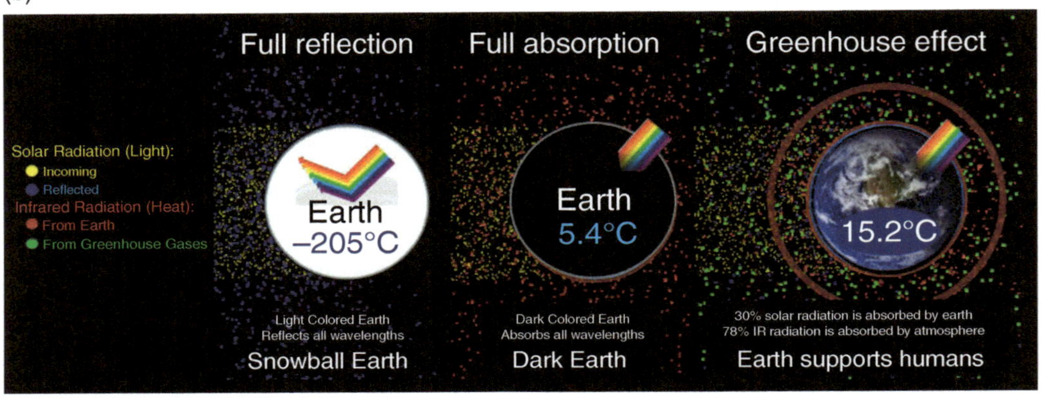

Figure 11.12 (a) The color of any object determines whether solar radiation is absorbed or reflected. (b) Differential reflection of incoming solar radiation depends on the color of the Earth. White ice and clouds and lighter colored sand and vegetation reflect the incoming solar radiation and make the Earth cooler. Dark-colored surfaces like the ocean and forested areas absorb the incoming solar radiation and warm the Earth. (b) Image created by author.

out to space, whereas dark surfaces like forests and oceans absorb more of the radiation (Fig. 11.13).

Greenhouse Gases and Radiant Heat Absorption

The greenhouse gas molecules in the atmosphere absorb the heat energy as it radiates from the Earth's surface and become more vibrationally excited (Fig. 11.14). As they dissipate this vibrational energy, they reradiate the heat in all directions (or isotropically), first warming the atmosphere (increasing the kinetic energy of the surrounding gas molecules), but also redirecting about half of that heat energy back to Earth's surface. This recycling of radiant heat energy by the greenhouse gases results in twice the radiation at the surface of the Earth as compared with the energy we receive from the Sun (see 100% for the large red arrow compared with only 48% for the large yellow arrow at the Earth's surface in Fig. 11.10). This is why we refer to greenhouse gases as a blanket over the Earth. However, recall that the blanket on the globe is merely a metaphor, as blankets also function more

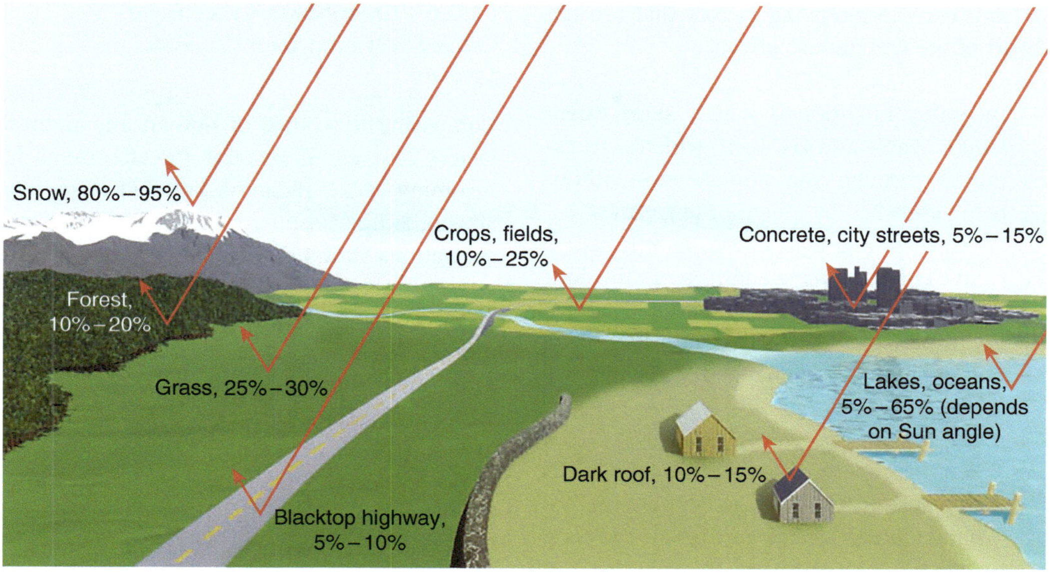

Figure 11.13 Variation of albedo depending on the brightness of ground cover. Lighter surfaces generally have higher albedo, or reflection, and darker surfaces have low albedo, or reflection. From Arbogast, 2017.

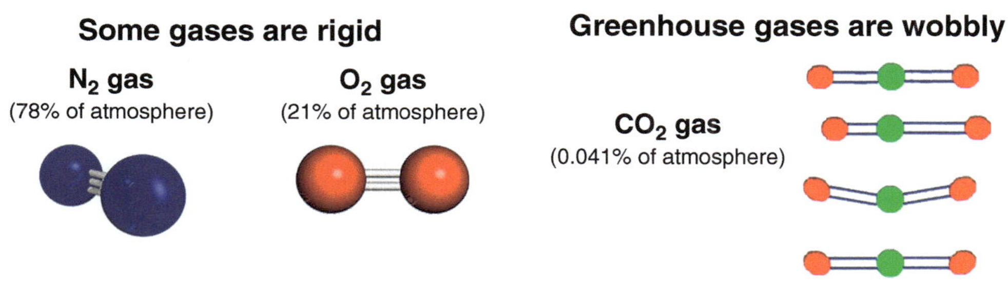

Figure 11.14 More complex systems of bonds within a molecule exhibit a larger variety of vibrational modes creating more wobbly bonds in greenhouse gases that act as shock absorbers, retaining the heat of intercepted heat waves and reemitting the absorbed radiation in all directions. Image courtesy of Harun Mbuvi/YouTube with a screenshot by the author.

as a physical barrier to convective and conductive heat losses and less via absorption and reradiating of radiant heat energy, which is the primary mechanism by which greenhouse gases warm the Earth's atmosphere.

Note that the Sun warms the Earth's surface, but the physical motion of the greenhouse gases absorbing and reradiating that energy slows the rate at which the heat is released to space, thereby warming the atmosphere sufficiently to allow for life on Earth. Preindustrial levels of greenhouse gases were perfect for keeping Earth warm enough for humans, but the increase in greenhouse gas concentrations that humans have added since the Industrial Revolution have resulted in an enhanced greenhouse effect resulting in global warming. Let's examine each factor that impacts the global temperature.

Consider the following factors that are the heart of the greenhouse effect:

1) Incoming radiation is mostly short wavelength visible and UV light, which is transmitted through the atmosphere without being absorbed by greenhouse gases.
2) When incoming light energy is reflected, it hasn't changed wavelength, and still does not interact with the greenhouse gases. We experience that when reflected solar UV radiation is unchanged when we get sunburned under our chin when we go skiing or boating.
3) When light energy is absorbed by the Earth it changes form to become heat energy.
4) When radiant heat energy is reradiated back to space it gets absorbed by wobbling the bonds of the greenhouse gases. That heat energy is then reradiated in all directions, so about half of it is directed back to Earth and acts to warm the planet.

11.2.6 Greenhouse Gases Have Caused Warming Over the Past Century

Comparing the range of human and natural factors that could account for alterations in incoming solar radiation, altered planetary albedo, and atmospheric greenhouse gas concentrations shows that human emitted greenhouse gases are the only factor that could account for global warming since the beginning of the Industrial Revolution (Fig. 11.15). More detail can be found in these articles, "Why Scientists Think 100% of Global Warming Is Due to Humans" (Pearce, 2017) or "How to Calculate the Impact of Carbon Emissions" (Goodwin, 2014). The main takeaways from these articles are listed here.

1) Scientists have examined all human and natural factors that could potentially account for global warming since 1850. Of all the factors examined, greenhouse gases

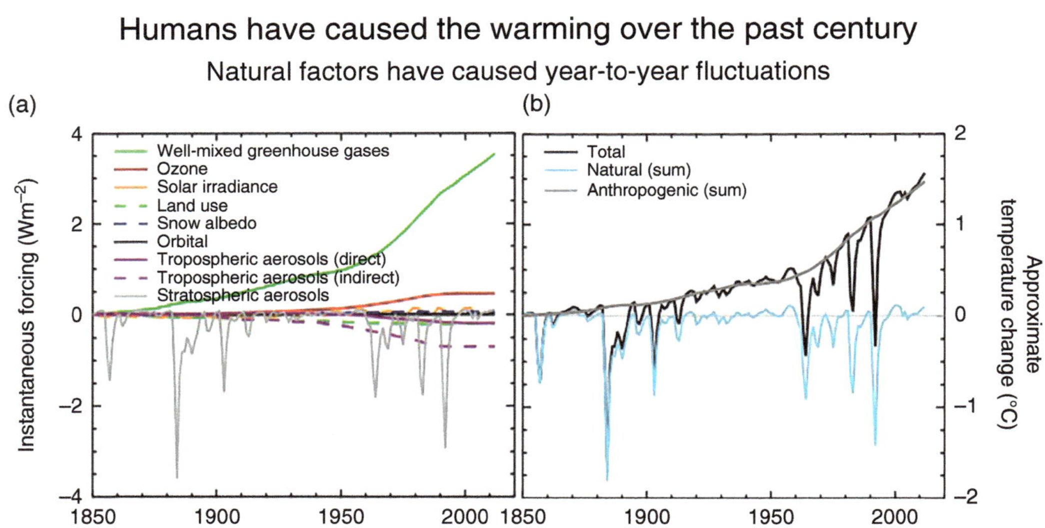

Figure 11.15 Comparison of the relative impacts of human versus natural radiative forcers since 1850. Radiative forcing calculated from model simulations (CMIP5, E2-R NINT ensemble). (a) Individual forcings. (b) Total forcing along with separate sums of natural (solar, volcanic, and orbital) and anthropogenic forcings. Note that natural factors account for the year-to-year variation, but human factors are what account for the warming trend. Image https://data.giss.nasa.gov/modelforce/courtesy NASA under the public domain with axes edited by author.

and tropospheric ozone pollution (which is a short-term greenhouse gas since its atmospheric lifetime is on the order of hours to months) are the only factors that could account for the rise in temperatures since 1850.

2) The cooling caused by tropospheric aerosol pollutants has offset the overshoot in warming that would otherwise have occurred due to greenhouse gases. While greenhouse gases absorb and reradiate heat with their wobbly gaseous bonds, aerosols, or "soil of the air," are particles that scatter and reflect incoming solar radiation. Note that black carbon, the soot from burning coal, is one exception that does absorb incoming solar radiation thereby acting to warm the atmosphere. But the remainder of aerosols generally reflect incoming solar radiation, which has offset the warming of the planet on the order of 0.5–1.0°C (Masson-Delmotte et al., 2021) over the past century in a process called *global dimming*.
3) Human and natural factors are both needed to explain past temperatures. However, greenhouse gases and ozone are the only factors that could account for the warming of the planet (Fig. 11.15a), while natural factors account for the year-to-year variation (Fig. 11.15b).

Additional factors that were considered, but that could not account for warming over the past century are the following.

1) Sunspots. Climate scientists must consider the roughly 11-year cycle of solar activity which describes the oscillation in the number of observed Sunspots on the Sun's surface, and the number of solar flares emitted. Variations that occur during this 11-year cycle cause a slight but measurable shift between higher and lower amounts of light and heat reaching the Earth's surface. Reductions in the number of Sunspots is thought to be one factor accounting for the Little Ice Age in the 1500–1700s. These oscillations in solar activity are far too small and of too high frequency to account for the warming observed during the past century.
2) Milankovitch cycles. The dynamic positioning of Earth relative to the Sun and were named the Milankovitch cycles (Fig. 11.16) after their discoverer, Serbian astrophysicist Milutin Milankovitch (1879–1958). These cycles alter the tilt of the Earth on its axis (obliquity), the relative distance of the North and South Poles from the Sun in summer and winter seasons (precession), and the shape of Earth's orbit around the Sun from more circular to more elliptical (obliquity) (Fig.11.16). Together, these cycles alter the angle of incidence of sunlight hitting the poles in various seasons which initiates the melting or growing of polar ice caps. It is important to note that while these cycles can account for the initiation of the comings and goings of the Ice Ages, the effects – whether considered individually or collectively – fail to account for the observed warming of the climate during the past century because they act over tens of thousands of years and have negligible impacts over decadal time scales (described below).
3) Volcanic Activity. Another impact on global climate doesn't involve greenhouse gases, but it does involve how the atmosphere interacts with solar radiation. During large volcanic eruptions, tiny aerosol particles can be ejected high into the atmosphere where they reflect the Sun's rays back out to space. This has a global cooling effect, which can also be mimicked by anthropogenic (human-produced) aerosols such as the sulfate aerosols that may be released by coal burning electrical utilities, but which are now mostly removed from flue gas in countries with strong environmental regulations. The net effect of volcanoes on the climate system during the recent geologic past (and certainly since the beginning of the Industrial Revolution) is that of global cooling.
4) Evolution of the Sun. We tend to think of the Sun as a very consistent source of light

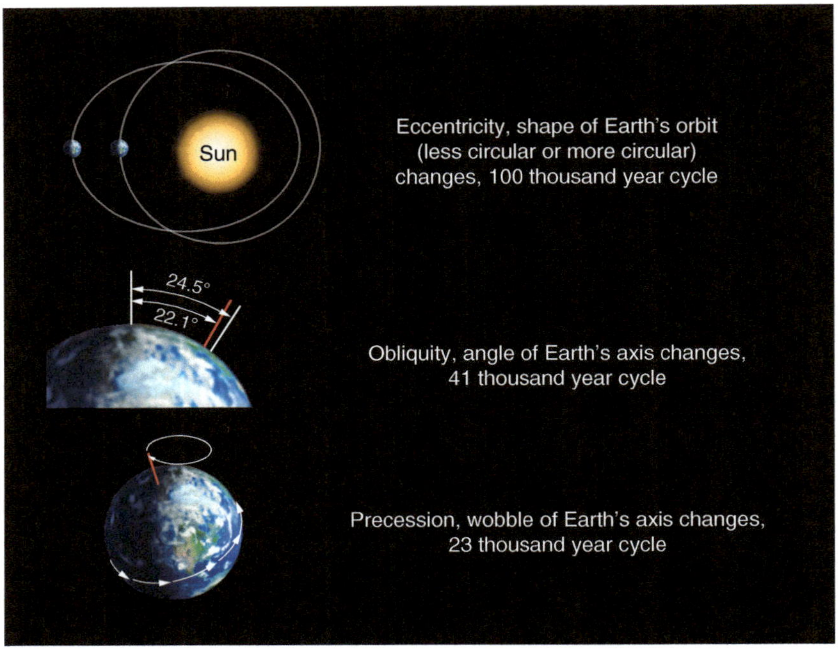

Figure 11.16 The Milankovitch cycles of eccentricity, obliquity, and precession. From Fletcher, 2014.

and heat, but if we think about the Sun across very long time periods the amount of energy it emits and the amount that reaches the earth has not always been the same and will change over time into the future. We know from observing other solar systems in our galaxy that stars (like our Sun) undergo a process of formation and destruction while in between a long and slow evolution occurs. In its current phase of evolution our Sun is becoming gradually brighter and smaller. In fact astronomers estimate that the Sun becomes approximately 10% brighter every billion years. This comes to only 0.000,000,010x (10^{-8}x) brighter per century, a rate far too slow to account for observed climate warming over the past century.

5) Earth's Interior Heat. While we have seen that energy from the Sun and its interactions with the material of Earth's systems represent the primary source of heat for Earth's climate, there are two additional factors that must also be considered to complete our understanding. The radioactive decay of heavier elements in the Earth's interior and throughout the crust and residual heat from Earth's formation provide a source of heat quite apart from energy inputs from the Sun. However, heat emission from the Earth's interior (0.09 W/m^2) is very small compared to the 341.3 W/m^2 coming from the Sun, so makes only a small 0.03% contribution to the warming of the planet.

11.2.7 Greenhouse Gases as the Primary Driver of the Earth's Temperature

The final planetary temperature model used in the Paris Agreement shows greenhouse gases as the sole predictor for the Earth's temperature (IPCC, 2013, Fig. 11.17).

Read the exciting story behind Figure 11.18 here: https://www.weforum.org/agenda/2014/12/how-to-calculate-the-impact-of-carbon-emissions/ (Goodwin, 2014).

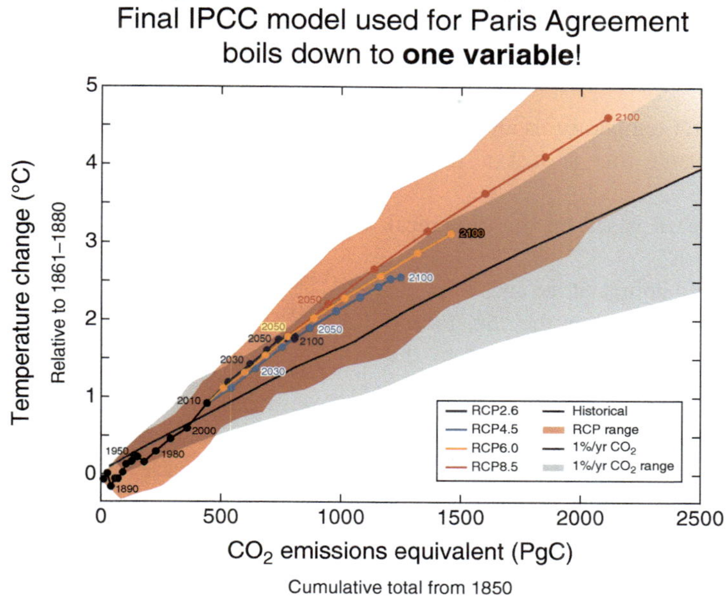

Figure 11.17 Final planetary temperature model used in the Paris Agreement showing greenhouse gases as the sole predictor for the Earth's temperature. Image courtesy of IPCC (2013) Fig. SPM.10. IPCC Caption: Global mean surface temperature increase as a function of cumulative total global CO_2 emissions from various lines of evidence. Multimodel results from a hierarchy of climate-carbon cycle models for each RCP until 2100 are shown with colored lines and decadal means (dots). [See RCP discussed in Fig. 11.31.] Some decadal means are labeled for clarity (e.g., 2050 indicating the decade 2040–2049). Model results over the historical period (1860 to 2010) are indicated in black. The colored plume illustrates the multimodel spread over the four RCP scenarios and fades with the decreasing number of available models in RCP8.5. The multimodel mean and range simulated by CMIP5 models, forced by a CO_2 increase of 1% per year (1% year, 1 CO_2 simulations), is given by the thin black line and grey area. For a specific amount of cumulative CO_2 emissions, the 1% per year CO_2 simulations exhibit lower warming than those driven by R.C.P.s, which include additional non-CO_2 forcings. Temperature values are given relative to the 1861–1880 base period, emissions relative to 1870. Decadal averages are connected by straight lines. For further technical details see the IPCC 2013 Technical Summary Supplementary Material. [Fig. 12.45; T.S. TFE.8, Fig. 1] .

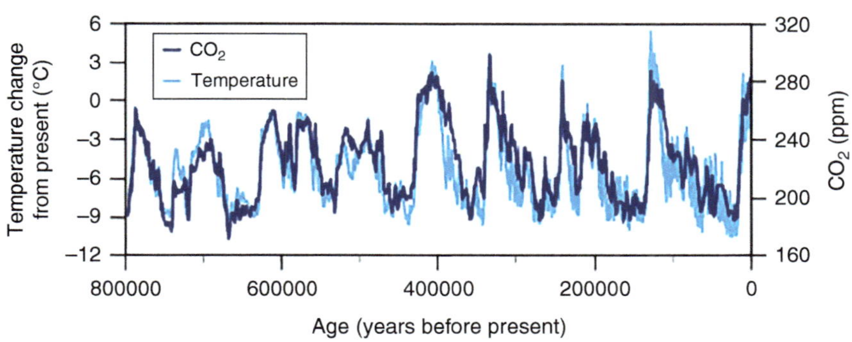

Figure 11.18 Temperature and CO_2 were tightly coupled during the comings and goings of glacial and interglacial periods during the Pleistocene ice ages. Image courtesy Wikimedia Commons.

11.2.8 Earth's Temperatures Have Also Been Correlated With CO_2 Concentrations Over Geologic History

People ask, Haven't temperatures always been fluctuating and isn't this is just a natural process?

What can we learn based on the geological history of the Earth?

The comings and goings of ice ages over the past 2 million years during the Pleistocene epoch have been shown to be initiated by the Milankovitch cycles. The first of the Milankovitch cycles has a periodicity of 23,000 years and describes a pronounced wobble in the Earth's rotation referred to as *precession*. If you've ever watched a top spin with near perfect stability with a vertical axis, but gradually begin to tip outward as its rotation slows with an axis of rotation that gradually rotates around the center, this is precession. The effect that Earth's precession has on climate is that sometimes northern winter solstice occurs as Earth is furthest from the Sun (aphelion), sometimes northern winter solstice occurs as Earth is closest to the Sun (perihelion). Between these two extremes we see a spectrum of intermediate states. Currently, the Earth is closest to the Sun during the Southern Hemisphere summer, so the Southern Hemisphere has the most pronounced seasonal variation in solar radiation. Yet 13,000 years from now, the Earth will be closest to the Sun during the Northern Hemisphere summer. Since there is significantly greater land mass in the Northern Hemisphere and land mass can heat up much more rapidly than the oceans, the Northern Hemisphere may be expected to experience extreme hot summers and cold winters, though this seasonal temperature variation would be tempered by the eccentricity (circular versus elliptical) of the Earth's trajectory around the Sun and tilt of the Earth on its axis at that time, as described in the following sections.

The second Milankovitch cycle deals with how pronounced Earth's axial tilt is, or its *obliquity*. While this has relatively little impact on the total annual input of solar radiation, it profoundly impacts the seasonality of this input especially when we consider the freeze-thaw cycle occurring at high latitudes. The period of this oscillation is 41,000 years and represents a shift from 22.1° to 24.5° of tilt. We are currently about midway through one of these cycles at 23.5° of tilt and decreasing. As the Earth progresses toward minimal tilt, there will be less solar radiation at the poles in their respective summers. In the absence of the current warm temperatures on Earth due to the high concentration of greenhouse gases, less solar radiation at the poles would lead to an increase in the polar ice caps, which would in turn reflect solar radiation and lead the Earth into the next ice age. Note that this pattern is in stark contrast to the rapid ice loss we are currently experiencing as the globe warms in response to greenhouse gases (Fig. 11.3).

The last Milankovitch cycle that we'll discuss is *eccentricity*. It deals with how elliptical, or eccentric, Earth's orbit around the Sun is. This cycle has a periodicity of about 100,000 years. Currently, this causes about 6.8% difference in solar radiation between aphelion and perihelion, while at maximum eccentricity we'd expect a 23% difference between aphelion and perihelion.

Each of these cycles, based upon astronomical observations, is supported by the paleoclimate record (Figs. 11.18, 11.19). Taken collectively, the Milankovitch cycles represent important triggers for the onset of glaciation and thawing that describe the alternation of ice ages and warm periods during recent geologic time. In the absence of human-caused climate change, the astronomical forecast is of nearly 100,000 years of gradually reduced eccentricity, and insolation (the amount of solar radiation reaching a given area) that is dominated by changes in obliquity and a slight decrease in axial tilt. This suggests that Earth will undergo slight summertime warming in the Northern Hemisphere for about 25,000 years followed by about 25,000 years

(a)

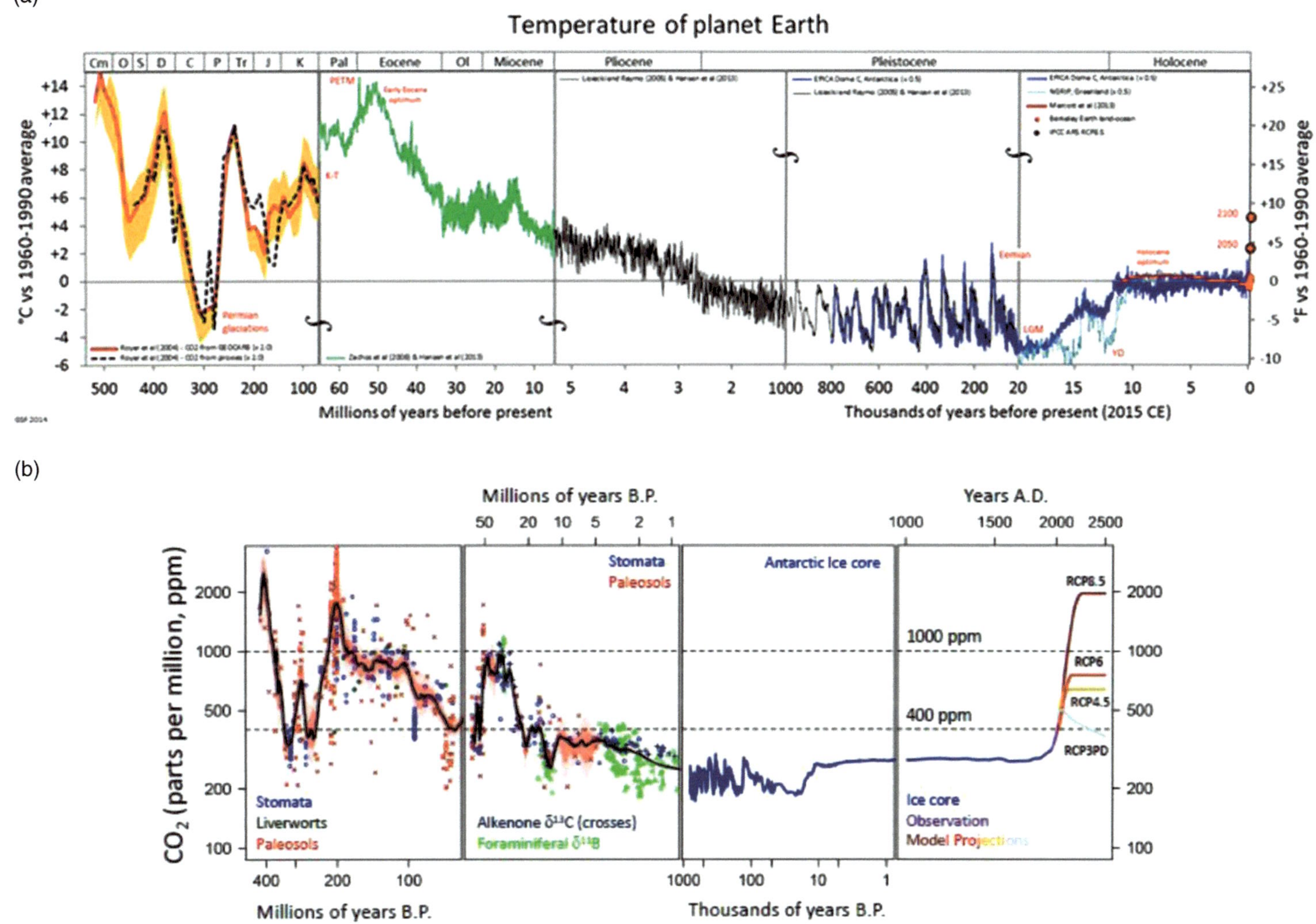

Figure 11.19 (a) Average planetary temperature and (b) CO_2 concentrations of Earth over the last 540M and 400M years, respectively. Multiple confounding factors affect interpretation of samples this old, so data are best viewed as a qualitative indication of temperature (warmer/colder) and CO_2 concentrations. Temperature image by Wikipedia under CC-BY-4.0. CO_2 image by Foster et al. (2017) in *Nature Communications* under CC-BY-4.0; https://www-nature-com.oregonstate.idm.oclc.org/articles/ncomms14845 also in https://skepticalscience.com/Past-and-Future CO2.html

of summertime cooling, but not significant enough cooling to trigger glacial advance into the midlatitudes for at least 50,000 years. Most long-term climate models suggest that even beyond this time-frame, human-caused climate change will be severe enough to delay, or possibly prevent entirely, the next glacial advance.

Despite the importance of the Milankovitch cycles for initiating the onset or decline of an ice age, this change in the Earth's orientation to the Sun turns out to be insufficient to account for the 10°C temperature difference between these glacial and interglacial periods. Instead, the ice core data show that fluctuations in global temperatures through the ice ages were directly correlated with the atmospheric CO_2 concentrations (Fig. 11.18). In a glacial to interglacial transition, the initial melting caused by the relatively minor Milankovitch warming, appears to have been accentuated by melting polar ice, decreased albedo, increased respiration rates, and release of dissolved CO_2 from the warmer oceans, resulting in higher atmospheric CO_2 concentrations that acted as the primary driver to warm the planet. In an interglacial to glacial transition, an initial increase in polar ice mass triggered by Milankovitch cooling would then be accentuated by higher albedo, increased dissolution of CO_2 in the cooler oceans, and greater primary productivity on the newly exposed landmass, resulting in lower atmospheric CO_2 concentrations and further planetary cooling. These phenomena of slightly warming or cooling of the planet leading to more significant warming or cooling by another mechanism is referred to as *climate feedback*. Climate feedbacks can be positive (destabilizing or reinforcing) or negative (stabilizing or counteracting). The temperature needed to trigger a climate feedback-loop irreversibly is often referred to as a *tipping point*. So, Milankovitch warming/cooling, by slightly changing the light and heat reaching Earth, is thought to have been sufficient to reach the tipping points needed to trigger dramatic positive warming or cooling climate feedbacks observed during glacial/interglacial transitions. Climate was then stabilized by negative feedbacks during the periods between transitions.

Proxy data for paleoclimatic temperature and CO_2 records over the past 540 and 400 million years of Earth's history, respectively (Fig. 11.19), show a similar pattern where warm periods are associated with higher CO_2 concentrations. Multiple confounding factors affect interpretation of samples this old, so data are best viewed as a qualitative indication of temperature (warmer/colder) and CO_2 concentrations (higher/lower). However, qualitative comparisons of these proxies for past climate indicate the global temperatures were correlated with atmospheric CO_2 concentrations.

11.2.9 Earth's Current Rise in CO_2 Is From Humans Burning Fossil Fuels Not From Natural Factors

While the rise in temperature and CO_2 over the past century are correlated and these factors also changed in concert over the waxing and waning of the ice ages, correlation does not indicate causation. The greenhouse effect explains the mechanism by which CO_2 warms the planet, and we have examined the details of other impacts of altered solar inputs or albedo influences, narrowing down to CO_2 and other greenhouse gases being the primary cause for altering the global temperatures over the past century.

The final question that remains then, is whether humans are responsible for the higher CO_2 concentrations, the primary greenhouse gas that has the greatest impact on global warming. We saw previously that CO_2 concentrations fluctuated naturally in the ice ages as a positive feedback in response to warming or cooling temperatures. Here, we examine four key figures that document that humans have been responsible for the higher CO_2 concentrations since the beginning of the Industrial Revolution.

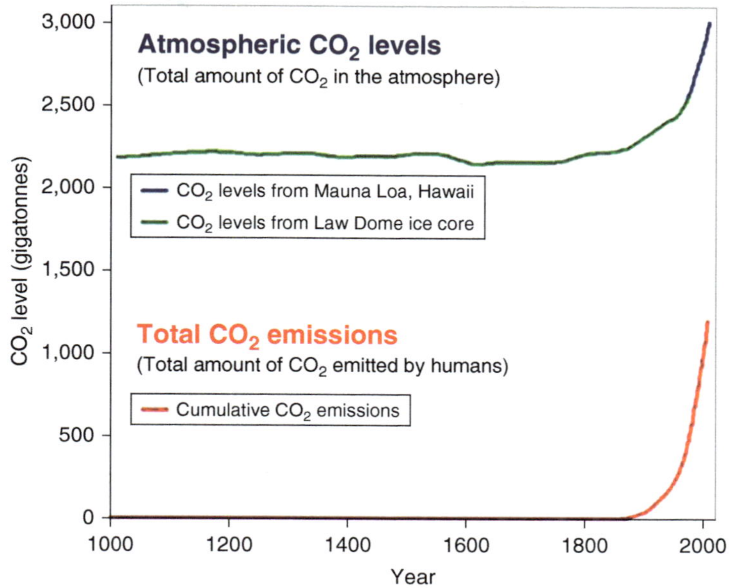

Figure 11.20 Gigatons of CO_2 in the atmosphere (green and blue lines) and cumulative CO_2 emissions (red line) from 1000 to 2000 AD as calculated by the CDIAC. Image by Cook, 2020, in *Skeptical Science* under CC BY 3.0.

Figure 11.20 demonstrates the similarity in the timing of human CO_2 emissions compared with the rise in atmospheric CO_2 concentrations. This figure also shows the twofold greater increase in CO_2 emissions compared with that accumulated in the atmosphere. These data indicate that the timing and magnitude of emissions match closely the rise in CO_2 concentrations.

Prior to the Industrial Revolution, atmospheric carbon dioxide exhibited a near constant ^{14}C relative abundance ($\Delta^{14}C$) owing to the fact that carbon dioxide concentrations and cosmic radiation have both been stable over the long term. ^{14}C is a radioactive isotope of carbon with a half-life of 5,730 years and is incorporated into biomass via photosynthesis by the same biochemical pathways as the other stable isotopes of carbon ^{12}C and ^{13}C. Because fossil fuels derive from ancient biomass, their ^{14}C has decayed entirely to ^{14}N. Upon burning, the CO_2 emitted contains little or no $^{14}CO_2$. Meanwhile, CO_2 emitted by the natural C cycle through respiration and even biomass burning has nearly the same $\Delta^{14}C$ as the atmospheric CO_2 taken up by the growing plants (and animals for that matter). By looking at the isotope ratios in tree rings and ice cores, we can reconstruct past atmospheric $\Delta^{14}C$. Figure 11.21 shows the decline in the radioisotope ^{14}C signatures in tree rings before 1950. This shows that the CO_2 incorporated into tree rings came increasingly from CO_2 emitted from fossil fuels that are depleted in ^{14}C. The organic materials have been buried in the ground for ~350–370 million years (since the Carboniferous period), so all the radioactive ^{14}C has decayed. These analyses are obfuscated after 1950 due to the $^{14}CO_2$ emitted to the atmosphere from nuclear weapons testing.

Figure 11.22 shows the steady rise in CO_2 concentrations through the years in contrast to the periodic eruption of volcanoes. Despite many calculations indicating that human CO_2 emissions are 100–1,000 times greater than that from volcanoes, it is still reassuring

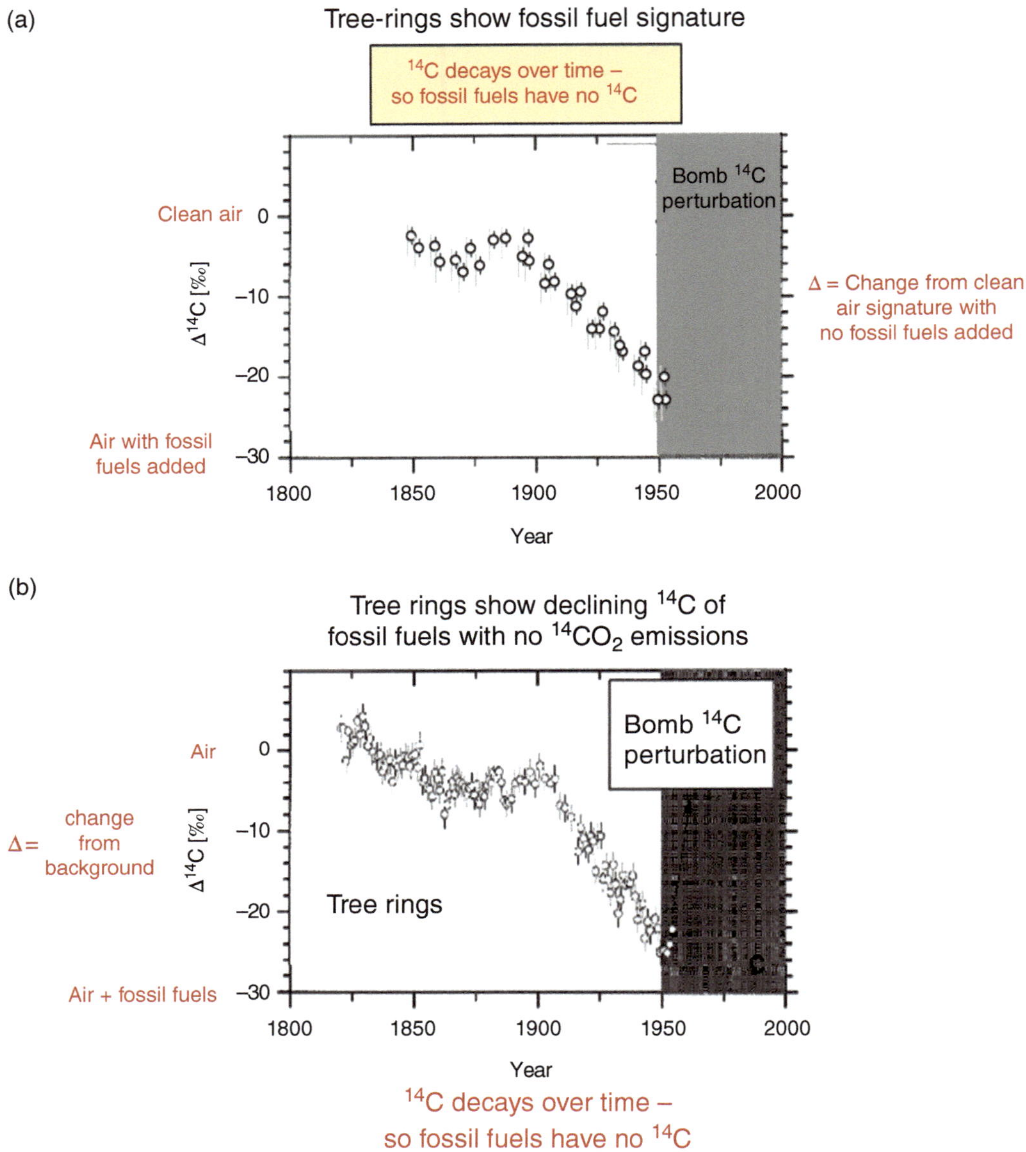

Figure 11.21 The relative abundance of ^{14}C in atmospheric CO_2 from 1825 to 1950 as recorded in tree rings. The dramatic decrease from 1900 to 1950 is due to the dilution of naturally occurring $^{14}CO_2$ by adding CO_2 from fossil fuels that has no ^{14}C because it is so old (see text). Data cannot be examined past 1950 due to the release of ^{14}C from extensive atomic bomb testing. Note that ^{14}C is a tiny fraction of total C in the atmosphere ($^{14}C/^{12}C$ is approximately $1/10^{12}$). The ppt (‰) units used for the graph tell us if the relative abundance of ^{14}C is higher (positive values) or lower (negative values) than an established reference point. Image by author, after Stuvier & Quay, 1981.

to see that recent large volcanic eruptions have had no impact on the steady rise in CO_2 concentrations.

The final figure documenting that the current rise in CO_2 concentrations could not be from volcanoes is the lockstep decline in global oxygen (O_2) concentrations as CO_2 rises (Fig. 11.23). This shows that O_2 was consumed as the CO_2 rose, so the CO_2 has to have come from some combination of respiration and

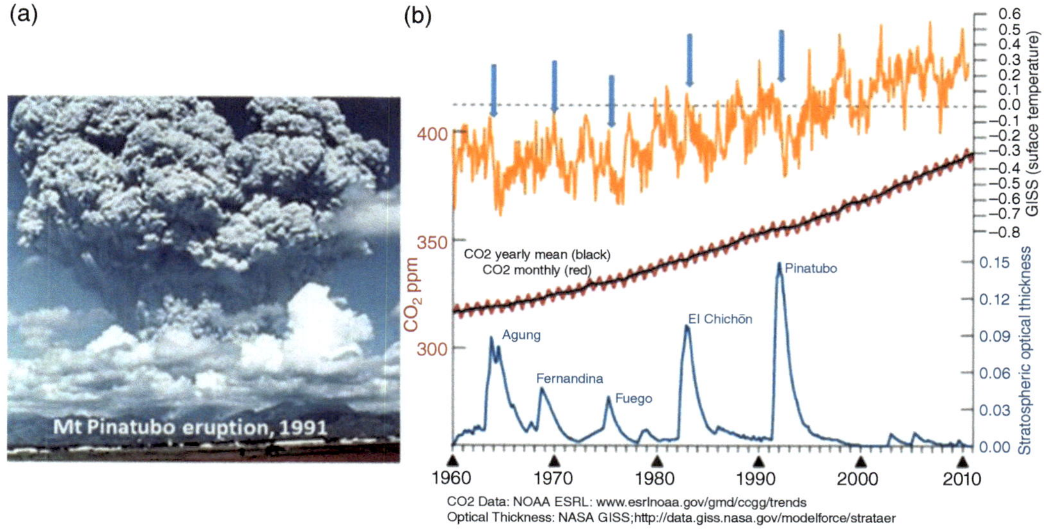

Figure 11.22 (a) Photo of Mt. Pinatubo erupting. (b) The rise in global mean CO_2 concentration (red and black lines, left y-axis), the change in stratospheric optical thickness due to periodic volcanic eruptions (blue line, right y-axis), and the rise in global temperature (orange line, top right y-axis) from 1960 through 2011. Arrows show the years the volcanoes erupted and the dip in temperatures after the eruptions, but no impact on the global CO_2 concentrations. Image (a) courtesy USGS under the public domain. https://pubs.usgs.gov/fs/1997/fs113-97/. Image (b) courtesy Skepticalscience.com under CC BY 3.0, edited by author. https://brightstarswildomar.blogspot.com/2012/05/volcanoes-co2-and-temperatures.html

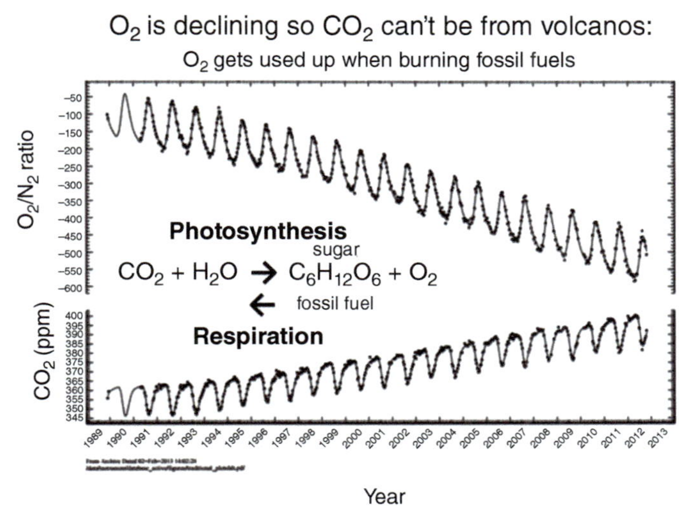

Figure 11.23 The decline in O_2 (shown as the O_2/N_2 ratio) in concert with the rise in atmospheric CO_2 concentrations shows that respiration reactions were involved in the release of CO_2 to the atmosphere. Photosynthesis can also be seen in the annual drawdown in CO_2 concentration, together with the release of O_2 to the atmosphere. Image courtesy UCAR by permission, edited by author. National Center for Atmospheric Research Staff, 2018.

combustion rather than emitted from volcanoes. This figure, combined with the timing and magnitude of increased atmospheric CO_2 versus emitted CO_2 documented in Figure 11.20, the postindustrial decline in $^{14}CO_2$ documented in Figure 11.21, and the lack of appreciable volcanic emissions signal within well-documented rising CO_2 concentrations in Figure 11.22, all combine to provide the evidence that the higher CO_2 concentrations are the result of burning fossil fuels.

11.2.10 Projected Average Global Temperatures by 2100 and Beyond

People ask, can someone please explain to me how my local weatherman can't get his weather prediction right a day out, but when it comes to global warming, we know what's going to happen in 10 years?

Because climate represents the average weather over many years, atmospheric models used to forecast the long-term global climate turn out to be more accurate than weather predictions since they are not attempting to estimate individual outcomes of daily weather events in specific locations. An analogy might be that of predicting the outcome for one flip of a coin or roll of the dice, representative of weather, compared with the average over many replicates, which is representative of the climate. The average of all possible outcomes is much more easily calculated than the prediction of any individual event. Over the past century, scientists have thus become increasingly adept at calculating the human impact on the average global and regional climate and how this could change based on the increased concentration of greenhouse gases.

The current predictions for global warming by 2071–2100 are shown in Figure 11.24 and are highly dependent on which shared socioeconomic pathway (SSP1-5) we take. Note that the models indicate that if we were able to take the SSP1-1.9 or SSP1-2.6 pathway (detailed in lower panel), we could maintain the average global temperatures below the 2.0°C threshold called for in the Paris Agreement (Box 11.1).

Note that while many climate models stop at 2100, most greenhouse gases remain in the atmosphere for millennia. Since there is currently no economically efficient mechanism to remove them (though see current attempts in section 11.4), the greenhouse gases we emit today are expected to remain and accumulate in the atmosphere (Fig. 11.2 and Table 11.1). Therefore, the socioeconomic pathway we take in the next decade will have important implications for global temperatures for thousands of years.

11.3 Natural Hazards Caused Directly and Indirectly by Greenhouse Gas Emissions and Climate Change

11.3.1 Climate Change Impacts

Now that we have a broad overview of the data and mechanisms accounting for global warming, the goal for the rest of the chapter is to assess the impacts of climate change on environmental disasters. To this end, we provide a broad overview linking climate change to impacts on the environment, exacerbation of climate disasters (Table 11.2), and direct impact on humans (Fig. 11.25). Fossil fuels, agriculture, deforestation, and many amplifying feedbacks all lead to the increase in greenhouse gases, global warming, and climate change. Climate change in turn has impacts on habitat destruction, ecosystem collapse in the Arctic, and the Great Barrier Reef (for example), species migrations and extinctions, sea level rise and coastal submersion, and disease carrier and pest propagation. There are also many direct impacts on humans and ocean ecosystems including ocean acidification and warming leading to coral bleaching and fish stock declines.

The remainder of this chapter will, therefore, discuss the climate change impacts on the

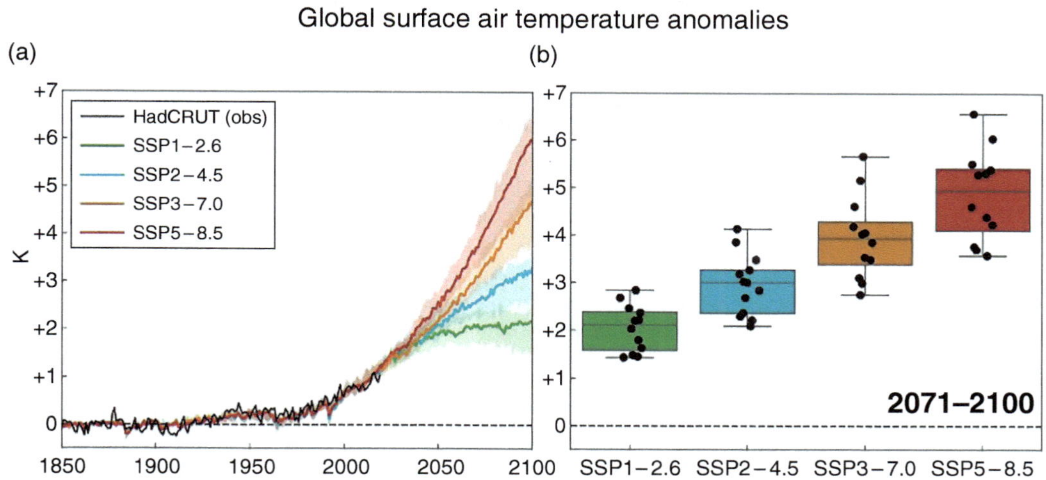

(c)

Shared Socioeconomic Pathways in the IPCC Sixth Assessment Report[15]:SPM-14 V·T·E

SSP	Scenario	Estimated warming (2041–2060)	Estimated warming (2081–2100)	Very likely range in °C (2081–2100)
SSP1–1.9	Very low GHG emissions: CO_2 emissions cut to net zero around 2050	1.6 °C	1.4 °C	1.0–1.8
SSP1–2.6	Low GHG emissions: CO_2 emissions cut to net zero around 2075	1.7 °C	1.8 °C	1.3–2.4
SSP2–4.5	Intermediate GHG emissions: CO_2 emissions around current levels until 2050, then falling but not reaching net zero by 2100	2.0 °C	2.7 °C	2.1–3.5
SSP3–7.0	High GHG emissions: CO_2 emissions double by 2100	2.1 °C	3.6 °C	2.8–4.6
SSP5–8.5	Very high GHG emissions: CO_2 emissions triple by 2075	2.4 °C	4.4 °C	3.3–5.7

Figure 11.24 Global, annual average surface air temperature (SAT) anomaly changes by 2100 (baseline 1851–1880) for four Shared Socioeconomic Pathway (SSP) scenarios. (a) Time series, showing the SSP ensemble medians (solid lines) and the interquartile ranges calculated across models (colored shading). Black line shows actual data through 2017 using the same 1851–1880 baseline. Light gray shading is 2071–2100, the time interval used for the box and jitter plots. (b) Box and jitter plots for all models (median SAT anomaly, 2071–2100) in each SSP scenario. Individual model values are indicated by the black dots. (a & b) From Cook et al., 2021. (c) Details of the various SSPs, by Wikipedia under CC-BY-4.0; https://en.wikipedia.org/wiki/Shared_Socioeconomic_Pathways

Table 11.2 Climate distasters in 2022.

Time frame	Climate disaster
January–September 2022	Many rivers in Europe have run low or dried up partly because of the worst drought in 500 yr and intense heat waves. Climate change has likely played a significant role in this crisis by increasing the frequency and intensity of droughts and heat waves.
February 2022	La Niña and climate change contributed to record-breaking rainfall on the east coast of Australia. This led to flooding that damaged thousands of properties and killed eight people.
February–March 2022	Record-breaking flooding occurred along the northeastern coast of Australia, leading to standing water, which, in turn, promoted the spread of mosquitoes that carry the

(Continued)

Table 11.2 (Continued)

Time frame	Climate disaster
	Japanese encephalitis virus. Such flooding is likely becoming more common because of climate change.
February–July 2022	The number of people affected by drought in Kenya, Somalia, and Ethiopia who have limited access to safe water increased from 9.5 million to 16.2 million. This increasing drought severity may be at least partly due to climate change (Ghebrezgabher et al., 2016)
March 2022	A severe drought in the southern plains of the United States put the winter wheat crop at risk. Although droughts are complex phenomena with many possible causes, increasing drought intensity has been linked to climate change (Mukherjee et al., 2018).
March–April 2022	A deadly heat wave occurred in India and Pakistan, killing at least 90 people and contributing to widespread crop losses and wildfires. It was estimated that climate change made this event 30 times more likely to occur.
April 2022	Climate change likely contributed to extreme rainfall in eastern South Africa, which triggered flooding and landslides that killed at least 435 people and affected more than 40,000 people.
April–June 2022	Widespread dust storms in the Middle East led to thousands of people being hospitalized; such dust storms may be increasing in frequency because of climate change.
May 2022	Extremely heavy rainfall in northeastern Brazil resulted in landslides and flooding that killed at least 100 people. Climate change may be responsible for the increasing frequency of extreme rainfall.
June 2022	A severe storm in Yellowstone (United States) caused the Gardner River and Lamar River to overflow, destroying parts of various roads in Yellowstone National Park. Such extreme flooding could be increasing in frequency because of climate change.
June 2022	Several countries in Western Europe experienced a record-breaking heat wave. This heat wave contributed to major wildfires in Spain and Germany. Many other parts of the Northern Hemisphere also experienced extreme heat; for example, temperatures reached 104.4° F in Isesaki, Japan, an all-time record for the country. Similarly, a heat dome in the United States contributed to record-breaking temperatures. Other affected countries include Finland, Iran, Norway, and Italy. In general, extreme heat is becoming more common because of climate change (Luber & McGeehin 2008).
June 2022	Following extreme heat, China experienced record-breaking rainfall, which may be linked to climate change.
June 2022	Bangladesh experienced the worst monsoon flooding in 100 years, killing at least 26 people. This flooding is likely at least partly due to climate change causing monsoons to become more variable.
June–July 2022	Extreme rainfall led to flooding in some parts of New South Wales, Australia. Sydney which ultimately led to the ninth wettest year on record since 1900. It is likely that climate change contributed at least partly to this rainfall and flooding.
June–August 2022	Deadly floods in Pakistan ultimately submerged one third of the country, killed 1,839 people and affected roughly 33 million people, including 16 million childrenfrom mid-June until Oct. Impacts include surging rates of dengue fever, gastric infections, and malaria. These floods may be at least partly related to climate change causing monsoon rainfall to become more intense.
June–August 2022	China experienced an extraordinary heat wave, which may be the most severe that has ever been recorded globally. Such events are likely becoming more common because of climate change. The extreme heat contributed to large-scale crop failures and wildfires, in addition to exacerbating a major drought that caused 66 rivers to dry up and led to a significant decline in hydroelectricity generation.

Table 11.2 (Continued)

Time frame	Climate disaster
August–September 2022	California and other parts of the western United States faced extremely hot temperatures because of a heat dome, which caused seven firefighters to be hospitalized with heat-related injuries. The effects of the heat dome may have been worsened by climate change.
September–October 2022	In the United States, Hurricane Ian caused damage across many parts of Florida and the Carolinas, killing more than 100 people and leaving at least 2.5 million without electrical power. Ian is one of the costliest and strongest hurricanes to ever hit the United States. Climate change is likely causing strong and rapidly intensifying storms such as Ian to become more common.

Source: Ripple et al., 2022, Bioscience, Open Source.
Notes: Here, we list numerous recent disasters that may be at least partly related to climate change. This list is not intended to be exhaustive. Because of the recent nature of these events, our sources often include news media articles. For each event, we generally provide references indicating that the likelihood or strength of such an event may have increased because of anthropogenic climate change. References to scientific articles are given directly in the table, and links to news articles are provided in supplemental file S1 in Ripple et al., 2022. Note, some of these climate disasters may be at least partly related to changes in jet streams (Stendel et al. 2021; Rousi et al., 2022). These references have been added to the bibliography, but they are not the source of the table.

intensification of the range of extreme weather events and disasters including heat waves, hurricanes, floods, droughts, freshwater loss and desertification, wildfires, and glacial retreat that can cause massive mudslides and flooding.

Changes to the natural environment that derive from greenhouse gases and climate change include melting of the cryosphere, sea level rise, changes in rainfall patterns, longer fire seasons, more extreme weather, and loss or degradation of marine habitat and productivity. Each of these have corresponding impacts on humans and society. We outline the range of climate disasters that can be intensified by climate change.

Heat Waves

Heat waves have been occurring at triple the long-term average rate since 2010. This trend is expected to continue in a warming climate. These warmer temperatures are intricately tied to the increase in precipitation as the warmer air temperatures draw more moisture from land and oceans, thereby enabling megastorm events. Globally, large urban population centers that exhibit urban heat island effects and have large populations of poor and homeless will be at particular risk of increased heat related health emergencies and deaths.

Perhaps counterintuitively, global warming may be leading to incursions of extreme cold into midlatitudes in the Northern Hemisphere. More extensive melting of sea ice in the Arctic Ocean is thought to lead to a more wavy circumpolar jet stream that allows extremely cold air masses previously confined within the Arctic to be pushed further south by a nearly sinusoidal (S-shaped) high-altitude wind field. Whatever the ultimate cause, these wavy jet streams, can also prevent temperate zone westerlies from moving out hot air masses resulting in record heat waves in many regions. Ultimately, extreme heat waves will also cause droughts that are expected to become more frequent, more intense, and more difficult to recover from.

Drought

Periods of drought have long been one of the greatest impediments to agricultural production. *Drought* is defined as a period of below average rainfall in a given region and may encompass periods as short as 14 days, months, or even several years. Even short-term

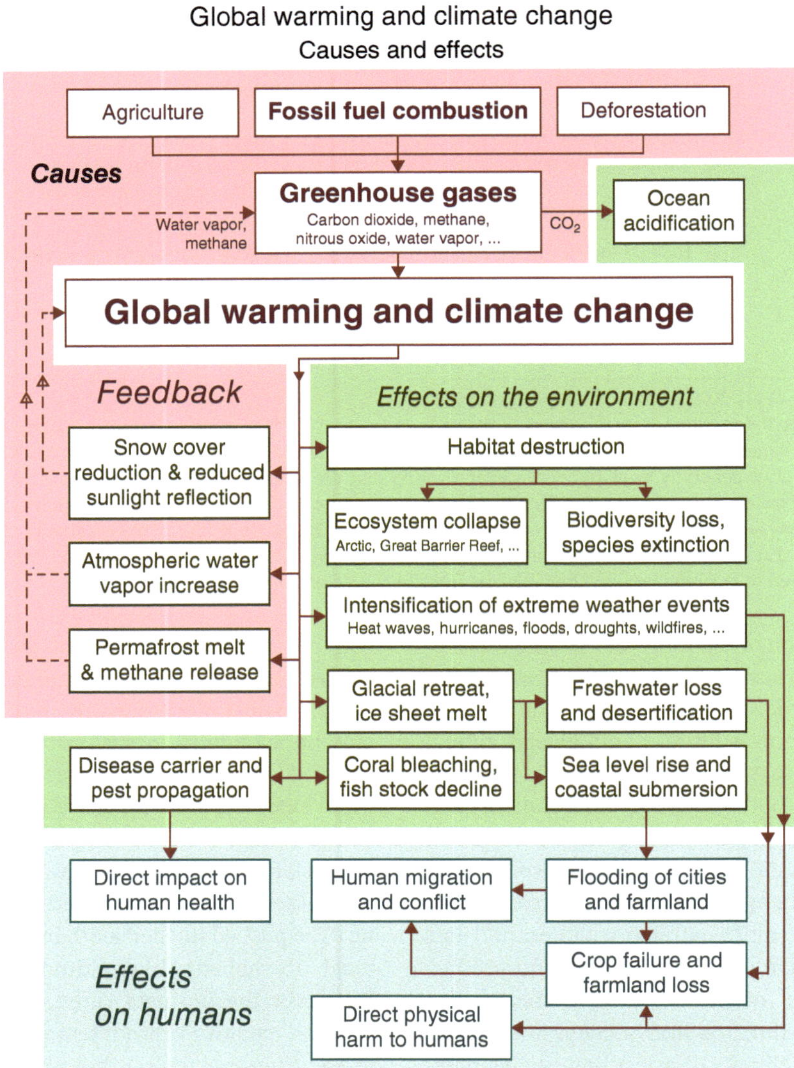

Figure 11.25 Block diagram indicating primary causes and wide-ranging effects on the environment and humans. Including amplifying feedbacks. Image by RCraig09 File:20200118, Global warming and climate change, vertical block diagram, causes effects feedback.png under CC-BY-4.0; https://creativecommons.org/licenses/by-sa/4.0/deed.en

droughts can be problematic for agricultural production if the timing is such that water or heat stress limits the development of plants or impacts ripening. Droughts that persist long enough to significantly impact water supplies (lake/reservoir level, water table depth, river flow, etc.) are of greater concern, as they can often impact a region's productivity over multiple growing seasons. However, agricultural productivity is not the only risk associated with drought. Droughts also increase the risk of fire; damage wild ecosystems, forests, and wildlife populations; trigger dust storms; lead to malnutrition, mass migration, war, and other forms of social and political unrest; cause economic losses from a variety of sources; and threaten municipal water sources.

One of the largest-scale droughts in the modern history of North America was the Dust Bowl period of the 1930s. The region most impacted was the High Plains region centered around the Oklahoma panhandle and encompassing parts of New Mexico, Texas, Kansas, Nebraska, and Colorado. Drought conditions generally occurred in three waves throughout the region in 1934, 1936, and 1939–1940 with some areas experiencing persistent drought for 8 years or more. While the drought was acute and persistent, the negative consequences of the drought were badly exacerbated by human activities. Thanks to the fact that the previous decade had enjoyed unusually high rainfall, farmers in the region abandoned or simply failed to implement dry-land farming techniques that help minimize wind-borne soil erosion. Farmers of the time typically practiced deep plowing of the topsoil, which stripped away the native plants of the short grass prairies and eschewed practices such as cover cropping. Overgrazing of livestock had been typical in the region since the late nineteenth century. Cotton farmers left fields barren during windy winter months and burned fields prior to planting, stripping the soil of organic aggregates, moisture, and root structures. These practices coupled with drought conditions and a period of strong thunderstorms led to some of the worst dust storms ever witnessed in North America. One of the worst was nicknamed the Black Sunday Storm, which displaced over 300,000,000 tons of prairie topsoil in 1935 (Fig. 11.26). Other storms transported dust as far as the east coast of the United States. After the widespread dust storms of 1933–1935, the U.S. government began to strictly regulate agricultural practices in the Great Plains by limiting deep plowing and encouraging crop rotation, cover cropping, and grass seeding. These attempts at mitigating the impacts of the drought reduced dust storms by 65%, but the "Dirty Thirties" didn't truly end until drought conditions eased in the autumn of 1939.

(a)

(b)

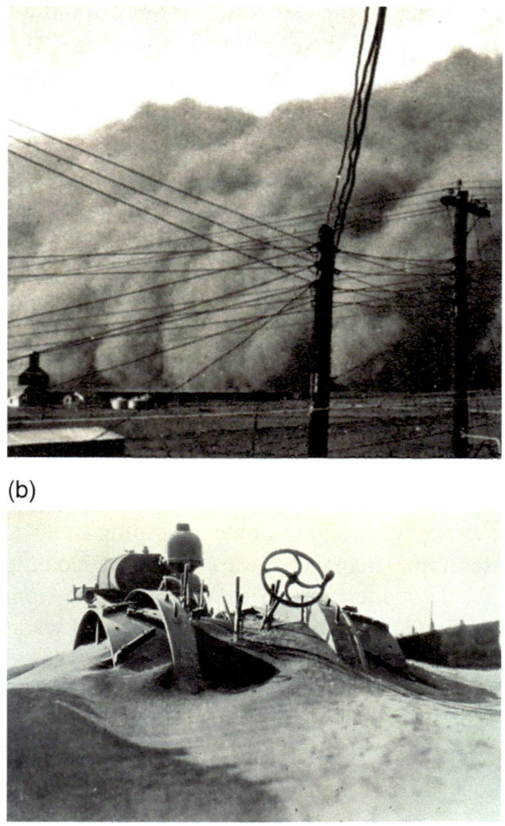

Figure 11.26 (a) Black Sunday storm approaching a Texas town during the Dust Bowl 30s. U.S. National Oceanic and Atmospheric Administration, National Weather Service. Photo from https://en.wikipedia.org/wiki/Black_Sunday_(storm)#/media/File:Wea01422.jpg. (b) Dust storms often buried farm equipment, cars, and even small structures such as houses. Photo from National Weather Service/National Oceanic and Atmospheric Administration; https://www.weather.gov/images/oun/wxevents/19350414/wea01408.jpg

Neither extreme droughts in the United States nor human activities intensifying their consequences are limited to history books. The California drought of 2011–2019 was a massively consequential phenomenon, extending fire seasons, killing more than 100,000,000 trees, damaging a wide variety of crops, decimating multiple winter runs of salmon and other fish, and sparking intense political debates about water rights and land use. The drought is not currently thought to be part of a long-term trend, and a return to greater annual rainfall occurred in 2019. However, increasing temperatures, which *are* part of a long-term trend are thought to have exacerbated the drought conditions and will significantly delay recovery following this and future droughts.

Mitigation efforts were extensive and novel. Outdoor water use for automotive and residential cleaning and watering of lawns was severely restricted for nearly the entire period. The water allocation for farmland was cut by 50% during 2014. Los Angeles County began a proactive program of cloud seeding in 2016. Reservoir managers even deployed floating plastic shade balls to prevent sunlight from heating and evaporating highly stressed water supplies. Los Angeles County alone deployed 96 million of the 4 inch diameter polyethylene balls.

During roughly this same time frame, 2000–2022 (and likely on-going as of this writing), nearly the entire Colorado River watershed experienced drought conditions. The two largest reservoirs, Lake Powell and Lake Mead, were drastically depleted (Fig. 11.27) while the estuary, where the Colorado River once met the Gulf of California, has remained mostly dry for decades. Problems in the Colorado River Basin are perceived largely as caused by natural variation in rainfall, but have deep roots in overly optimistic forecasts, growing demand for water in the developing Sun-Belt, rising temperatures, and unreasonable land use for a desert landscape. In short, human activities in the American West have impacted both supply and demand for freshwater in the watershed.

The importance of the Colorado River to the people living in the western United States cannot be overstated. The river and its reservoirs provide drinking water to more than a tenth of the human population in the United States, irrigate 5.5 million acres of farmland, generate over 4,000 megawatts of hydroelectric power annually, and generate over 1 billion dollars in tourism revenue. The especially intense 2011–2019 period of drought appears to be

Figure 11.27 Lake Mead before and during the ongoing millennial drought. From https://www.nps.gov/lake/learn/drought.htm

the worst since the 1100s, but even if rainfall in the upper Colorado River Basin returns to normal levels, the increased demand due to continued development in the region means that the water levels in major reservoirs are unlikely to recover without drastic conservation and mitigation measures (See Boxes 11.2 and 11.3). Human-caused climate change has led to higher temperatures in the region, which in turn has led to a phenomenon recently coined *hot drought*. When hot drought conditions develop, the higher air and soil temperatures mean that a greater amount of rainfall is needed to recover from drought conditions due to the higher rate of evaporation and transpiration. The added water burden of hot drought recovery is particularly concerning when the warming climate is considered. The challenges facing federal, state, and local planners and policy makers surrounding water infrastructure in the desert Southwest are daunting. With the need for innovative new conservation practices and better technological solutions comes unprecedented opportunity to engineer greater community resilience.

Box 11.2 Water scarcity in western North America.

In 2022, Park Williams along with Benjamin Cook and Jason Smerdon published a study in the journal *Nature Climate Change* that showed that the megadrought of 2000–2022 in the southwestern United States represents the most intense/prolonged soil moisture deficit since at least 800 AD. This assertion is based upon an analysis of tree-rings providing a soil moisture reconstruction that goes back to 800 AD. This unprecedented dry period in North American history demonstrates the powerful synergy that can develop between human activities and events that we might otherwise think of as natural disasters far beyond our control. This synergy has the potential to mitigate risks, but as we'll see in this chapter it often represents an exacerbation of risk factors.

Much of western North America is composed of arid and semiarid ecotypes including landscapes as diverse as the paleodesert of the Nebraska Sandhills, the Okanagan Desert of British Columbia, the Badlands of South Dakota and New Mexico, and the hot sand dunes and salt pans of California's Death Valley and Mexico's Chihuahuan Desert. Despite the seeming inhospitality of the so-called desert southwest, there has been a marked increase in human population in urban centers such as Las Vegas, Phoenix, Sante Fe, Albuquerque, and Salt Lake City throughout the 20th and 21st centuries. Similarly, there has been an expansion in industrial and agricultural activities within the region. This large-scale settlement and development brought with it an intense need to exploit increasingly scarce freshwater resources. To provide drinking water, irrigation, and hydroelectric power, the United States began a period of ambitious large-scale dam building during the middle of the 20th century, with 8 of the country's 10 largest reservoirs (by area) being established west of the Mississippi River.

During this same time period, human activity has brought about global-scale changes to Earth's atmosphere. Through a variety of means, chief among them the burning of fossil fuels, humans have rapidly increased the amount of CO_2 in the atmosphere, which has led to an amplification of the greenhouse effect and a corresponding increase in global temperature and disruptions to the large-scale flow and distribution of atmospheric moisture and precipitation patterns. Earth's atmosphere has so far warmed an average of 1°C (1.8°F) globally. CO_2 is now at a concentration exceeding 415 ppm that is higher than at any time in human history and is now comparable to a period more than 4 million years ago when sea level was on average 23.8 m (78 ft) higher than today and the average global surface air temperature was 3.9 °C (7°F) higher than today.

Box 11.3 Twenty-first century megadrought.

What is a megadrought? A megadrought is a multidecadal (lasting 2 or more decades) period of drought. Megadroughts have been responsible for mass migration of humans and wildlife and are suspected as the root cause of several collapsed civilizations (e.g., the Anasazi and Mayans of North America and the Khmer Empire and Yuan Dynasty in East Asia). Western North America and the southwestern United States in particular have experienced drought conditions since 2000 that continued through 2022). Thanks to human-caused climate change, these conditions are expected to persist and ultimately establish a new normal of less rainfall/snow and higher temperatures in the region. This new normal will impact urban planning and likely limit agricultural activity throughout western North America.

By manipulating the flow of the region's rivers and inadvertently causing more frequent and more intense periods of drought, humans have brought about startling ecological devastation in the lower reaches of several western freshwater systems. For example, the Colorado River no longer reaches the Gulf of California, depriving northern Mexico of a once thriving fishery as well as freshwater. The Salton Sea, a once thriving lake ecosystem in Southern California, is now in rapid ecological decline. Even Lake Mead and Lake Powell, two of the region's largest reservoirs, impounded by the Hoover and Glen Canyon dams along the Colorado River respectively, have declined dramatically during the early decades of the 21st century.

The importance of these two reservoirs to the regions economic well-being is hard to overstate. Lake Powell provides water for 30 million people, irrigates 2 million ha (5 million ac), and produces 5 billion kilowatt-hours of electricity. Similarly, 25 million people get at least some of their water from Lake Mead while it irrigates nearly 253,000 ha (600,000 ac), and produces 4 billion kilowatt-hours of electricity.

In late 2021, regional stakeholders began implementation of the 500-plus plan, which aims to conserve or otherwise reclaim 500,000 ac/ft of freshwater within Lake Mead during 2022–2023. The plan includes 8.3 billion US dollars in federal infrastructure investments as well as promised cuts to water usage by utilities in Arizona, California, and Nevada. However, the plan does little to alleviate continued drought conditions downstream of the Hoover dam and fails to identify sustainable or long-term solutions to problems of water scarcity within the region.

Meanwhile, throughout southwestern North America more broadly, climbing temperatures have sharpened the crisis brought about by reductions in rainfall, annual snowpack, and groundwater recharge. Higher air and soil temperatures result in a higher demand for precipitation to recover between periods of drought because of higher evaporation rates and increasing the risk factors associated with wildfires. This combined with a growing tendency for precipitation events to occur in less frequent but heavier downpours creates a perfect storm scenario in areas prone to landslides and other mass wasting events. If humanity is to develop a better relationship with physical surroundings and natural resources, it will have to find ways to collectively mitigate rather than intensify these climate extremes. Yet the next section explains that alterations in the timing of the start and end of growing, migration, reproductive, and dormant periods have already begun.

Phenology is the study of the timing of biological responses to the seasons: when spring flowers emerge, when bears hibernate, when birds migrate, when autumn colors and defoliation happen, and so on. Phenological observations represent clear evidence that the climate is warming. From a biological perspective, spring is coming earlier and fall is happening later. Examples of changing phenology abound. As of this writing, peak cherry blossoms in Washington, D.C., have shifted 5 days earlier since 1921. One of several annual Azalea Festivals in the southern United States has had to shift from the second week of April to the first week of April in order to time the festival to coincide with fresh blooms that have begun to emerge as early as March. Similar concerns are being addressed by planners of tulip, dogwood, and other spring festivals all over Europe and North America. The melt of Arctic sea ice has led to changing nutrient dynamics such that fall phytoplankton blooms are intensifying. Changing animal migrations are leading to pressure to reschedule hunting and fishing seasons. Forest insects are expanding their ranges thanks to the longer active seasons. Migrating birds have shifted average arrival dates to northern breeding grounds as much as 6 days earlier since record keeping began. While in many regions this means that growing seasons are longer, changing phenology can be problematic for migrating species that rely on the coincidence of physical habitat (open water in the case of waterfowl for example) and food sources. While many plants emerge and flower based on soil temperatures, others time their flowering-fruiting based upon day length (photoperiod). Nearly all plants are sensitive to day length for biomass production (i.e., photosynthesis can be either light, moisture, or nutrient limited). A potential mismatch between temperature and photoperiod cues (and/or the timing of precipitation) highlights the danger of reduced productivity, health, and disease resistance for both wild and commercially, agriculturally important species, and represents a growing area of botanical, horticultural, and agronomic research. It is anticipated that some migrating species may alter their migrations and arrive before food availability becomes sufficient or that species that retain current migration timing will be dampeningly impacted by changing conditions.

Melting Cryosphere

The melting of the cryosphere includes the melting of alpine glaciers (glaciers in a mountainous region) and is observed as summer glacial retreat that is not matched by winter glacial advance, or in extreme examples as year-round glacial retreat. Between 1303 and 1860, there was a series of three cool periods comprising the so-called Little Ice Age. The Little Ice Age was a relatively cold period in the Northern Hemisphere caused by changes to orbital forcing, solar activity, ocean circulation (melting of land ice during the medieval warm period), volcanic activity, human population, and land use. Following this period, glaciers began a period of retreat that, under natural conditions, should have stabilized around the middle of the twentieth century. Instead, mountain glaciers have continued to retreat and disappear. The pace of melting has accelerated since 1980. Figure 11.28 shows a pair of photographs taken as part of a large-scale project by the U.S. Geological Survey to replicate the location, angle, field of view, and seasonal conditions of previous photographs to compare the extent of glaciation in U.S. national parks. Glacier National Park in Montana, USA, is an informative example of the fate of alpine glaciers in our current climate warming paradigm. Of the 150 glaciers larger than 25 ac at the time of its founding in 1910, only 26 remain at or above the 25 ac threshold. The remaining named glaciers lost as much as 85% of their ice over the 50 year period ending in 2017. Glacier National Park is projected to be ice free within the next few decades. This will obviously be a great loss to the millions of tourists who visit the park each year, but much more important, there are freshwater habitats and human communities that rely upon the springtime melting

(a)

(b)

Figure 11.28 (a) Grinnell Glacier in Glacier National Park, Montana, circa 1900. (b) The same glacier in 2008. Note the change in the size of the glacier. This is representative of the change seen elsewhere in the park. (a) Photo credit F. E. Matthes photo, courtesy of Glacial National Park Archives. (b) Photo by Lisa McKeon, U.S. Geological Survey. Photos from Strahler, 2017.

of these glaciers. There are rivers flowing into the Gulf of Mexico, Pacific Ocean, and Hudson Bay that currently or historically receive melt water from the glaciers of the region. There will also be a continued decrease in albedo, which is a amplifying warming feedback accelerating the pace of warming in the coming century.

Glaciers retreat, fragment, and dissipate by two main processes. In many cases the annual melt begins to outpace the annual accumulation leading to a net retreat or thinning. In other cases, such as with the main ice pack atop Mt. Kilimanjaro in Africa, changes in atmospheric moisture and/or snowfall lead to the complete loss of an ice accumulation zone. In the second case, glaciers tend to fragment with each summer season leading to smaller and shorter-lived glaciers in an alpine region.

Sea Level Rise

In addition to its impact on alpine glaciers, global warming has caused a decrease in the size (both geographic extent and thickness) of the world's largest ice sheets, the Antarctic Ice Sheet and the Greenland Ice Sheet. It has also caused serious decreases in the size and extent of smaller glaciers located in northern Canada, Siberia, Alaska, and southern Patagonia. The ice sheet on Greenland (including its outlet glaciers) covers more than 80% of Greenland's surface or roughly 660,000 mi^2 (1.7 million km^2). The ice sheet is built up by deposition of snow during the Arctic's frigid winters with the oldest ice found to be as old as 1 million years old. It is around 1.2 mi (2 km) thick on average with a maximum elevation of 2 mi (3.2 km) above sea level. The thickness of Greenland's ice has been decreasing steadily since at least 1958. The process by which ice moves from the center of the island to the coasts is speeding up largely due to vast amounts of melt water lubricating the interface between the ice and the continental crust beneath. If the entire mass of ice currently covering Greenland were to melt, it would lead to around 24 ft (7.2 m) of sea level rise. Climatologists have projected that the atmospheric temperatures needed to trigger this melting will likely be reached within the next century, but the pace of that melting remains unclear with projections ranging from one hundred to a few thousand years needed once the tipping point is reached and complete melting becomes inevitable.

The Antarctic Ice Sheet represents the largest single mass of ice on Earth and covers an area of nearly 5.4 million mi^2 (14 million km^2). It is estimated that 61% of the planet's fresh water currently exists as part of the ice sheet and its associated outlet glaciers. Complete melting of the Antarctic Ice Sheet would account for around 190 ft (58 m) of sea level rise. In some areas, the ice sheet extends beyond the continent resulting in structures known as ice shelves. Where seawater undermines the ice shelves and causes thinning by melting from underneath, the intruding seawater is referred to as the *basal cavity*. As the subsurface waters of the Antarctic Ocean are warming, above the growing basal cavities increasingly large areas of ice shelf are *calving*, or breaking off the ice sheet, and ending up in the Southern Ocean (Fig. 11.29). In 2017 one of the largest icebergs ever observed separated from the Larsen-C ice shelf on the Antarctic peninsula. The resulting mass of ice was roughly the size of the U.S. state of Delaware and as thick as 1,000 ft (300 m).

The observed and projected thinning and shrinking of the major ice sheets and the accelerated calving and melting of land ice into the oceans has several consequences relevant to climate change and natural hazards. As glaciers retreat, the albedo of the bare continents or open waters are much lower than the ice, leading to an amplifying warming feedback (more melting leads to more warming, more warming leads to more melting, etc.). Melting land ice also means that the water previously stored on land will end up in one of two other reservoirs of the global water cycle. Evaporated meltwater (or sublimed ice) results in more water vapor in the atmosphere, which we have already discussed as a warming climate feedback, and also allows for more intense rain and snowstorms. Calved land-ice and glacier and icesheet meltwater also contribute to sea level rise. As the melting continues, the reduced weight on the Antarctic and Greenland landmasses is expected to cause continental uplift referred to as rebound that we are already seeing in Alaska. Because of continents rebounding in some areas and ocean floors sinking in others, sea level rise is not expected to be uniform along all coastlines, which has prompted the importance of using satellites to measure the absolute change in sea level. Calving and melting have also contributed to navigational and shipping hazards in the Southern Ocean, South Pacific, and South Atlantic Oceans, as well as in the Arctic. In the case of calving of land ice, that risk is due to the presence of icebergs, while in the case of melting,

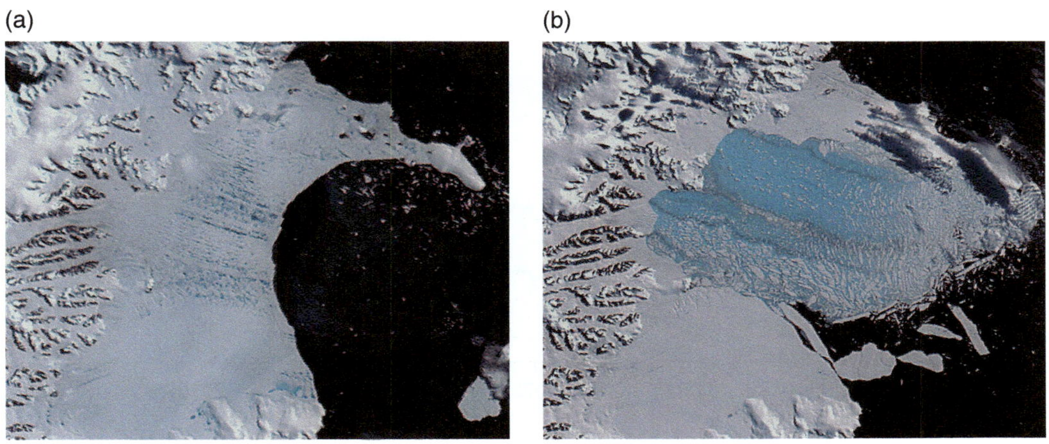

Figure 11.29 Ice calving from the Antarctic Peninsula, creating massive icebergs. These icebergs can cause hazards in shipping lanes throughout the Southern Ocean. (a) Larsen B ice shelf in January 2002. Blue patches on the ice indicate that the ice shelf is deteriorating. (b) The same area in March 2002 where the solid iceshelf is very near the shoreline and the light blue area is now free-floating icebergs. The ice shelf has lost more than 40% of its ice volume since 2002. From Fletcher, 2014.

the freshening of the water adjacent to Antarctica is causing the annual sea ice pack to grow in size (unlike the Arctic sea ice) because lower salinity seawater freezes at higher temperatures than normal salinity seawater.

Sea ice in the Arctic Ocean has been showing a more direct trend with respect to global warming. The warmer summers in the Arctic Circle have been leading to significantly smaller sea ice packs by the start of fall. This reduction of sea ice pack has made ice navigation more economically viable for shipping along the formerly inaccessible arctic coasts of Siberia, Canada, and Alaska. In fact, the long-sought-after Northwest Passage has been navigated at least 14 times between 2000 and 2018 via ice-free routes, which represents more open water passages than were accomplished during the preceding two centuries of intensive exploration and effort. The Northwest Passage and other potential trans-Arctic trade routes may spur some economic development in the form of cheaper shipping and tourism (expedition cruising in the Arctic is growing rapidly). However, the loss of year-round sea ice pack is problematic for a number of Arctic species such as polar bears, several seal species, narwhals, Pacific walruses, and beluga whales. There is also pressure on governments to permit commercial offshore petroleum exploration and production in these extremely sensitive habitats, which can only worsen climate change. As with the loss of land ice, reductions in sea ice coverage reduce Earth's albedo in the Arctic, which amplifies the effects of atmospheric warming.

We have discussed the potential impact of the melting land ice on sea level rise, but note that melting sea ice is already in the ocean so has no impact on sea level rise. Yet the melting land ice is only one aspect of sea level rise (just the tip of the iceberg so to speak!). As the ocean warms the average density of seawater decreases, meaning it has more volume or takes up more space for the same mass.

As the globe warms, 90% of the heat energy is added to the oceans, which causes them to expand. The melting land ice previously held in glaciers and Greenland and Antarctic ice sheets have also contributed to sea level rise in response to the warmer temperatures. Figure 11.30 shows breakdown of sea level rise

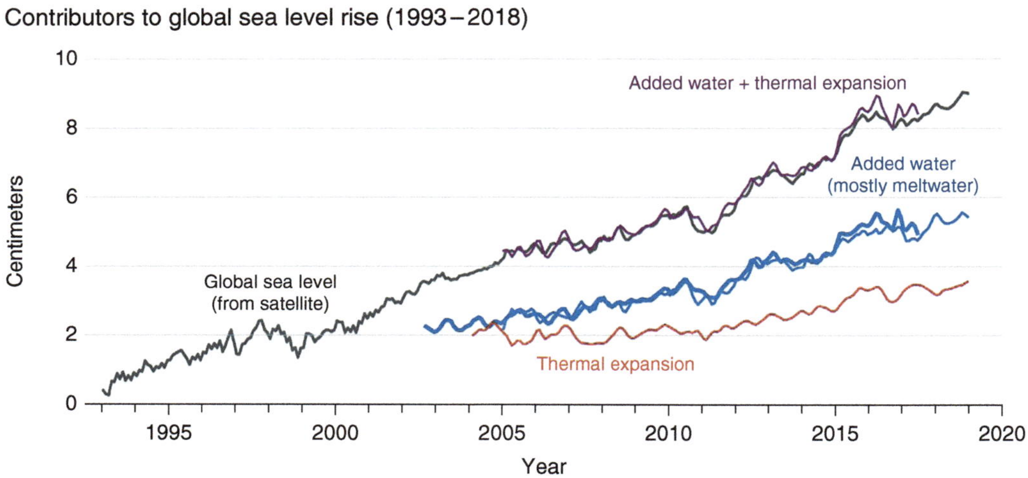

Figure 11.30 Global mean sea level rise (GMSL) from 1993 to 2020 as measured by the combined GMSL time series from satellite altimetry (black). Comparisons are made with increased ocean mass due to melting land ice (blue, measured via GRACE/GRACE-FO satellite), increased volume due to the warmer ocean temperatures (red, measured via the ARGO floats), and the combination of increased mass and volume (purple). Comparison of methods shows the general agreement so we can "close the GMSL budget." Image from Lindsay, 2022.

as calculated by two different methods. The first is satellite altimetry that measures the height of the ocean surface compared with the Earth's center-of-mass, thereby eliminating any potential complications of changing land height due to continental uplift or subsidence due to colliding land masses or rebound following ice loss. The figure also shows the amount of sea level rise due to thermal expansion, as measured by an array of ~3,800 ARGO floats that circulate throughout the world's oceans, and the change in ocean mass as a result of added meltwater as measured by NOAA's GRACE satellite. Perhaps the most fascinating aspect of the figure is that the combined measurement of thermal expansion and increased water mass fall right in line with the satellite altimetry data. As such, it appears that we have an excellent understanding of the impacts of global warming on sea level rise. Future projections based on this understanding show that, depending on which representative concentration pathway we take, sea levels could rise by over a meter by 2100 and nearly four meters by 2300 (Figure. 11.31).

More than an eighth of the world's population lives in low lying coastal regions, putting them at a growing risk of coastal flooding especially during astronomical high tides, which occur with new and full Moon phases, and hurricane and tropical storm surges. The increased risk to coastal communities from storm surge is particularly disturbing when we consider the projections for extreme weather as climate change proceeds (Fig. 11.32). While 1.6 m of sea level rise may not occur until after 2100, there are two U.S. towns that have already had to be relocated due to sea level rise: Isle de Jean Charles, Louisiana (Fig. 11.33), and Newtok, Alaska (Beaumont 2023).

Another risk associated with sea level rise includes *saltwater incursion* into fresh groundwater, in which saltwater invades freshwater sources at the coast, leaving coastal residents with no source of freshwater. Additional hazards include *salinization* (increasing salt

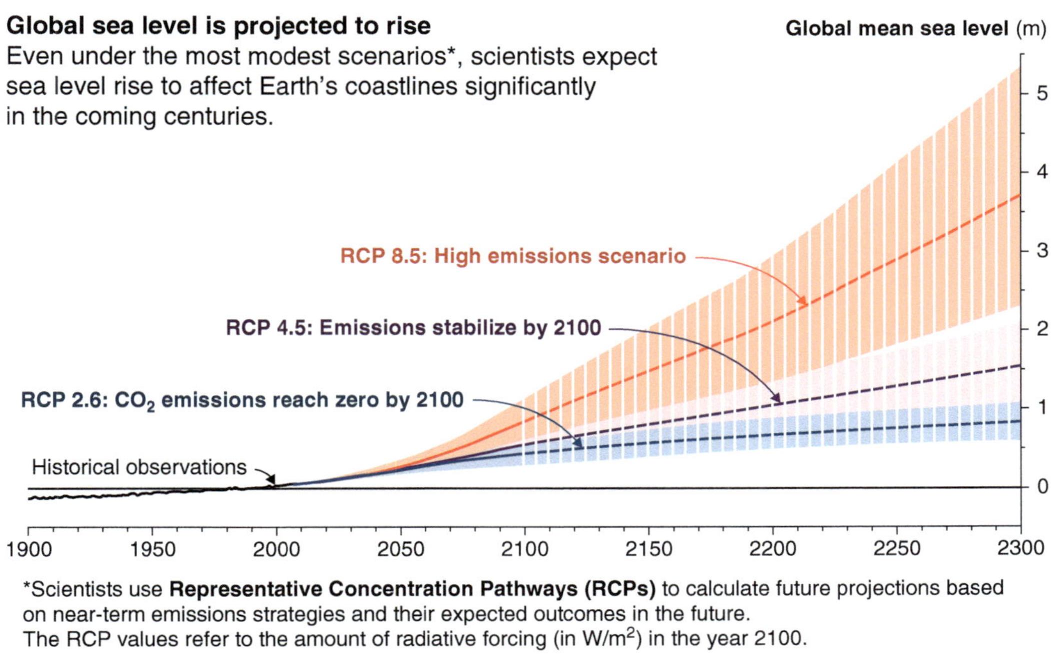

Figure 11.31 Sea level rise projections through 2300. Image from NASA Earth Observatory under the Public Domain. Voiland, 2021.

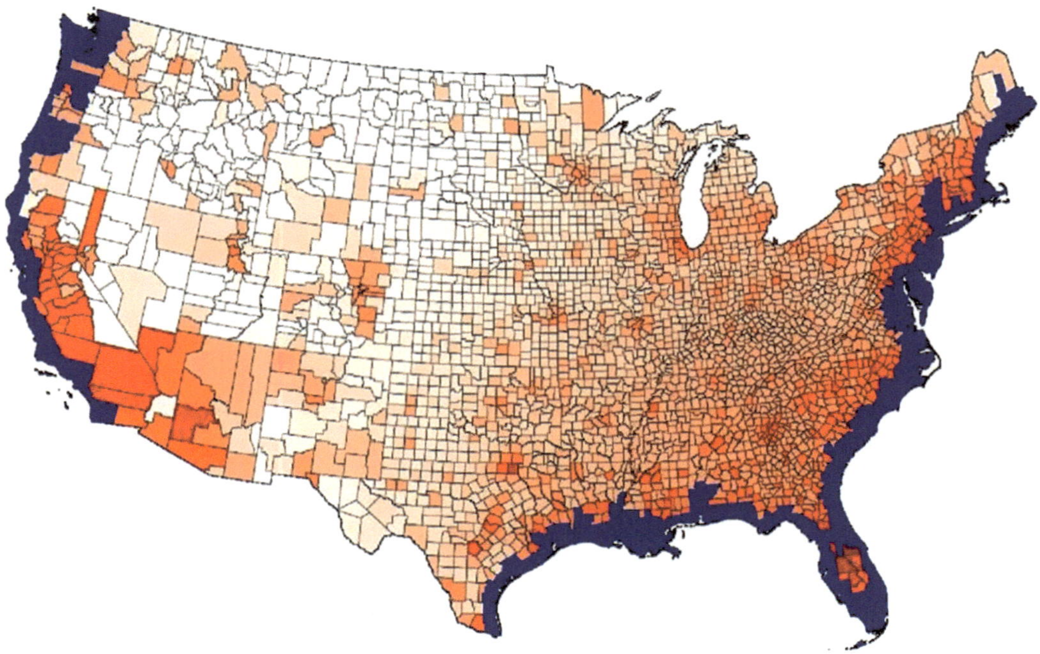

Figure 11.32 U.S. map showing counties impacted and those migrated to as a result of sea-level rise of 1.6 m. Image from Robinson et al., 2020.

Figure 11.33 Images showing land loss surrounding Isle de Jean Charles over a recent 45 year period. In 2022, the U.S. government provided funding to purchase land and build new homes for the residents (Watch: Isle de Jean Charles (10 min). USGS via U.S. GAO.) In Florida, one in eight homes, worth over $400 billion in current housing value, could be under water by 2100. Image from Simms et al., 2021.

content) of soils and drinking water supplies, enhanced *erosion*, and wetland habitat degradation as freshwater marshes and swamps struggle to adapt during the transition to saltwater marsh.

Extreme Weather

According to the 2014 U.S. National Climate Assessment, there are at least six types of extreme weather that are projected to become either more intense, more frequent, or both during the remainder of the 21st century. Since high quality satellite data became available in the early 1980s, almost every measure of Atlantic hurricane activity has steadily increased. The length of the hurricane season, the number of strong cat 4–5 storms, rainfall amounts, and the intensity of storms have all grown and are projected to continue to grow in a warming climate. As discussed in Chapter 10, the destructive power, potential death toll, and economic impact of large tropical and subtropical cyclonic storms are difficult to overstate.

Tropical Cyclones

A 2°C anthropogenic global warming is projected to impact tropical cyclone activity via (1) higher sea levels will lead to higher storm surge, (2) increased tropical water vapor will lead to increased precipitation amounts, (3) increased hurricane intensity; about 5% increase in wind speed (range: 1%–10%), (4) 13% increase in proportion of tropical cyclones that reach category 4–5, (5) a further poleward expansion of the latitude of maximum tropical cyclone intensity in the western North Pacific; and (6) potentially a slower translation speed, allowing for greater precipitation inundation.

With increasing moisture in the atmosphere, winter storms are also expected to bring larger snowfall amounts and higher wind speeds. There are several projected changes to rainfall patterns including to the timing and spatial variability of rainfall. However, the most notable change for most of the world's population is the greater proportion of rainfall that will come from heavy downpours. While agriculture is deeply dependent upon precipitation, heavy downpours can damage crops, erode soil, and cause mineral leaching or nitrogen loss in soils, while failing to provide sufficient water for crops when compared with smaller less intense rainfall events. Heavy downpours also increase the likelihood of mudslides, landslides, and other mass wasting events along with their associated risks. As we might expect with more intense storms, more frequent heavy downpours, and sea level rise, floods of all types are expected to become more problematic as a result of climate change. That includes coastal flooding, fluvial or riverine flooding (whether overbank- or flash-floods), and pluvial or surface floods brought about by heavy rains inundating low-lying areas or flowing down hillsides. In the case of pluvial flooding in particular, urbanization (covering a landscape with impervious surfaces such as concrete and buildings) can have a compounding effect by decreasing the amount of porous surface and leaving nowhere for flood waters to escape once storm sewers are overwhelmed.

Melting Permafrost

Much like the melting of the cryosphere discussed previously, warmer temperatures in the Arctic are problematic for a number of reasons. Melting permafrost in tundra and boreal forests are causing structural damage to plant life, forests, riverbanks, and shorelines, as well as human infrastructure such as roads, buildings, power transmission lines, and pipelines (Fig. 11.34).

Increasing the Range of Insects

Increased insect ranges means that mosquito-borne and tick-borne pathogens are spreading more rapidly as are forest and crop pests. Anthrax spores that had been frozen in the Siberian tundra for decades recently thawed and caused an outbreak of the disease in a herd of reindeer. Cholera outbreaks are expected to become more frequent with higher water temperatures and heightened risk of flooding. Tropical diseases are beginning to spread to

(a)

(b)

Figure 11.34 (a) Permafrost is permanently frozen soil with a seasonally active layer on top that can support plant life. (b) A house in Dawson City, Yukon Territory, Canada, that was built on Permafrost. The house has been severely damaged due to melting permafrost beneath the structure. (a) Photo credit AccentAlaska.com/Alamy. From Fletcher, 2014. (b) © imageBROKER/Alamy. From Murck & Skinner, 2016.

higher latitudes. Warmer water temperatures and higher salinity estuaries are blamed for the spread of oyster parasites along the U.S. Atlantic coast. Warmer water temperatures as well as the pulses of nutrients entering coastal waters from heavy downpours (not to mention poor agricultural and water quality policies) are causing an increase in the frequency, severity, and persistence of toxic algal blooms and the expansion of eutrophic dead zones leading to large losses in coastal fisheries and diminished biodiversity along our coasts.

11.4 Ocean Impacts

11.4.1 Coral Bleaching

Warmer water temperatures are blamed for most of the large-scale coral bleaching events (Fig. 11.35) that have been increasingly common and increasingly damaging during the last several decades. Coral bleaching occurs when symbiotic phytoplankton that live within coral polyps, the tiny animals responsible for reef building, are expelled from the polyps in response to a stress. Since corals depend on the phytoplankton to survive, bleaching often leads to coral death. Surviving polyps may or may not repopulate and produce a healthy reef if the stressful conditions are alleviated. Australia's Great Barrier Reef supports 1,625 species of fish, 215 species of birds, 6 of the world's 7 species of sea turtles, more than 30 species of marine mammals, around 1,300 species of crustaceans, at least 133 species of rays and sharks, and several reptiles.

At lower *trophic levels* (of the food web) the reef is also home to coralline algae (a photosynthetic reef building organism), sea grasses, sea lettuce, seaweeds (green and brown algae), thousands of microalgae (phytoplankton) and zooplankton species, over 3,000 mollusk species, hundreds of echinoderm species, more than 450 known hard coral species, and many varieties of sponges, jellyfish, and other marine invertebrates.

(a) (b)

Figure 11.35 When water gets too warm, corals expel the symbiotic algae that provide food for the coral to survice and also give them their green color. This is called coral bleaching. This pair of images from American Samoa shows relatively healthy coral reef on the left, compared to a reef that experienced severe bleaching in 2015 on the right. Photo credit Richard Vevers/The Ocean Agency. https://www.coralreefimagebank.org/

The Great Barrier Reef has been a World Heritage Site since 1981. Since that date, the Great Barrier Reef has lost over half its corals, with over two thirds of these losses occurring since 1998. Similar bleaching damage has been observed at all of the major tropical and subtropical coral reefs. Bleached coral structures support much less diverse populations of fish and marine invertebrates. Though corals can recover after minor bleaching events, they become more susceptible to diseases and invasive species. These large reefs are extremely important to maintaining the world's stock of fish that sustain subsistence, recreational, and commercial fisheries. Tiny coral polyps may be beginning to build new reefs even now at higher latitudes. However, it takes millennia to build the large structurally sound and biologically diverse ecosystems that are at risk of disappearing at lower latitudes within just decades.

11.4.2 Ocean Acidification

While coral bleaching, or the loss of symbiotic algae, can be attributed primarily to higher water temperatures, corals and most other marine organisms that produce calcareous shells are also impacted by ocean acidification. Ocean acidification is a consequence of more carbon dioxide being dissolved in seawater and converting the primary buffering agent in seawater, carbonate, into bicarbonate (equation (11.1)).

$$CO_2 + H_2O + CO_3{}^{2-} \rightarrow 2HCO_3. \tag{11.1}$$

This decreases the pH of seawater and reduces the carbonate ($CO_3{}^{2-}$) concentration, so the solubility of calcium carbonate increases. Organisms that have calcium carbonate body parts ranging from shells (see Fig. 11.36) to bones now find that these features readily dissolve and/or are much more difficult to maintain.

Other impacts of ocean acidification include sensory impairment in marine vertebrates, significantly lower primary productivity in marine environments due to the stress on calcareous phytoplankton, less nutritional value in krill for marine mammals such as whales, and lower biodiversity as acid-resistant species such as jellyfish outcompete more sensitive species. As ocean warming and acidification

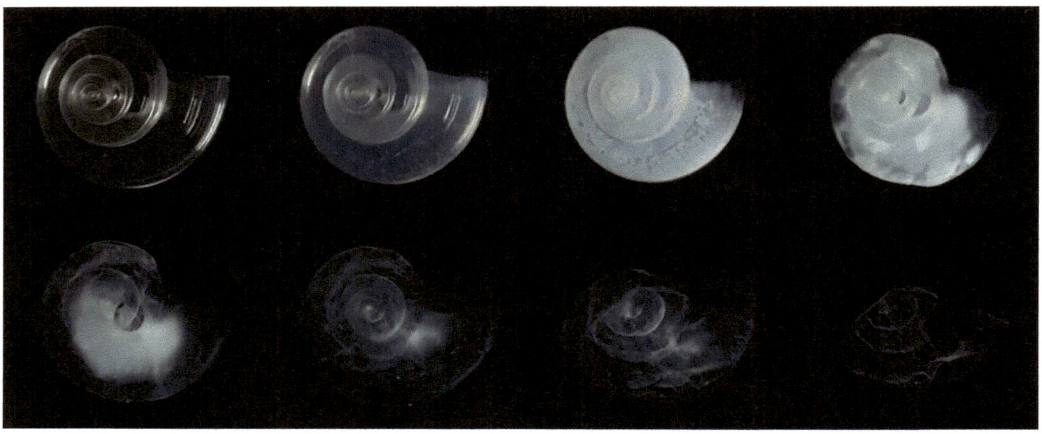

Figure 11.36 A pterapod (sea snail) shell dissolved over the course of 45 days in seawater adjusted to an ocean pH projected for the year 2100. Photo courtesy of NOAA via public domain; https://upload.wikimedia.org/wikipedia/commons/b/b1/Pterapod_shell_dissolved_in_seawater_adjusted_to_an_ocean_chemistry_projected_for_the_year_2100.jpg

threaten biodiversity and the viability of many organisms at lower trophic levels, ocean productivity from a fisheries standpoint is likely to be threatened as well, which will compound other threats to fish stocks such as overfishing, coastal development, water pollution, and habitat loss.

11.4.3 Slowing of the Great Ocean Conveyor Belt

Since the Sun shines more directly upon lower latitudes near the equator and less directly on the higher latitudes nearer the poles, the higher intensity of the direct solar radiation at low latitudes warms a larger volume of ocean water to a greater depth. This warmer water is less dense and more buoyant, compared with the colder, denser water at the poles. If one were to imagine a planet without continents, the warmer, more buoyant equatorial waters would remain at the surface while the colder, denser polar waters would sink, thereby initiating the process of thermohaline circulation. "Thermo" refers to the temperature gradient just described, and "haline" refers to the more saline polar waters due to evaporation of water during poleward transport and saltwater rejection in the process of ice formation, which adds to their density. High density cold-saline downwelling in the North Atlantic is the primary site of initiation of what is often referred to as the Great Ocean Conveyor Belt (Fig. 11.37).

One reason the flow of surface waters is not strictly poleward with a counterflow at depth is that the Earth is rotating and the flowing fluid hydrosphere and atmosphere experience a deflection referred to as the Coriolis Effect resulting in flowing air and water turning right in the Northern Hemisphere and left in the

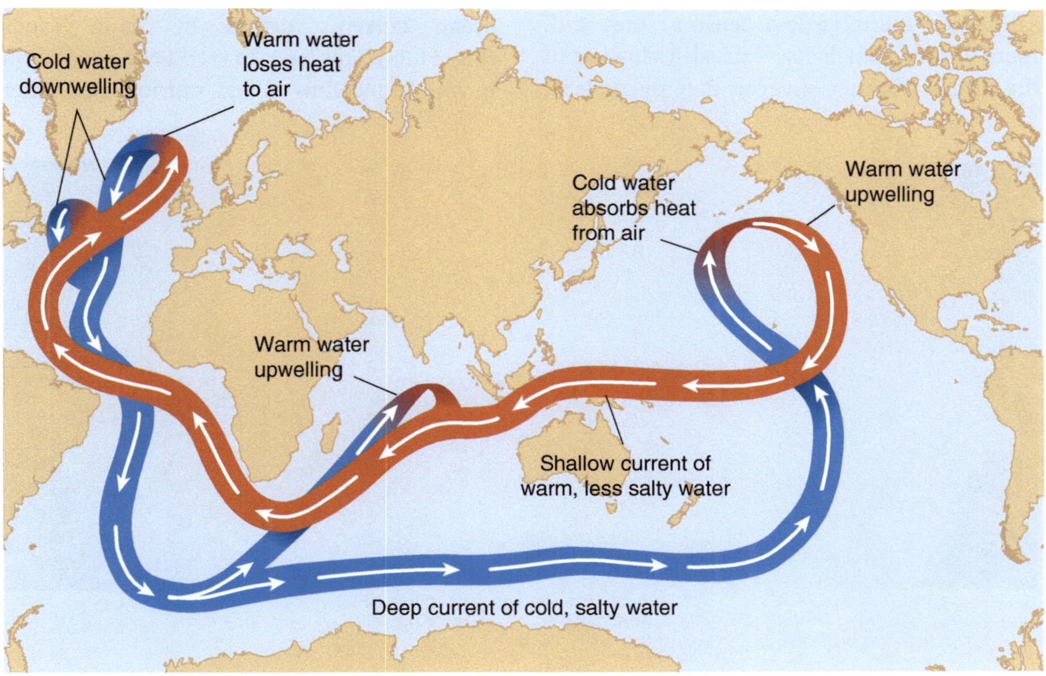

Figure 11.37 Thermohaline circulation of warmer surface waters (red) and colder, more saline deep waters (blue) throughout the Earth's oceans. High density cold, saline downwelling in the North Atlantic is the primary site of initiation of what is often referred to as the Great Ocean Conveyor Belt. These temperature and salinity gradients drive ocean currents around the world. From Arbogast, 2017.

Southern Hemisphere. Much of the behavior of the Great Ocean Conveyer Belt is owed to the indirect influence of the Coriolis Effect in the form of prevailing winds dragging water along the surface and the direct influence of the Coriolis Effect diverting the flowing water to form the major ocean gyres. Another, perhaps more obvious, reason this poleward surface flow is limited to specific regions of the world's ocean is quite simply because there are continents in the way. In fact, throughout geologic history, the efficiency with which the ocean has transported heat from low latitude regions with net warming to high latitude regions with net cooling has partly, and at times largely, been determined by the amount and distribution of landmasses. The size and location of landmasses is controlled by plate tectonics (see Chapter 2).

Recent data show that the rapid melting of Greenland ice has diluted the ocean waters in the North Atlantic, causing them to be less saline and therefore less dense so they sink more slowly, which has caused a slowing of the Great Ocean Conveyor Belt by at least 15% since 1950. Slowing of the conveyor, in turn, allows increased heat buildup at the equator, which can lead to more intense hurricanes as well as a variety of weather impacts such as colder temperatures in northern Europe.

11.4.4 El Niño and La Niña

In early 1997, satellite-based measurements of sea surface temperature began to indicate a marked and persistent warming in the equatorial Pacific (Fig. 11.38). Based upon many years of observation and atmospheric modeling, scientists knew that this indicated the onset of the warm period of a phenomenon known as *El Niño Southern Oscillation* (ENSO). The name derives from the fact that each year just after Christmas, fishermen in coastal Peru would notice a warm northerly current, hence "El Niño de Navidad." In the ocean, the phenomenon manifests as a periodic slowing of ocean currents driven by trade winds (Fig. 11.38). This leads to the El Niño condition in which upwelling of cold, nutrient-rich water

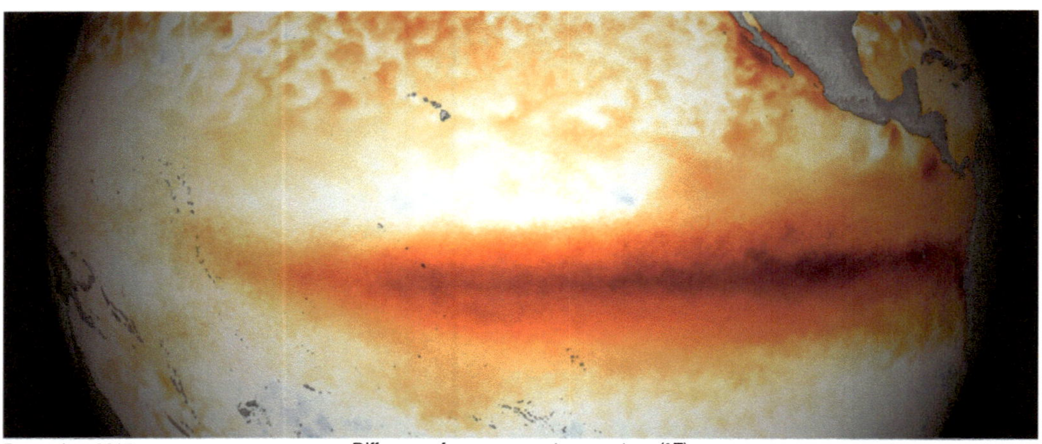

Figure 11.38 A map of the equatorial Pacific Ocean with South America visible to the right and Australia and New Guinea visible to the left. Warmer colors represent more positive temperature anomalies (above average temperatures), while cooler colors represent more negative temperature anomalies (below average temperatures). Source: National Oceanographic and Atmospheric Administration; https://www.noaa.gov/education/resource-collections/weather-atmosphere-education-resources/el-nino

along the Pacific coast of Ecuador and Peru slows considerably. This upwelling normally fuels a highly productive marine ecosystem and an abundant fishery. Every 2 to 7 years, this warm countercurrent persists well into the spring or even summer months and eventually the term *El Niño* came to refer only to these atypical years in which warming of the equatorial Pacific is more pronounced or more prolonged than normal.

By May of 1997, water temperatures near the Pacific coast of South America were 7°C (12.6° F) above normal, and a large pool of abnormally warm water lay off the coast of North America. Quite apart from the impact on local fisheries, the warming of these equatorial waters has a profound effect on atmospheric conditions above and thus leads to rather unusual weather conditions. Westerly winds begin to blow these air masses inland to the Americas and trade winds carry them toward Asia and Oceania. One manifestation of these impacts on weather can be seen in the record-breaking typhoon season with 11 storms reaching super typhoon status. The hurricane season was also intensified with 24 named depressions and 9 hurricanes, 7 of which were major storms (Category 3 or higher) and included only the second December tropical storm ever recorded in the Pacific. Other weather impacts of what was, at the time, the most intense El Niño event in history included a record-breaking drought in Papua New Guinea as well as droughts in Indonesia and other parts of southern Asia (and wildfires in many of these areas); extreme rainfall, flooding, and landslides in Peru; record rainfalls in California (Fig. 11.39); extreme heat in China; and wet and cool weather in the southeastern United States. In eastern Africa, Kenya had its wettest year on record while Mozambique to the south experienced drought. Even as far away as central Europe, flooding attributed to the El Niño phenomenon killed 55 people.

Figure 11.39 Flooding along the Russian River in Northern California during the 1997 El Niño. Credits FEMA/Dave Gatley; https://www.nasa.gov/feature/jpl/2015-and-1997-el-ni-os-d-j-vu-or-something-new

The previous record holder for sea-surface temperature anomaly was the 1982 El Niño event, which caused 2.2 billion dollars in damages in California alone. Thanks to improvements made to the observing systems and investments made in basic atmospheric research and climate modeling, the 1997 El Niño was the first such event that climatologists and meteorologists were able to predict and extensively warn the public about; the public and private sectors, in turn, were able to prepare and mitigate damages. While California saw even greater rainfall amounts in 1997, the early warnings resulted in a halving of the monetary damages compared with the 1982 event. The 1997 El Niño was so strong that, by January of 1998, water temperatures were still increasing and reached 11°C (19.8°F) above normal in some areas. Fortunately, cooler than average water was beginning to spread outward from the western Pacific Ocean signaling that the opposite phase of ENSO known as *La Niña* would soon emerge.

La Niña is named to simply denote the opposite condition of El Niño. Some choose instead to refer to the condition as anti-El Niño. La Niña refers to colder than average surface waters in the equatorial Pacific Ocean driven by relatively strong trade winds and strong upwelling. When neither El Niño nor La Niña extreme is observed, it is referred to as the neutral (or simply normal) phase or condition of the oscillation. As one might expect, the distribution of unusual weather patterns during La Niña is largely the opposite of those seen during El Niño. La Niña produces wetter than normal conditions in southern Africa and drier than normal conditions in equatorial eastern Africa, causes heavy rains in Malaysia and Indonesia, yields drought across much of California along with above average precipitation across much of the northern United States, and cold, snowy winters in Canada. It also generates higher than average rainfall in Brazil and some other parts of South America. It is worth noting that even a relatively weak El Niño/La Niña event can have important consequences for regions where its impacts are amplified by ongoing weather patterns or where other factors lead to an intensification of existing conditions such as drought, extreme heat or cold, or heavy precipitation. As we will see, this is also true for longer term climate change. Whether El Niño frequency or intensity will be impacted by global climate change is an open area of inquiry, as is whether El Niño events will lead to sudden intensification of ongoing global oceanic and atmospheric trends.

11.4.5 Distribution, Current Status, and Projections

Whereas wildfire, hurricanes, and monsoons are expected in specific but widespread locations, hot extremes, heavy precipitation, and drought are expected to impact most regions of the world. Table 11.1 shows the 22 climate disasters experienced in 2022 while we are at 1.1°C of global warming and Figure 11.40 shows the trajectory of expected increases in these events as the globe warms.

11.5 Mitigation of Hazards

By far the most important thing we can do to mitigate the impact of global climate change is to curtail our emissions of greenhouse gases (Fig. 11.41). This includes breaking our dependence on fossil fuels through continued development and adoption of renewable energy sources and alternative fuels; efforts to improve the efficiency of our power grid; improving the efficiency of our transportation systems; cutting production of Portland cement; decreasing the chemical intensity of our agricultural production; and reducing meat consumption, ruminant meat in particular. These actions can help minimize our so-called carbon footprint. On 5 October 2016, the Paris Agreement was adopted as part of the United Nations Framework Convention on Climate

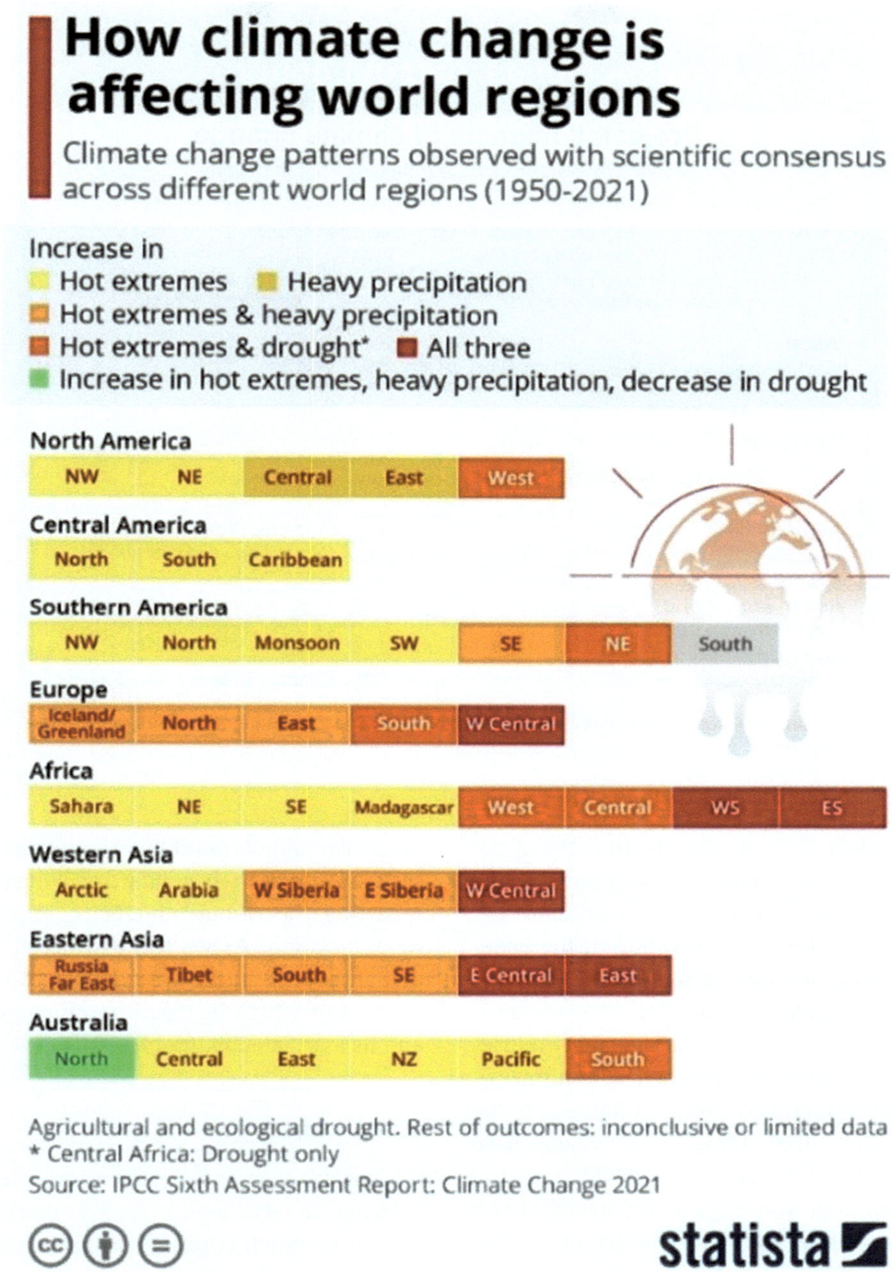

Figure 11.40 How climate change has already impacted world regions. Image from Statista under Creative Commons.

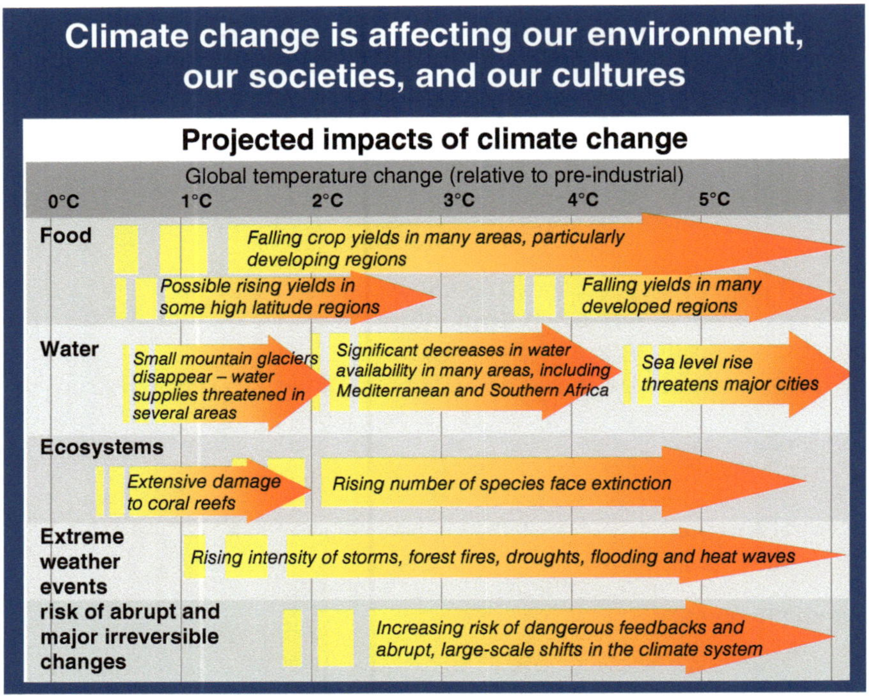

Figure 11.41 Temperatures at which we expect even greater impacts on climate disasters. Image from IPCC, 2007, by permission.

Change (Box 11.1). As of Feb 2023, 195 of the UN's 198 nations have officially ratified the agreement, which deals with greenhouse gas emissions mitigation, adaptation to the changing climate, and how to finance these efforts. It should be noted that the Paris Agreement is nonbinding and it's up to individual countries to make progress toward their commitments. Only three relatively small countries in the Middle East (Iran, Libya, and Yemen) have not yet ratified the Paris Agreement.

While a single country's choice to seek its own path in dealing with these issues may seem inconsequential, we should recognize that during the first decade of the 2000s, the United States represented just 5% of Earth's population but consumed nearly a third of the world's resources, including a quarter of all fossil fuels. Ultimately, developed countries must seek individual and societal lifestyles that are less resource intensive and more sustainable. We must also support developing countries as they seek to improve their standard of living without repeating the resource-intensive mistakes of modernization. As we learn to become more carbon neutral, there are many steps that we can take to reduce the risks from climate change and its associated natural hazards.

11.5.1 Mitigation

Renewable energy and *alternative fuels* are growing in the world energy marketplace despite entrenched resistance from politically connected and well-financed oil, coal, and gas industries and their allies. During four of the six years between 2012 and 2017, new utility-scale electrical generating capacity from wind and solar surpassed new generating capacity from non-renewable sources in the United States. Other strategies to decarbonize energy use include energy recovery from solid

Figure 11.42 Geothermal power station near Taupo, New Zealand. Chrispo/Shutterstock.

waste and sewerage in the form of biogas recovery from sewage digesters and landfills; use of the nonrecyclable solid waste directly as a fuel source; using geothermal hot spots to produce steam heat and electricity (Fig. 11.42); using geothermal heat exchange to heat and cool indoor spaces; burning sustainably harvested biofuels; and developing wave, tide, and hydroelectric generation methods that forgo the practice of building large dams and reservoirs. By far the most effective strategy to minimize the effects of climate change is to mitigate its primary cause and limit the increase in climate forcing by greenhouse gas emissions.

11.5.2 Carbon Sequestration via the Biological Carbon Cycle

Another approach to mitigating the amount of CO_2 in the atmosphere is to increase the amount of carbon that is stored in the environment. The relative abundance of oxygen, carbon dioxide, and methane in the atmosphere is controlled also in large part by biological processes including photosynthesis, respiration, *methanogenesis* (anoxic production of methane by bacteria from organic acids or carbon dioxide), and *methanotrophy* (microbial usage of methane as food to produce energy). For the sake of simplicity, we will describe only the biogeochemical cycle of carbon.

The *biogeochemical cycle* is a conceptual model that describes (1) the transport of material through different environmental regimes (such as the flow of carbon-rich soil particles from the continents to the oceans via rivers and airborne dust plumes) and (2) the transformation of carbon from one chemical form to another (such as microbial respiration of a solid carbohydrate to dissolved or gaseous carbon dioxide). A removal mechanism in a biogeochemical model is referred to as a *sink*, so mechanisms that sequester CO_2 from the atmosphere by converting it to another form or transporting it to a different reservoir are also referred to as *carbon sinks*.

The human or anthropogenic impact on the *global carbon cycle* (Fig. 11.43) has been ongoing since at least the agricultural revolution and has been intensifying since the beginning

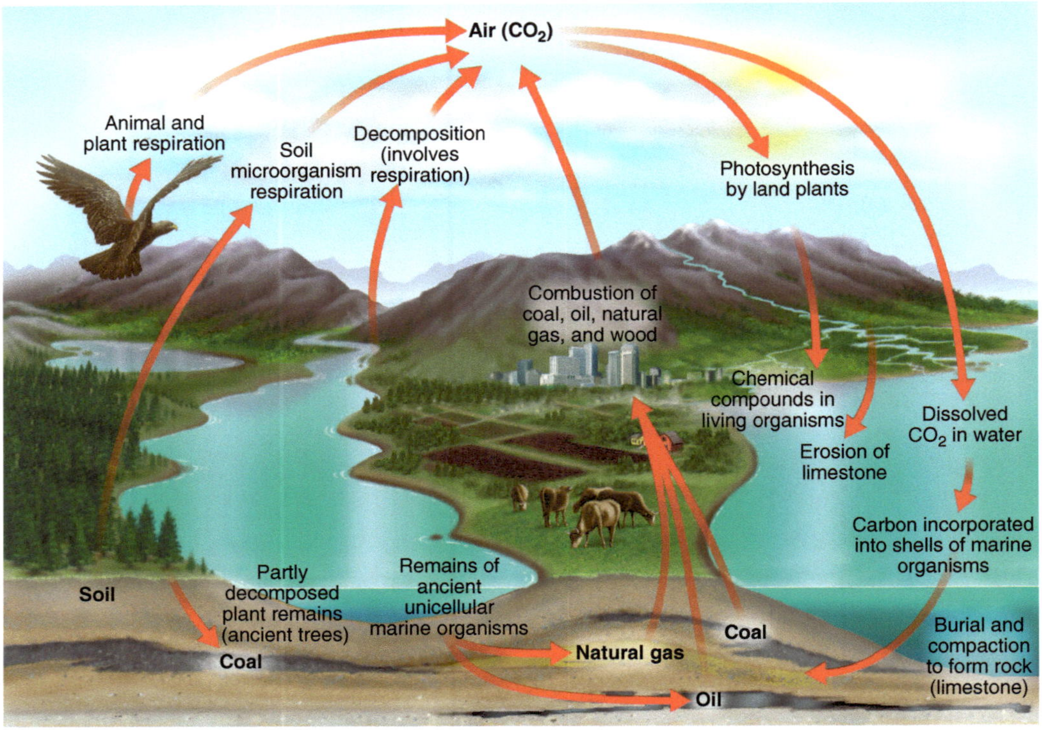

Figure 11.43 Earth's carbon cycle describes how carbon atoms flow between various reservoirs in the Earth system. From Strahler, 2013.

of the Industrial Revolution. These impacts range from changes in land use to the mass burning of *fossil fuels* (coal, oil, and natural gas). Agricultural land, for example, stores less carbon and releases more carbon dioxide than forest or native grassland. Fossil fuels take carbon stored within the Earth's crust that is the remnant of plants and animals from hundreds of millions of years ago, and humans convert it through burning to the most abundant greenhouse gas, CO_2. By contrast, biomass burning generates CO_2 from recently formed plants, which requires substantially less time to regenerate. While both biomass burning and fossil fuel burning affect the amount of CO_2 in the atmosphere, fossil fuel burning adds carbon to the active carbon cycle so is a much greater challenge to mitigate, whereas biomass burning releases only recently fixed carbon that could be resequestered by current ecosystems so long as they remain intact.

The strategy of combating climate change by enhancing carbon sinks range from conservation agriculture and traditional conservation efforts to the development of new and innovative technologies. Currently the most effective and widespread examples of carbon sinks involve the uptake and storage of CO_2 as biomass by forests, which are being replanted in many regions.

According to the IPCC's Fourth Assessment Report on mitigation of climate change, sustainable forest management practices that include sustained harvesting for wood or fiber as well as reforestation and forest conservation efforts will yield the greatest carbon sequestration benefit currently at our disposal. If we can slow and ultimately stop deforestation and drastically increase reforestation efforts on a global scale, we can eliminate a major source of atmospheric carbon dioxide and even begin to sequester some of the fossil carbon that we are releasing as forest biomass.

A related approach is wetland restoration, which leads to greater soil carbon storage than dry land ecosystems. In fact, soil conservation practices can also be thought of as carbon sequestration efforts. Traditional open till agriculture gradually strips the organic carbon from the soil as it oxidizes. On the other hand, no-till methods, use of cover crops, and other farming techniques that retain soil carbon can prevent the loss of carbon as CO_2 and even increase the amount of organic carbon in the soil over time leading to carbon sequestration. Likewise, allowing low-productivity agricultural land used to grow cereal grains for livestock to revert to native grasslands has a twofold mitigating impact, native grasslands represent carbon sinks whereas row cropping typically represents a net annual carbon source to the atmosphere. Native grasslands also have higher albedo than either comparable cropland or forest, leading to increased reflection and reduced absorption of solar radiation. In many cases, the grassland can still be used as pasture supporting the nutritional needs of livestock but with a much-reduced climate impact. Fire management using controlled burns in many grassland and savannah ecotypes and some forest biomes can reduce the overall amount of carbon released during seasonal wildfires. This has been shown to be true in the sub-Saharan African savannahs where patchy early dry season burning, which is naturally limited by the cooler evening temperatures, higher moisture content in soils and vegetation, and occasional rainfall drastically decreases the amount of biomass that burns during the late dry season.

11.6 Geoengineering

11.6.1 Radically Altering the Carbon Cycle

Carbon Capture and Storage

Similar to the biological carbon cycle discussed previously, there is also a much slower, geologic carbon cycle that could help restore carbon in the form of rocks over a geologic timescale (too slow to help us in the moment). Yet it is still useful to understand this cycle, as this is the cycle that carbon capture and storage techniques are attempting to emulate. Before the Industrial Revolution and ignoring biological influences for a moment, there were two primary geologic controls on the amount of carbon dioxide in the atmosphere, both of which are related to the recycling of material between the outer mantle and the crust. These geologic controls include volcanic activity, a CO_2 source, and chemical weathering of silicate minerals to sequester CO_2 in the formation of carbonate rocks such as limestone and dolomite, where carbonate rocks act as a CO_2 sink, or reservoir, for CO_2. When carbonate rocks are subducted back into the mantle, the carbon is largely converted back into CO_2, which can then be released back into the atmosphere through volcanic activity and the geologic carbon cycle begins again.

New and novel carbon sequestration methods are being tested, proposed, and in some cases implemented on an increasingly ambitious scale. Collectively these techniques are referred to as *geoengineered sequestration* or *artificial sequestration*. *Carbon capture* technology refers to a collection of techniques that allow the removal of carbon dioxide from flue gas and other industrial sources. In each case, the goal of carbon capture is either to make use of the captured CO_2 for another coupled process (such as chemical manufacture or production of algal biofuel) or to store it long-term through geological sequestration, or mineral storage. *Geological sequestration* refers to the potential long-term storage of CO_2 deep within the Earth's crust in subterranean structures such as saline aquifers, exhausted oil wells, or unmineable coal seams. The idea originates with the practice of injecting CO_2 into oil wells to boost production. Since 2012, the Southeast Regional Carbon Sequestration Partnership (SECARB) and the National Energy Technology Laboratory of the U.S. Department of Energy (DOE) have been investigating the viability of large-scale deep injection of carbon dioxide at test sites in Alabama, Kentucky, and Mississippi. By 2022, the project is expected to provide extensive monitoring data and, if the results are promising, a set of guiding principles for successful implementation on a larger scale.

Another geochemically based carbon sequestration scheme called *mineral carbonation* involves taking waste stream or atmospheric CO_2 and converting it to stable carbonate minerals. The process occurs naturally but is sped up by increasing the surface area of the metal oxide minerals or by increasing the temperature and/or pressure. A plant in Iceland, which injects CO_2 dissolved in water into hot basaltic rock underground began test operations in 2017 and became fully operational by September 2021. It sequesters an estimated 1,300 tons of CO_2 per year as carbonate minerals and also removes H_2S emissions. The primary benefit of mineral carbonation over more conventional geological sequestration is that it is not prone to leaks or other accidental releases. Once the carbonate minerals form, there is no longer any gaseous carbon dioxide that can escape. The primary block to expansion of this method is the current cost of \$300/ton CO_2 removed, while C offset pricing stands at \$50/ton in the European Union and as low as \$15/ton in other locations.

Deep Ocean Storage

The oceans have also been suggested as possible storage reservoirs for carbon. Until the early 1990s, it was considered plausible to store CO_2 in the deep ocean where the low temperature and high pressure causes carbon dioxide to remain stable as a liquid or a hydrate. Concerns over ocean acidification, potential harm to benthic organisms, and unpredictable stability led to the abandonment of such plans. Others have proposed deep-ocean dumping or burial of biochar or agricultural wastes such as corn fiber as a way of storing organic carbon on time scales similar to the production of fossil fuels.

Iron Fertilization of the Ocean

More recently, several would-be geoengineers have proposed iron-fertilization of the high nutrient, low chlorophyll (HNLC) regions of the ocean. In these remote open-ocean regions, primarily the North Pacific, Equatorial Pacific, and Southern Ocean, marine primary productivity is not limited by nitrogen or phosphorous nutrients, but rather iron (a nutrient needed for photosynthesis), which has low solubility at ocean pH (8.1). The natural source of iron to these regions are windblown dust or volcanic ash, both of which are episodic (or at most seasonal). When iron becomes abundant in these regions, it leads to short-lived, but intensely productive blooms of phytoplankton, which consume carbon dioxide at the water's surface as they grow and then transport much of the now solid organic carbon to the ocean floor as they die and sink.

In 1988 oceanographer John Martin famously stood up after a public lecture on climate change and stated (reportedly with a fake accent to mimic Dr. Strangelove), "Give me a half tanker of iron, and I'll give you an ice age!" Shortly after Martin's death in 1993, his careful but controversial laboratory studies and iron hypothesis were shown to have merit during the Iron Enrichment Experiment (IRONEX), which used a research vessel to systematically spray dissolved iron across a 64 km^2 patch of ocean near the Galapagos Islands in the Pacific Ocean. The iron enrichment caused a tripling of algal biomass. Later experiments generated even more dramatic results, though scientists still question whether such an endeavor would result in a net draw down of atmospheric carbon, if one accounts for the fossil fuels consumed in preparing the iron and transporting it to the remote ocean via boat or aircraft, and analyzes the efficiency with which the bloom biomass is transported to the ocean floor to escape recycling.

Expanding this practice to encompass larger spatial and greater temporal scales continues to garner interest from a number of countries and corporations. However, most scientists have espoused application of the precautionary principle since the possible unintended consequences of such a practice (harmful algal blooms, benthic anoxia, food-web disruption, etc.) remain poorly understood. The *precautionary principle* states that a new product or process whose ultimate effects are disputed or unknown should be resisted.

Dimming the Sun and/or Brightening the Earth

Stratospheric sulfate (SO_4^{2-}) injection is another widely discussed geoengineering approach to climate mitigation. The idea is to mimic the effects of a large volcanic eruption by periodically releasing sulfate or one of its precursors (i.e., H_2S or SO_2) or products (i.e., H_2SO_4) from high altitude aircraft, weather balloons, or artillery shells. The resulting small, reflective particles would increase the albedo of the stratosphere and decrease the amount of solar radiation reaching Earth's lower atmosphere and surface. Proponents suggest that based on the more optimistic models, this approach combines the greatest cooling efficacy among the technologically feasible approaches. Concerns include the potential to whiten the sky and change the colors of sunsets and sunrises and potential ozone depletion, but the most serious concerns have centered around the relatively short stratospheric lifetime of the particles and the potential consequences once they settle below the tropopause. These consequences range from acid rain, respiratory illness, possible humidification of the atmosphere, and changes to weather patterns. These have proven difficult to predict and depend largely on where and how the particles are released. Moreover, stratospheric aerosol geoengineering also fails to address ocean acidification.

Mirrors

Many science fiction writers as well as more serious and practically minded physicists have proposed the use of ground or space-based mirrors to either alter the amount of incoming radiation (space mirrors) or artificially increase Earth's albedo (less exciting earth-bound mirrors). To date only one space mission has been used to deploy a space mirror. In 1992, the Russian Federal Space Agency launched Znamya-2, a 20 m wide space mirror, that created a 5 km diameter bright spot that moved across Europe on 4 February 1993 before the craft was de-orbited and allowed to disintegrate during atmospheric reentry. While space mirrors are unlikely to deflect enough radiation to offset the warming from the greenhouse gases of their manufacture and launching into orbit, both space mirrors and Earth mirrors remain a climate engineering strategy that are actively discussed.

11.7 Adaptation

Because we have already documented significant climatic changes and have yet to effectively curtail our consumption of fossil fuels, much of what will need to be done will be in the form of human adaptation to new and changing climatic conditions. As extreme weather events become more frequent, early warning systems and increased rescue and shelter capacities will become increasingly critical for protecting communities. As sea levels continue to rise during (at least) the next century, densely populated low-lying coastal regions will most likely need to be evacuated.

Many coastal communities are already grappling with these issues. The Bayside neighborhood of Portland, Maine, has begun to experience significant street flooding and property damage during astronomical tides (i.e., tides during the full moon and new moon) even in the absence of storm surge or rainfall. Such sunny day flooding is becoming increasingly common in many coastal American cities such as Annapolis, Maryland, Norfolk, Virginia, and Miami, Florida. Miami Beach has announced plans to spend at least $400 million to raise its streets, install pumps, and build higher seawalls. Smaller and less affluent coastal communities are unlikely to be able to commit similar resources to mitigation of flooding.

11.7.1 Building Prohibition and Buyouts

In cases where storm surge or heavy rains coincide with high tide or astronomical high tide, the effects are additive and flooding may be catastrophic. In southern Louisiana, where land subsidence is accelerating the effects of sea level rise, state officials are developing a program known as LA SAFE (Louisiana Strategic Adaptations for Future Environments) that will prohibit new home building in at-risk areas, provide buyouts for current residents, and impose tax hikes on residents that refuse to leave. Understandably, many residents have had very dampening reactions to the draft plan, and many state officials are concerned what the creation of a new class of climate change refugees will mean for regional communities further inland. Other coastal states are watching the rollout and gauging the effectiveness and feasibility of these programs very carefully as they will have to be duplicated or adapted in many parts of the Gulf Coast and Atlantic seaboard.

11.7.2 Storm Surge Barriers

Some cities may be able to engineer protective or resilient approaches to living in what are becoming flood prone areas, but the scale and scope of engineering needed to meet these challenges should be considered. Already, the largest moving human-made object on Earth is a storm surge barrier built (Fig. 11.44) to protect the Dutch city of Rotterdam. The coming decades will find many coastal cities and commercial ports around the world in need of similar protections for which financial resources may or may not be available.

Figure 11.44 The world's largest moving human-made structure, the supercomputer-controlled Maeslantkering surge barrier, in its closed position, was built to protect the city of Rotterdam, Netherlands, from storm surges and related flooding that is worsening due to sea level rise. https://commons.wikimedia.org/wiki/File:Maeslantkering_closed.jpg#/media/File:Maeslantkering_closed.jpg

11.7.3 Water Restrictions

In the interior of the continents, people will have to deal with the opposite end of the climate adaptation spectrum with respect to water. As drought conditions for many areas become more common and in extreme examples become the new norm, residents will have to devise strategies to deal with dryer climatic conditions. Water restrictions will become increasingly common and increasingly restrictive. Many cities in the southwestern United States have begun to consider banning turf grass lawns and many are already incentivizing removing existing turf grass for replacement with rock gardens and landscaping with drought tolerant plants. Most states in the region are updating building codes to specify water efficient fixtures and appliances. While restrictions on household water use are important tools for combating drought, public water supplies and domestic withdrawals account for only around 14% of water use. According to data compiled in 2005 by the U.S. Geological Survey, about 42% of U.S. freshwater resources were used by electrical utilities, more than 38% were used for irrigation, and more than 5% were used for industrial and mining applications. These findings highlight the need for careful planning when locating new industrial development and power infrastructure.

Adapting to the new challenges and risks of our changing climate and doing what we can to limit our continued damaging impact on Earth's climate and environment will be humanity's greatest challenge for the foreseeable future. With these daunting challenges, though, come new and exciting opportunities to improve the sustainability and resiliency of our communities and institutions with respect to both environmental quality and social welfare. By evolving our understanding of how human economic and political constructs interact with Earth's natural systems, we can optimize continued development to minimize risks, alleviate suffering, and conserve and protect resources.

11.8 Chapter in Review

Throughout the entirety of Earth's history, the planet's climate has oscillated widely between extremes of warm and cold. In general, a cold period in geologic history is called a glacial period or ice age, and a warm period is called an interglacial period or interglacial. We are currently in an interglacial period and expect to be for at least another 50,000 years. Variations in the Earth's orbit, volcanic emissions, and plate tectonics are overarching historical causes for the advance and retreat of glaciers, and thus the oscillation between glacial periods and interglacials. Yet Earth's temperature, CO_2, and sea levels have all remained relatively constant since the end of the last ice age and the rise of human civilization based on agrarian lifestyles that depend on a stable climate. This stable climate is now threatened by human emissions of greenhouse gases, which are altering the stable climate system.

The Earth has both weather and climate. Weather is an instantaneous picture of the state of the atmosphere, and includes such readings as temperature, humidity, cloud cover, and atmospheric pressure. Climate is a long-term average of weather conditions, typically over a period of more than 30 years.

The Earth's climate is controlled by the overall balance of incoming and outgoing radiation. If there is more incoming than outgoing radiation, this causes global warming. Incoming solar radiation is absorbed by the landmasses and oceans or reflected by snow, ice, and clouds. Albedo is a measure of a surface's reflectiveness. High albedo means high reflectance, and low albedo means more absorption of the Sun's incoming solar radiation. Clouds and ice have high albedo, and land surfaces and the ocean have low albedo. Greenhouse gases in the atmosphere also prevent the loss of heat energy from reradiating back to space. They recirculate heat in the Earth's atmosphere, causing the temperature to rise. The greenhouse effect is beneficial in that it provides sufficient heat for life on Earth. But as humans have added greenhouse gases to the atmosphere, this has caused global warming.

The ongoing warming trend in Earth's climate can be attributed to greenhouse gases from human activity. Carbon dioxide, the most abundant human influenced greenhouse gas, has increased from a preindustrial average concentration of 278 ppm to 417.2 ppm in 2022, a full 50% above preindustrial levels. Sources of atmospheric carbon dioxide include burning fossil fuels and damage or destruction of ecosystems where primary producers use photosynthesis to draw CO_2 out of the atmosphere. The most potent greenhouse gases are synthetic and made by humans and are thousands of times stronger, more potent, than CO_2. The strongest of these is SF_6, which is currently rising unabated. The intensifying use of fossil fuels since the onset of the Industrial Revolution and especially during the 20th and 21st centuries has led to greenhouse warming unprecedented during human history.

Short-term climate variations can occur for a variety of reasons. During El Niño events (every 2 to 7 years), warm water replaces the normally cool currents off the Pacific coast of Ecuador and Peru. The warm water spreads west near the equator and causes wide-ranging weather effects. Some normally humid areas experience droughts, some dry areas experience downpours, there is an increase in the intensity of tropical storms, and food for fish disappears in the warm current, causing a collapse of the fishery and of populations of animals that rely on those fish for food. The opposite condition, when unusually cold water replaces the normally cool water along the west coast of South America, is called a La Niña event. Areas that experience one type of weather may experience the opposite during a strong La Niña. For example, an area that experiences intense rainfall during an El Niño year may experience a drought during a La Niña year.

Droughts are periods in which an area experiences much less rainfall than usual. They can last for as little as 2 weeks or as long as several years. The worst U.S. drought of the nineteenth and twentieth centuries was the Dust Bowl period of the 1930s, which lasted for 8 years. The drought was exacerbated by poor farming practices and led to changes in how the United States provides guidance to farmers. During the first part of the 21st century, thus far, the western United States, including the Colorado River watershed, has experienced prolonged drought conditions. In addition to the lack of rainfall, demands on the water within the river basin rose considerably due to rising temperatures and population growth. This has led to a hot drought in which higher air and soil temperatures delay recovery. Governments and stakeholders will need to think creatively about how to reduce demand on scarce water resources if the urban, residential, and farming communities in the American southwest are to become sustainable.

Changes in Earth's temperature are altered not only by greenhouse gases, but also by amplifying and dampening feedbacks to the Earth system. An example of an amplifying feedback loop for global cooling would involve an expanding glacier, or snow fields increasing the Earth's albedo. This would lead to greater reflection of incoming solar radiation, greater cooling, so more ice and snow, and further increasing the amount of glacial cover. An example of an amplifying feedback loop for global warming involves higher temperatures, which melts snow and ice, which reduces albedo, which causes more absorption of heat from the Sun, which in turn causes higher temperatures and more melting. Warmer temperatures would also cause an increase in evaporation and hence more water vapor in the atmosphere, which is a greenhouse gas. The greenhouse gas absorbs outgoing IR radiation and increases atmospheric temperatures, thus ensuring more melting of ice and snow. While current climate models included these and many other amplifying feedback loops, they generally have not yet incorporated the amplifying impacts of CO_2 released from forest fires, or CO_2 and CH_4 released from melting permafrost, for example, so may be underestimating the rate of warming. There are 27 documented amplifying feedback loops thus far, and it will be important for modelers to take all of these into account (Ripple et al., 2023).

During the 800,000 years prior to the Industrial Revolution, CO_2 levels in the atmosphere ranged from 180 ppm to 278 ppm. Yet by 2023 we reached an average annual CO_2 concentration of 420 ppm, 50% above preindustrial levels. This drastic increase is largely because of the burning of fossil fuels. This level of radiative forcing is unprecedented in all of human history. As a result, global and regional weather patterns are changing. Desertification is occurring in new areas. Alpine glaciers are retreating. Ice sheets on Greenland and Antarctica are melting and calving. Melting of these ice sheets together with ocean warming are causing sea level rise. The melting ice also reduces ocean salinity and density near those ice sheets, which has resulted in 15% slower circulation of the Great Ocean Conveyor Belt.

Other effects expected from ever-increasing temperatures include increased incidents of downpour events, resulting in more cases of extreme flooding; increase intensity of tropical storms; more frequent and hotter heat waves; reduced stream flow in the spring and summer months as alpine glaciers that once melted in the spring and summer disappear completely; likely destabilization of the circumpolar jet stream extenuating both cold and warm extremes in the normally temperate regions of the Northern Hemisphere; poleward shifting of plant and animal habitat ranges (including pest species); changes to the timing of animal migrations; melting of permafrost, which disrupts animals, plants, streams, forests, shorelines, and human infrastructure; expanded ranges for pathogens, parasites, and crop pests; coral bleaching, which leads to loss of biodiversity and animal populations in the oceans; and ocean acidification.

The best way to mitigate the effects of climate change is to significantly reduce emissions of carbon dioxide and other greenhouse gases. This would include reducing the use of fossil fuels, developing and adopting renewable energy and alternative fuels, improving the efficiency of the power grid and transportation systems, reducing production of Portland cement, and reducing fertilizer use. Unfortunately, despite knowledge of these needs for several decades, greenhouse gas emissions and atmospheric CO_2 concentrations are still on the rise, although we are encouraged by recent trends showing the increased production and low cost of alternative energy, and declarations of many municipalities and nations to achieve 100% renewable electricity and/or carbon neutrality (more carbon sequestered than released) in the coming decades.

Despite these efforts toward carbon neutrality, adaptation measures are already needed. Climate-driven sea level rise has already forced many in the Pacific islands and even two U.S. towns to relocate. Other locations have already built well-engineered barriers that keep floodwaters at bay. A significant investment in climate adaptive infrastructure is needed.

Many geoengineering approaches are also being studied. While these are less desirable solutions since they may have unintended consequences, it is important to explore these options in the event that global temperatures keep rising uncontrollably. Carbon capture and storage has already begun to be deployed, though these practices are not yet energetically or financially tractable.

12

Wildland Fires

Bethany D. Hinga

University of Nebraska at Kearney, USA

12.1 Introduction

12.1.1 The Camp Fire, 2018

At 6:29 a.m. on 8 November 2018, a brush fire was reported to the California Department of Forestry and Fire Protection station at Jarbo Gap in Northern California. Captain Matt McKenzie attempted to visit the site to assess the situation. He found the fire was in a nearly inaccessible area of Butte County, off of Camp Creek Road. Camp Creek Road was a narrow, steep, and winding mountain road just outside the community of Concow. He knew this was a dangerous situation, and radioed his fellow firefighters, "This has got potential for a major incident." He knew this was likely to bloom into a dangerous wildland fire because the area had not seen appreciable rainfall since May of 2018, and the heavily vegetated area of mountains and canyons was tinder-dry. In addition, warm winds were blowing strongly across the landscape at speeds up to 52 mi/hr (84 km/hr). The combination of steep landscape, low humidity, dry vegetation, and high winds was a recipe for disaster.

Within 1 hr, the fire had moved into its first inhabited area 3 mi (4.8 km) from its origin, and had begun burning structures. The Butte County Sheriff's Office ordered evacuation of the town of Pulga at 7:23 A.M. It was followed shortly thereafter with an order to evacuate the community of Concow. The first home reportedly caught fire at 7:35 A.M. near the town of Concow. Though people were ordered to evacuate it was already too late for some to get out of the way. A tree fell across Concow's main escape route, Hoffman Road. Trapped on the wrong side of the roadblock was a state fire crew and twenty residents. They sought refuge in a park, but the fire was closing in. Some people jumped into a creek to escape the flames. Others sought refuge inside firefighters' emergency shelters. Unfortunately, four people were severely injured and eight people died. Of the injured, three people were burned over half their bodies and required extensive medical intervention, and an elderly man who jumped in the water suffered from hypothermia.

In the next hour, the fire traveled another 3 mi (4.8 km) southwest to the community of Paradise. Paradise was a city of 27,000 people located on top of a ridge. Fire tends to spread upward along slopes quite quickly. Since Paradise is located in an area at a higher elevation than the surrounding landscape, the fire spread quickly upward and began burning structures in town. The first evacuation call came at 7:46 a.m. as fire swept into the city, yet the evacuation order only covered the eastern quarter of the city. Despite the incredible speed of the fire, evacuation of the rest of the city was not ordered until nearly an hour later.

Earth's Natural Hazards and Disasters, Advanced Textbook 7, First Edition. Bethany D. Hinga.

Adventist Health Feather River Hospital in Paradise began evacuating patients, moving the last patients out as the hospital's human resources building caught fire.

Paradise had four main evacuation routes out of town. The roads normally handle a few thousand cars per day each, but in the evacuation, all 27,000 residents were trying to leave town at the same time. Roads became jammed, and people were stuck on the highway as the forest burned around them. Car tires exploded and melted, leaving cars unusable. Stalled cars blocked the road so others could not pass. The heat was intense. One evacuee watched the fires burn around her while noting that her steering wheel was melting and sticking to her hands. With cars rendered useless, many people got out of their cars and ran for their lives. This was incredibly dangerous as well, not only because of the intense heat from fires burning on either side of the road, but also because of the heavy smoke and lack of visibility. In many cases, people were picked up on the road by civilians whose tires were still intact, or by emergency workers. Unfortunately, many people died in their cars along the evacuation route. Firefighters worked desperately to evacuate people all day long, so they had no chance to actually fight the fire. By 10 a.m., nearly the entire town of Paradise had been engulfed by fire, and more than 9,000 homes were destroyed. Wind gusted to 72 mi/hr (116 km/hr) and blew flames sideways. People took refuge anywhere they could find. Some found empty fields, others found intact stores in which to seek refuge. At the end of the day, the hospital was still standing, so it became a shelter for many who were unable to escape the city. By nightfall, the fire had burned 20,000 ac (8,094 ha).

Two days later, the fire had expanded to 100,000 ac. Smoke from the fire caused the air quality to fall to dangerous levels, prompting closure of schools as far away as the San Francisco Bay Area. Smoke drifted eastward and was detected across the country, even in New York City.

By the time the Camp Fire was finally extinguished on 23 November, 15 days after it started, it had consumed more than 153,000 ac (40,469 ha). There were 85 fatalities. The deceased ranged in age from 39 to 99. Of those who died in the fire, 60% were in their 70s, 80s, or 90s. Age and infirmity were clearly complicating factors in the evacuation process. More than 13,750 homes were destroyed, along with about 525 commercial buildings and nearly 4,300 other structures. Insurance claims for wildland fires in California in November 2018 exceeded $12 billion. Most of those economic losses were from the Camp Fire, making it the most expensive year in California history to date in terms of insured losses from fire. One company, Merced Property & Casualty Company, which insured some homes and businesses in the area, went bankrupt because of all of the losses in this fire. The state took over the insurance claims, guaranteeing payment to those who carried insurance and lost property and possessions.

What caused this fire? There is strong circumstantial evidence that damaged power lines sparked the blaze. Pacific Gas and Electric (PG&E), California's largest utility, is the electricity carrier in the area affected by the fire. Despite a company policy to shut off power in extreme fire hazard conditions to avoid sparks igniting flames, PG&E did not take measures on 8 November 2018 to protect the community from rogue sparks, and therefore fires. Thousands of homeowners filed claims against the utility. On 29 January 2019, PG&E filed for bankruptcy, citing $7 billion in claims from the Camp Fire. The company was directed to pay $25.5 billion in fire-related claims and emerged from bankruptcy in 2020 with $38 billion in debt, considerably more than the $22 billion in debt it had going into its bankruptcy in 2019. The company also enacted new policies regarding shutting off power to fire-prone areas during times of high fire danger to prevent sparks from power lines from starting new fires.

In addition to concerns about how the fire was started, people are asking questions about the timeliness and accessibility of evacuation orders, as well as the evacuation plans for communities in the area. Residents of the area decided several years ago to enact an opt-in emergency warning system rather than a more universal emergency warning system. Less than 40% of the residents in the Paradise area opted in, which raises questions about the effectiveness of the system. During the Camp Fire, only about 30% of those who had opted in actually received the evacuation warnings. Cell towers were destroyed in the first few hours of the fire, and land lines were not capable of handling the volume of calls necessary to warn the entire population of Paradise at the same time. Better systems are available, such as one similar to that which sends Amber Alert style warnings to all cell phones in the area. It is likely that the city councils in many fire-prone areas of California will be looking into upgrading warning systems after this fire.

The last time fire encroached on Paradise, in 2008, the after-incident report very clearly indicated that escape routes and evacuation plans needed to be improved. In the years since, one of the major arteries that leads through downtown Paradise was actually narrowed from four lanes to two. This was done to provide a safer environment for pedestrians and encourage downtown commerce. Arguments have been made that the highways would become jammed no matter how many lanes were available because of the volume of people trying to escape, all at the same time. During the fire, all lanes in town were turned into outbound traffic lanes, turning two-way highways into one-way roads. It is likely that the narrowing of streets downtown was not a major factor in the deaths, injuries, and harrowing evacuations on 8 November 2018. However, the narrow roads leading out of Paradise became blocked with stalled cars, and were therefore inadequate to handle evacuation traffic safely. Evacuation routes clearly still need to be improved.

The city of Paradise was nearly uninhabitable immediately following the fire, and for more than a year afterward. Thousands of families were displaced. Many fled to the nearby city of Chico to find temporary housing. Several people want to rebuild their homes in Paradise, and the State of California is working with those individuals to be sure that they have a safe space in which to do so. Much of the city was contaminated with the toxic remains of household materials and substances. Older homes had asbestos in roofing materials or siding, and most homes contained chemicals that leaked out during the fire since many household items become toxic when they burn. Cleanup crews with the California Office of Emergency Services and other state agencies removed hazardous waste (batteries, pesticides, lead, asbestos, and structural debris) from more than 11,000 properties by April of 2021. As of November 2022, more than 1,400 homes were rebuilt in Paradise, and the population of the city was nearly 10,000.

Wildland fires are common in California and, at least in the past, occurred in a somewhat predictable season each year. However, the wildland fire season has grown considerably in length over the last 30 years, and is now almost year-round. As climate changes, California is becoming warmer and drier, making conditions ideal for more fires in the future. How California responds to this growing threat has the potential to become a model for other areas of the world experiencing similar changes to their climates and therefore their fire hazards.

The Camp Fire burned in a forested area at the *wildland/urban interface*. This is an area where people are encroaching on wildland areas, and where they are most at risk from wildland fires. While rare catastrophic fires in forested areas, such as the Camp Fire, make headlines and can displace thousands of people, forests are not the only areas in which wildland fires occur. Fires are just as common in grassland environments. On 30 December 2021, a devastating grass fire, the Marshall Fire, burned in the wildland/urban interface in Boulder County, Colorado.

A months-long drought in the region had left the grass, shrubs, and trees in the area extremely dry. On 30 December, conditions were perfect for a wildland fire. Wind speeds were high and variable, with some gusts of up to 115 mi/hr (185 km/hr) recorded. The fire quickly spread from uninhabited areas toward cities, towns, and homes in unincorporated areas of the county. Three towns were evacuated, and fire entered two of them. In the city of Superior, a hotel was destroyed by the fire, and a shopping center containing Target and Costco stores was damaged. Superior, Colorado, saw 378 residential structures destroyed and 58 structures damaged for residential losses of $152.8 million. The city of Louisville, Colorado, lost 550 residential structures and 43 structures were damaged for a total of approximately $229.2 million in residential losses. One hundred fifty six residential structures in unincorporated sections of Boulder County were destroyed and 48 damaged, resulting in $138.3 million in residential losses. Total losses topped $513 million.

Snow on the night of 1 January 2022 extinguished the Marshall Fire, but not before it burned more than 6,000 ac (2,428 ha) and claimed two lives. The cause of the fire was never determined with any certainty, but downed power lines, a fire at an outbuilding, and an underground coal fire have been cited as possible ignition sources. The Marshall Fire was the most costly in Colorado history in terms of residential losses.

Climate change has also resulted in severe fires in other places in the world. In March of 2023, wildland fires in Spain consumed over 9,884 ac (4,000 ha) of land near Valencia. Officials there cite higher than normal temperatures for the springtime and a winter drought for this fire. However, Spain has seen an extreme uptick in wildland fires in recent years due to extreme temperatures, low humidity, strong westerly winds, and lack of rainfall. In 2022, Spain saw 757,000 ac (306,000 ha) burn in more than 500 wildland fires. The largest fire that year was in Sierra de la Culebra in northwest Spain, which alone burned over 148,000 ac (60,000 ha).

Terrible droughts are also plaguing Australia, leaving the landscape parched and ripe for wildland fires. In recent years, bushfires in Australia have threatened populations of koalas, an iconic Australian species. Koala populations in some Australian states are declining rapidly. Koalas are endangered in New South Wales, and bushfires there in 2020 killed approximately 70% of some koala populations within that state. Scientists fear that koalas will go extinct within 20 years in New South Wales if more is not done to protect them from bushfires and other disasters.

As climate change brings hotter and drier conditions to areas that were previously cooler and received more precipitation, wildland fires are expected to consume more acres each year. However, it is also likely that climate change will cause some areas to experience a decrease in fire hazard because of a decrease in fuel availability. As some areas get hotter and drier, fewer plants will grow, and thus less fuel will be available to burn.

12.2 Fire Basics

While fire can certainly occur in any environment, rural or urban, our discussion in this chapter will be focused on wildland fires. *Wildland fires* are fires occurring in natural vegetation and fuels. They can occur in forests, grasslands, parks, highway medians, or any other outdoor space with vegetation. Wildland fires have the ability to burn thousands, even hundreds of thousands of acres during their burn times, which can last up to several weeks. They can also, of course, burn human-made structures. Places that are particularly at risk for destruction of human-made structures are found at the *wildland/urban interface* where residential development is occurring in wildland areas (Fig. 12.1).

12.2.1 Combustion

Fire is caused by the chemical process of combustion. *Combustion* is a self-sustaining process in which there is rapid oxidation of fuel. (*Oxidation* is the process of a substance combining chemically with oxygen.) It produces

Figure 12.1 Images of the Jesusita Fire in Santa Barbara in May 2009. (a) Satellite image of the fire as it burned on 8 May 2009. Smoke plumes billow southward across the city of Santa Barbara toward the Channel Islands. (b) Photo of the Jesusita Fire as it moves into Santa Barbara. (c) Satellite image of the burn scar from the Jesusita Fire. The burned area is in reddish-purple tones. The city of Santa Barbara is visible in the lower right. (a) Photo courtesy Jeff Schmaltz MODIS Land Rapid Response Team, NASA GSFC; (b) From Alamy; (c) NASA image created by Jesse Allen, using data provided courtesy of the NASA EO-1 Team. From Strahler, 2013.

both light and heat. Rapid oxidation takes place in two forms. One is glowing, or *smoldering fire*, in which a material burns without a flame. The second is free-burning fire, steady-state fire, or *flaming combustion*. As the name implies, flaming combustion is a state in which material burns with a flame.

Combustion of plant material, in a sense, is the reverse of the process that created the plant material in the first place. *Photosynthesis* is the chemical process in which carbon dioxide from the atmosphere, water, and the Sun's energy combine to produce cellulose and other chemical components of plants. Fire and decay reverse the process. Combustion takes that *cellulose*, an insoluble substance that is the main constituent of plant cells walls and vegetable fibers (and other chemical components of plants), oxygen from the air, and heat energy, and breaks the plant material down into carbon dioxide, water, and energy in the forms of heat and light.

Photosynthesis equation,

$$6\ CO_2 + 5\ H_2O + \text{solar energy} \rightarrow C_6H_{10}O_5 + 3\ O_2.$$

Carbon dioxide plus water and solar energy yields cellulose and oxygen.

Combustion equation for plant material,

$$C_6H_{10}O_5 + 3\ O_2 + \text{ignition temperature} \rightarrow 6\ CO_2 + 5\ H_2O + \text{heat and light energy}.$$

Cellulose in plant material plus oxygen and an ignition source yields carbon dioxide, water, and heat and light energy.

Burning starts with a process called preignition or preheating, in which a pilot source of heat acts on the organic material. The organic material is warmed and dehydrated. As the vegetation warms, materials within the plant such as waxes and oils are brought to a vapor phase (volatilized). As the temperature rises and volatile products are brought to the surface, the process of *pyrolysis* begins. Pyrolysis is a chemical change brought on by heat.

Fuel exists in all forms of matter in nature: solid, liquid, and gas. However, the only material that burns is gas. Why, then, do we see wood burning? In reality, the wood undergoes a chemical reaction, pyrolysis, before burning takes place. Through pyrolysis, wood fuel is converted to its numerous components. This includes not only water and carbon dioxide, but also semivolatile tars and *lignins* as well as waxes and flammable oils and a solid material called char. (Lignin is a complex organic material in the cell walls of many plants, making them rigid and woody.) During the process of pyrolysis, the volatile products (gases) migrate to the surface of the wood. The gases then ignite in the process of combustion, and burn on the surface of the wood, giving the appearance of the solid wood burning.

The process of pyrolysis requires heat in order to begin. After it is underway, it produces more heat, which helps ensure further pyrolysis. This produces more combustible gases that burn. It is through this loop that pyrolysis and combustion can become self-sustaining. Pyrolysis continues to prepare new sections of the wood by breaking it into its components, then the zone of combustion moves on and consumes those wood components.

12.2.2 Heat Transfer

Heat is transferred from one substance to another in three different ways: radiation, convection, and conduction. *Radiation* transfers heat from one place to another through radiant energy. Radiant energy travels outward away from a source in all directions. The energy travels until it encounters something that absorbs it. One example of this type of energy is the heat and light energy we get from the Sun. The radiant energy of the Sun travels through space until it reaches the Earth and its atmosphere. The radiant energy is absorbed by the atmosphere and the Earth's surface. It warms the planet and makes the planet habitable for us. In wildland fires, radiant energy preheats objects ahead of an advancing fire, contributing to preignition and pyrolysis.

Convection is the transfer of heat via movement of gas or liquid. The Earth's atmosphere redistributes heat through the process of convection. Hot air tends to rise, and cools as it does so. Cool air tends to sink, and warms as it reaches lower elevations. This creates a vertical circulation pattern in which warm air rises and is cooled, and then the cool air sinks and is warmed. This is true in fire situations as well. Warm, buoyant air circulates near fires and has the ability to preheat fuels that exist at some level above ground surface, including the tops of trees and vegetation at progressively higher elevations on slopes. This produces *updrafts* around fires, where the heated air rises and flows outward away from the fire. Once the air is sufficiently far away from the fire, it can cool and sink. The cool air may then be pulled back into the fire and heated, completing the convection cycle.

Conduction is the transfer of heat from one object to another through direct contact. This is most evident to those who burn themselves by touching hot pans or burners while cooking. Metals are good conductors of heat, which means they are very effective at transmitting heat. Vegetation, however, is a poor conductor of heat. Therefore, conduction is not a significant factor in the spread of wildfires. However, conduction can be an effective means of igniting fires. For example, if a hot welding torch from a fence building crew is dropped in a dry, grassy field, the torch can become a source of ignition for the grass. The same is true for *firebrands*, or embers that escape from the fire. Hot, glowing firebrands are often lofted above the fire and blown downwind. As they meet flammable vegetation in front of the fire, they can start new fires. This process in which firebrands start new fires is known as *spotting*.

12.2.3 The Fire Triangle

Combustion requires three elements in order to exist. These three elements are fuel, oxygen, and heat. Together, they make up the fire triangle (Fig. 12.2). If you take away any of the three elements required to sustain combustion, a fire can no longer exist. In firefighting, many techniques are used to eliminate at least one leg of the fire triangle. The most common material used to fight fires is water. Dousing a fire with water cools the fuel and removes heat from the environment, therefore eliminating one side of the fire triangle. Removing fuel from the fire is another effective means of control. This can be accomplished by creating vegetation-free space in front of an advancing fire. Firefighters do this by creating fire lines, in which they cut and remove vegetation all the way down to *mineral soil*, which is the lower soil horizons that do not contain appreciable organic material. This removes flammable materials from the fire's path in an effort to starve the fire of fuel. A third method of controlling fire is to remove the oxygen supply to stop the chemical reaction of combustion. This is done by smothering the fire. Since oxygen is present everywhere in the atmosphere, removing oxygen is challenging on a large scale. However, firefighters have tools in their arsenal to do this as well. In small fires, they can spray it with water or throw dirt on a fire to smother it, making oxygen unavailable. A second method for limiting oxygen availability is to spray flame retardant material on the fuel, which coats the fuel in nonflammable material so oxygen cannot reach it to undergo the process of oxidation. For large fires, water can be dropped by helicopter, and fire retardant sprayed by tanker plane.

12.3 Fire Behavior

Fire behavior is determined largely by three different factors present at the location of the fire. These are fuels, weather, and *topography* (the shape of the landscape). Together, these influences can be expressed as a fire behavior triangle (Fig. 12.3). Changes in any of these three environmental components can radically change the behavior of a fire.

Figure 12.2 The fire triangle. A fire requires heat, fuel, and oxygen in order to burn. From https://www.powerandcables.com/how-to-label-cables-safely-in-zone-1-zone-2-hazardous-areas/fire-triangle/

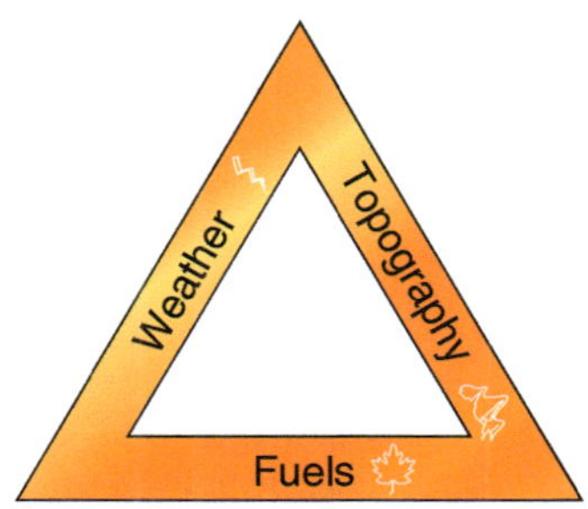

Figure 12.3 The fire behavior triangle. A fire's behavior is influenced by topography, weather, and fuels. Weather and fuel can change rapidly, but topography is a constant. From https://www.nps.gov/articles/wildland-fire-behavior.htm

12.3.1 Fuels

Firefighters classify fuels in many ways, which helps them predict fire behavior. First, fuels can be described by their vertical arrangement. *Ground fuels* are found below the surface of the Earth. They consist of root systems and *duff*, which is decaying plant material such as leaves and twigs in the upper soil horizons. *Surface fuels* are those that are found at and near the surface of the ground. This includes grass, fallen branches or twigs, pine cones, small trees, or shrubs found less than about 6 ft (1.8 m) above ground level. *Aerial fuels* are found higher than 6 ft (1.8 m) above the surface of the ground. It includes the canopies of trees with their leaves, branches, and limbs. Fires in forested areas can burn the aerial fuel in the *crowns*, or tops of trees. For this reason, they are called *crown fires* when they burn the entire tree. In addition to the three basic heights of vegetation, this classification integrates the concept of *ladder fuels*. Ladder fuels lead from the surface environment to the aerial environment. They can cause a fire to spread from the surface of the ground to the canopies of trees in very short order.

Next, fuels can be described as light, medium, and heavy. *Light fuels* are also termed flash fuels, flashy fuels, or fine fuels. Examples include grass, light brush, and other vegetation that is less than 2 ft (0.6 m) tall. It also includes pine needles, twigs, and other small-diameter vegetation or vegetative debris near the surface of the ground. Fires in light fuels tend to burn intensely and spread rapidly.

Medium fuels are those that are taller and have a slightly greater diameter than light fuels. This includes shrubs and saplings up to 6 ft (1.8 m) tall, as well as the grassy understory. Medium fuels burn at a moderate to high intensity, but fires in this type of fuel do not tend to spread as rapidly as those in light fuels.

Heavy fuels are trees more than 6 ft (1.8 m) in height, continuous brush, and slash. *Slash* (Fig. 12.4) is the material left over after logging a patch of forest. It includes branches, tree tops, stumps, and logs. Heavy fuels also burn at a moderate to high intensity, but tend to spread less rapidly than those in light or medium fuels.

Another way fire professionals describe fuel is by moisture content as a percentage of the weight of the same vegetation, dehydrated (Table 12.1). Fresh, young plants may have a moisture content of 300% the dry vegetation weight. This means that it is full of moisture and requires more preheating in order to

Figure 12.4 Slash is the material left over after logging a patch of forest. It includes branches, tree tops, stumps, and logs. Photo courtesy of the Oregon Department of Forestry; https://www.flickr.com/photos/oregondepartmentofforestry/14329363829

Table 12.1 Moisture content of various types of wildland fuels.

Moisture content (%)	State of vegetative development
300	Fresh foliage, annuals developing early in the growing cycle
200	Maturing foliage, still developing
100	Mature foliage, new growth complete and comparable to older foliage
50	Entering dormancy, coloration starting, some leaves may have dropped from their stems
30	Completely cured, treat as dead (dehydrated) fuel

Source: After Goodson & Adams, 1998.

become flammable enough to ignite. Mature foliage from late in the summer may have a moisture content of 100%. Moisture content falls to 50% for plants entering dormancy in the fall, as well as for plants that are suffering through a drought. Any plant with a moisture content of 30% or less is likely dead and is dehydrated. It takes less preheating before it ignites.

Distribution of fuel is also important in predicting how a fire will behave. *Fuel load* is a description of how much fuel (both alive and dead), exists. It is reported in tons per acre (or tonnes per hectare). Low fuel load may be less than 1 ton of fuel per acre (2.2 tonnes per hectare) if there are only patchy grasses on a rocky surface. High fuel load could be more than 500 tons per acre (1,100 tonnes per hectare) in a dense stand of shrubs or a mature forest. Embedded in this description of fuel load is the *fuel continuity*, which is how well connected fuel is within an area. Patchy areas of vegetation with low fuel continuity burn differently than do continuous areas of vegetation with high fuel continuity.

12.3.2 Weather

Weather, as described in Chapter 7, is the set of atmospheric conditions that exist in an area at a specific time. Weather is variable across time and space, and is monitored very closely by fire officials both before and after fire breaks out. Temperature, relative humidity, and wind speed and direction are the factors most important to the behavior of a fire. In general, the higher the temperature, the more flammable material becomes. This is because as the temperature increases it dehydrates vegetation, leaving it with a lower moisture content. This makes the vegetation more susceptible to ignition. Temperature is variable throughout the day. That means fire danger is also somewhat variable throughout the day. Fire danger is greatest during daytime hours when temperatures are highest and least during hours of darkness when temperatures are at their lowest.

Humidity is also important in assessing fire danger. The less humid the atmosphere is, the more likely vegetation is to lose moisture. As the vegetation loses moisture, it becomes more susceptible to ignition. Humidity varies daily, and just as with temperature, humidity influences the likelihood of ignition. In general, humidity is lowest during the day and highest at night. Therefore, it reinforces the effect of temperature in making daytime hours the most likely hours in which a fire will start or will experience its greatest rate of spread.

Temperature and relative humidity are often used to forecast fire danger. Fire danger is lower if relative humidity is high, temperature is low, and the area has received plenty of precipitation. Fire danger is extreme if the temperature is high, the humidity is low, and the area has not received appreciable rainfall in the preceding weeks or months. During times of extreme fire danger, red flag warnings can be issued if the added element of strong winds is present.

Winds are produced by a number of different sources: fronts moving through the area, convection, daytime heating, or thunderstorms. They are characterized by their speed, direction, and variability (such as gusts or rapid changes in direction). Often, topography, or the shape of the landscape, both creates local wind patterns and affects regional wind patterns. Various types of winds can affect fires in diverse ways.

Frontal winds are those produced by a strong atmospheric pressure gradient (change in air pressure over a distance). For example, consider the case of a cold front moving through an area. The cold air has higher air pressure than the warmer air mass it is encroaching upon. The result is a strong air pressure gradient across the frontal boundary. The greater the change in air pressure between the two masses, the stronger the winds will be. Arrival of a front can also cause wind direction and speed to change abruptly. This in turn can cause a shift in the direction and rate of fire spread.

Some wind is produced by differences in temperature across an area. Land breezes and sea breezes are examples of these types of winds (Fig. 12.5). In coastal areas, the land has great capacity to heat up during the daytime and cool down at night. The water, on the other hand, remains a relatively constant temperature. As the land heats up during the day, it becomes warmer than the water. Air above the land heats and rises. Cooler air is pulled inland from over the water to replace the rising air. The result is a convection cell in which the warm air over the land rises and moves seaward, while cooler air over the sea sinks and moves inland. This is a sea breeze.

In late afternoon, as the Sun is going down, the process is reversed. The cooler temperatures are now over land, so cooler air over land sinks and moves seaward. Warmer air over the ocean rises and moves inland. This is a land breeze. Together, land breezes and sea breezes moderate the temperature of coastal environments. Fires in coastal areas can change direction and speed with the time of day because of the effect of the land and sea breezes.

Another weather phenomenon that can affect fire ignition and spread is thunderstorms. Unstable air in the atmosphere often leads to cloud formation. That, in turn, can lead to the formation of thunderstorms. Thunderstorms carry with them strong and variable

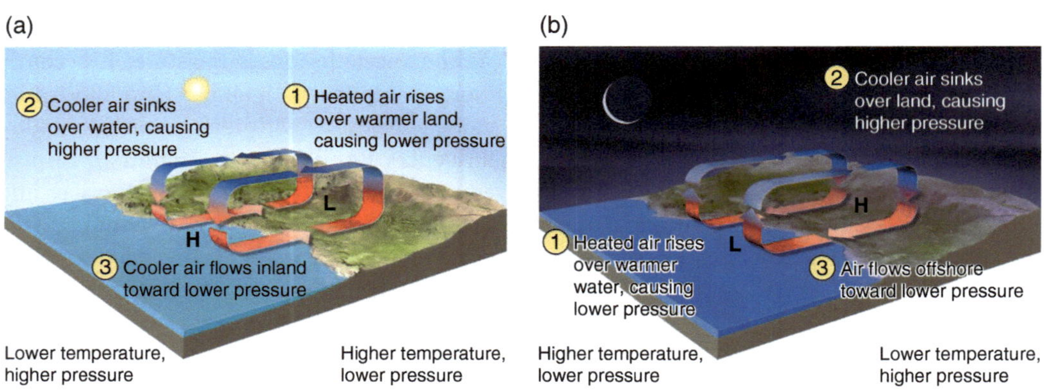

Figure 12.5 Land and sea breezes are caused by changes in temperature and atmospheric pressure in coastal areas. (a) Sea breezes occur in the daytime. The land is warmer than the ocean. Warm air rises over the land and moves over the water. There, the air cools and sinks toward the water. This creates a high pressure area over the water. Wind blows from the water toward the land during the day. (b) Land breezes occur at night. The ocean is warmer than the land. As a result, warm air rises over the water and moves over land. The air cools over the land and creates a high pressure area. Wind blows onshore as a result. From Arbogast, 2017.

winds. Those winds, generally speaking, move toward a thunderstorm as it is forming, and away from a thunderstorm as it is dissipating. As thunderstorms form, the wind speed increases and blows toward the approaching thunderstorm. Convection carries surface winds higher into the clouds via updrafts. When a storm becomes mature, it produces both updrafts and downdrafts. This means wind speed and direction change often since wind is moving toward the storm and rising in updrafts, and away from the storm and sinking in downdrafts toward the storm. Dissipating storms produce mostly downdrafts. Wind direction will largely be away from the center of the storm, and wind speed can be quite high. Thunderstorms have the additional danger of producing lightning. It is not uncommon for lightning from a strong, well-developed thunderstorm to move over an area and ignite dozens of fires.

Other atmospheric conditions can also lead to devastating fire. One of the worst grassland fires of recent years occurred in March of 2017 on the Great Plains as a result of a passing dry line. A dry line is a situation in which warm, dry air from Mexico meets moist air from the Gulf of Mexico. Grass on the plains had grown tall thanks to rains in the fall of 2016. However, the area had received very little rainfall for almost 6 months so it was extremely dry by the end of the winter in March 2017. On 6 March, a dry line developed above Texas, Oklahoma, and Kansas. The resulting weather included high winds and low humidity. Conditions were just right for a major wildland fire. As the dry line moved across the area, winds blew at 50 mi/hr (80 km/hr). Fires started in multiple locations. One fire started when a metal chain being dragged behind a truck produced sparks that ignited roadside grass. Multiple fires were started by electrical discharge from power lines that had been damaged in an ice storm in January. The causes of many of the fires are unknown. Fires moved quickly, fanned by the winds produced by the dry line (Fig. 12.6).

On 6 March 2017, firefighters across the panhandle of Texas arrived to fight multiple fires near Amarillo and Perryton. Seven people tragically lost their lives in those fires. Among them was a young cowboy named Cody Crockett, his girlfriend Sydney Wallace, who was a nurse at a hospital in Amarillo, Texas, and rancher Sloan Everett. They were killed as they worked to free livestock on the Franklin Ranch northeast of Amarillo so the horses and cattle had a chance to escape the flames. While thousands of cattle and horses died in the fires, their efforts saved hundreds of head of livestock. In all, over a million acres burned across the southern Great Plains that day. Oklahoma and Kansas each had about 400,000 ac affected by fire. Texas saw 325,000 ac burn.

Intense local heating creates a situation in which winds become variable and utterly unpredictable. In extreme cases, and when conditions are right, they can cause a phenomenon called a fire whirl, or fire tornado. Fire whirls are spinning columns of rising hot air and gases. They carry flames, smoke, and debris into the air and resemble small tornados made of fire. Fire whirls range from just a foot or two (0.3 to 0.6 m) in diameter to the size and wind speed of a small tornado. They can form on flat, open ground, but often form on the leeward (far) side of a ridge where convergent air currents cause a vortex to form.

Another hazard of wind, which cannot be ignored, is its ability to move firebrands or embers away from the fire. Firebrands can start new fires in front of an existing fire. They can help an existing fire jump fire lines and other barriers, causing the fire to spread very quickly. Firebrands were a major factor in the rapid movement of the Camp Fire.

Foehn winds are another major factor in some wildland fires. Foehn winds are sometimes called gravity winds. These can occur over relatively small areas, but are perhaps best known as regional wind patterns. The best known of the Foehn winds in the United States are produced in late summer and early fall as a result of a high pressure system over the Great

(a)

(b)

Figure 12.6 Images of the fires near the Franklin Ranch (March, 2017) near Amarillo, Texas, taken by Hoover, Texas volunteer firefighter Nathan Stillwagon. (a) The fire was fueled by high grasses that grew in the area after a particularly wet fall and winter. (b) Tragically, the fire burned quickly, overtaking wildlife, livestock, and people as well. Photos by Nathan Stillwagon.

Basin in Nevada. High pressure over the Great Basin forces air up and over the surrounding mountains. When the air is forced up in elevation, it cools and the atmospheric pressure decreases. Most moisture is dropped on the Great Basin side of the mountains. When the air moves over the crest of the mountains, it is dry and cool. It starts moving downhill. As it does, it becomes warmer and compresses, causing a strong atmospheric pressure gradient. The resulting winds flow outward in all directions away from the Great Basin. The winds are hot, dry, and persistent. The most famous of these winds are the Santa Ana winds of Southern California, the Diablo winds of Northern California, and the Chinook winds on the eastern side of the Rocky Mountains.

The onset of the Santa Ana winds each fall corresponds with the peak of the Southern California fire season. The area typically goes without appreciable precipitation during the hot summer. In the fall, vegetation such as the ubiquitous chaparral shrubbery (Fig. 12.7) is dry and primed for ignition. Any poorly placed spark from a power line, unextinguished campfire, carelessly discarded cigarette, or bolt of dry lightning can ignite a fire that quickly grows and spreads thanks to the powerful Santa Ana winds. Conditions on the ground are hazardous for firefighters because the Santa Ana winds are so strong. These winds also make it nearly impossible to use aerial methods for firefighting, such as water drops from helicopters or spraying of fire retardant by air tankers, because atmospheric conditions are too unstable and winds too strong and unpredictable. In addition, liquid droplets evaporate in the hot, dry conditions before they can hit the ground. Therefore, fires fanned by the Santa Ana winds and other similar Foehn winds are extremely difficult to fight.

12.3.3 Topography

Topography is the third major component of the fire behavior triangle. Topography includes variation in elevation and the steepness of slopes, both of which have an influence on how a fire is likely to behave. Topography is the most predictable of the three aspects of

Figure 12.7 Much of the wildland areas of southern California and northern Mexico are covered in chaparral, a shrub-dominated ecosystem. These areas burn readily when the Santa Ana winds and local dry, hot conditions merge to create the right conditions. This photo was taken at Cajon Pass, California, after a wildland fire swept through the area. Photo © Marli Miller, used with permission.

the fire behavior triangle because it does not change with time like the weather, and it does not have the ability to change moisture content and flammability like vegetation.

The most basic aspect of topography is the elevation above sea level. Each elevation has its own characteristic vegetation and weather. Temperature and relative humidity vary with position on the slope. The direction a slope faces is called its *aspect*. Daytime heating is greatest on slopes that face the Sun, causing those slopes to have higher temperatures and lower humidity. This makes them more susceptible to ignition. In the Northern Hemisphere, the south and southwest aspects are most likely to face that increase in temperature.

In mountainous areas, the temperature typically drops with increasing elevation. However, there are often inversion layers in the mid-slope, which means that there may be a band of relatively warm temperatures and lower relative humidity at a higher elevation. The presence of an inversion layer can influence fire behavior on the slope by creating a zone of vegetation that is already somewhat preheated and more susceptible to ignition.

In the absence of significant wind, fires tend to burn much faster as they move up slopes than when they burn downhill. This is partially because of the nature of fire to climb upward. This is easy to demonstrate in a simple experiment. When you light a match and hold it upright, the flame will move down the match, but it will do so quite slowly. However, if you hold the match at an angle so the match head is downhill, the fire will climb up the match rather quickly.

Topography can also help create air currents that help drive a fire, even when there is no large influence from regional winds. A fire heats the atmosphere around it, causing hot air to rise due to convection. On a steep slope, this drives the hot air up the slope, creating upslope winds and updrafts. The hot winds moving quickly up the slope preheat the vegetation above through the process of radiation. This spreads the fire quickly upward, continuing all the way to the crest of the slope where cooler air and a change in vegetation and slope slows the movement of the fire at those higher elevations.

Narrow canyons and ravines can affect fire behavior in a multitude of ways. If one side of a narrow canyon catches fire, radiant heat from the fire will preheat the other side of the canyon and make it highly susceptible to ignition. Ignition will occur quite rapidly thanks to the upslope winds blowing firebrands. In addition, canyons can create a chimney effect that causes the fire to move rapidly. Hot air is funneled upward very quickly through the narrow gap between the sides of the canyon. This causes rapid preheating and incredibly fast ignition of the fuels on both sides of the canyon.

12.4 Firefighting in Public Lands

The United States is no stranger to wildland fire. Each year it seems that fires become more severe and destroy more property than ever before. The causes for the rising toll of fires are many. Because of climate change, much of the western United States is becoming warmer and drier. This makes the West more prone to wildland fires. The southeastern United States has also seen an uptick in the number of fires, and in losses caused by those fires.

Many more people are also choosing to move into areas that were once reserved as natural areas. Their homes and businesses now sit at the wildland/urban interface, where homes are now at risk when wildland fires burn. This makes fires more expensive, and more hazardous to human life because many more people are in harm's way than in the past.

In addition, many decades of firefighting policy on public lands have created a situation in which there is an overabundance of fuels. The lack of regular fires in forested areas has led to a proliferation of ladder fuels, which makes it possible for fires to become catastrophic, burning fuels from ground level to the crowns of trees, and spreading quickly from one tree to the next.

Native Americans used fire for thousands of years to control their environment. They understood that fire was a vital tool in management of the ecosystem. They engaged in controlled burns and kept forests clear of debris, kept grasslands productive and well suited to the grazing herds they hunted, and reduced the likelihood of catastrophic wildland fires. Termed *cultural burning*, Indigenous people routinely set fires to support the growth of certain types of vegetation they used for medicine and food, as well as materials for string and baskets and those used for shelter. Prior to the late nineteenth century, these practices kept the environment healthy.

Fires that were set naturally through lightning strikes and by Native Americans via cultural burning kept wildland areas healthy. In forests, leaf litter and discarded pine needles on the forest floor burned, as did small trees and scrub brush in the understory of the forest. This regular burning of low-level fuels eliminated much of the vegetation that could grow into ladder fuels. This eliminated the likelihood of more intense, fast-burning and catastrophic forest fires. The natural life cycle of vegetation in the forest was also preserved. Many plants also require fire in order to allow growth of a new generation. Several species of trees require fire in order to release seeds from cones, or to weaken the hard outer shell of a seed. In essence, fire has always been part of the natural order of things, and for centuries was managed effectively by Indigenous people.

Starting in the late nineteenth century, the United States adopted an official policy to fight wildland fires wherever and however they began. In doing so, the government fought not only fires that threatened populated areas, but also the natural fires that were vital to the health of the ecosystem, including those that burned far from population centers. The result of the government's decades-long policy was a dangerous buildup of fuel. Fire professionals often speak of the pay-me-now or pay-me-later nature of fire. In essence, this means that fire is inevitable. The choice then becomes either allowing natural-set, cultural burns, and prescribed fires to burn within safe boundaries that firefighters work to establish and maintain, or suppressing all fire until the fuel load is immense and any fire becomes too large and severe to control.

It was not until the late 1960s that the United States changed its policy on firefighting in wildlands. Currently, many wildland fires that are set by lightning strikes are allowed to burn, within certain parameters (Fig. 12.8). These fires that are allowed to burn are termed *prescribed natural fires*. Firefighters often determine a perimeter that they need to keep the fire within. They clear fire breaks and monitor the perimeter to be sure the fire does not burn outside the intended area. Areas that would not typically be allowed to burn would include homes, businesses, national park or U.S. Forest Service buildings, and historic sites. Many human-caused fires are still fought, since they are often set in nonideal conditions, and near vital infrastructure.

Yellowstone National Park is a prime example of a landscape affected by the lack of natural fires for nearly 100 years. Fire had been suppressed within the park from the late 1880s until the early 1970s. As a result, much of the forested area within Yellowstone was overloaded with fuel. Yellowstone had more than 200 prescribed natural fires in the years between 1972, when the park first adopted the policy of letting prescribed natural fires burn, and 1988. In the mid-1980s, Yellowstone had about 5 years of above-average precipitation, allowing vegetation to grow tall and healthy. As late as spring of 1988, precipitation levels were near normal. However, drought conditions took over in the early summer of 1988. Many individual fires started. Nine fires were human-started, and 42 were caused by lightning. Some fires in the park caused by

Figure 12.8 Fire is part of the life cycle of many trees. Photo by Kent Dennen/Photo Researchers, Inc. From Strahler, 2013.

lightning were allowed to burn, and were not suppressed until later in the summer when the fires became dangerously close to visitor areas. Spot fires occurred throughout Yellowstone where firebrands were carried on the wind to new patches of forest, at times up to a mile from the main fire. Some spot fires combined into much larger fires. In September of 1988, Yellowstone closed to the public for the first time in its history as upwards of 9,000 firefighters battled the flames that had burned out of control. A total of about 794,000 ac (321,320 ha) were burned (Fig. 12.9).

About 36% of Yellowstone National Park burned in 1988. However, several fires also started outside but adjacent to the park. Those constituted about 63% of the total area burned, or approximately 500,000 ac. Many patches of forest within Yellowstone National Park and its surrounding landscape were severely damaged and even today host stands of dead, burned trees. Several areas were rejuvenated by the fires and have since come back healthier than ever.

In addition to allowing naturally set fires to burn, government fire scientists also participate in actively burning areas on public lands in an effort to safely reduce fuel load. These are called *prescribed fires*. They are carefully planned and intensely monitored in an effort to keep them contained. Prescribed fires can restore ecosystems to their natural state, resulting in healthier plant and animal communities. They also reduce the amount of fuel available to an unplanned fire later on.

Unfortunately, there have been a small number of incidents in which prescribed fires have gone terribly wrong. In May of 2000, a prescribed fire was set at Bandelier National Monument in northern New Mexico. After the fire was set, the wind changed speed and direction, and firefighters lost control of the fire. It became known as the Cerro Grande Fire (Fig. 12.10). The fire burned approximately 150,000 ac and 400 homes. It also threatened the Los Alamos National Laboratory, the premier research facility of the Department of Energy. People were rightfully concerned about the threat to the national laboratory because a great deal of research on nuclear materials is still done there. The facility is secure, and much of its research is classified,

Figure 12.9 Photos of burned areas in Yellowstone National Park. Forty-five individual fires raged in Yellowstone National Park and the surrounding area in the late summer of 1988. They were fueled by dense vegetation in areas that had not burned in many decades. (a) Many acres of lodgepole pines were killed by an intensely hot crown fire. (b) Ten years after the fire, a new pine forest was beginning to grow. (a) Photo courtesy Jonathan Blair/© Corbis; (b) M. P. Kahl/Photo Researchers, Inc. From Strahler, 2013.

Figure 12.10 Aerial image of the Cerro Grande fire near Los Alamos, New Mexico in 2000. This fire started as a prescribed fire in Bandelier National Monument, but quickly grew out of control when weather conditions changed. Photo © *Albuquerque Journal,* used with permission.

so the firefighters involved in battling this fire were uncertain of the type and volume of fuels they might encounter should the fire enter the campus of the laboratory. In the end, a few outbuildings on the campus were burned, but the fire did not breach the main part of the property. Thankfully, there was no loss of life.

Wildland fires can burn almost anywhere there is vegetation. California often garners most of the country's attention since fires there tend to be large and grow out of control quickly. However, wildland fires occur in nearly every state each year, even near populated areas. A large fire in 2011, for example, burned on the outskirts of Austin, the capital city of Texas, near the suburb of Bastrop. Dry vegetation in the late summer was ignited by sparks, and the result was a wildland fire that burned for over a month. The fire destroyed over 1,600 homes and killed two people.

12.5 Mitigation of Fire Hazards

There are a number of things federal, state, and municipal governments can do to reduce the risk of catastrophic wildland fires. For instance, after decades of fire suppression, prescribed burns and cultural burns are again becoming more commonplace. By setting fires under controlled circumstances, stewards of the land have the ability to reduce the hazard of catastrophic wildland fires by clearing out unwanted ladder fuels. In addition, prescribed burns can kill invasive plants and help reestablish native plants that require fire to reproduce effectively. The Bureau of Indian Affairs is assisting national parks and other public land managers within the United States to understand Indigenous use of fire and implement many of those policies to the benefit of the ecosystem.

Fire scientists have also designed and implemented fire modeling. The most advanced of these models use digital topographic maps and real-time temperature, humidity, and wind data to forecast the path and speed of a fire. This allows firefighters to position their resources in optimal locations and create fire breaks ahead of the advancing fire. It also provides local officials with enough specific information about the fire and its likely path to call for evacuations in a timely manner. These fire models have great potential to save lives and property.

Local officials in areas prone to fire should also keep citizens aware of the dangers of fire by supplying citizens with information about their fire risk, communication they can expect during fire season, and evacuation routes. Governments should also assure that there are sufficient evacuation routes leading away from populated areas. The city of Paradise, California, learned this lesson in the most tragic way possible when narrow evacuation routes became jammed with traffic and people in their cars perished as the fire overtook them (see opening case study). Assuring that neighborhoods in fire country have access to fire hydrants, wide streets, and functional warning systems in place can save lives. Wide streets with 100 ft (30.5 m) radius turns in neighborhoods allow plenty of space for fire trucks to reach the neighborhood and maneuver into the best position for fighting fires. It is critical for a city to assure that there are working fire hydrants close enough to homes to be useful.

There are things homeowners can do to make their homes somewhat safer in the face of wildland fire as well. Since hazards are highest in high-fuel areas, the first line of defense is to not build in the wildland/urban interface where fuel loads are high, roads are steep and narrow, and few escape routes are there. Even in relatively fuel-rich areas, it is possible to make a home safer. One way to do so is to reduce the fuel load around your home. Fire officials urge homeowners to create a defensible space around their homes, which significantly lessens the fire hazard. In the area closest to the home (30 ft or 10 m from the home), plant short grass and keep it mowed and watered. Eliminate any dry or dead brush and plant trees and other larger plants far apart so fire does not readily spread from one to the next. Remove any firewood to at least 100 ft (30.5 m) away from the home. Keep at least 10 ft (3.3 m) between trees and any chimneys at the house. Keep a spark arrestor (cap) on the chimney at all times. In the area between 30 ft and 100 ft (9 and 30.5 m) from the home, make sure trees are an appropriate distance apart, and that there are no shorter shrubs planted near trees. This eliminates ladder fuels and prevents fire moving from one tree to the next.

In addition, it is a good idea to ensure there is appropriate and easy-to-read signage on your street and around your home. Large, reflective numbers on the home or in the yard allows firefighters to readily locate your home in the event of a fire.

Home construction practices are also important. Roofs should be made of nonflammable material. Wood shingles may look attractive, but they are also easy to preheat and ignite. Fire-resistant siding such as brick, stucco, or masonry make the home safer. Do not leave open space under the eaves that leads into the interior of the house. Firebrands can be blown under the eaves and can enter the attic of a home if these areas are left open. Use double-pane windows in the home. The heat from a fire can shatter single-pane windows. The extra pane in the windows affords a modest amount of protection from the fire breaking through the window barrier and reaching the inside of the home.

Do not build a home overhanging a slope. Both convection currents and radiant energy from fires below cause these structures to become primed for ignition. As hot wind sweeps the fire uphill, the house is in danger from almost certain incineration. Do not build decks overhanging a slope. As fire burns up a slope, it can catch decking material on fire. If the deck is attached to the house, this will lead immediately to fire in the home. If you are building on a sloped lot, it is always best to build a home at the top of the lot, on a level space that is set back 100 ft (30.5 m) from the edge of the slope. This will prevent both convection and radiation from preheating the home and pushing it closer to the ignition point. Trim vegetation on the slope below the home as well, to prevent fire from climbing so easily uphill. Creating wide spaces between trees and shrubs can be an effective ways of slowing down fires that are burning up a slope.

12.6 Conclusion

Wildland fire is a major hazard that varies with season, weather, fuel availability, topography, and climate. Wildland fire has always been a part of the natural world, and is part of the life cycle of a many types of vegetation. Hazards to people exist when they build homes within the wildland, or at the wildland/urban interface where fuel is plentiful and the topography and weather are conducive to fire at least part of the year. Climate change is causing many areas to become warmer and drier, leading to increased risk of catastrophic fires. Changes to climate in California, for example, have led to historic droughts and record-breaking fires for each of the last several years. To protect themselves and their homes, homeowners must make themselves aware of mitigation strategies that limit the fuel available to fires that may burn in their neighborhoods. As climate continues to change, the potential increases each year for deadly and widespread fires in regions previously not affected.

12.7 Chapter in Review

The Camp Fire in Northern California was a devastating fire that burned for a month in November 2018. It raged through multiple communities, including the city of Paradise, which had 27,000 residents before the fire. It spread and moved incredibly quickly, trapping many people who tried to escape. The fire burned more than 180,000 ac and caused 85 deaths. The fire was likely caused by sparks from a power line. The power company, PG&E, declared bankruptcy after it was clear that it would be faced with billions of dollars in lawsuits from those affected by the fire.

Wildland fires are fires occurring in natural vegetation and fuel. They burn hundreds of thousands of acres each year. Wildland fires are becoming larger and more deadly in some areas due to climate change that leaves many areas prone to drought, higher temperatures, and lower humidity for long stretches of the year.

Fires burn by combustion. Combustion of vegetation is in many ways the opposite of photosynthesis. Photosynthesis converts water, carbon dioxide, and sunlight into cellulose and oxygen. Combustion converts oxygen, cellulose and other plant components, and heat into carbon dioxide and water, and solid residue called char. The process of burning an object starts with preignition or preheating. During this phase, the fuel is dehydrated. In addition, waxes and oils are volatilized (turned into a vapor). During the process of pyrolysis, volatiles migrate to the surface of the fuel and then are ignited so they burn.

Heat transfer occurs in three ways. The first is radiation, which is when heat is transferred by radiant energy. As radiant energy moves in advance of the fire, it preheats fuel and dehydrates it, making it more likely to burn. The second way heat is transferred is by convection, which is heat transfer via the movement of gas or liquid. Convection can create wind, updrafts and downdrafts, and sea and land breezes. Air movement helps spread fires. Conduction is the third method of heat transfer. Conduction involves the direct transfer of heat by contact. This is less of a factor in the spread of wildland fires than radiation and convection, but is important in the case of firebrands. Firebrands are glowing embers that are lofted away from a fire and blown downwind. They can ignite new fires, which is called spotting.

Fire requires fuel, heat, and oxygen to burn. Scientists use a diagram called a fire triangle to describe the interdependence of these elements in creating and sustaining a fire. If one leg of the fire triangle is removed, the fire can no longer burn. Heat can be removed by dousing a fire with water. Fuel can be removed by creating a fire line, an area cleared of vegetation down to the bedrock or mineral soil layers. Oxygen can be removed by coating fuel with fire-retardant material or throwing dirt on a fire.

The fire behavior triangle is used by scientists to describe the factors that control how a fire behaves: weather, topography, and fuels. Of these three, only topography does not change as the fire burns. Weather can occur on a large scale, such as with Foehn winds, which are hot, dry winds that blow outward from the Great Basin. It can also occur on a small scale, as when the temperature and humidity of a slope change throughout the day due to the effect of the Sun. Fuels affect fire behavior in many ways. The amount of moisture in fuels controls how long it takes to preheat the fuel to the point where pyrolysis and ignition take place. Whether the fuel is patchy or dense affects how quickly the fire can spread. Ladder fuels cause a fire to burn quickly as well. Grasses that lead upward to shrubs then to trees create fuel at every level within a fire, causing it to burn very intensely and spread quickly. Topography, while it does not change, can affect fires because weather is different at various elevations on a slope, and convection can carry hot wind up a slope and preheat fuel quickly. Fire also naturally tends to burn uphill very quickly, and much more slowly downhill. This makes communities on slopes particularly vulnerable.

Prior to the late 1800s, Indigenous people managed the ecosystem across the United States through the use of cultural burns. These cultural burns were prescribed fires set under controlled conditions that cleared underbrush from forests and kept prairies healthy. These cultural burns encouraged the growth of native vegetation the Native Americans used for medicine, food, fibers, and shelter, and which animals used for food.

From the late 1800s to the 1960s, the policy of the U.S. government was that all wildland fires should be fought, no matter the circumstances or location. Unfortunately, this had two negative effects. One was that it interrupted the natural life cycle of trees that depend on fire to free seeds from cones or weaken seed casings. The other is that it allowed a dangerous amount of fuel to build up in forests, creating conditions that were favorable for intense fires that did not help the ecosystem but harmed it. In the 1960s, the government changed its stance on wildland firefighting, and now allows prescribed natural fires (those started naturally in isolated public lands) to burn within certain parameters. They are fought only when approaching human-made structures, historical sites, or critical infrastructure. Fire scientists also set prescribed fires, which clears out underbrush and promotes a healthy ecosystem. Indigenous people are also being invited into the conversation to talk about cultural burns and traditional management of ecosystems.

Mitigation of fire hazards takes many forms. First, prescribed fires and cultural burns should be used to reduce fuel loads and ladder fuels. Fire modeling can help determine the likely speed and direction a fire will burn, allowing firefighters to position resources and create fire breaks ahead of an advancing fire. Models also give municipalities actionable data that allow them to warn and evacuate people.

Cities and towns can mitigate fire hazards by assuring that citizens have wide streets. These wide streets can be used more effectively than narrow roads for both evacuation and for firefighters maneuvering fire trucks to most effectively fight fires. Working fire hydrants close enough to homes to be useful are also important.

There are a number of steps individuals can take to mitigate fire hazards in their homes and communities. First, if possible people should avoid building homes in the wildland/urban interface. For those that do build in these areas, they should not build homes or decks that overhang slopes, since fires tend to burn quickly up slopes. Homes should be set back at least 100 ft (30 m) from the edge of a slope to discourage radiant heating, and the slope should be cleared of as much vegetation as possible to slow the spread of a fire. Roofing should be made from fire-resistant material. A home site should be cleared of tall vegetation a certain distance away from the home itself, and only widely spaced vegetation should be planted slightly further away from the home. This reduces both the amount of fuel and the distance between fuels so wildland fires have a chance to bypass the homesite.

Part III

Other Hazards

13

Biological Hazards With COVID-19 Case Study

Anni Moore[1] *and Austin Nuxoll*[2]

[1] *Morningside University, USA*
[2] *University of Nebraska at Kearney, USA*

13.1 Biological Disasters

Many of Earth's natural disasters, such as earthquakes or tornadoes, can occur with minimal or no warning, last for a short time, and the destruction and devastation they cause can be seen almost immediately. On the other hand, the beginning of many biological disasters can remain undetected for quite a while, they can last for long periods of time, and the death toll can take years to finalize. Depending on where you live, you may never get to see a volcanic eruption or experience a hurricane, but every person in the world recently experienced several years of a global pandemic, the effects of which will be felt for years to come. Biological disasters differ from other natural disasters in several ways, but most importantly because they are caused by living microorganisms that can travel from one part of the world to another by the same means and in the same speed as people, having the capacity to replicate and cause disease anywhere they go. The pandemic of COVID-19 is an excellent example of how human interaction with nature, global travel, and insufficient preparedness resulted in a biological disaster that has left a lasting effect on society.

13.1.1 Case Study: COVID-19

In December 2019, reports of a mysterious pulmonary disease began surfacing out of China. This unknown disease was initially characterized by a dry cough, shortness of breath, and fever. Occasionally, gastrointestinal symptoms were also associated with this new illness. Fever was the most consistent symptom followed by cough with ~90% and ~80% of patients initially demonstrating these two symptoms, respectively. Some progressed to viral pneumonia, which was recognized by the World Health Organization. As this disease spread across the world, we would later find a much wider array of associated symptoms (discussed later in this introduction). By the end of 2019, the Chinese government reported 44 cases largely associated with the city of Wuhan. It is assumed many early cases were missed and the number of illnesses caused by this emerging virus was underestimated.

On 7 January 2020, scientists identified that these mysterious pulmonary infections were caused by a novel coronavirus and named the virus *SARS-CoV-2* (Fig. 13.1). The virus causes the disease *COVID-19*. *Novel* means new or not previously known: it had not previously been

Earth's Natural Hazards and Disasters, Advanced Textbook 7, First Edition. Bethany D. Hinga.

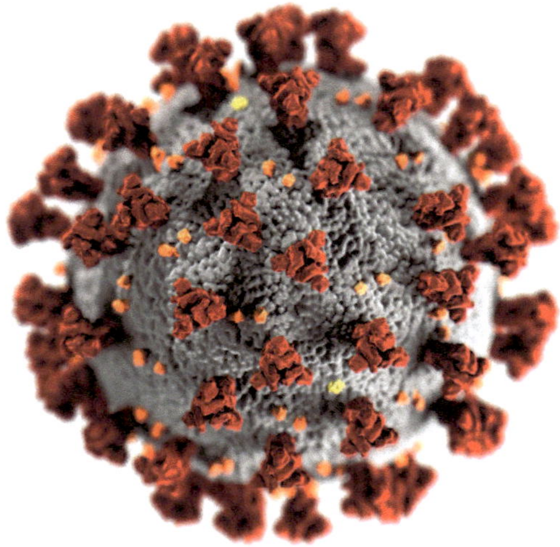

Figure 13.1 Colorized image of a coronavirus, as might be seen under a scanning electron microscope. Image credit Centers for Disease Control Image Library; https://www.cdc.gov/media/subtopic/images.htm

seen as a pathogen in humans. *Coronavirus* is a family of viruses, some of which are responsible for respiratory illnesses. A few days later, on 11 January, the first death was reported in Wuhan, China, a 61-year-old man. A couple of weeks following the first death associated with this pandemic, Wuhan, a city of 11 million people, was cut off from the rest of the world. Public transportation within the city was suspended. This included buses, ferries, and trains. Additionally, no travel into or out of the city via trains or by air was allowed. These actions were taken to stop the spread of this unknown disease. However, by the time these preventative actions were taken, at least 570 people were confirmed to have the disease in Thailand, South Korea, the United States, Taiwan, and Japan.

On 15 January 2020, a 35-year-old man from Snohomish County, Washington, returned from visiting family in Wuhan, China. Four days after his return, he visited an urgent care clinic presenting with a cough, fatigue, nausea, and vomiting. Based on his travel history and symptoms associated with SARS-CoV-2, the patient underwent a nasal swab. Following a positive test, the patient was admitted to an isolation unit for further observation. On day two of the hospital stay, the patient developed diarrhea and abdominal pain. During his hospital stay, treatment was largely supportive. He received saline, acetaminophen, ibuprofen, guaifenesin (medication for cough and congestion), and ondansetron (used to alleviate nausea and vomiting). Five days post hospitalization, the patient's condition worsened and he developed pneumonia with oxygen saturation levels falling to 90% (above 95% is considered normal). The patient was given supplemental oxygen and antibiotics vancomycin/cefepime to prevent secondary bacterial infection. Remdesivir, an antiviral drug that inhibits RNA-dependent, RNA polymerase, was administered on day seven. The following day his condition started to improve and by day ten, only a cough remained. Though the first known U.S. patient recovered, this was the beginning of one of the worst pandemics in modern history, second only to the Spanish Flu pandemic in the early 1900s.

13.1.2 What Are Coronaviruses?

Coronaviruses belong to the Orthocoronavirinae subfamily of viruses. These viruses get their name from the crown-like spikes (glycoproteins) on their surface. The ribonucleic acid (*RNA*, structurally similar to DNA but is single-stranded and contains the base uracil instead of thymine) genome is an important feature, as this genome type is more susceptible to random mutations that can alter virulence and transmission properties of the virus. Furthermore, new mutations make it harder for the host immune system to recognize the virus, which results in reinfection occurring. This was observed in COVID-19 because a number of *variants* (mutated forms) began appearing, with increased levels of transmissibility. Two of these are the Delta variant and the Omicron variant, both of which were responsible for more illnesses and deaths than the original virus. As SARS-CoV-2 persists in high numbers in the population, the likelihood for further variants remains high.

There are hundreds of known coronaviruses; however, most do not infect humans. Occasionally a *spillover* event will occur where these viruses will jump from an animal reservoir such as pigs, bats, cattle, and so on, to humans and cause disease. There are seven coronaviruses known to infect humans, with three known to cause moderate to severe respiratory illness. In 2002, Severe Acute Respiratory Syndrome (SARS) appeared and infected over 8,000 individuals, killing 774 by the end of 2003. No new cases were reported since 2003 and the World Health Organization (WHO) declared an end to the outbreak. Middle Eastern Respiratory Syndrome (MERS) emerged in 2012. It is often associated with travel to the Arabian Peninsula, and has been responsible for over 850 deaths since it emerged. Both of these coronaviruses have a relatively high mortality rate compared with SARS-CoV-2, which causes COVID-19, although transmission from person to person is much lower. SARS and MERS are estimated to have around 10% and 35% case fatality rate, respectively. This is much higher than the current estimated mortality rate of COVID-19, which ranges from less than 1% to 5% depending on the region examined.

13.1.3 What Symptoms Are Associated With SARS-CoV-2?

One aspect of COVID-19 that makes it challenging to diagnose without molecular testing is the wide array of symptoms that the infected individual can display and the similarity to influenza, allergies, or the common cold. The average incubation time tends to be 4–7 days but can be over 10 days in a small number of individuals. Infected individuals can range from being asymptomatic or with mild symptoms, to presenting with life threatening respiratory issues that may progress to death. Roughly 80% of infected individuals do not display symptoms or display only mild symptoms (low fever, coughing, fatigue, etc.). Commonly seen symptoms include but are not limited to cough, fever, muscle pain, sore throat, gastrointestinal symptoms, fatigue, headache, loss of smell, loss of taste, congestion, chest pain, and difficult or labored breathing. The high percentage of mild or asymptomatic cases contributes to the high rates of community transmission. Many individuals are unaware they have the virus and continue to go on with their normal daily routine, thus contributing to the spread of the virus. This also leads to an underreporting of the disease, and the number of cases is likely much higher than confirmed cases being reported.

13.1.4 Epidemiology

Since the beginning of the pandemic, the United States remains the country hit hardest by the virus and leads the world in confirmed cases. By March of 2023, there were over 104 million confirmed cases and approximately 1.12 million deaths attributed to the virus in the United States. China reports 99 million confirmed cases. India and Brazil have also experienced widespread infection with nearly 45 and 37 million cases, respectively. Within the United States, California, Texas, and Florida have reported the most cases of SARS-CoV-2, which largely follows population trends. A more accurate picture is illustrated by looking at per capita cases. The Midwest, followed by the southeastern United States, tend to have the highest rates of confirmed SARS-CoV-2 cases per capita. A number of factors contribute to these trends, including lower vaccination rates, relaxed mask policies, and comorbidities in these populations. Age is a key factor associated with the mortality rate of the virus with most deaths observed in those individuals over 75-years-old followed by the 65–74 age group. Males have an increased risk of dying from the virus compared to females. Men have approximately 1.5 times the death rate than that of women. Other risk factors leading to increased mortality rates include cardiovascular disease, diabetes, chronic respiratory disease, hypertension, and cancer. This is by no means a comprehensive list of risk factors. More research will likely uncover more factors. For example, researchers have identified a reduced risk of mortality associated with blood type O compared with other blood types.

13.1.5 Origins

The origins of the SARS-CoV-2 virus remain a point of debate among experts. Most evidence points to a market in Wuhan as being the source of the novel coronavirus, and where patient zero was infected. The SARS-CoV-2 virus was suspected to have originated in bats. From there, it hopped to an intermediate host (originally thought to be a pangolin) before jumping to a human host. Recent research suggests the pangolin was not the intermediate host due to low binding affinity of the ACE2 receptor. It still remains unknown how the virus made its way to humans. However, the prevailing hypothesis is the virus indeed emerged from a bat reservoir. Bats are a natural reservoir of coronaviruses and genome sequencing of coronaviruses in bats provides support to this hypothesis. There is a fringe hypothesis that the SARS-CoV-2 virus originated in a lab in China although evidence for this is lacking. Most scientists believe this unlikely. However, most hesitate to completely discredit it until details of the spillover event are confirmed.

13.1.6 Vaccines

As of January 2022, three COVID-19 vaccines have been FDA approved. The Pfizer-BioNTech and Moderna vaccines are *mRNA vaccines*. The mRNA (messenger RNA) is a type of RNA that is used as a template by the cell to synthesize proteins. The Johnson & Johnson's Janssen vaccine is a *viral vector vaccine*. All three vaccines work by using the host cells' machinery to produce a viral protein called the *spike protein*. Our immune systems will recognize the spike protein and subsequently produce antibodies to fight off future infections. None of the COVID-19 vaccines have the potential to interact with our DNA. There have been concerns raised by the public regarding the mRNA technology, as this is the first vaccine that has been developed and approved for human populations using this technology. The primary advantage of this technology is the increased speed of vaccine development compared with other traditional methods while producing the same outcome, stimulation of the immune system. The mRNA vaccines also tend to be one of the safer ways to produce vaccines, as there is no risk for infectious virus particles to be produced. The mRNA from the vaccine never enters the nucleus, therefore does not encounter our DNA. Concerns and claims about the vaccine staying in our body for extended periods of time are also unsupported. The mRNA is degraded within a few days and the spike protein is eliminated from the body after a few weeks. Many concerns regarding the speed at which these mRNA vaccines were generated have been voiced. The mRNA vaccines have been investigated and studied for decades and for several other viruses including coronaviruses, influenza, rabies, Zika, and cytomegalovirus. The first of these was tested with influenza virus in the 1990s. These mRNA viruses had potential, but the mRNA transcript was degraded prior to being translated into protein. Recent advances in nanotechnology stabilized the SARS-CoV-2 mRNA transcripts by protecting them in fatty droplets until the mRNA could be translated into spike protein.

Vaccine development takes 7 years on average. Due to the nature of the pandemic and the critical need for a vaccine, the timeline for vaccine approval was altered. A vaccine usually progresses in a series of steps. Approval must be granted at each step before moving forward to the next. With the COVID-19 vaccines, these approval steps were completed in a parallel fashion. It is important to note that no steps were skipped and, instead, were just altered to shorten the time frame for approval. In December 2020, the FDA granted the Pfizer and Moderna vaccines Emergency Use Authorization

(a) (b)

Figure 13.2 Cyanobacteria have been present on the Earth for billions of years. (a) A scientist exposes cyanobacteria to ultraviolet light in the laboratory. This causes the photosynthetic chlorophyll in the specimen to fluoresce red. (b) Stromatolites are mats of cyanobacteria and sediment. Most stromatolites are known from the fossil record, but these rare living specimens are found in Shark Bay, Australia. From Babcock, 2009.

(EUA). The vaccines continued to be evaluated for safety following approval and have been generally safe aside from a small number of allergic reactions and side effects often associated with other vaccines (injection site soreness, fatigue, headache, fever, and muscle pain). The vaccines were highly effective in preventing illness early in the pandemic but, as variants arose, the vaccine had a decrease in preventing illness. However, the vaccines have significantly reduced severe disease, hospitalizations, and death from COVID-19 infections. There are currently variant-specific vaccines available, which are further reducing the number of severe illnesses from COVID-19. The pandemic was declared over in early 2023, though the disease has not disappeared.

13.2 Introduction to Microbes

Microbes have existed on the Earth longer than any other living organism and helped to shape the living conditions that we have today. For example, it is only thanks to cyanobacteria (small photosynthetic bacteria, which used to be called blue-green algae) that we have oxygen in the atmosphere in the first place (Fig. 13.2). In fact, it is estimated that 20% of atmospheric oxygen, or one of every five breaths a person takes, is produced by just one genus of cyanobacteria, *Prochlorococcus,* which is one of the most numerous organisms on the Earth.

In addition to oxygen, bacteria are also responsible for producing most of the biologically available nitrogen for all other organisms. Nitrogen is a building block for basic biomolecules of life (such as DNA and proteins). Nitrogen is also abundant in the Earth's atmosphere. About 78% of the Earth's atmosphere is made of nitrogen, but it is useless for most cells in its gas form. The only organisms that can take atmospheric nitrogen and fix it into a biologically available form (ammonia) are nitrogen-fixing bacteria in the oceans and in the soil.

For the first 2 billion years after life first arose approximately 4.2 billion years ago, small unicellular organisms (bacteria and archaea) ruled the Earth. They depended on the physical world for their energy needs and synthesized their own biomolecules. However, as larger multicellular life forms (plants, animals, and multicellular fungi) appeared, many microbes evolved to obtain nutrients from these new organisms instead, which is simpler and requires less energy. Some new hosts did not appreciate microbes scavenging their nutrients. As a result, they reacted adversely to microbes. This adverse reaction is what we call a *disease*.

In this chapter we will look at biological hazards and disasters caused by microbes. It is important to understand the different types of microbes, how they cause harm to people, animals, and plants, and what the consequences are. We will also look at how microbes interact with Earth systems: the solid Earth, hydrosphere, and atmosphere. Since microbes themselves are part of the biosphere, we will explore the factors that make rest of the biosphere, especially human populations, vulnerable to diseases, and finally look at strategies that can be used to mitigate these biological disasters.

13.3 Types of Microbes

It is important to understand the basic differences between different kinds of microbes. There are seven basic types of microscopic agents: small multicellular *parasites*, *protozoans* (including algae), *fungi*, *archaea*, *bacteria*, *viruses*, and *infectious molecules*. While all microbes are capable of causing biological disasters (whether it is a human disease outbreak, crop failure, or economic loss due to some other factor), the mechanisms and dynamics of how, where, why, or how fast these events happen and how to control them are specific to each type of microbe.

Before we look at each type of microbe and the disasters they cause more closely, there is some common terminology that needs to be understood. Most of these terms pertain to disease outbreaks.

13.3.1 Pathogen, Host, and Vector

A *pathogen* is any organism that causes disease. A *host* is the organism that will harbor the pathogen; the host can, but does not necessarily, get sick. If an animal naturally harbors a pathogen, we call that animal a reservoir host. For example, bats are natural reservoirs for many viruses, including SARS viruses, various encephalitis viruses, and rabies virus, but these viruses usually do not cause a serious disease in them.

Sometimes a pathogen needs two kinds of different host animals to complete their life cycle. In this case the hosts are usually called primary hosts and secondary hosts. For example, *Plasmodium*, the malaria parasite (Fig. 13.3), needs both the mosquito as well as an animal (such as a human) to complete its life cycle.

Some disease agents depend on another organism to transport them from one host to another. The transporting organism is known as a *vector*. The vector itself does not get sick, and the pathogen does not use the vector to complete its life cycle. A vector is frequently a blood-sucking insect, such as a mosquito or a tick. For example, the common house mosquito is a vector for West Nile virus, transmitting it between individual birds (the natural reservoir hosts). Occasionally, the infected

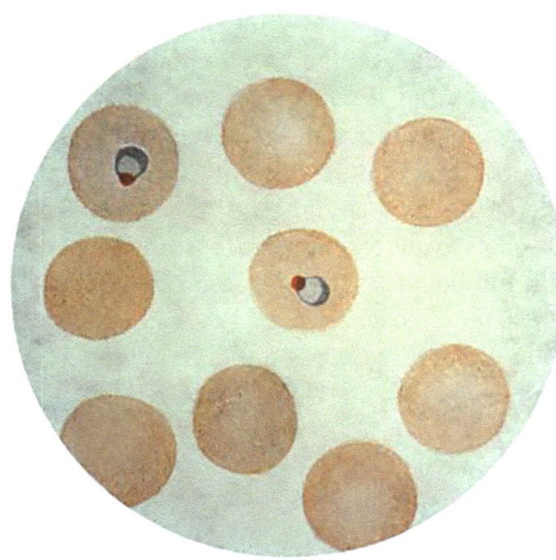

Figure 13.3 A photomicrograph of *Plasmodium*, the microbe responsible for malaria. The parasite *Plasmodium falciparum* is seen as dark, circular structures within red blood cells (the larger red circles). From Black & Black, 2015.

mosquitoes will also feed on people or horses. These are the accidental, or *dead-end hosts* who will get sick with the disease but cannot transmit it to others.

13.3.2 Endemic, Epidemic, and Pandemic

Some diseases will always be present in the population at a low level. This natural low-level presence is called *endemic*. For example, there are always people in the population who will have the common cold. Occasionally some diseases will suddenly arise in population, which is then called an outbreak. Periodically there are national news stories about several cases of food poisoning from *E. coli* or *Salmonella*-infected meat, fruit, or lettuce; these are called outbreaks. If an outbreak persists over longer period of time and spreads to a wider area, it is called an *epidemic*. A large-scale epidemic that spreads between continents is called a *pandemic*. The H1N1 swine flu that started in 2009, spreading to every continent and killing over 14,000 people, is an example of a pandemic. SARS-CoV-2 (COVID-19) that spread throughout the world in 2020 is another example of a pandemic. During the COVID-19 pandemic more than 761 million people were infected worldwide and the virus caused approximately 6.9 million deaths.

Epidemiologists sometimes use the terms *outbreak* and *epidemic* interchangeably, and there really is no concrete line between the two. Sometimes the term outbreak is used instead of epidemic so as to not scare the public.

13.3.3 Multicellular Parasites

Multicellular *parasites* are microscopic animals that have coexisted with humans and other vertebrates for a long time. These types of parasites include mainly *helminths*, or parasitic worms such as tapeworms, roundworms, hookworms, whipworms, flukes, and schistosomes (Fig. 13.4). Many (although not all) of them have adapted well to life in the human or animal body, and thus cause little damage and only minor discomfort. In fact, since the human immune system evolved during the time when our ancestors were heavily infested with various worms, it has developed an array of strategies to keep these parasites in check. While sanitation and hygiene practices have reduced the occurrence of helminths in developed countries, approximately 30% of world's population still carries the most common roundworm, *Ascaris lumbricoides*.

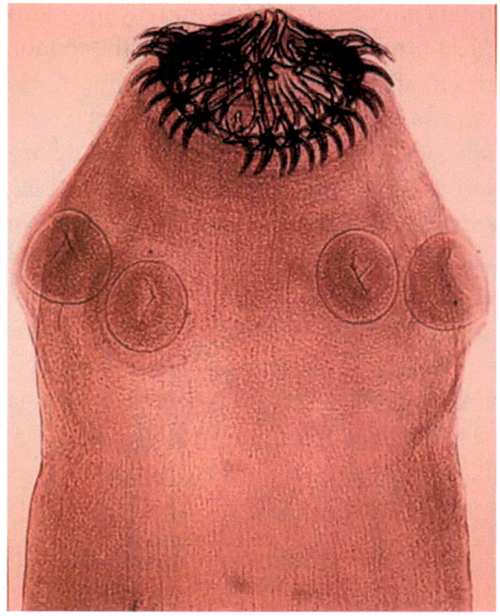

Figure 13.4 Colorized scanning electron microscope (SEM) image of the head of a tapeworm. There is a ring of hooks and a series of suckers that help the worm attach to the intestinal wall of its host. From Black & Black, 2015.

While some parasitic infections are relatively benign, others can be deadly, or cause debilitating illnesses and require costly treatment. The World Health Organization estimates that 3.8 billion people will contract a parasitic worm infection annually, leading to over a million deaths.

People are not the only ones suffering from parasitic infections. Because parasitic worms are endemic to grass and soil, they add a significant burden to the livestock industry, costing farmers billions of dollars annually. In addition to sickening and killing livestock, the economic burden also comes from the reduced milk, meat, and wool production as well as other associated costs, such as anthelmintic treatment or disposal of infected carcasses. (*Disease burden* is the impact of a health problem on a given population in terms of its health, as well as social, political, environmental, and economic factors.) Fisheries are also vulnerable to parasitic infections. For example, an estimated 20% of farmed fish are lost each year due to parasites, amounting to a loss of \$1–\$9.5 billion annually.

13.3.4 Case Study: The Comeback of a Hookworm

Many of the parasitic worm infections are classified as Neglected Tropical Diseases because it is the poorest and least-developed areas of the world that bear over 90% of the disease burden, frequently lacking the funds for treatment and prevention. However, these diseases can also occur in poor areas of developed countries, especially as the inequality between the rural poor and the rest of the population keeps widening. This is exactly what is happening in Lowndes County, Alabama, where 34% of people tested were positive for hookworm infection in 2017. Hookworms are small parasitic worms that hook their mouthparts to the lining of the small intestine, where they puncture small blood vessels and suck blood.

Lowndes County is located in rural Alabama where an estimated 80% of residents do not have access to a municipal sewer system. Instead, they rely on poorly functioning septic tanks that overflow and flood the houses and mobile homes, crawl spaces beneath homes, and yards. In these conditions, both children and adults frequently come in contact with raw sewage, which can be teeming with these parasites. Hookworms enter the body through small cuts in the skin – usually through the bare soles of the feet. After entering the body, the worms will travel to the small intestine, where they attach and suck blood. Hookworm infection can lead to diarrhea, fatigue, weight loss, and anemia, as well as mental impairment in children.

The hookworm, which was rampant in the southern United States throughout the nineteenth and early twentieth centuries, was thought to be eliminated in 1980s as the sanitation in rural areas improved greatly. However, not all areas saw these improvements, and various socioeconomic factors together with economic recession have forced many residents back into unsanitary living conditions. Considering the number of Americans living in similar conditions to those in Lowndes County, the National School of Tropical Medicine estimates that as many as 12 million Americans may currently be infected with hookworm.

13.3.5 Protozoans

Protozoans are a diverse group of small unicellular organisms that can be parasitic or free-living. While parasitic protozoans by their nature require a host to feed off, usually leading to a disease, the free-living protozoans can also cause diseases, usually through toxin production. Parasitic protozoans include *Plasmodium* spp. (agent of malaria), *Trypanosoma brucei* (agent of sleeping sickness), *Entamoeba histolytica* (agent of amoebic dysentery), *Toxoplasma gondii* (agent of toxoplasmosis, Fig. 13.5), *Cryptosporidium parvum* (cryptosporidiosis), and many others. Other free-living protozoans that can cause problems include dinoflagellate algae, which can cause red tides (red-colored algal blooms, Fig. 13.6). The shellfish that feed on these toxin-producing organisms will concentrate the toxins in their bodies. When people eat the shellfish that grew in the presence of these dinoflagellates, they can get shellfish poisoning.

Cats are the primary vector for toxoplasmosis. When cats consume infected rodents, birds, or other raw meat, the *T. gondii* reproduces in their reproductive tract, and they leave behind reproductive structures in their litter boxes,

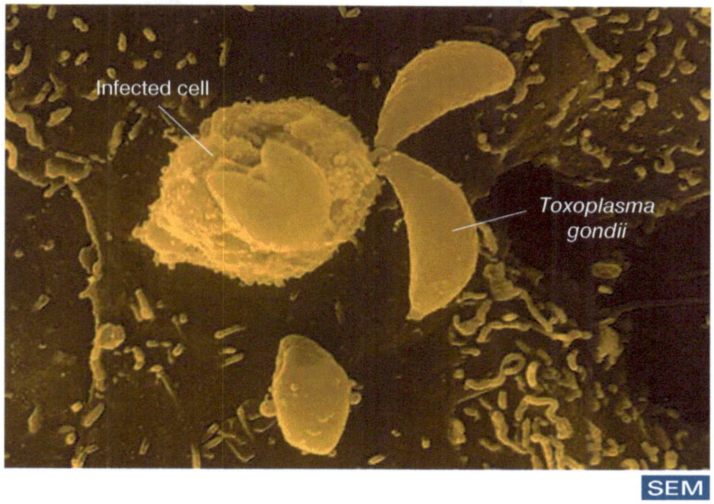

Figure 13.5 Crescent-shaped protozoans, *Toxoplasma gondii*, leaving an infected cell in which they have multiplied. These organisms cause the illness Toxoplasmosis. From Black & Black, 2015.

Figure 13.6 Red tide is caused by an algal bloom of dinoflagellates. They produce a toxin that can kill organisms in the water, and cause illness in people who eat fish and shellfish from affected waters. From Black & Black, 2015.

sandboxes, or soil. For most people, if they ingest the reproductive structures of *T. gondii* through the act of cleaning a litter box or entering a sandbox, they may develop minor flu-like symptoms. However, if a pregnant person ingests *T. gondii*, they can pass the protozoan through the placenta to the fetus. The fetus can develop severe congenital defects or die as a result. For this reason, women are advised not to clean litter boxes while pregnant.

In general, the illnesses caused by protozoans are far more deadly and numerous than the illnesses caused by the helminthic worms. The World Health Organization reported that in 2022, malaria alone infected 249 million people, causing 608,000 deaths. In the same year, \$3.5 billion was spent to prevent and treat malaria. The World Bank and World Health Organization estimate the combined disease burden of protozoan and parasitic infections worldwide to be about 12% of all infections, accounting for 1.154 million deaths per year.

13.3.6 Case Study: Leishmaniasis

Leishmaniasis is a disease most people in developed nations have never heard of. However, it is the ninth highest disease burden for tropical and subtropical areas around the world, putting an estimated 350 million people at risk (Fig. 13.7). The World Health Organization estimates that 12 million people have the disease, with 1.5 to 2 million new cases occurring each year.

(a) (b)

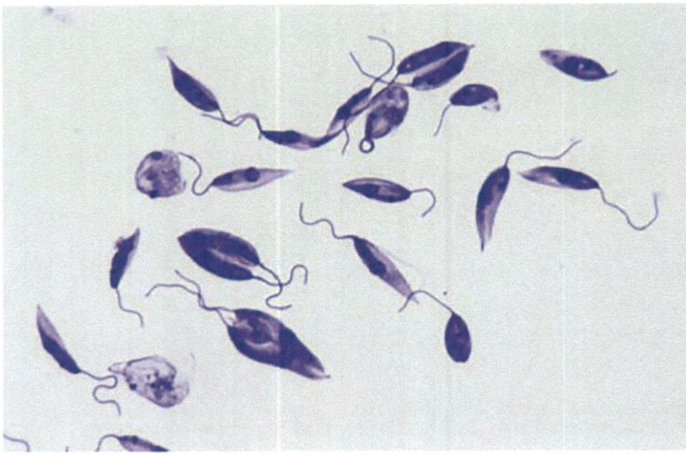

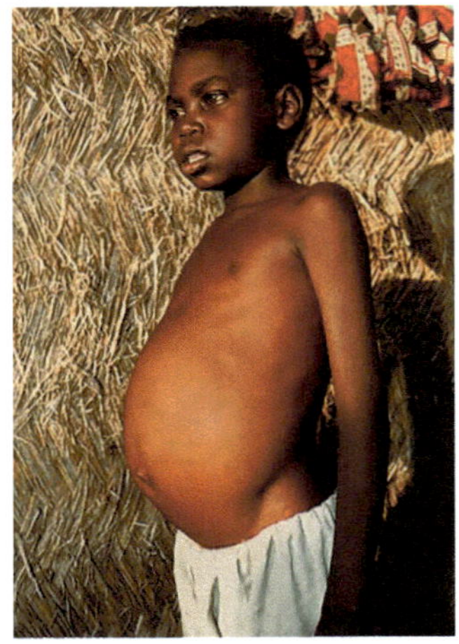

Figure 13.7 (a) *Leishmania donovani*; (b) a child infected with leishmaniasis. From Black & Black, 2015.

Leishmaniasis is caused by several species of a protozoan called *Leishmania* and spread by a bite of infected sand flies. Gerbils and other rodents act as reservoir hosts, but people and dogs can also serve as hosts for the parasite. There are two forms of leishmaniasis: the cutaneous form that causes disfiguring and slow-healing ulcers on the skin and mucous membranes in the mouth and nose, and the deadly visceral form that causes damage to red blood cells, liver, and spleen.

Afghanistan has currently the highest number of leishmaniasis cases in the world. Several factors have contributed to this epidemic in that country. Though they reported their first case in 1964, occurrences increased greatly in the 1980s. The first reason was the construction of irrigation canals in 1980s. The sandy soil removed to construct the canals was piled on the embankments, attracting a very high number of gerbils, the natural host of *Leishmania*. Gerbils came in contact with humans, and *Leishmania* was spread to humans as a result. The second factor that contributed greatly to the current epidemic is the war in Afghanistan that started in 2001. As a result, large numbers of people who were displaced from their homes ended up in densely populated refugee camps, which made the transmission of leishmaniasis very efficient: the sand flies would transmit the parasite from one infected individual to another without the need of a rodent host. As a result, since 2002, roughly 225,000 new cases have been reported in Afghanistan every year.

The third factor that has contributed to this large-scale outbreak in Afghanistan is that the relatively high cost of the medicine was not included in the standard primary health care packages. For a country where per capita yearly income is less than $2,000 and where 42% of people live below poverty line, $1 to $2 per dose is often not affordable. The same is true for the simplest prevention method, insecticide treated bed netting. While the treatment and prevention of this disease are relatively simple and inexpensive, they are still outside the means for many people in Afghanistan, and it has not been a priority disease for international disease fighting efforts.

13.3.7 Fungi

For most people, fungi are just mushrooms, but in reality, fungi are a large and diverse group of organisms. They range from unicellular yeasts (such as *Saccharomyces cerevesiae*, the brewers yeast, or *Candida albicans*, which is part of the natural flora of human beings but can cause yeast infections) to filamentous molds (such as *Penicillium*, which produce the antibiotic penicillin and grow as green mold on bread, fruits, and vegetables, to rusts and smuts, plant pathogens that cause billions of dollars' worth of damage to crops across the world. It is estimated that there are over 8,000 species of fungi that are plant pathogens, while about 300 species cause disease in humans and animals.

While dealing with the fungal pathogens is important, it is equally important to recognize the role of *symbiotic* fungi. Fungi form symbioses (meaning "living together") with an estimated 80% of all plants, usually (but not always) forming a network with plant root systems called *mycorrhizae* (Fig. 13.8). Many plants, such as orchids, could not survive without the mycorrhizal association; many others would grow but have considerably lower productivity and growth rate.

Fungal diseases in people range from relatively benign, such as yeast infections of the skin (e.g., athlete's foot and ringworm), to potentially lethal, such as histoplasmosis and coccidiomycosis, both of which cause pneumonia-like illness and are acquired by breathing in fungal spores from dust. Fungal diseases in people are also much more difficult to treat than bacterial diseases, since fungi are much more closely related to animals than are protozoans, and consequently the drugs that harm fungi can also be toxic to people.

In addition to invasive fungal diseases, people are also affected by *mycotoxins*, poisonous compounds produced by fungi that can have acute toxicity and potential carcinogenic effects. The most common toxins include aflatoxins from *Aspergillus* (mainly affecting corn and peanuts); ochratoxins from *Aspergillus* and *Penicillium* (affecting cereal crops, coffee, and dried fruit); ergotamine from *Claviceps* (affecting rye, barley and wheat); trichothecenes from various fungi (affecting corn, oats, and wheat); and fumosins from *Fusarium* (affecting mainly corn). The United Nations Food and Agriculture Organization estimates that 25% of crops are affected by mycotoxins each year, resulting in the loss of 1 billion tons of food and food products. In the United States and Canada, the loss from mycotoxin-contaminated crops can be as high as $5 billion per year. The losses due to mycotoxins are even higher in tropical regions and developing nations. The aflatoxin-related losses in

(a)

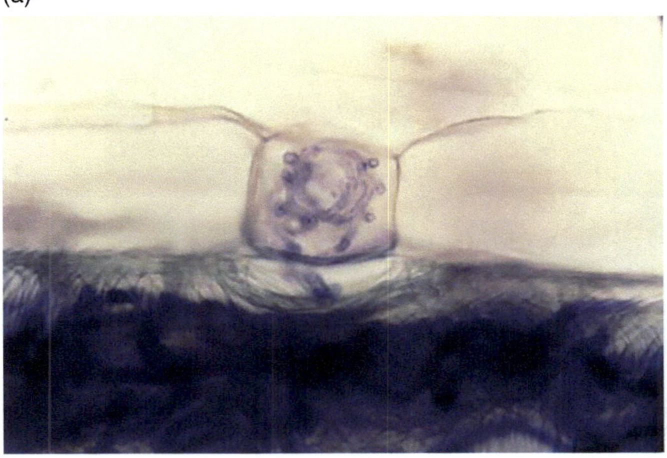

(b)

Figure 13.8 (a) Photomicrograph of mycorrhizal fungi growing in association with a root. (b) Orchid blooming, showing its roots. From Black & Black, 2015.

Indonesia, Philippines, and Thailand amount to roughly $900 million per year.

13.3.8 Case Study: Chestnut Blight

At the beginning of the twentieth century, the American chestnut was a prevalent tree in the eastern part of the country. It is estimated that in the Appalachian region, chestnuts made up 25%–30% of all trees. That is, until 1904, when chestnut blight (a fungal disease caused by *Cryphonectria parasitica*) was introduced to the country from a Japanese tree nursery. The Japanese chestnuts were resistant to the blight; the American chestnut turned out to be highly susceptible. The fungus grows and spreads under the tree bark, producing toxins and killing the living tissue. Within 40 years, the disease wiped out an estimated 4 billion American chestnut trees, leaving just several hundred chestnuts standing in other parts of the country where the fungus did not reach.

The chestnut blight did not just kill the beautiful trees, but deprived people and animals of food as well as income from the popular chestnut lumber for construction, furniture, and even musical instruments. In 1912, the standing chestnut timber was estimated to be worth over a billion dollars in current value. As this revenue was lost, rural communities suffered greatly. It is argued that the death of the chestnut trees greatly contributed to the decline of Appalachian culture and self-subsistence.

13.3.9 Archaea

All living things are classified into one of the three branches in the tree of life. So far, we have looked at organisms classified as *eukaryotes*, meaning cells that have a nucleus and complex organelles, and that did not arise until about 2 billion years ago. The organisms that ruled the Earth before that were much simpler in their cellular composition, although highly adaptive to varying environmental conditions. These organisms were bacteria and archaea.

While bacteria have been known for over 200 years, mainly as disease-causing organisms, it was not until 1970s that scientists had the tools to distinguish archaea (which look very much like bacteria at first glance) as an entirely separate branch in the tree of life. Archaea (Fig. 13.9) possess some novel abilities

Figure 13.9 Many organisms of the kingdom Archaea have properties that allow them to survive in extreme environments. *Thermoacidophile archaea*, such as those in this image, thrive in hot, acidic environments. From Babcock, 2009.

that are not seen in either bacteria or eukaryotes, such as the capacity to live in extremely hot conditions and to produce methane. Interesting enough, to date not a single species of archaea has been found to be pathogenic to eukaryotes, although they do parasitize on other archaea.

13.3.10 Case Study: Did Archaea Cause the Permian Mass Extinction?

The Earth history shows some well-documented mass extinction events, the best known of which is probably the Cretaceous-Paleogene extinction event 66 million years ago that killed the dinosaurs. However, much bigger extinction events have occurred in the past, such as the Permian-Triassic extinction event 252 million years ago in which 90% of marine and 70% of terrestrial organisms went extinct, including the last trilobites. The proposed mechanisms that contributed to this event include large meteorite impact, mass volcanism, and release of massive amounts of methane over a short period of time, which contributed to global greenhouse effect. One proposed source of the methane is the hyperactivity of methanogenic archaea in the oceans.

The hypothesis that archaea are at least partially responsible for the mass extinction event is based on the fact that conditions for methane mass production were just right. Methanogens, such as *Methanosarcina*, were (and still are) present throughout the oceans, and genetic analysis shows that they had just then acquired the ability to produce methane. There was abundant organic carbon (the starting point for biological methane production) available in the ocean sediments because the oceans were teeming with life. And to enhance the growth of the microbes, massive volcanic eruptions had added nickel into the ocean waters, just the element that *Methanosarcina* needs for growth. As a result, a large amount of methane was released into the atmosphere over a short period of time, causing average global temperatures to rise by 14.4° F (8° C) and the CO_2 levels to rise by 2,000 ppm (currently at 410 ppm). This led to anoxia and acidification of the ocean waters, all of which made it unsuitable for many life forms. Although there is not a complete agreement on the exact causes, it was probably the combination of all these factors that led to the biggest mass extinction event in Earth's history.

13.3.11 Bacteria

Bacteria are probably the most diverse living organisms on the planet. Despite their incredible diversity, only relatively few bacterial species are capable of causing disease, and therefore biological disasters. Bacteria were also the earliest known organisms to inhabit the Earth, and still today they are the most numerous organisms, making up about one third of Earth's biomass, or about 250 billion tons of carbon. There is a huge diversity within this domain of life, with most of the bacteria being harmless environmental microbes. Only about 1% of the known bacteria are human pathogens. It is also important to note that the true diversity of bacteria is still not known, since to date most of them cannot be grown and studied in the lab. It is only within the past 30 years that molecular tools have been widely available to study the microbes without culturing them.

Life on the Earth would not exist without bacteria. Bacteria are *primary decomposers* (they recycle nutrient back into the environment), they form the basis of the food chain, and they live symbiotically with other organisms, including people, protecting us from various diseases and producing many necessary chemicals for our bodies. In fact, the number of bacterial cells in the human body is on the same order of magnitude as the number of the human cells! However, bacteria are still mainly known for the diseases they cause, probably because everybody has a firsthand experience of getting sick.

It is hard to estimate the economic impact of overall bacterial infection because bacteria and their diseases are highly diverse and hard to track. Instead, the impact is calculated separately for each microbe or group of diseases. For example, according to the U.S. Department of Agriculture, human illnesses from six common food-borne bacteria (*Salmonella, E. coli* O157:H7, *Listeria, Campylobacter, Staphylococcus aureus,* and *Clostridium perfringens*) cost the United States $2.9 to $6.7 billion annually.

While bacterial diseases cause many illnesses and billions of dollars' worth of economic losses, it is important to note that thanks to improved sanitation, hygiene, vaccinations, and antibiotics, the overall disease burden and mortality from bacterial diseases (and infectious diseases overall) was reduced throughout the twentieth century. According to the Centers for Disease Control (CDC), in 1900, 73.5% of the 10 leading causes of death were caused by mainly bacterial infections. In 1997, the same number was 3.2%. Infectious diseases do contribute more to overall mortality in developing countries where life expectancy and vaccination rates are low, sanitation and hygiene are not well developed, and life-saving drugs are expensive.

13.3.12 Case Study: Plague Epidemics

Plague is a disease caused by a bacterium *Yersinia pestis*, a tiny rod-shaped organism that is only 1–2 μm long. Reservoir hosts for this bacterium are various rodents, primarily marmots, which harbor the bacterium but do not get sick from it. Other rodents, such as rats, chipmunks, and prairie dogs also carry the microbe, and can get the disease. Mammals other than people and rodents that can occasionally also get the disease include house cats, rabbits, goats, sheep, chickens, and pigs.

Yersinia pestis needs fleas as a vector for transmission from one host to another. The disease starts with a bite from an infected flea. The bacterium has several mechanisms to hide from the immune system, which allows it to multiply in the blood and lymphatic systems. Eventually the bacteria end up in lymph nodes, where they cause acute infection, hemorrhage, and necrosis. The swollen, darkened lymph nodes (Fig. 13.10) are known as buboes, giving the disease the name "bubonic plague." When the lymph nodes get overwhelmed, bacteria spill into the bloodstream, where they secrete various toxins that cause small, localized blood clots all over the body. This is known as "septicemic plague" and is manifested by patchy rash, necrotized tissue, and vomiting of blood. When *Yersinia pestis* gets into the lungs, it causes "pneumonic plague," which can be transmitted through air like common cold. The last two forms of plague can progress rapidly (within hours to 1–2 days) and are fatal without treatment.

Plague has "plagued" society for a long time, causing many epidemics throughout history. In fact, some of the most devastating and deadly pandemics in human history were caused by

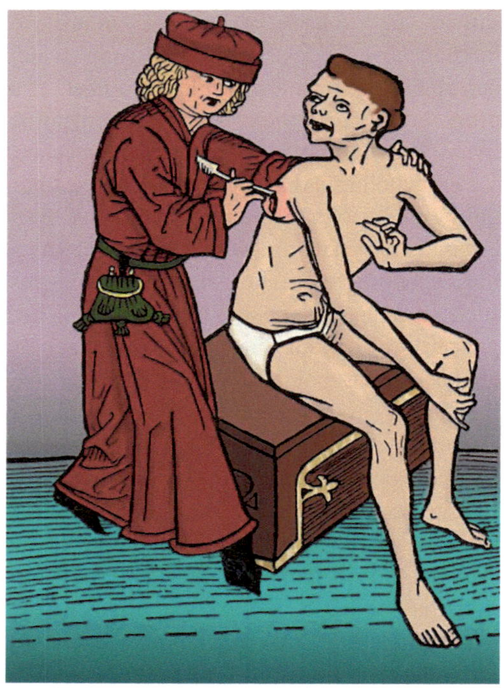

Figure 13.10 In this drawing from a fourteenth century Flemish manuscript, a physician lances a bubo. From Black & Black, 2015.

Yersinia pestis. In the sixth century, the Plague of Justinian in the Byzantine Empire killed between 25 and 50 million people, which amounted to about 13% of the world's population at that time. In the late Middle Ages, during the Renaissance period in 1300s, the Black Death (also known as the Great Plague) killed between 75 and 200 million people across the Eurasian continent. In Europe, about 30% to 60% of the population was killed by the plague. It took 200 years for the population to recover to its previous numbers. Outbreaks of plague in Europe continued throughout the 19th century. Another pandemic started in China in 1855 and lasted until the middle of the 20th century, killing over 12 million people, mainly in India and China.

The plague is still around in the 21st century, although thankfully, these days *Yersinia pestis* can be controlled with timely antibiotic treatment. According to World Health Organization, 3,248 cases, including 584 deaths, were reported from 2010 to 2015, mainly from countries such as Democratic Republic of Congo, Madagascar, and Peru that have high numbers of indigenous flea-carrying rodents. In the United States, plague still occurs in the grasslands of the western part of the country, with between 1 and 17 human cases reported by the Centers for Disease Control (CDC) every year.

13.3.13 Viruses

So far we have examined living microbes as disease agents. The last two we will investigate, viruses and infectious molecules, are technically not living. Viruses have been described as a "piece of bad news wrapped up in protein." This is definitely a good way to describe an organism that is not made of cells (as all living things are), but instead have their genetic information in a form of either DNA or RNA surrounded by a protein coat. Viruses cannot do many things that living things can do: they cannot grow or reproduce on their own and they do not take in nutrients from the environment or generate their own energy. However, as organisms that rely completely on their hosts for food and reproduction, they can take over the host cell and use its materials and mechanisms to replicate themselves over and over, killing the host cell in the process.

Viruses are known to parasitize every type of living organism: bacteria, archaea, plants, fungi, protozoans, and animals. Viruses have also evolved to be highly specific to the cell type they infect: a virus that infects liver cells generally cannot infect skin cells, and a virus that infects plants cannot infect animals. However, occasionally some viruses can jump species. For example, a virus that had evolved to infect bat cells gains an ability (usually through a mutation) to infect human cells. This event of a virus jumping species is called *spillover,* and most emerging infectious diseases, such as Ebola and SARS (including SARS-CoV-2, the virus that causes COVID-19), that seem to appear out of nowhere and cause serious outbreaks occur as a result of a spillover from an animal host to a human host. In most cases, the animal reservoir is identified, which helps to curb the disease. In the case of one outbreak of SARS, the animal reservoir was first misidentified as the civet (a cat-like mammal), leading to the mass killing of civets. The actual reservoir was later identified as the bat. In other cases, such as Ebola (Fig. 13.11), the natural reservoir is still unknown although bats are suspected here as well.

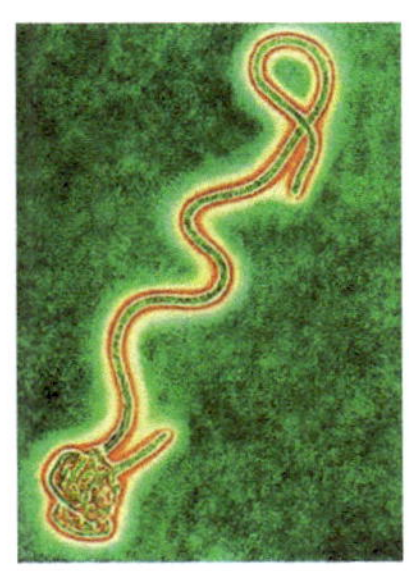

Figure 13.11 Photomicrograph of the Ebola virus. From Black & Black, 2015.

Historically, viruses have caused innumerous outbreaks, epidemics, and pandemics, which have had a devastating effect on societies. For example, smallpox has caused numerous epidemics throughout history on every continent. The most notable epidemic was the introduction of smallpox to the North American continent by sailors from Europe. They first accidentally and later deliberately infected Native Americans who lacked natural immunity to the disease. As a result, an estimated 90% of the North American Indigenous people died of smallpox.

Other viral diseases have a huge impact on populations as well. An unknown hemorrhagic fever virus killed about 80% of Aztec population in the 1500s. Various flu pandemics have killed millions throughout history; the best-known example was the outbreak of Spanish Flu in 1918, which killed an estimated 75 million people worldwide (Fig. 13.12). HIV, which made its debut into broader population in 1960s, has since killed over 35 million people. Other viruses that have caused deadly but more localized epidemics and outbreaks include measles virus, mumps virus, Ebola virus, Dengue Fever virus, Yellow Fever virus, SARS coronaviruses, poliomyelitis, hepatitis B virus, Chikungunya virus, Zika virus, and several others.

13.3.14 Case Study: Rabbit Myxomavirus as a Population Control Method

Because viruses are highly specific to an organism, they have been cautiously used as a control method to kill invasive species. This was attempted in Australia in the 1950s to control the rabbit population. The rabbits had been introduced to Australia in 1788 as a food animal but were mainly kept in cages or enclosures. In the middle of the 1800s, some estate owners started releasing the caged rabbits for hunting purposes. As a result, the rabbit population in Australia exploded due to lack of natural predators or diseases. In addition, rabbits could breed in the mild Australian climate year-round. Within 10 years of the release, the rabbits had become so numerous that shooting millions of them every year did not make a significant difference in the population size. Rabbits are thought to be the major contributors to native species loss by overgrazing, leading to the death of native plants and animals.

Several control measures have been implemented since the 1800s, including hunting, rabbit-proof fences, and various biological control measures, including chicken cholera, rabbit scab, rabbit myxomavirus, rabbit calicivirus, and rabbit hemorrhagic disease. Of these, the

(a)

(b)

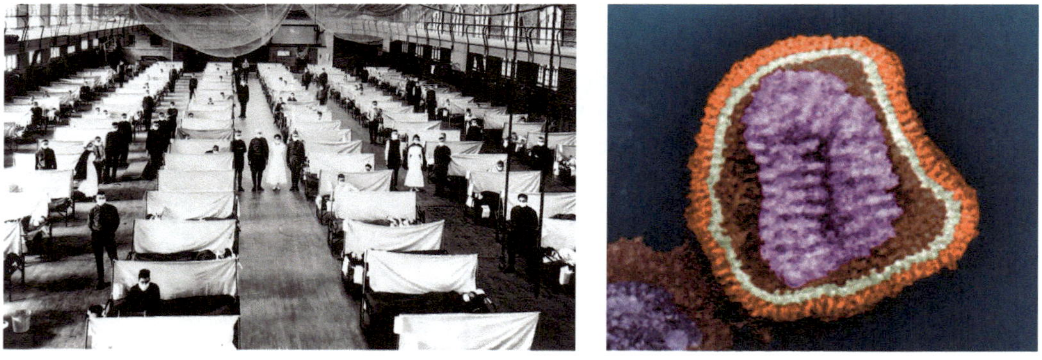

Figure 13.12 (a) During the height of the Spanish flu pandemic in 1918, this gymnasium at Iowa State University was temporarily converted into a hospital ward. (b) A colorized electron microscope image of an influenza virus. From Black & Black, 2015.

introduction of myxomavirus has been one of the most successful biological control examples against invasive species.

Myxomavirus, which naturally infects rabbit species in South America, has no other host than rabbits, although there is varying susceptibility among different rabbit species. The South American rabbits get a relatively mild disease from the virus; the European rabbits that were introduced to Australia are highly susceptible to the virus, which causes skin tumors, followed by fever, conjunctivitis, and blindness, and leaving the rabbits susceptible to various bacterial illnesses.

When myxomavirus was first introduced to the Australian rabbit population, it was fatal in 99% of the cases, killing the rabbits within 48 hr to 2 weeks. Upon introducing the virus to wild rabbit population in 1950, the number of rabbits dropped from an estimated 600 million to 100 million within 2 years. Unfortunately, the surviving rabbits developed partial immunity to the disease, and while the myxomavirus is still circulating in the rabbit population, the fatality rate is now only 35%, prompting the need for other control measures. However, there is no doubt that introduction of the myxomavirus was a huge success in controlling the rabbit population in Australia.

13.3.15 Infectious Molecules: Viroids and Prions

The smallest known infectious agents are simple molecules that can cause disease. The examples of infectious molecules are viroids and prions. *Viroids* resemble viruses in a sense that the infectious part is a nucleic acid, and in the case of the viroid, the nucleic acid is always RNA. However, viroids lack the protein coat that protects the nucleic acid in viruses. To date, viroids are only known to be plant pathogens, such as the potato spindle tuber viroid that causes stunted growth in potatoes and tomatoes (Fig. 13.13).

Prions are infectious protein molecules that can cause fatal neurodegenerative diseases. Proteins are long polymers of amino acid molecules that fold into their final three-dimensional structure in a way that takes least energy. For most proteins there is only one such way to fold; for prions, there are more. Normally folded healthy prions are present in the brain and spinal cord, although their function is not exactly known. The misfolded prions are infectious: upon contact they will cause the normal prions to misfold as well. These misfolded proteins stick together, causing amyloid plaques (dense aggregates of certain proteins) in the brain. These proteins that no longer

Figure 13.13 The tomato plant on the left is healthy. The plant on the right has been infected with a viroid, causing tomato apical stunt disease. From Black & Black, 2015.

function normally surround the nerve cells in the brain and disrupt the normal function of the cells and tissues, leading to neurodegenerative disease that always ends in death.

The misfolding also makes prions highly resistant to heat, radiation, various chemicals, and enzymes that usually degrade proteins, making prion-contaminated materials extremely hard to disinfect. This contributes to the transmission of prion diseases. Usually, prion disease is transmitted between the same species of animals, but occasionally they can jump species, creating a new prion disease in a new animal. To date, prion diseases have been found in sheep (scrapie), cattle (bovine spongiform encephalopathy (BSE), also known as mad cow disease), deer (chronic wasting disease), cats (feline spongiform encephalopathy), people (various forms of Creutzfeldt-Jacobs disease, Kuru, fatal familial insomnia, and few others), and in some other animals, such as mink, camel, ostrich, and antelope (Fig. 13.14).

13.3.16 Case Study: BSE

Creutzfeldt-Jacobs disease (CJD) has been known since the early 1900s. It is a rare neurodegenerative disease that occurs roughly one in 1 million people, usually late in life when it generally kills a person within one year of the first onset of symptoms. It usually starts with failing memory, behavioral changes, and lack of coordination, then progresses to mental deterioration, blindness, and occasionally coma before the death. Because it is so rare, it came as a great surprise when in early 1996, a cluster of 10 CJD cases were diagnosed in the United Kingdom, mainly in younger people. As more cases were diagnosed over the following years (177 total in the United Kingdom, 52 cases elsewhere in the world), the epidemiologists tied this to the epidemic of BSE (mad cow disease) in cattle that had struck the United Kingdom a decade earlier. Apparently the cow prion had crossed the species and started to infect the human brains.

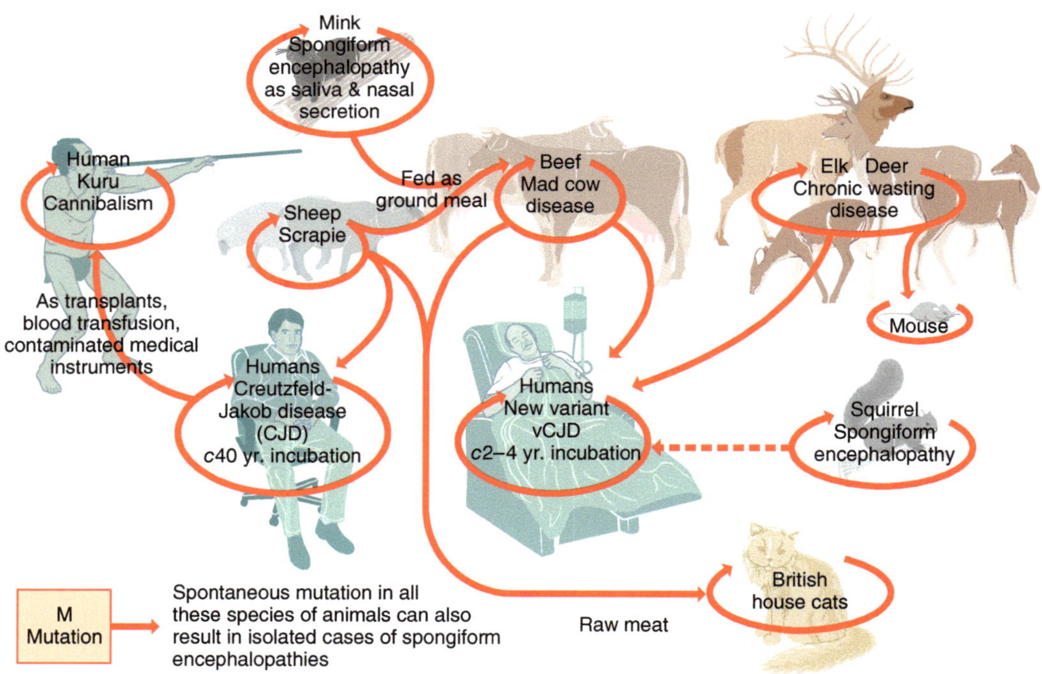

Figure 13.14 Prion-caused spongiform encephalopathy diseases occur in many species and can be transmitted from one to another. Spontaneous mutations also produce a few cases each year without involving transmission from another animal. From Black & Black, 2015.

The BSE epidemic hit the United Kingdom in 1980s. It is suspected that BSE was caused by misfolded sheep prion that mutated to infect the normal prions in cows. (Scrapie is a prion disease in sheep that has been known since 1732.) Cows came into contact with the sheep prions through protein supplement in their feed. In most countries, soy meal is used as protein supplement in feed; in the United Kingdom (where soybeans do not grow well) the protein supplement was made from ground meat and bones that are left over from meat processing, as well as sick animals (cows and sheep) that could not be used for human food. The ground meat and bones received high temperature treatment, which killed pathogens, such as worms, protozoa, fungi, bacteria, and viruses, but not prions, which have high heat resistance (up to 600°C, or 1,112°F). As a result, cows frequently consumed the sheep prions, increasing the likelihood for it to mutate to infect the cows.

Even before it was known that the cow prions could mutate to infect people, the economic hit from this epidemic was devastating. Over 180,000 cases of BSE in cattle were confirmed within 20 years after the first known case, and over 4 million cows were slaughtered as a precautionary measure to stop the infection. The European Union banned the import of British beef, as did several other countries. Domestic beef consumption dropped by 40% and the prices fell by 25%. Implementation of precautionary measures in the 1980s and 1990s cost the British taxpayers over 6 billion dollars.

13.4 Microbes and the Earth System

While microbes are usually associated with diseases that occur in people and animals, most microbes actually live independently in the soil and water. The majority of these organisms are harmless; however, they can occasionally cause a disease, either by getting into a wrong place of the body, or by producing toxins that can make people and animals sick. Almost all environments and components of the Earth System (the geosphere, atmosphere, hydrosphere, and definitely the biosphere) harbor microbes that can cause a disease or other natural disaster in the environment.

13.4.1 Solid Earth (Geosphere)

Even though microbes are mainly associated with soil and the upper levels of Earth's surface, microbes have been found living as deep as 1.9 mi (3 km) within the Earth in African gold mines. Specific microbes can actually live inside minerals, or on mineral surfaces. Sometimes they secrete acids to dissolve the rock around them to gain trace amounts of minerals from the rock. These microbes, known by their energy requirements as *chemolithotrophs* (they use an inorganic substrate, such as a rock to provide them with electrons for energy production), resemble the early Earth colonizers during the first 2 billion years and do not have the ability to exploit another living system for nutrients or cause a disease.

However, that does not mean these microbes cannot cause disastrous events for people. Some of these microbes, such as *Acidithiobacillus ferrooxidans,* will produce acids to dissolve minerals for their energy production, which can lead to corrosion of metal, rock, and concrete, known as *MIC* (microbially induced corrosion). This is a serious problem for infrastructure, especially in sewer pipes, where microbes under right conditions can corrode up to 10 mm of concrete per year.

These acid-secreting microbes can also cause environmental disasters in areas where mine tailings are piled near mine entrances. *Mine tailings* are leftover crushed rocks after most of the valuable minerals have been extracted from the ore. Microbes attach to the rocks, secreting acids to access necessary minerals, which leads to a rapid drop of environmental pH. Since the mine tailings are exposed to the rain, the runoff water from this pile of rocks will now contain both the acids as well as the dissolved minerals, usually sulfur, iron, and copper. This runoff frequently finds its way to mountain streams and can be spotted as water that has turned brown with iron oxide. The combination of very low pH and high concentration of metals is usually toxic to all living organisms in these waterways, except the microbes that created these conditions firsthand.

In addition, the microbes that dissolve iron in rocks can also dissolve other minerals, including heavy metals like arsenic. Some areas in which arsenic has been mobilized and contaminated drinking water are Bangladesh and Vietnam, though this is far from an exhaustive list. Poisoning from naturally occurring materials such as arsenic and lead can occur as a result of microbial activity that mobilizes elements.

Much more common than having microbes on the rock are microbes in the soil. Soil is one of the most microbe-rich environments on the planet. It is estimated that one teaspoon of soil contains between 100 million and 1 billion bacteria! Microbes that live in the soil are not seeking to cause a disease, but when presented with rich nutrient source of human or animal body, they can take advantage of it. Good examples of these advantage takers are bacteria that are *spore-forming bacteria*. Spores are tough little structures that bacteria form as a protection from adverse environmental conditions. Soil is a tough environment that challenges microbes with temperature extremes (summer-winter), moisture extremes (drought-flood), and in the case of agricultural fields, also occasional exposure to UV radiation, pesticides, and herbicides.

Several well-known pathogens are spore-forming bacteria from the soil, including *Clostridium botulinum* and *Clostridium tetani*, the agents of botulism and tetanus (Fig. 13.15), and *Bacillus anthracis*, the etiological agent of anthrax (Fig. 13.16). Anthrax, which is primarily a disease of cattle, can kill a cow in the field, and the spores of *B. anthracis* will remain viable in the soil for over 60 years or even more. These spores can be inhaled by anyone walking in the field, kicking up dust.

13.4.2 Case Study: Anthrax From Melting Permafrost in Siberia

In the summer of 2016, a mysterious illness struck the residents of Yamal Peninsula, Russia. In this remote northeastern corner of Siberia, dozens of people were hospitalized, and one person was killed by what turned out to be *Bacillus anthracis*, the bacterium that causes anthrax. Because anthrax is mainly cattle disease, the officials investigated the local reindeer herds, and eventually identified a reindeer carcass that was teeming with these bacteria.

However, the officials discovered that this particular reindeer had died of anthrax 75 years ago and was buried in the permafrost. The unusually hot temperatures in the summer of 2016 had melted the frozen soil and exposed the carcass where the spore-forming bacteria had germinated and were now actively spreading. Reindeer grazing near the carcass had picked up the bacteria and transferred it to their owners. While the outbreak was contained by vaccinating the reindeer herds against anthrax, scientists are concerned that as the global temperatures continue to rise, more frozen microbes may escape their permafrost graves.

13.4.3 Atmosphere

Like any living organism, microbes influence atmospheric composition through metabolic processes. As mentioned earlier, microbes are responsible for the oxygen in our atmosphere, but their respiration also adds carbon dioxide (CO_2) to it. Microbes also release methane, hydrogen sulfide, nitrogen gas, and many other gases that can affect the environment.

However, it is not just the gases produced by microbes, but also microbes themselves that are present in the air. There is a whole branch of biology (*aerobiology*) dedicated to studying such organisms. Microbes from water, soil,

(a)

(b)

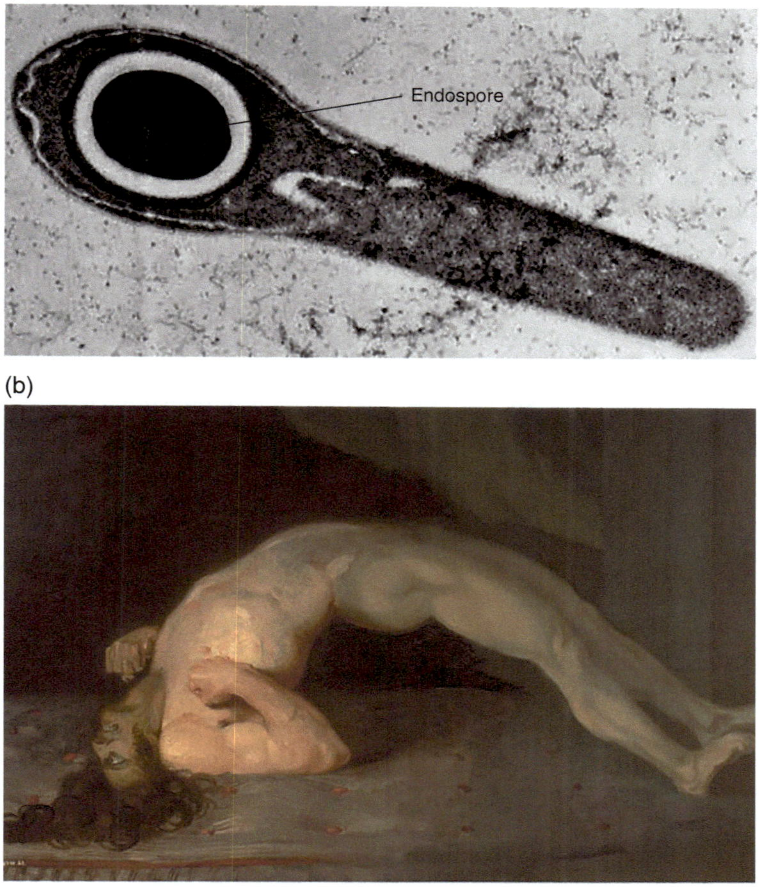

Figure 13.15 (a) Electron microscope image of *Clostridium tetani* with a large, dark endospore. (b) A soldier dying of tetanus. Tetanus causes extreme contraction of all muscles, from those of the face to those of the toes. It used to be a common cause of death in the days of cavalry troops. From Black & Black, 2015.

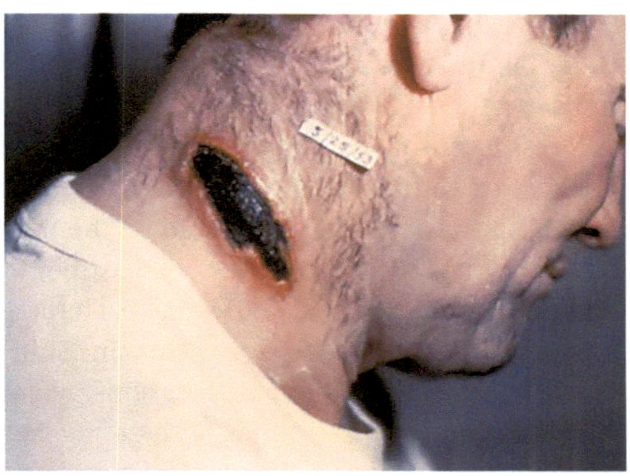

Figure 13.16 A photo of a man with cutaneous anthrax (anthrax in the skin). Anthrax spores entered the skin through abrasions or cuts. The organisms multiply and release a toxin into nearby tissue, and damage spreads. Other forms of anthrax include respiratory anthrax and intestinal anthrax. From Black & Black, 2015.

and human activities get aerosolized and carried over long distances and can stay airborne for days or weeks. Microbes have been found as high as 50 mi (80 km) above sea level, and actually contribute to atmospheric processes, like precipitation. Water molecules use bacteria as nucleation points, contributing to cloud formation and precipitation. All precipitation (rain, snow, and sleet) are full of microbes, which is why it is not wise to eat snow.

When it comes to disease causing microbes, air does not seem to be a place for them to hang out. However, due to their small size, all kinds of microbes can be carried by air currents. Before people knew that diseases were caused by microbes, "bad air" was often blamed for illness. In fact, this is how the disease malaria got its name: translating from Italian, *mala aria* literally means "bad air."

Many respiratory pathogens, such as influenza, common cold viruses, and the coronavirus that causes COVID-19 are transmitted through airborne transmission. This is usually a localized transmission, such as picking up microbes from someone coughing and sneezing in the same room. However, some microbes can be dispersed over significant distances. For example, industrial cooling towers in Pas-de-Calais, France, where the water was contaminated with *Legionella pneumophila* (the agent of Legionnaire's Disease, a form of pneumonia), the aerosolized bacteria were carried nearly 4 mi (at least 6 km) from the source, sickening 86 and killing 21 people. Even more extreme examples include the big dust storm events, created by desert winds in Africa. These can carry microorganisms across continents and even across the Atlantic Ocean and have been investigated as potential dispersal methods for novel pathogens.

13.4.4 Hydrosphere

Much like soil, water is also rich in microbes. For example, 1 drop (1 ml) of seawater contains an estimated 1 million microbes. Most of these microbes are nonpathogenic. Instead, their numerous functions include being the basis for aquatic food chain, driving the nutrient cycles, decomposing organic matter, and reducing it into inorganic matter, acting as carbon sinks, and many more.

However, water also harbors microbes that cause problems for people, animals, and fish. Sometimes the microbes themselves are nonpathogenic but produce illness-causing toxins. *Karenia brevis* and *Alexandrium fundiyense* are dinoflagellates (unicellular algae) found in the eastern coast of the United States and Canada that frequently cause red tides, algal blooms that look reddish brown in color. While the organisms themselves do not cause disease, they do produce toxins that affect people, marine animals, and fish either through consumption of contaminated water or by consumption of toxin-contaminated shellfish. Red tides and associated massive fish kills have been recorded in the shores of the Gulf of Mexico since Spanish explorers arrived in the 1500s.

Some microbes naturally present in the water can be straightforward pathogens. While careless campers drinking unsafe water can get infected with these organisms from time to time, large-scale outbreaks are usually associated with failure in drinking water treatment process. In the spring of 1993, a water purification plant in Milwaukee, Wisconsin, failed to eliminate oocysts (tough spore-like structures formed by protozoans) of *Cryptosporidium parvum*, resulting in over 400,000 people becoming ill with cryptosporidiosis, characterized by stomach cramps, nausea, vomiting, and diarrhea. One hundred and four elderly and immunocompromised people died as a result.

Although ingestion of waterborne pathogens seems to be the most logical way of acquiring cases of gastroenteritis, other routes work too. Some bacteria do their damage upon contact with skin. Skin frequently has small cuts and abrasions, which opens the portal for various flesh-eating bacteria, such as *Vibrio vulnificus*. *V. vulnificus* can cause a variety of illnesses depending on the place of entry, ranging from mild gastroenteritis and invasive septicemia (bacteria in the bloodstream) from ingestion, to necrotizing fasciitis and blistering dermatitis from contact with skin and underlying tissues.

People can also acquire pathogens from water through a respiratory route. *Naegleria fowleri*, aptly nicknamed "the brain eating amoeba," is an especially heinous example. *N. fowleri* has been found present in many freshwater lakes, ponds, and rivers (and swimming pools) throughout the world. When people inhale droplets of contaminated water during swimming, the amoeba attaches itself to respiratory epithelium inside the nose. From there it follows the olfactory nerve into the brain, where it feeds on brain tissue and blood cells, causing primary amoebic meningoencephalitis (PAM). Even with treatment, PAM is usually fatal. In rare cases, *N. fowleri* infections leading to PAM have also been obtained from the use of tap water in a neti pot or sinus rinse. It is important to use purified water rather than tap water for these uses.

13.5 Factors That Put Populations at Risk

While nobody is immune to all the diseases that can occur in the population, several factors can make either individuals or the whole populations more susceptible to various outbreaks. On the individual level, the factors that make populations more susceptible include high number of vulnerable people (very young, very old, immunocompromised). On the population level, the density and diversity of the population play a significant role in transmission of diseases. Socioeconomic factors that contribute include the levels of poverty and hygiene standards, but also increased global traveling. An environmental factor that contributes is global climate change. And finally, the pathogens themselves are capable of mutating and adapting to changing conditions, contributing to new outbreaks and epidemics.

13.5.1 Population Vulnerability

The medical advances throughout the 20th and 21st centuries have increased newborn survival rates, contributed to the increase of the elderly population, extended the life expectancy of immunocompromised people, and created a relatively significant population of people with compromised immune systems (people with organ transplants as well as cancer patients undergoing chemotherapy). All these people are more vulnerable to various diseases, especially to opportunistic infections, which are caused by microbes that usually live in peaceful symbiosis with our bodies but can cause serious disease in people with weak immune systems.

Some of these diseases are rarely seen in healthy individuals but show up in clusters among immunocompromised people. Such was the case in 1982, when doctors in California noticed an unusual number of rare pneumonia cases, caused by *Pneumocystis*, a fungus that lives in the lungs. All the patients were young gay men with severely weakened immune systems. The publications of these findings were the first step in discovering the Human Immunodeficiency Virus (HIV).

Sometimes a pathogen can cause a different disease in people with weakened immune systems. For example, *Listeria monocytogenes* is a food-borne pathogen that usually causes regular food poisoning symptoms in otherwise healthy individuals. However, in immunocompromised people, especially in pregnant women, *L. monocytogenes* can cause much more serious effects, including sepsis, meningitis, encephalitis, miscarriage, stillbirth, and life-long complications for the child.

13.5.2 Population Density

On a population level, it is clear that when there are more people together in a small area, the transmission of diseases from one person to another will happen faster and more efficiently than in sparsely populated areas. The main problem with high population density is overcrowding, which can lead to a decrease in quality of life, often accompanied by poverty and inadequate sanitation. Overcrowding usually refers to conditions in big cities but crowding also happens in social settings where many people get together, such as schools or shopping malls, which make these places great facilitators of disease transmission.

Most diseases spread well in overcrowded conditions, especially those that require airborne droplet transmission, and those that use parasite vectors to travel from one host to another. An example of a disease that takes advantage of overcrowding is typhus. This is a disease caused by a bacterium *Rickettsia prowasekii* and is transmitted through contact of infected body lice. Typhus causes flu-like symptoms and a rash, and it can lead to various neurological symptoms and multiple organ failure. Without treatment, it is fatal in 60% of the cases. Before the vaccine was developed in 1940s, this disease frequently plagued military troops, prisons, and refugee camps. It is estimated that during Napoleon's retreat from Russia in the winter of 1812, more French soldiers were killed by typhus than by Russians. Improved sanitation and hygiene, together with aggressive pesticide use against lice, brought typhus epidemics under control in Europe and North America after World War II. However, typhus epidemics have still been recorded in refugee camps in war zones of the Middle East and Africa.

13.5.3 Population Diversity

Another factor that contributes to disease spread in the population is how diverse the population is. This pertains both to the diversity within the species (genetic diversity) as well as how many different species make up a population. In both cases, individuals who cannot get the disease act as a buffer between the individuals who can get the disease, slowing down disease progression. In the best-case scenario, this buffering allows the disease to run its course in one individual before they can transmit it to the next individual.

Genetic diversity within species determines how the individual reacts to the disease agent, ranging from getting very sick to not getting the disease at all. The difference often comes from just a single mutation. For example, an estimated 1% to 10% of Europeans have a mutation in their CCR5 receptor molecules on white blood cells, making them immune to HIV. Several other important mutations throughout the immune system allows another 0.3% of people, the so-called elite controllers, to control the HIV infection without help from antiretroviral drugs.

Because people (and most other animals) have a diploid genome (two copies of each chromosome, one inherited from the mother and one from the father), difference within the genome can offer protection from various diseases as well. For example, *Plasmodium*, the malaria parasite that kills over 400,000 people every year, completes part of its life cycle in red blood cells. People with healthy, normal red blood cells are highly susceptible to malaria. In the same time, sickle cell anemia that occurs frequently in the same area of the world as malaria is a genetic disease that causes abnormal shape of red blood cells for people who possess two copies of the sickle cell gene. *Plasmodium* cannot live in these abnormally shaped cells, and therefore people with sickle cell anemia are immune to malaria. However, sickle cell anemia, too, is a debilitating disease. People who carry just one copy of sickle cell trait and one healthy copy end up with slightly warped blood cells. This is enough to ward off the most severe symptoms of malaria, and not enough to bring on the sickle cell disease.

Interspecies diversity is also extremely important in controlling disease transmission. This can be seen in the example of Lyme disease (also called *borreliosis*), which was first identified as an infectious disease in 1981. It is caused by a bacterium *Borrelia burgdorferi* and is transmitted by a tick vector (Fig. 13.17). This is another case where humans are just an accidental host: the main reservoir are white-footed mice. While other rodents

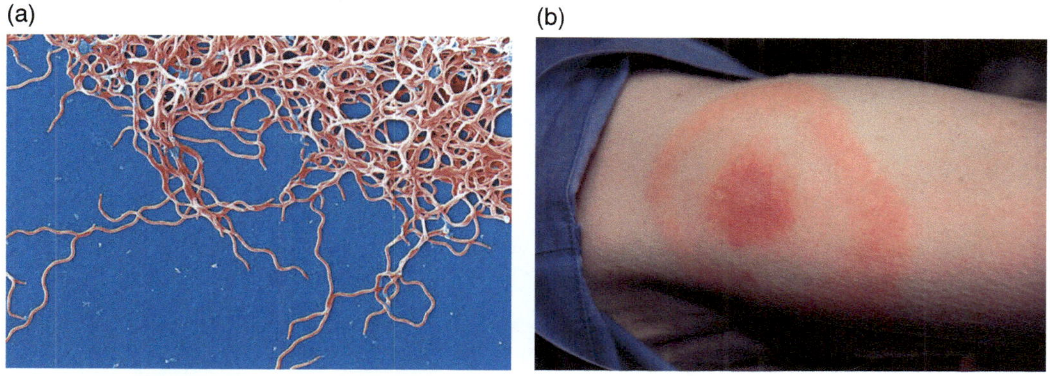

Figure 13.17 (a) Photomicrograph of *Borelia burgdorferi*, the bacterium that causes Lyme disease. Image from Black & Black, 2015. (b) Typical bulls-eye rash seen in a Lyme disease infection. Image by James Gathany, Wikimedia Commons.

can get infected by *Borrelia* and serve as a tick feeding ground, mice are poor groomers, meaning that they do not clean the ticks off their bodies, allowing the parasites to transmit or pick up enough bacteria from the blood. As long as there were enough predators to the mice (especially foxes and owls), and diversity of other rodents that were better groomers, the bacterium was also kept in check.

However, as people started building in natural woodland areas, disrupting the ecosystem, mice became the predominant rodents in the proximity of people. With mice came also the infected ticks that had no problem biting people and transmitting the bacterium and starting the fastest growing vector-borne disease epidemic in the United States with an estimated 300,000 new cases of Lyme disease in the United States every year.

13.5.4 Hygiene and Sanitation

Improved *hygiene* and *sanitation* throughout the twentieth century are responsible for a huge drop in the number of infectious diseases. This was mainly accomplished through implementation and improvement of drinking water quality, waste and sewage disposal, food safety, and public education on hygiene practices. While these practices have definitely improved the quality of life in developed countries, there is still a lot of room for improvement, especially in developing countries. The WHO estimates that 9.1% of the global disease burden could be alleviated through basic access to clean water and sanitation.

Access to clean water is indeed the main improvement that has made several diseases almost obsolete in certain parts of the world. Diseases like cholera and dysentery are no longer common in developing countries, where many people no longer even recognize the names of these diseases. However, it was not so long ago that both caused devastating epidemics in Europe as well as in the United States.

Cholera is a disease that is caused by a bacterium (*Vibrio cholerae*) that is transmitted through contaminated water (Fig. 13.18), and where the main symptoms include persistent watery diarrhea. The bacterium itself originates from the warm waters of South China Sea, where it was introduced to Europe in the 1800s through the ballast water of the ships. Several long cholera epidemics were recognized in Europe throughout the 1800s as well as epidemics throughout Asia and Africa in the twentieth century, killing hundreds of millions of people. Today, cholera infects 3 to 5 million people per year, causing 21,000 to 143,000 deaths per year. The latest notable cholera outbreaks occurred after the Haitian

Figure 13.18 London physician John Snow, who produced a study that proved cholera was being spread by contaminated water in the area of the Broad Street Pump. He plotted the locations of cholera cases in the city and found they were clustered around the Broad Street Pump. He discovered that the well was contaminated by sewage. When the handle of the pump was removed, the outbreak subsided. From NIH National Library of Medicine Digital Collections, https://collections.nlm.nih.gov/catalog/nlm:nlmuid-101434295-img

earthquake from 2010 to 2017, where it arrived with infected U.N. peacekeepers, and the epidemic in war-torn Yemen from 2016 to 2018, where Saudi-led air strikes demolished water purification and wastewater treatment facilities.

13.5.5 Poverty

Poverty remains one of the main risk factors for disease development throughout the world. It contributes to many of the factors discussed above, such as overcrowding and lack of sanitation, but also to malnutrition, lack of safety and personal protective equipment, environmental exposure to various chemicals, lack of education about risk factors, and inadequate access to health care. Millions of people in the world die every year as a direct result of poverty. Diseases that correlate strongly with poverty are called *poverty-related diseases* (PRDs). The main PRDs are AIDS, malaria, and tuberculosis, but also preventable childhood diseases such as measles, polio, and pertussis, as well as various diarrheal diseases and pneumonia. Neglected tropical diseases, such as leishmaniasis, Chagas disease, and leprosy also fall under the category of PRDs. In addition to infectious diseases, the WHO estimated that 98% of neonatal and maternal deaths in the world occur in poor countries.

One of the examples of poverty-related diseases in the 21st century is poliomyelitis (polio). Polio is caused by poliovirus, a virus that can infect only people and has no other reservoir (Fig. 13.19). It is spread from person to person through contact with contaminated fecal matter, mainly through sewage contaminated drinking water. While polio can be asymptomatic in 70% of infected people, in about 30% of the infected it will cause mild symptoms, such as fever, sore throat, and headache. However, about 0.5% will experience rapid muscle weakness and paralysis that can

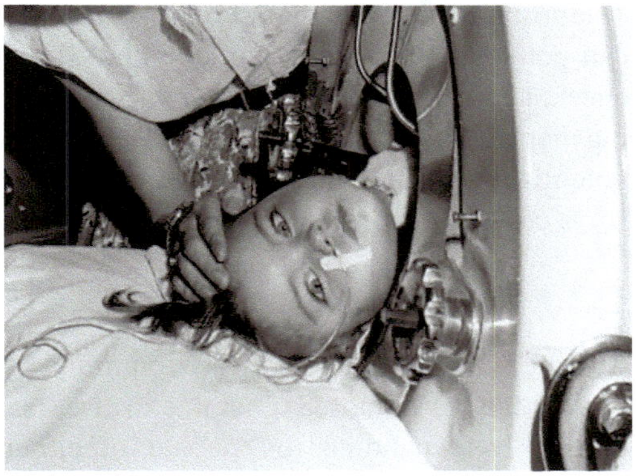

Figure 13.19 During polio epidemics, which occurred before the vaccine became available in 1955, patients were treated in iron lungs. Polio paralyzes muscles, including those that allow a person to breathe. Infected patients with paralyzed breathing muscles had to live in iron lungs for years, or until their deaths. From Black & Black, 2015.

be permanent, although many people recover at least partially. Childhood polio is fatal in 2% to 5% of the cases; fatality rate for infected adults is 15% to 30%. The disease infected 58,000 children per year at its peak in 1930s through 1950s in the United States, killing thousands and leaving tens of thousands disabled.

The polio epidemic was brought under control after two polio vaccines, injectable and oral, were introduced in the 1950s. From 1958 to today, the cases have dropped by 99.9%, and the effort to eradicate polio was announced by WHO in 1988. Currently, polio caused by wild poliovirus still occur in two countries: Afghanistan and Pakistan, where it is endemic. However, a new concern about oral polio vaccine-derived poliovirus type 2 (VDPV2) has been on a rise with over 600 cases reported globally in 2023.

In some developing countries, delivering polio vaccine (that needs refrigeration) is complicated, and travel may be restricted due to ongoing wars, even though the government supports vaccination efforts. In Pakistan and a number of northern African countries, several militant groups have declared polio vaccination to be a plot by Western countries to make their children sterile, which has subsequently hindered vaccination efforts.

13.5.6 Global Travel

When Captain Kirk of "Star Trek" described the mission of the *USS Enterprise* as "to boldly go where no man has gone before," he probably did not bear in mind that travel also comes with a risk of bringing back a disease that no man has had before. Developing technology and transportation have made global travel faster and easier not only for people, but also for various microbes that people pick up in locations far away from home. Many disease agents and other invasive species have made their way from one corner of the world to another, causing a devastating outbreak or epidemic in their destination. Chestnut blight, which killed billions of American chestnuts, is one example. More recently in 2003, a SARS coronavirus (Fig. 13.20) made its way from China to Hong Kong, from where it traveled to Canada, Singapore, Vietnam, United States, Philippines, and other countries in 2003, sickening 8,273 people

Figure 13.20 An electron microscope image of SARS virus particles (in orange) on the edge of an infected cell (green). Note the similarity to Figure 1, the SARS-CoV-2 virus, another SARS virus. Image by NIAID, Public domain via Wikimedia Commons.

and killing 775. COVID-19 became a pandemic because people traveled from China, the presumed source of the virus, to every other continent, bringing the SARS-CoV-2 virus with them.

Of course, diseases have traveled globally before air travel was even a dream. In 1846, a ship from Denmark made a stop at Faroe Islands, isolated islands in the North Atlantic, whose population had limited contact with outsiders due to its location. One of the passengers was infected with measles, a relatively common respiratory virus in Europe at that time (Fig. 13.21). As a result, 6,000 of the 8,000 Faroe islanders who had never before come in contact with the virus quickly contracted the disease.

The WHO along with the local health ministries have recognized this threat of easy dispersal of diseases through air travel. While it is hard to restrict traveling people, many countries protect their agriculture from potential crop diseases by restricting fruits and vegetables that can be brought into the country. Some countries that have been geographically isolated from others for a long period of time take extreme measures to keep certain pathogens out. For example, in Australia and New Zealand there is a strong effort to keep out soil organisms, such as fungi and protozoans, that can potentially harm their unique flora. They check, and disinfect if necessary, visitors' hiking boots. In a world where any disease is potentially a 24 hr plane ride away, these measures constitute only a minor discomfort.

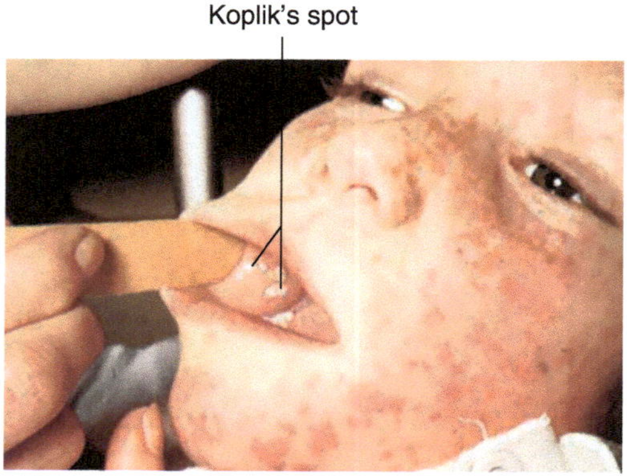

Figure 13.21 Photo depicting the typical appearance of a skin rash and white Koplik's spots in the mouth in a patient infected with measles. From Black & Black, 2015.

13.5.7 Climate Change

Climate change makes most communities around the world more vulnerable to diseases, usually by expanding the habitat of either pathogens or the disease vectors. Changes in climate that have a direct effect on pathogens include warming or cooling temperatures and increased or decreased precipitation. A good example is again malaria. *Plasmodium*, the malaria parasite, is spread by a mosquito with habitat range of tropical and subtropical climates: warm, humid regions with high rainfall. As the climate changes, so does the habitat. Some areas in the world are seeing more rainfall and warmer temperatures, and consequently more malaria cases, while other areas are seeing less rainfall and humidity, and fewer cases of malaria.

Warming oceans have already been blamed for driving many species of cold-water fish north, away from where fishermen have depended on the catch for centuries. However, the warming water also brings with it harmful bacteria and algae that thrive in tropical waters. One of these organisms is *Vibrio vulnificus*, the same bacterium that causes food poisoning or flesh eating disease. Water samples taken in different locations of the North Atlantic since 1958 show a threefold to fourfold increase in the amount of *V. vulnificus*, which correlates both with rising water temperatures and numbers of documented infections.

13.5.8 Evolution of Organisms

While most of the risk factors pertain to individuals and populations, the pathogens themselves can also contribute to the risk, sometimes with the help of people. Most microbes evolve much faster than hosts, which is mainly due to their simpler cellular structure. As a rule, the simpler and smaller the organisms, the faster they can adapt to their environment through evolution. Bacteria evolve faster than eukaryotes, and viruses evolve even faster as they do not have the constraint of being alive.

A good example of viral evolution is the ever-changing Influenza A virus. One of the reasons the virus is mutating is that when it uses the host cell to make copies of its genome, the enzymes that copy the nucleic acid make mistakes. These mistakes are then reflected on the surface molecules of the virus (called *viral antigens*), which in turn are no longer completely recognized by our immune systems. This is why people can get flu over and over even though they have already had it or were vaccinated for it the year before (Fig. 13.22). However, since the immune system can recognize some parts of the viral antigens, it can start fighting against it immediately, making the person sick, but unless they have a weakened immune system (elderly, very young, or immunocompromised), seasonal flu has relatively low mortality rate. However, it is still estimated that during a normal flu season, 250,000 to 500,000 people worldwide are killed by the virus.

Occasionally, more serious changes can occur in the virus, which make it deadly. Sometimes these changes happen in the antigen molecules, which make the virus almost completely unrecognizable for the body. Other times several different Influenza A viruses can recombine (or mix and match) inside one cell, which can lead a virus from one organism (such as birds) to infect another organism (such as people). This kind of mixing and matching of influenza viruses usually happens in pigs. When these kinds of big changes occur, the result is usually a global pandemic, with deaths generally exceeding 1 million. The latest such pandemic occurred in 2009, although luckily the new H1N1 swine flu strain (actually a recombination of human, swine, and bird influenza strains) had low case fatality rate with less than 0.1%, which is comparable to the seasonal flu. As a comparison, the Spanish flu strain that caused a global pandemic from 1918 to 1920 had case fatality rate of about 2% to 3%, causing 40–75 million deaths worldwide.

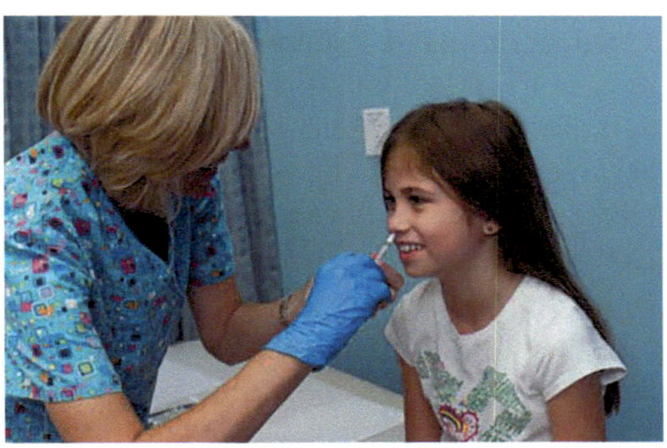

Figure 13.22 A child receiving the Flu Mist vaccine to guard against new strains of influenza. From Black & Black, 2015.

While flu viruses mutate on their own accord, people and their behavior cause the evolution of some bacterial strains. This is particularly the case with the rise of *antibiotic resistance* among common pathogens. Antibiotic resistance is coded in the bacterial DNA. Bacteria have many ways to exchange genetic information with other bacteria, spreading the ability to resist antibiotic treatment. In nature, antibiotics are mainly produced by soil bacteria and fungi, which act as their chemical weapons against other bacteria in a fight over space and nutrients. Antibiotic resistance developed as a means to withstand this attack.

In health care settings, antibiotics are used to kill pathogenic bacteria, but *antibiotic resistance* develops and is spread for the same reason as in the soil, for survival. Human and veterinary medicine, together with prophylactic antibiotic use in agriculture, use vastly more antibiotics than what is produced naturally by small-scale production in the soil. Therefore, antibiotic resistance in medical, veterinary, and agricultural settings has become a huge and widespread threat. Doctors and other scientists worry for some strains of bacteria, we may progress to a point in antibiotic resistance that we will no longer be able to treat infectious bacterial diseases.

In part we are already there. Resistance to penicillin (Fig. 13.23), the first antibiotic discovered in 1928, is common among most microbes that were routinely treated with it in the middle of the twentieth century. Tuberculosis, which in the pre-antibiotic era accounted for 10% to 25% of all deaths in Europe, was almost eradicated after streptomycin was utilized as an effective treatment in the 1950s. However, it started to return in the 1980s, now resistant not only to streptomycin, but also many other antibiotics, including fluoroquinolones, rifampicin, and isoniazid, the strongest antibiotics approved for tuberculosis treatment. Various antibiotic-resistant bacteria, such as methicillin-resistant *Staphylococcus aureus* (MRSA), vancomycin-resistant enterococci (VRE), and carpabenem-resistant Enterobacteriaceae (CRE), make up a significant portion of hospital-acquired infections.

Figure 13.23 Amber-colored drops on the surfaces of these cultures of *Penicillium commune* and *Penicillium chrysogenum* are the antibiotic penicillin, which it produced. From Wikimedia Commons, https://commons.wikimedia.org/wiki/File:Penicillia_on_Petri_dish.jpg

A most concerning antibiotic resistance case occurred in the fall of 2017, when a Nevada woman died of a *Klebsiella pneumoniae* infection where the bacterium was resistant to all 26 antibiotics that were approved for its treatment.

13.6 Diseases in the Food Supply

So far we have mainly examined diseases that affect people or animals. However, as was discussed in the case of the chestnut blight, plant diseases can have a significant impact not just to the plant communities, but also to people who depend on the plants as part of their livelihood. This is especially acute when diseases devastate food crops. Any plant is vulnerable to a disease, and there are two factors that increase the likelihood of large-scale plant disease outbreaks. The first is introduction of a novel pathogen, which the plants have no resistance to. The other is growing and depending on a single crop species (see Population Density, section 13.4.2, and Population Diversity, section 13.4.3). A great example of how both of these factors brought on a food crop failure that drastically changed a society is the Irish Potato Famine of 1845–1849.

In the 1800s, when Ireland was under British rule, the Irish, who were mainly Catholic, were not granted many rights under the Protestant British government. For example, laws prohibited the Irish Catholics from owning land. The Catholics instead rented out farm plots that belonged to British landowners. The farm plots rented to the tenants were limited to 1–5 ac, and there was only one crop that could feed a family from such a small area, the potato. As a result, most of the farmland was planted with the monoculture of potatoes, which were mainly of just one variety. This left the food staple vulnerable to various potato diseases, and occasional potato crop failure was not uncommon.

However, none of the crop failures was as great as the ones brought on by *Phytophthora infestans*, a parasitic protozoan that was brought over from North America in 1844, causing potato blight throughout Europe. Because potato was their main crop, Ireland suffered especially harshly. It is estimated that in 1845, almost half of potato crop failed; in 1846, the estimated crop loss was 75%. In the spring of 1847, the seed potatoes were scarce, since most of the small harvest of 1846 potatoes had been eaten for survival. Most of the people were starving. An estimated 1 to 1.5 million people died. Most families could not afford to pay the rent for the farmland.

As a result, the landlords, who had to pay taxes for every family living on their property, were now faced with massive payments and no income from the tenants. Since the families could not pay, they were evicted from the farm plots, leaving them with no means to earn income or feed their families. This left many of them with one viable option, to emigrate. The most common destinations included England, Scotland, North America, and Australia. It is estimated that during the famine years, about a quarter of a million of people left the island every year. The people who reached their destination established the vibrant Irish communities seen in the United States, Canada, and Australia these days.

13.7 Mitigation of Hazards

While microbes are small and the prevention of diseases can be hard, or sometimes almost impossible, there are several steps that can be taken to mitigate biological hazards. Some of these actions are taken to prevent the disease outbreak from developing in the first place. These include the improvement of sanitation assuring the basic cleanliness and safety of food and water, but also personal hygiene, vaccinations, control or elimination of disease vectors, prevention of antibiotic overuse, and varied agriculture. Other measures are taken to contain an outbreak that is in its first stages. These include culling of animals and quarantine.

13.7.1 Sanitation

It is the basic improvement of sanitation that is responsible for reduction of most diseases in the world. Sanitation is a broad term that includes drinking water purification as well as proper treatment and disposal of sewage. Several diseases already covered in this chapter, such as hookworms and cholera, arise from the lack of sanitation. The World Health Organization estimates that 4% of all deaths in the world result from the lack of sanitation, including 1.5 million children and infants. Diseases that arise from lack of sanitation usually involve diarrhea, anemia, and stunted growth and development.

While overall sanitation has improved considerably throughout the twentieth century, improvements are not evenly distributed in the world. While most people in the developed countries take clean drinking water and working sewers for granted, things can be dire in other regions of the world. For example, the WHO estimates that 2 billion people in the world still lack clean drinking water. Even more concerning is that in middle- and low-income countries, 38% of health care facilities do not have access to clean water, 19% lack properly working sewers, and 35% do not have water and soap for hand washing. The total economic losses from inadequate water supplies and sanitation are estimated to be $260 billion annually.

The World Health Organization together with local governments are working to improve sanitation in vulnerable areas. The estimated cost for attaining universal coverage for safe drinking water and sewage disposal is about $400 billion. While this number is high, the WHO estimates that for every dollar spent on sanitation improvement, there will be $4.30 return from reduced health care costs and increased productivity.

13.7.2 Hygiene

Improvements in personal hygiene have been yet another factor in reducing the number of diseases and deaths around the world. The term *hygiene* refers to a set of practices that help to maintain health and prevent the spreading of diseases. Hygiene is directly linked to sanitation, the availability of clean water and working waste disposal, but also to the actual habit of keeping the body and surroundings clean. Hygiene is a powerful practice when it comes to reducing the number of diseases. CDC estimates that hand washing by itself reduces the spread of diarrheal diseases by 50%, potentially preventing millions of deaths.

Hygiene practices such as hand washing, which seem self-explanatory for us, have routinely been practiced in the Western world for a just little over 100 years. The first person in modern history who proposed hand washing as a means of preventing diseases was Ignaz Semmelweis, a physician working in the obstetric clinic in the Vienna General Hospital in the 1840s. He noticed that maternal deaths from puerperal fever were very high in the clinic where medical students went straight from dissecting cadavers to tending their patients, while the deaths were low in the clinic where childbirths were attended by midwives. He proposed that the medical students should wash their hands with disinfectant between the cadavers and patients, which immediately and sharply reduced maternal mortality. Unfortunately, the practice of hand washing was abandoned as Semmelweis left the clinic, and it took decades to become a widely practiced precaution in health care settings, and even longer to become routine for the general public.

13.7.3 Food Safety

Assuring the safety of foods is exceptionally important to reducing foodborne diseases, several of which can be more severe than just the classical food poisoning symptom of diarrhea. Food safety depends on availability of clean water and sanitation as well as the practice of hygiene. To keep up with increasing population, global food production has increased 145%, but the number of foodborne illnesses as well as the number of foodborne pathogens has increased too. In developed countries, it is estimated that 20% to 30% of the population will experience a food-related illness every year. In the United States, about 76 million foodborne illnesses occur yearly with about 5,000 deaths occurring as a result. While the risk of dying from a foodborne illness is relatively low (about the same chance as a pedestrian dying from being hit by a car), USDA estimates that the economic cost from just the top seven foodborne pathogens is between $6.6 and $37.1 billion per year.

The need for food safety and monitoring has been in a race to keep up with increased food production for the past 100 years. These days, food microbiology, which used to be regarded as just an applied science, is using the latest, basic scientific approaches and techniques to solve food safety problems, including predicting and preventing future outbreaks from occurring. Several high-profile food poisoning cases, such as the 1993 *E. coli* O157:H7 outbreak in Jack-in-the-Box restaurants that killed 4 children and left 178 permanently disabled with kidney and brain damage, have created more public awareness for food safety issues as well as helped to develop prevention measures to keep these tragedies from occurring again.

13.7.4 Vaccinations

Vaccinations have been one of the most successful methods of infectious disease control, giving the body its own ability to fight the disease. Vaccines are mainly developed against viral infections (measles, mumps, polio) or diseases that are caused by bacterial toxins (tetanus, diphtheria). Sometimes one series of vaccinations early in life is sufficient to protect the person against the disease for the rest of their life, as it is in the case of polio or hepatitis B. However, when the disease agent mutates rapidly or the body can partially lose the "memory" of the vaccination, frequent booster shots will be necessary. Influenza virus that mutates fast requires yearly new vaccinations; for tetanus, which is caused by a bacterial toxin, booster shots are recommended every 10 years.

While improvements in sanitation, hygiene, and pest control have rendered many diseases obsolete, vaccination strategies have been used to actually eradicate two diseases from the nature. One of them is Rinderpest, a viral disease of cattle and other ungulates, which was declared eradicated in 2011 after a successful vaccination campaign in the 1990s. The other eradicated disease is smallpox.

The global success story about eradication of smallpox is well known. This debilitating viral disease with 25% fatality rate had plagued people since prehistoric times, causing high fever and the characteristic fluid-filled painful blisters that often left the victim severely scarred for life (Fig. 13.24). An estimated 2 million people every year died of smallpox in the middle of the twentieth century.

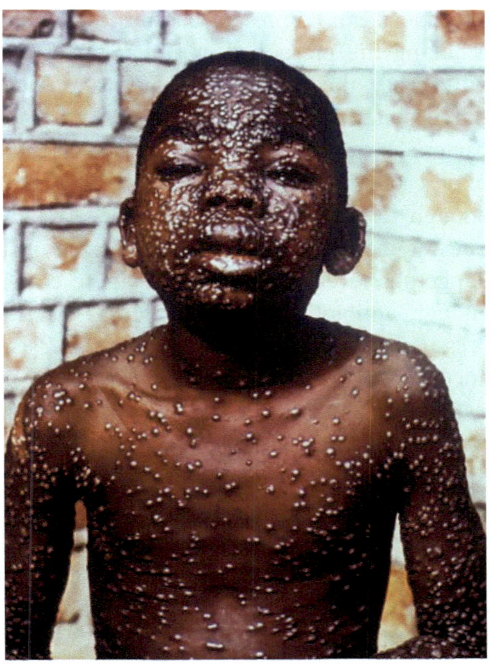

Figure 13.24 One of the last cases of smallpox in the world. This boy shows the blister-like lesions. As the disease progresses, the lesions become opaque and eventually form crusts and fall off, leaving scars. From Black & Black, 2015.

However, smallpox, caused by the *Variola* virus, had two weak spots that made its eradication possible: a successful vaccine had been developed against it, and it could only inhabit the human body. The World Health Organization started the global smallpox eradication program in the 1950s and spent about $23 million annually on effective vaccination strategies until 1979 when the last naturally occurring smallpox case was recorded. After waiting almost 2 years for the disease to reappear, the World Health Assembly declared smallpox to be the first disease eradicated from nature in May 1980.

Encouraged by this success, several other diseases were placed in the shortlist of global eradication: polio, Guinea worm disease, and yaws. Others were placed on a list for local eradication, such as hookworm, measles, and rubella. To date, none of these have been as successful, even though a great deal of effort has been expended. Diseases caused by pathogens that are exclusively human (polio, measles, rubella) are easier to eradicate, but even in those cases, conspiracy theories, political unrest, and wars can rapidly undo decades worth of medical success.

While vaccines are one of the most successful methods of combating diseases, they are relatively expensive to develop, produce, and deliver. To fund and coordinate the efforts, in January 2000, the World Health Organization together with the United Nations Children's Fund, the World Bank, various private philanthropic organizations such as the Bill and Melinda Gates Foundation, the International Federation of Pharmaceutical Manufacturers Associations (IFPMA), various research and public health institutions and national governments have created the Global Alliance for Vaccines and Immunization (GAVI). To date, this organization has raised billions of dollars to combat diseases through vaccinations, vaccinated an estimated 700 million people, and prevented over 10 million deaths.

13.7.5 Vector Control

Vector control operates on the principle that by removing the disease vector, the pathogen cannot spread. It is especially important in case of infectious diseases that have no cure, such as Dengue Fever or Zika virus disease, but equally effective for other vector-borne diseases, such as malaria, Lyme disease, and typhus. The most common vector is the mosquito; other arthropod vectors include lice, ticks, sand flies, and the tsetse fly.

Four basic strategies are used for vector control, which are frequently combined for maximum effect. The first is vector habitat control, which will usually inhibit their breeding. For example, removing all standing water containers (old tires, cans, tubs, stagnant ponds) will limit mosquito breeding places. The second is limiting contact with the vector by physical barriers, such as bed nettings, window screens, and protective clothing. Chemical control is the third option, where pesticides (insecticides, larvicides, rodent poison) are applied into the environment. While pesticide control has been shown to be highly effective in many cases, in some cases the chemicals have shown to do damage to nature (DDT), or the vectors have grown resistant to the chemical (rats growing resistant to rat poison).

The final strategy is biological control. In this case the vectors are controlled by application of bacterial or plant toxin, introduction of organisms that prey on the vector (use of fish that eat mosquito larvae), or by releasing genetically engineered sterile vectors, such as mosquitoes into the vector population. The latter has been successfully used in different parts of the world but comes with a caution about the ethics and biological safety of these organisms.

13.7.6 Antibiotic Overuse Prevention

The overuse of antibiotics is one of the examples where a great initial success in combating infectious diseases has turned into an arms race with some bacteria that develop and acquire antibiotic resistance genes soon after new antibiotics are developed. Penicillin, the first antibiotic, was discovered in 1928 and first mass-produced in 1943. By 1955, many common bacterial strains that were treated with it were showing resistance. The same happened to almost all antibiotics that were discovered, or later modified or developed. Bacteria started to develop resistance to multiple antibiotics at the same time, and in time, several "superbugs" evolved that are resistant to most of the antibiotics routinely prescribed to treat them. The World Health Organization estimates that about 1 million people every year die from antibiotic resistant infections.

While some bacteria are naturally resistant to certain antibiotics, most bacteria in the medical setting acquire the ability to resist antimicrobial agents by picking up genetic material that contains the antibiotic resistance genes. Bacteria pick up the genes more readily when there are antibiotics present in their environment. If the environment is the human body, antibiotics can be present because the person is taking the prescribed antimicrobial drugs, or because the person has eaten antibiotic-raised meat or washed their hands with antibacterial soap. In the latter two cases, the antibiotic is present at subtherapeutic level: enough for the bacteria to feel the presence but not enough to kill them. This is the most dangerous level, as at this level there is just enough antibiotic present to put a selective pressure on the bacteria to acquire the resistance genes.

There are other factors that contribute to the spread of antibiotic resistance, including the increased availability of different antibiotics, overprescribing of antibiotics, incorrect dosage, antibiotic misuse and overuse in agriculture, release of antibiotics into wastewater, and casual antimicrobial agent use, such as in the antimicrobial soaps. The standard answers to growing antibiotic resistance has been (1) to develop new antibiotics, (2) to prescribe a combination of antibiotics, and (3) to produce antibiotics that have added resistance inhibitor drugs. While all these are viable solutions, it has not stopped the development of new antibiotic resistant strains of bacteria (although it may have slowed them down).

The overuse of antibiotics in agriculture is one contributing factor to growing antibiotic resistance. Antibiotics are used in agriculture to treat infections (like in human medicine), but in some cases also prophylactically in the feed to prevent infections from developing. Until recently they were also used as growth promoting factor to allow poultry, cattle, and pigs to gain more weight rapidly. The latter practice has been discontinued in most countries. Although most antibiotics used in agriculture are not used in human medicine, they can be similar, and the ability to withstand animal antibiotics can also be used by bacteria to withstand human prescriptions.

One possible solution against so-called superbugs, which are resistant to most antibiotics, is the use of bacteriophages, employing something known as phage therapy. Bacteriophages (or just phages) are viruses that infect bacteria. Like all viruses, they are highly specific to the type of cell they infect, and thus bacteriophages are highly specific when they attack bacteria. This is beneficial, because it allows the phages to target only the pathogen and not harm other beneficial bacteria in the human body, which is what many antibiotics do.

Can bacteria develop resistance to the phages as they do to antibiotics? Of course, but the phages themselves are excellent at evolving and mutating, naturally keeping up with the bacterial mutations. In this case, the medicine has treatment that evolves together with the pathogen! Phage therapy was successfully developed in the former Soviet Union and has been used in the former Soviet countries for decades. Luckily, as the global antibiotic resistance is rapidly increasing, phage therapy is gaining more international recognition, with the first experimental treatments now taking place in the United States.

13.7.7 Quarantine

Quarantine is probably the oldest measure for isolating the sick, or people who have come in contact with the sick and may develop the disease. Quarantine can be applied to both people and animals. For animals, quarantine is frequently used as a routine prevention measure for traveling purposes.

The term *quarantine* comes from Italian *quaranta giorni*, meaning "forty days," the period of time required by Venetian government that all ships had to be isolated before the passengers could come on land in order to prevent more plague epidemics from occurring in the fourteenth and fifteenth centuries. Historically, quarantine has been successfully implemented to keep out plague, leprosy, yellow fever, typhoid, influenza, smallpox, and cholera. It is also applied to infected farms to keep infectious animal diseases, such as foot and mouth disease, from spreading.

Animals are frequently quarantined, too, especially in these days of global trade and travel. Most geographically isolated countries or areas (generally located on the islands that naturally limit disease spreading) impose quarantine on any animals that people may bring to the country. In addition to farm animals, this is also applied to people traveling with their pets. For example, Australia is free from many common cat and dog diseases, such as rabies. Because export of these diseases can potentially be devastating for its unique wildlife, the Australian government requires all the pets entering the country undergo a minimum of 10 days of quarantine. Similar rules are in place for pets entering Hawaii, New Zealand, Japan, and other islands.

One of the most notable quarantines of the modern times happened in 1969, when the returning Apollo 11 astronauts were quarantined for 3 weeks to prevent a potential spread of any infectious agents from the Moon, even though the existence of this kind of agent was highly unlikely.

13.7.8 Culling as Disease Control

One of the most drastic measures that can be taken to curb an epidemic is the killing of not just the infected animals, but also those that have come in contact with infected animals. Clearly, this is a solution for diseases that involve animals and crop plants and cannot be applied to human populations. While culling as a disease control is an effective measure of stopping the disease spread, it does raise questions about the ethics of such practices.

Culling for disease control has successfully been applied for both wild and domestic animals and birds. In wild animals, overpopulation frequently leads to spreading of diseases. For example, the overpopulation of white-tailed deer in the Midwest has facilitated the spread of chronic wasting disease (CWD), a prion disease in deer that resembles mad cow disease. In Illinois and Wisconsin, two states hit hard by CWD, the government culling programs have kept the disease relatively under control. When Wisconsin discontinued the culling program in 2007, the number of CWD cases started steadily increasing.

In domestic animals, culling has been prescribed for rapidly spreading diseases, such as foot and mouth disease and African swine fever. An epidemic of foot and mouth disease that swept the United Kingdom in 2001 led to the culling of over 6 million cattle, sheep, and pigs; the economic cost was estimated to be over $10 billion. It also is frequently used in the case of highly pathogenic avian influenza outbreaks, where culling of all infected flocks of domestic birds as well as the susceptible flocks surrounding the infected farms takes place. Depending on population density, the farms surrounding the immediate neighbors to the infected farm may need to be culled, too (in densely populated country), or undergo emergency vaccination (sparsely populated country). The latest large-scale avian flu outbreak in the United States happened in the Midwestern states in 2015, leading to the culling of over 3 million chickens and turkeys. The price of eggs doubled. In the case of the highly pathogenic avian flu, the concern is not only the birds, but also the real possibility that the virus would mutate and become a highly pathogenic human strain.

Obviously, culling for disease control is not a viable long-term strategy, and cannot replace preventive measures, such as vaccination. However, when vaccine and/or drugs for treatment are in short supply, rapid culling in early stages of an epidemic does help with containment of the disease.

13.8 Conclusion

Though it may seem from this chapter that there are dangers everywhere from parasites, viruses, bacteria, fungi, viruses, and infectious molecules, only a small percentage of microbes are capable of causing disease in humans. Microbes inhabit all areas of the Earth System, including rocks, soil, water, and even the air. Many microbes engage in symbiotic relationships with each other, as well as with plants and animals. In fact, we have as many bacteria cells in our bodies as human cells! The vast majority of these bacteria are there to help us in the body's everyday functions, and we couldn't live without them.

Many pathogens, however, can be quite devastating to the human body. Helminths, or worms, infect both animals and humans, and can cause malnutrition, blood loss, and even death. Both parasitic and free-living protozoans come in many varieties that can cause severe illness in people and can even create toxins that kill other organisms. Fungi come in both unicellular and multicellular forms, causing disease in plants and humans. Prions are becoming an increasing part of the conversation in North America because they are causing disease and death via chronic wasting disease that effects members of the deer family.

Viruses have been in the news and become the focus of much of the world's news since late

2019. COVID-19, caused by the SARS-CoV-2 virus, plunged the world into a pandemic that has sickened hundreds of millions of people, and killed well over 5 million people. As a result, we have all become better educated about viruses, how they are spread, and how to mitigate these biologic hazards to protect ourselves. Thorough hand washing, face masks, social distancing (leaving at least 6 ft or 2 m of space between you and the next person), and vaccines are all mitigation strategies used to fight the spread of COVID-19 and other viruses. These are common mitigation strategies for pathogens of all types.

13.9 Chapter in Review

Disease-causing pathogens can exist in soil, air, and water. Most pathogens need a host, or an organism in which to take up residence. The host may or may not get sick from the pathogen. For example, bats are reservoir hosts for a number of microorganisms, but do not typically get sick from them. Some organisms need two hosts, a primary and a secondary host, to complete their life cycles. Accidental or dead-end hosts do not transmit disease to another organism. For example, the protozoan that causes malaria needs both a mosquito and an animal to complete its life cycle. The mosquito acts as a vector, which means it transports the malaria parasite to its new, dead-end host. The mosquito is not affected by the parasite, but people can get very ill and die from malaria.

Some diseases are endemic in the population, which means they always exist in the population at a low level. When a disease develops in a population, this is called an outbreak. When that disease spreads to a wider area, it called an epidemic. If the disease then spreads to other continents, it is called a pandemic. The most recent pandemic to affect the world is the SARS-CoV-2 virus, which causes COVID-19. The most severe pandemic prior to COVID-19 was the Spanish Flu, which spread across the world in 1918.

There are many types of organisms and molecules that can cause disease in humans, animals, and plants. Parasites can be single-celled organisms or multicellular organisms. The largest group of multicellular parasites are the helminths, or worms. This includes roundworms, hookworms, and tapeworms. Worms have coexisted with larger organisms for millions of years. Many do not cause severe disease, but others can cause severe anemia, malnutrition, painful sores, distended abdomens, and death. Many worms are relatively easy to eliminate with treatment, but for much of the world's population the treatment is unavailable or too expensive.

Protozoans can cause disease by acting either as a parasite or as a free-living organism. Malaria, sleeping sickness, amoebic dysentery, toxoplasmosis, cryptosporidiosis, leishmaniasis, and the potato blight that ravaged Ireland in the 1800s are examples of diseases caused by parasitic protozoans. Free-living protozoans can create toxins that have deadly effects on other organisms. Certain dinoflagellate algae that live in seawater produce a toxin that is red-brown in color. This is the cause of red tides, which kill fish and sicken other marine animals. Red tides have been recognized in warm seas of southern North America and the Caribbean since the Spanish arrived in the 1500s.

Fungi can be both plant and animal pathogens. They also form symbiotic relationships with plants that allow those plants to survive and flourish. Orchids, for example, depend on mycorrhizal relationships between their root systems and a fungus. However, fungi can also cause disease. Ringworm and athlete's foot are caused by fungal infections of the skin. Some respiratory illnesses are also caused when people inhale fungal spores in dust. Many of these respiratory illnesses are difficult to treat. Drugs that work to eliminate the respiratory fungal infections can be toxic to humans, making the cure almost as bad as the illness. Fungi are also formidable plant pathogens. A chestnut blight in North America in the early 1900s nearly eliminated chestnut trees in the eastern United States.

Bacteria, like fungi, often form symbiotic relationships with other organisms, including humans. Bacteria form the basis of the food chain, and are responsible for turning atmospheric nitrogen into molecules that plants can use. Nitrogen-fixing bacteria are critical to life on Earth. However, as with other types of microbes, some bacteria cause disease. Salmonella, *E. coli*, and listeria outbreaks are caused by bacteria. Staph infections, clostridium, and Lyme disease are also the result of pathogenic bacteria. One of the deadliest pathogens in all of history was the bubonic plague, caused by the bacterium *Yersenia pestis. Y. pestis* is spread from rodents to humans via a flea vector. Outbreaks of bubonic plague in Europe killed as many as 200 million people between the sixth and sixteenth centuries. Additional outbreaks in China and India occurred more recently. Thankfully, though bubonic plague still exists today, it can be treated with an antibiotic and rarely causes deaths in the modern world. Recently, reindeer in Siberia were found dead of an unknown disease. When scientists investigated, they found that the deer had become ill from exposure to the bacterium that causes anthrax. These bacteria were released into the environment after the permafrost that contained them melted. With increases in temperatures worldwide, scientists anticipate that a great deal of permafrost in cold regions will melt, and more infections of people and animals will happen due to exposure to bacteria long-frozen in the soil.

Viruses consist of DNA or RNA surrounded by a protein coat, but are not living organisms. They cannot reproduce by themselves or gather their own nutrients. Instead, they invade other cells and use those cells' mechanisms to replicate and exploit the cell's resources, damaging or killing the host cells in the process. Ebola, measles, mumps, Dengue Fever, Yellow Fever, coronaviruses, poliomyelitis, hepatitis B, Chikungunya, and Zika are examples of viruses that have spread widely through the population. SARS, Severe Acute Respiratory Syndrome, is a group of viruses that cause acute respiratory distress, and can result in death. SARS-CoV-2, which causes COVID-19, is a member of this group.

Viroids are a type of pathogen called an infectious molecule. Like viruses, these are not living organisms. Viroids consist of RNA with no protein coat. Some viroids are plant pathogens, causing stunted growth in plants such as potatoes and tomatoes.

Prions are also infectious molecules, but they are composed of protein molecules. Most proteins fold in only one way, but prions can fold in multiple ways. These misfolded proteins are infectious, and spread throughout the nervous system. Diseases caused by prions are always fatal. Sheep can be infected with the prion disease scrapie. Bovine spongiform encephalopathy was recognized in the United Kingdom in the 1980s. Evidence points to scrapie-infected sheep byproducts in cow food as a possible step from scrapie to prion disease in cows. In North America, Europe, and most recently Korea, members of the deer family are being infected with a prion disease that causes chronic wasting disease. People can acquire prion diseases as well. Creutzfelt-Jacobs disease, Kuru, and fatal familial insomnia are neurodegenerative diseases that are always fatal.

Microbes exist in every part of the Earth System. Chemolithotrophs use enzymes to break down rocks and metals. They can become problematic when they corrode sewer pipes and worsen acid mine drainage. In the soil, many microbes cause plant diseases. The microorganisms that cause tetanus and anthrax are also present in the soil, as are worms and worm eggs. Waterborne pathogens cause a great number of diseases, including cryptosporidium, cholera, and invasive septicemia (an infection in the blood stream). In addition, necrotizing fasciitis, also known as flesh-eating bacteria, is usually spread as a waterborne pathogen. Primary amoebic meningoencephalitis (PAM), caused by the so-called brain-eating amoeba, is contracted when a person inhales water or water droplets from the air. It is usually fatal.

We have created many vulnerable populations thanks to increased levels of health care. Many people are immunocompromised, so they are more vulnerable to disease. Poverty is a driver of disease because sanitation and hygiene practices are poor. High population densities, both in large cities where people live close together and in large gatherings, can be places where disease spreads rapidly. Global travel allows diseases to spread anywhere in the world within 24 hr. Viruses and other pathogens evolve, often rather quickly. In the case of COVID-19, there have been a number of mutations to the original virus that have given us new variants that are more transmissible and deadly than the original virus. Overuse of antibiotics, often at subtherapeutic levels, have caused many bacteria to become antibiotic resistant. Therefore, many bacterial infections are harder to treat than they used to be. Climate change has already caused melting of permafrost that has released pathogens into the environment. Most scientists believe this will accelerate as the planet warms further.

There are things we can do to mitigate biological hazards. Improving hygiene includes accessibility to clean water and working waste disposal, as well as the habit of keeping the body and surroundings clean. Related to hygiene is sanitation, which is a broad term that includes drinking water purification as well as proper treatment and disposal of sewage. Poor sanitation often results in disease such as cholera and worm infestations. Food safety is a field that struggles to keep up with food production, but is vital to keeping the population free from foodborne pathogens. Outbreaks of *E. coli* and salmonella are not uncommon, and can be fatal, especially to the very young, very old, and immunocompromised among us.

Vaccines are a very effective means of mitigation of some diseases. Smallpox was a devastating disease that caused disfigurement and often death to those infected. Through vaccination campaigns, smallpox was eradicated and is no longer present in the world's population. People also have the ability to fight antibiotic resistance by eliminating the use of subtherapeutic antibiotics such as antibacterial soap. In addition, doctors can stop overprescribing antibiotics, prescribing them only when there is a clear bacterial infection that needs treatment.

Other mitigation measures include vector control, which seeks to control the population of vectors such as lice, fleas, mosquitos, and ticks, which carry pathogens from a primary host to a secondary host. Quarantine is a period of time where a person, plant, or animal must be kept in isolation before entering a new environment. Culling is a mitigation strategy of last resort. If a population of animals is infected with a disease, and the disease both spreads rapidly and cannot be treated, culling is a method that allows for removal of all members of the population so they cannot spread the disease to other populations in the area. Bird flu infections spread throughout poultry farms in the Midwest in 2015, and millions of birds were culled as a result.

14

Hazards From Space

Kelli Wakefield[1] *and Joel Berrier*[2]

[1] *Mesa Community College, USA*
[2] *University of Nebraska at Kearney, USA*

14.1 Space Hazards

This chapter, like the last, explores hazards that are neither geological or meteorological in nature. Here we will explore hazards that originate from outside of the Earth's atmosphere. When we examine other bodies in our solar system, we find that most planets and moons with solid surfaces are heavily cratered. Mapping and dating of craters on other planets tells us that in the early days of the solar system between about 4.6 billion and 2 billion years ago, collisions between bodies within the solar system were common. Asteroids and meteorites struck not only other planets and moons, but also the Earth. Large craters are still visible on Earth despite thousands or millions of years of erosion.

Thankfully, large and devastating meteor impacts are rare in our solar system today. That does not mean that the Earth and human populations are unaffected by events and objects originating outside of our atmosphere. Heightened solar activity can disrupt radio signals and satellite communications. Meteors still enter the Earth's atmosphere and some fall to Earth, occasionally in populated areas. As you will see in the case study that follows, even small meteors can cause significant damage. The Russian city of Chelyabinsk was the site of a rare event in 2013 in which a small meteor caused both injuries and property damage when it detonated in the atmosphere. This is illustrative of the type of event that is much more likely on Earth today.

14.1.1 Chelyabinsk, Friday, 15 February 2013, 9:20 YEKT (3:20 UTC)

Just after sunrise, residents of Chelyabinsk, Russia, and its surrounding area were witness to an unusual event. A 62 ft (19 m) diameter meteor entered the Earth's atmosphere over the Ural Mountains, along a trajectory with a shallow angle of 18.3 $\pm$ 0.4 degrees and a speed of 42,690 mi/hr (69,000 km/hr). This meteor was over 20 million lb (9.1 million kg). While passing over Chelyabinsk, the meteor detonated in the sky less than 19 mi (30 km) above the surface in the stratosphere. This type of explosion is classified as a meteor airburst, as the meteor breaks apart and much of its kinetic energy is converted into heat.

The first sign of impact was an incredibly bright flash of light and then a trail of smoke. The explosion created a shock wave that took 2 min to reach the surface. The shock wave was detected on seismic records in the area with the equivalent energy of a magnitude 3.7 earthquake. Surface waves were detected 2,500 mi (4,000 km) from the explosion site.

Earth's Natural Hazards and Disasters, Advanced Textbook 7, First Edition. Bethany D. Hinga.

The detonation of the meteor caused shock waves in the atmosphere, releasing most of the *kinetic energy* (energy of motion) of the meteoroid and was equivalent to the detonation of between 400–500 kilotons of TNT. Seismic energy released by the explosion and recorded on seismometers was used to measure the energy of the event. Numerous meteorite fragments were recovered in the vicinity of Chelyabinsk, and thousands more were probably created. In addition, the detonation created an atmospheric dust cloud that NASA was able to track for weeks after the event.

Thousands of residents of Chelyabinsk were injured due to secondary effects from the blast, and 1,491 of those people needed medical treatment. Most injuries were due to glass broken by the shock wave. Shattered windows in thousands of buildings across the area presented a particularly hazardous situation because they left people exposed to the cold winter temperatures in Russia. Seven thousand two hundred buildings were damaged in the region of Chelyabinsk and its neighboring cities. A zinc factory in Chelyabinsk was severely damaged when the roof of the building collapsed. In total, around $33 million in damages occurred as a result of this event.

The small object responsible for the Chelyabinsk meteor was undetected prior to the impact. How did the meteor approach the Earth undetected? One reason it was so difficult to detect was its size. It was only 62 ft (19 m) across, and objects that small are difficult to detect simply because they are so small. In addition, it arrived from the direction of the Sun and was therefore not visible relative to the Sun's light. After entering the atmosphere, however, numerous eyewitnesses observed the descent of the meteor and its explosion. Hours of video footage of the event were recorded by security cameras and dashboard cameras, which are quite common in Russia.

The Chelyabinsk meteor was classified as a *superbolide*. A *bolide* is a bright meteor typically called a fireball, while a superbolide must be at least 100 times brighter than the Moon. The flash of this explosion was momentarily brighter than the Sun and released around 30 times the energy of the nuclear bomb dropped on Hiroshima, Japan. Estimates suggest that similar events to the Chelyabinsk meteor are expected to occur on average about once every 60 years.

In an unusual twist, the Chelyabinsk meteor was not the only Near-Earth Object (NEO) of interest that day. A few hours later, Asteroid 2012 DA_{14}, later renamed 367943 Duende, passed 17,200 mi (27,700 km) above the Earth's surface, closer than the orbit of a geostationary satellite. This object was discovered in February of 2012, and made headlines due to its near-Earth orbit. Despite its close pass by the Earth the asteroid was deemed of no risk by NASA. After some investigation into the course of the Chelyabinsk meteor, it was determined that the two objects were unrelated. The coincidental arrival of both objects was just that, a coincidence.

14.2 Introduction

Our planet Earth is part of a dynamic solar system in which the Earth interacts with objects beyond our planet. In some cases, these interactions can become dangerous for the inhabitants of the Earth. There are two broad areas of concern in the realm of space hazards. The first comes in the form of potential impactors, which are objects that strike the Earth from space. The second category is Space Weather, which amounts to effects produced by our Sun that produce high energy interactions in our atmosphere, and interact with the Earth's magnetic field.

14.3 A History of Catastrophe

Since the beginning of our solar system, mergers and collisions of objects have driven the growth and evolution of planets. Our solar system formed through the collapse of a massive rotating cloud of molecular gas and dust. Over millions of years, electrostatic attraction between those gases and small pieces of dust caused them to collide with each other. The collision of these particles causes them to stick together, forming larger particles through accretion. Accretion is the process of creating larger bodies from several smaller ones through collisions. As these objects grow larger, they continue to draw more material to themselves through gravitational attraction. As the cloud of gas and dust collapses, the cloud flattens to a disk, and the disk radius shrinks. As it does so, its rotation rate increases. This disk is called a *protostellar disk*, and is the precursor to a solar system. Most of the material is pulled to the center of the disk by the incredible force of gravity. At a critical mass, gravity collapses the abundant hydrogen atoms within the cloud so that they begin to fuse together into helium atoms by nuclear reactions, a process called *fusion*. Once fusion begins, a star is born. The star begins to release light and energy, including electrically charged particles called the *solar wind*. Figure 14.1 is an artist's conception of the formation of a star system like our solar system.

In the protostellar disk, the planets formed from countless collisions between objects in the disk space. Dust, gases, and rocks within the rest of the disk continued to collide and small fragments became larger pieces called *planetesimals*. Some of these planetesimals eventually grew large enough through accretion that they became the planets of the solar system and their moons. Some planetesimals were heated due to impacts by other objects. As they entered a molten state, higher density material separated from lower density materials in a process called *differentiation*. Most of the gas in the inner portion of the solar system was removed from the inner disk after the Sun began to release solar winds. The rocky inner planets are Mercury, Venus, Earth, and Mars. These planets have low mass. As a result gravitational forces were unable to retain the original hydrogen and helium atmosphere on those planets. The solar wind and light from the Sun removed most of the remaining gas from the inner portion of the solar system. Of the four planets in the inner solar system, only Venus and Earth maintain a substantial atmosphere formed out of gasses released from the rock. Earth's atmosphere changes in composition due to the abundance of water and life on the

Figure 14.1 Artist's conception of the dust and gas surrounding a newly formed planetary system. Image Credit NASA.

planet. Mars maintains a thin atmosphere, less than 1% of the volume of Earth's.

Lighter gases that were far enough away to be protected from the Sun's solar wind coalesced into the gas giants Jupiter, Saturn, Uranus, and Neptune. Ice in the outer reaches of the solar system formed comets and smaller dwarf planets. The chaotic nature of solar system formation resulted in frequent collisions, first between gases and dust, and later between rocky, iron, and icy objects. These collisions were necessary to form the planets and moons.

The Earth's Moon is a result of a violent collision in the early solar system. A rocky planet named Theia, which was the size of Mars, collided with Earth shortly after its formation. The impact was so violent that it tore material from the Earth's mantle and crust away from the planet and ejected it into the Earth's orbit. The ejected material formed rings until it eventually coalesced and accreted to form the Moon. Mars is the only other rocky planet with moons, named Phobos and Deimos. Scientists believe they are likely asteroids that were captured by Mars's gravity.

As planets grew and accreted most of the material in the solar system disk, the rate of impacts slowed over time. Events like the Chelyabinsk meteor are a striking reminder of the violent nature of the Earth's place in the solar system. It is these leftover objects from the process of solar system formation (asteroids, meteoroids, and comets), which can still move into the Earth's path from the outer solar system to collide with the Earth. When they do, the impactors can form large impact craters with devastating effects. The extinction of dinosaurs is attributed to the collision of a meteor with the Earth.

14.3.1 Asteroids

Asteroids are moderate-sized rocky or metallic bodies within the solar system that are too small to be planets. Most of these are found within the Asteroid Belt between Mars and Jupiter. The largest body within the Asteroid Belt is Ceres with a diameter of approximately 588 mi (946 km), though its size and uniqueness in this region make it the only *dwarf planet* (an object too small to clear debris from its orbit) within the orbit of Neptune. Second largest in the asteroid belt is the asteroid Vesta. Small craters on the surface of Vesta are shown in Figure 14.2. Ceres is shown in a size comparison to Vesta in panel (b) of Figure 14.2. Two large impact craters in the southern hemisphere of Vesta are likely the source of a suite of stony meteorites that represent the different layers of the asteroid. Some asteroids are orbited by their own small moons, such as the asteroid Ida and its moon Dactyl.

Many asteroids are not single rigid objects, but rather a "rubble pile" of fragments held together by gravity. One example of this is the asteroid Itokawa (Fig. 14.3). Rubble pile asteroids had been theorized for some time, but imaging of Itokawa's surface by the Hayabusa spacecraft demonstrated that Itokawa was indeed composed of a collection debris held together by gravity. Samples of Itokawa were returned to Earth by Hayabusa in 2010.

14.3.2 Comets

Icy bodies that originate in the outer reaches of the solar system are *comets*. These "dirty snowballs" are composed of dust, water and carbon dioxide ice, ammonia, and hydrocarbons. Organic compounds found in comets suggest that they may have carried the building blocks of life to Earth. The most easily recognizable features of a comet are the comet's tails, which only develop as the comet reaches the inner solar system. Heating from the Sun causes a small percentage of the comet to be *sublimated*. Sublimation is the process of turning a solid into a gas. In this case, the *nucleus*, which is the solid main body of the comet composed of dust and ice, is heated as it approaches the Sun. Two cometary tails form due to the loss of dust and gas during the sublimation of the surface. The ion tail of a comet, which is composed of these gases, grows and becomes longer

(a)

(b)

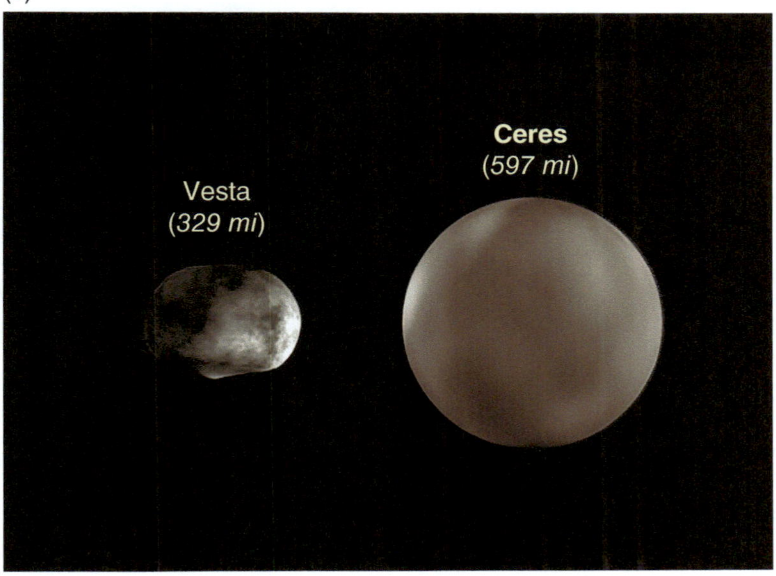

Figure 14.2 (a) The asteroid Vesta is the second largest object within the Asteroid Belt. It is pictured here with eight other asteroids, all of which have been visited by spacecraft. (b) Ceres is designated a dwarf planet and is the largest object in the main asteroid belt. (a) From NASA/JPL-Caltech/UCLA/MPS/DLR/IDA; (b) NASA, ESA, and A. Feild (STScI).

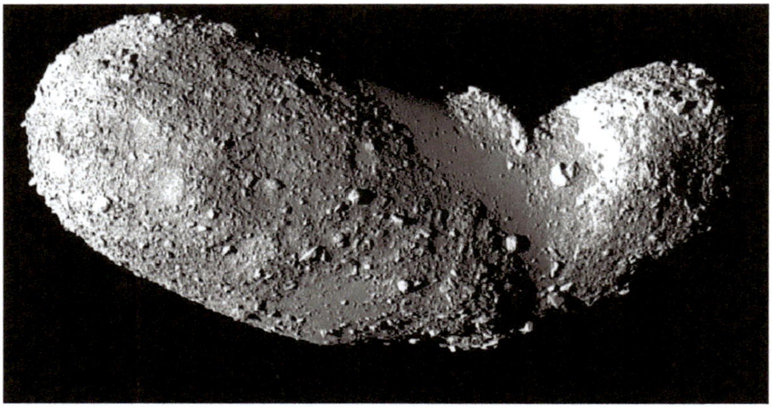

Figure 14.3 The asteroid Itokawa is an example of a "rubble pile" asteroid. The asteroid consists of small pieces of various sizes bound together by gravity, but not as a single large mass. From ISIS/JAXA.

as the comet travels nearer to the Sun. This makes the comet visible from the Earth. This tail points away from the Sun, since charged particles streaming outward from the Sun (the solar wind), push those gas particles away. Solar winds blow dust and gases away from the nucleus creating a second tail of dust as well. The ion tail is significantly dimmer than the dust tail. The dust tail is a trail of debris left behind in the wake of the comet. The twin tails of comet Hale-Bopp can be seen in Figure 14.4. Note that the two tails point in different directions.

Figure 14.4 The tails of comet Hale-Bopp on 23 July 1995. The long blue tail is the ion tail of the comet while the yellow-white tail is the comet's dust tail (Eckhard Slawik). From Bond, 2020.

14.3.3 Meteorites

Dust or other small particles in space due to comets or pieces of asteroids broken off during a collision or interaction are known as *meteoroids*. When one of these objects enters Earth's atmosphere it interacts with gas molecules, compressing and heating the air and the meteoroid. The heating of the meteoroid causes it to become incandescent, emitting light. This object is called a *meteor*, or a "shooting star," such as the one in Figure 14.5. During a meteor shower, the Earth moves through a trail of material that broke off from a comet or asteroid. For example, the Orionid meteor shower takes place during October and November every year as Earth passes through debris from Halley's Comet. Most meteors are very small particles, often dust- to sand-sized, that burn up due to the heat of friction in the atmosphere. Especially bright meteors are larger in size. They are often referred to as bolides or fireballs. These are large enough they may survive the friction of the atmosphere and fall to the surface. Once at the surface, these rocks are called meteorites.

Figure 14.5 A meteor, or shooting star, races across the sky during the Perseid meteor shower in 2016. Photo credit NASA/Bill Ingalls.

Types of Meteorites

There are three types of meteorites: stony, iron, and stony-iron. *Stony meteorites* look like accumulations of rocks and minerals. They tend to have low amounts of iron and nickel, but this varies based on where and how they formed. One type of stony meteorite is the *chondrite* meteorite, named after the round structures within them, known as *chondrules*. These meteorites, such as the Thuathe Meteorite in Figure 14.6a, are thought to have been formed early in the solar system's history and are composed of the dust that accumulated as the solar system first began to form through accretion. The other type of stony meteorite, the *achondrite* meteorite, is made of silica-rich minerals thought to have been part of the crusts of rocky bodies sent into orbit by impacts on the surface. These are often younger in age than the other types of meteorites. Specific achondrites have been identified as breccias (rocks composed of angular particles), plutonic rocks, and basalt flows that occurred on the Moon, Mars, and asteroids such as Vesta. Figure 14.6b is an achondrite formed due to impacts on the surface of Vesta.

Iron meteorites are composed of iron and nickel. The metal makes the meteorites more durable and recognizable on the surface than stony meteorites. Because of this, they are the most commonly identified meteorites, though they are a small fraction of observed falls. They likely represent the metallic core of differentiated (layered) asteroids and protoplanets. They are recognized by the crisscrossing Widmanstatten pattern that is displayed when a sample is sawed open, polished, and etched with acid as seen in Figure 14.6c. Widmanstatten patterns occur only in meteorites formed in space in the absence of an atmosphere. They cannot form at or near the surface of the Earth. The structures in a Widmanstatten pattern can form only when iron cools very slowly creating a pure iron crystal. This process does not occur on the Earth because iron combines quickly with other elements, such as oxygen, making it impossible to form a pure iron crystal.

Stony-iron meteorites, like the Fukang meteorite in Figure 14.6d, have both nickel-iron and silicate minerals. These may originate from the core-mantle boundary of protoplanets. Only larger objects stay warm long enough during their formation to differentiate and separate into layers of metal, stone and metal, and just stone. Therefore, this type of meteorite indicates the type and size of the protoplanet or parent asteroid where it originated.

Figure 14.6 (a) Meteorite NWA 869, an ordinary chondrite meteorite. Numerous small round chondrules can be seen on the cut and polished face of this specimen. (b) Bununu is an achondrite consisting of a breccia from asteroid Vesta. (c) Muonionalustav, an iron meteorite found in Sweden, displays a Widmanstatten pattern. (d) The Fukang meteorite, from China, is a stony-iron meteorite known as a pallasite and is composed of olivine crystals in a nickel-iron matrix. This likely represents the core-mantle boundary of an asteroid. (a) From Wikimedia; from Bond, 2020; (b) NASA/JPL-Caltech/Hap McSween (University of Tennessee), and Andrew Beck and Tim McCoy (Smithsonian Institution); (c,d) Daderot (CC0).

It is interesting to note that Earth is not the only location where meteorites have been identified. The rovers Opportunity, Spirit, and Curiosity have all imaged iron meteorites on the surface of Mars. Curiosity has even imaged more than one, including those pictured in Figure 14.7.

Airbursts and Impact Hazards

Most meteor entries into the atmosphere are relatively uneventful. Some meteors will produce airbursts similar to the Chelyabinsk event, while others will be able to directly impact the ground before a detonation can take place. In each case, the size of the meteor, its composition, its velocity relative to the Earth, and the angle of entry into the atmosphere determine whether the meteor produces an airburst or whether it directly impacts the surface without the airburst. These parameters also determine the altitude of the airburst if the meteor detonates, and the energy released in the impact or detonation.

As a meteor descends, air is compressed in front of the meteor, building up pressure and heating the air. As the meteor continues to

Figure 14.7 The Curiosity rover found this collection of iron meteorites on Mars in 2014. From NASA/JPL-Caltech/LANL/CNES/IRAP/LPGNantes/CNRS/IAS/MSSS.

move through the atmosphere, defects form in the surface allowing greater surface area to interact with the heated air. This interaction can eventually cause the meteor to break apart and release tremendous amounts of kinetic energy in the form of heat and light in an airburst. Thanks to better and more advanced monitoring of incoming meteors, as well as monitoring of the atmosphere for low-frequency atmospheric disturbances with greater numbers of infrasound stations, more data on the number and frequency of airbursts have been gathered than at any time in our history. Figure 14.8 shows the locations of known airbursts from 1994 to 2013 along with the energy released in the detonation. These events would still account for only a fraction of the total number of bolides over this time frame. New detection methods are being developed, which will more accurately detect bolides without direct observations of the event.

During the Earth's 4.6 billion year history, it has been impacted many billions of times by solid objects from within the solar system. These metallic, rocky, or icy bodies release vast amounts of energy and excavate craters on the Earth's surface. Analysis of the other rocky bodies of the solar system (Fig. 14.9), shows that impact craters play a significant role in shaping the surfaces of planets and their moons. The Moon has countless craters, ranging in size from micrometeorite craters that are less than a centimeter in width to the 1,600 mi (2,600 km) wide South Pole-Aitken Basin. Mercury's surface is dominated by impact craters due to its close proximity to the Sun and its gravity, which makes Mercury particularly vulnerable to impacts. The Southern Highlands of Mars record a long history of impacts, whereas the Northern Lowlands are far more sparsely cratered. This distribution of impact scars indicates that the Northern Lowlands are covered in younger rocks, deposited, or erupted in the more recent past. Older craters have been buried by younger rocks. The younger surface of Mars has experienced fewer impacts since those rocks were deposited. Still, cratering can be seen on Mars's surface as

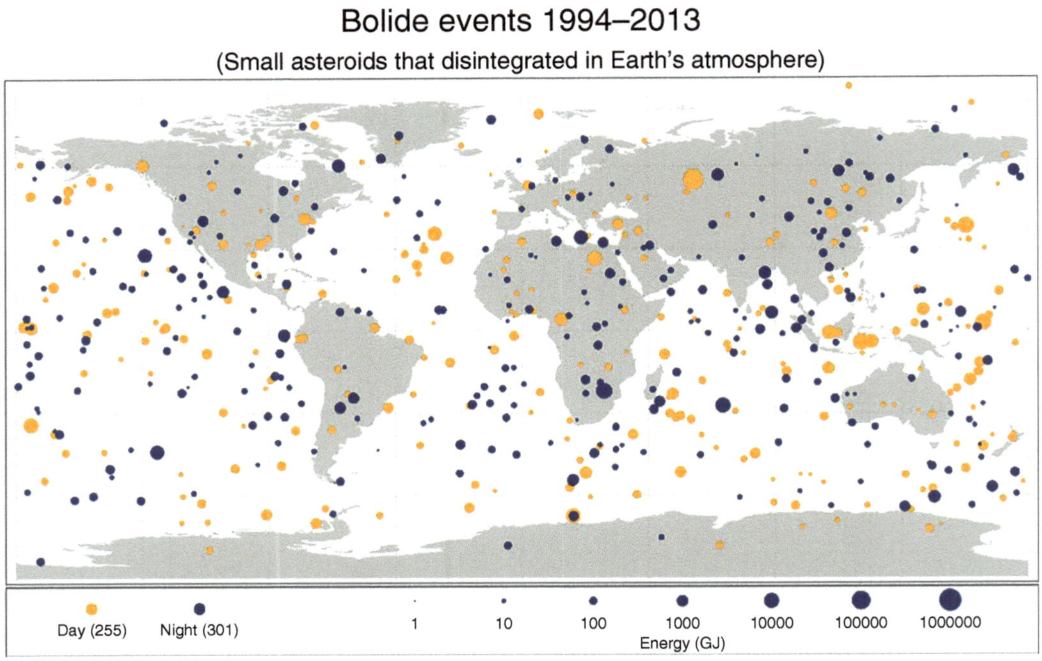

Figure 14.8 A map of the location of small bolide events from 1994 to 2013. The size of the dot indicates the energy of the detonation, and the color indicates whether the event took place during the day or during the night. From NASA/Planetary Science.

illustrated in Figure 14.10, which shows the Cassini crater in the Southern Highlands of Mars.

There are approximately 200 craters on Earth that have been identified as extraterrestrial in origin, far fewer than identified on the other rocky planets. Their locations are mapped in Figure 14.11. What makes Earth so different from these neighbors in the solar system? Earth is the only body within the solar system known to have active plate tectonics. This constant recycling of the Earth's crust means that much of the land surface is young in age. When a landscape is buried with new sediments or lava flows, these new materials cover and conceal impact craters. Craters, carried along on tectonic plates, can also be erased through the process of subduction. Most impact craters are found in older crustal material, within the old stable parts of the continental crust, called *cratons*. This is to be expected, since collisions of meteors and other bodies were more common millions to billions of years ago than they are today. Though impact craters have been identified on six of the seven continents, more have been identified on North America, Europe, and Australia.

Since Earth's surface is over 70% water, it is much more likely that any impact with a meteor or asteroid will hit the ocean rather than land. However, few impact craters are found on the ocean floor. Spreading centers and subduction zones are efficient at generating and destroying oceanic crust so the oldest oceanic crust is only around 180 million years old. This young crustal material in the oceans is much less likely to have been impacted than the older crust found on continents. In addition, the water provides a barrier between the impactor (also known as a bolide) and the ocean floor. Many impactors are simply not large enough to reach the ocean floor before

Figure 14.9 The surfaces of the other rocky bodies in the solar system have been heavily cratered. (a) The near side of the Moon contains large impacts called mare that are filled by basalt lava flows. (b) Mercury is heavily cratered. (c) The surface of Mars also contains numerous craters, but fewer than the Moon due to portions of the planets being more recently resurfaced. (a) From Bond, 2020; (b) From Bond, 2020; (c) From Fletcher, 2014.

their energy dissipates in the water. This limits the size of the objects that could generate a crater to only the largest of impactors.

Earth's atmosphere also plays a large part in the distribution of impact craters on the surface. The atmosphere and the presence of liquid water on the Earth's surface cause weathering and erosion. This erases impact craters from the surface or modifies them so that they are not easily recognizable. A planetary body such as the Moon has no atmosphere so there is no erosion by wind or water. Impact craters on the Moon are destroyed only by younger impacts. Venus

Figure 14.10 Cassini crater (the largest crater in the image) in the southern Martian highlands. Credit ESA/DLR/FU Berlin, CC BY-SA 3.0 IGO.

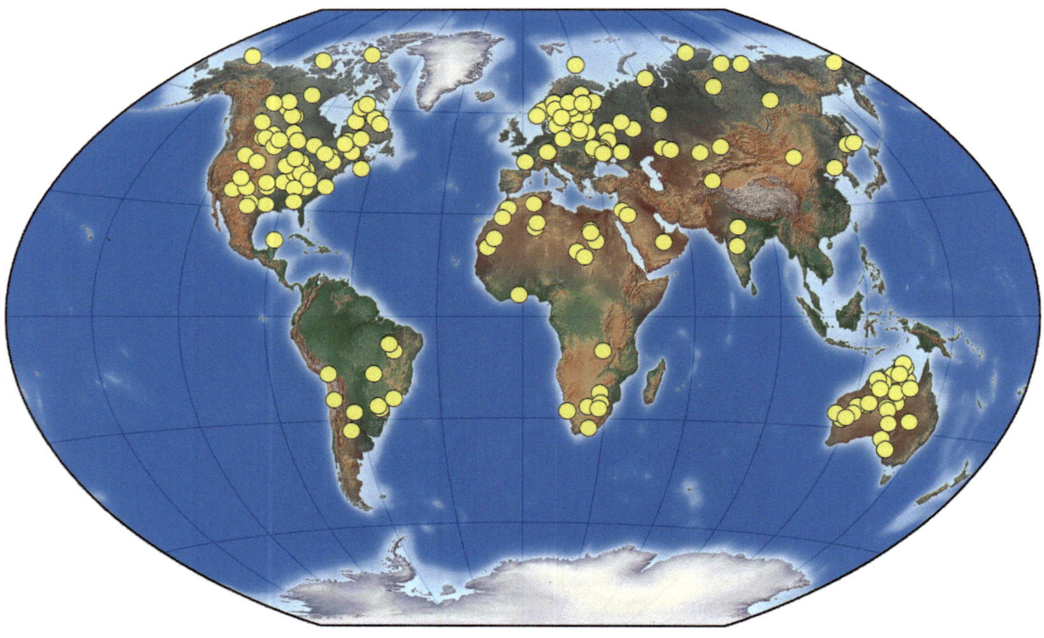

Figure 14.11 There are only ~200 verified impact craters on Earth, most identified in North America, Europe, and Australia. Courtesy Melinda Shimizu/data from Earth Impact Database.

has an atmosphere that is much denser than that of Earth, though it lacks liquid water. Small impactors are destroyed in the atmosphere because the density of the atmosphere produces extreme friction with and heating of the meteor. As a result, only the largest of impactors strike the surface of Venus. There are fewer small craters on Venus than on planets with less dense atmospheres.

14.3.4 Near Earth Objects

Near Earth objects (NEOs) include asteroids and comets that have an orbit that passes within 120 million mi (1.3 AU or 190 million km) of the Sun. These NEOs pose a higher threat to the Earth because of their closer proximity to us. It is not uncommon for NEOs to pass by the Earth at a distance closer than that of the Moon. NASA began tracking these near-Earth objects in 1998 as a response to the impact of a comet called Shoemaker-Levy 9 on Jupiter, which occurred in 1994. Initially, the goal was to identify 90% of NEOs that are 0.6 mi (1 km) or larger, representing the size of object for which a collision with Earth would be catastrophic. By November 2021, of 891 near earth objects out of an expected 920, over 97%, have been found.

Due to improvements in technological capability of detecting small objects, the current goal of the NEO- tracking program was revised in 2005, to identify 90% of objects that are larger than 460 ft (140 m) in diameter, which could potentially cause large-scale, though not worldwide, effects. By January of 2020, this goal had not yet been met, with only around 50% of the expected NEOs identified. Objects of this size are expected to strike the Earth on an average of only every 2,000 years, so the probability of an event is small, but work continues to attempt to identify and track these objects. Currently, NEOs are discovered at a rate of around 500 per year.

14.4 Measuring Risk

Approximately 44,000 tons (40,000 metric tons) of rocky debris strike the Earth each year. Studies of the frequency of meteorite impacts and airbursts versus the size of the impactor (or energy released by the event) show a clear trend. Smaller meteorites strike the Earth regularly. Events the size of Chelyabinsk are expected to occur roughly every 60 years. In general, meteorites of larger size have a greater average time between encounters with the Earth. Most meteorites we encounter are micrometeorites that do not even produce a visible meteor, but simply drift to the surface of the Earth. Small meteors around a millimeter in diameter are encountered roughly every 30 seconds, but meteors around 1 m in size are encountered only once a year. Objects greater than 1 km in size would produce catastrophic collisions with global effects. Meteorites of this size strike the Earth only around once every million years, and those 10 times larger once every 100 million years.

The *Torino Scale* was created as a way to communicate the threat of a possible impactor. The risk is determined based on both the probability of an impact occurring, and the size of the object. Values on the Torino Scale, shown in Figure 14.12, range from 0 (poses no threat to the Earth) to 10, indicating certain, catastrophic worldwide effects. All known objects are currently at level 0 on the scale. The first object to reach level 2 was Apophis, a Near Earth Asteroid (NEA), when it was first discovered in 2004. Early observations seemed to indicate a possible impact, which warranted the high value. Calculations suggested that the orbit might bring it into collision course with the Earth in 2036, at which time it reached a level 4 on the Torino Scale and shot Apophis into the spotlight with news coverage of the potential impact. But further study has ruled this out as a possibility and the Torino Scale level for Apophis has been reduced to 0 (Table 14.1).

14.4.1 Observed Collisions

In March of 1993, a team of comet hunters, Eugene and Carolyn Shoemaker and David Levy, discovered a comet on a trajectory to impact Jupiter. As the comet approached the planet in 1994, all eyes were turned toward Jupiter. For the first time, scientists were able to forecast a date and time of impact between another planet and a comet. The comet broke into numerous fragments due to the pull of Jupiter's gravity. Over the course of a week, 22 individual pieces of the comet (Fig. 14.13) entered Jupiter's atmosphere. The collisions

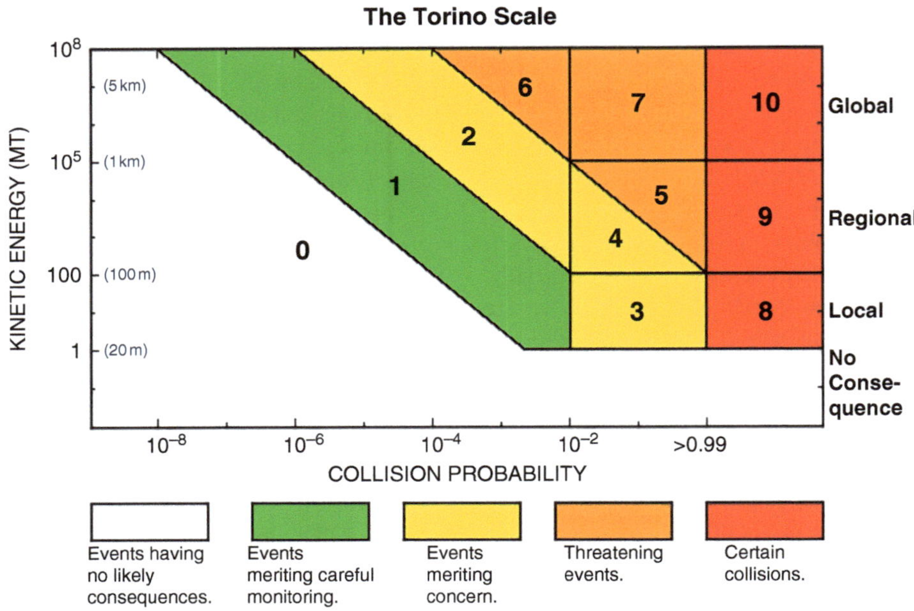

Figure 14.12 The Torino Scale is used to classify possible collisions. The highest level on the scale corresponds to an impending collision that will have global impacts. Courtesy of NASA.

Table 14.1 Terms that describe the relevant ratings of an NEO on the Torino Scale.

	NO HAZARD (white)
0	The likelihood of a collision is zero, or is so low as to be effectively zero. Also applies to small objects such as meteors and bodies that burn up in the atmosphere as well as infrequent meteorite falls that rarely cause damage.
	NORMAL (green)
1	A routine discovery in which a pass near Earth is predicted that poses no unusual level of danger. Current calculations show a collision is extremely unlikely with no cause for public attention or public concern. New telescopic observations very likely will lead to reassignment to Level 0.
	MERITING ATTENTION BY ASTRONOMERS (yellow)
2	A discovery, which may become routine with expanded searches, of an object making a somewhat close but not highly unusual pass near Earth. While meriting attention by astronomers, there is no cause for public attention or public concern as an actual collision is very unlikely. New telescopic observations very likely will lead to reassignment to Level 0.
3	A close encounter, meriting attention by astronomers. Current calculations give a 1% or greater chance of collision capable of *localized destruction*. Most likely, new telescopic observations will lead to reassignment to Level 0. Attention by public and by public officials is merited if the encounter is less than a decade away.
4	A close encounter, meriting attention by astronomers. Current calculations give a 1% or greater chance of collision capable of *regional devastation*. Most likely, new telescopic observations will lead to reassignment to Level 0. Attention by public and by public officials is merited if the encounter is less than a decade away.

Table 14.1 (Continued)

	THREATENING (orange)
5	A close encounter posing a serious, but still uncertain threat of regional devastation. Even though most objects of this level will be lowered to Level 0, critical attention by astronomers is still needed to determine conclusively whether a collision will occur. If the encounter is less than a decade away, governmental contingency planning may be warranted.
6	A close encounter by a large object posing a serious but still uncertain threat of a global catastrophe. Critical attention by astronomers is needed to determine conclusively whether a collision will occur. If the encounter is less than three decades away, governmental contingency planning may be warranted. Some objects of this level may be lowered to Level 1 or 2, if the chances of collision decrease.
7	A very close encounter by a large object, which if occurring this century, poses an unprecedented but still uncertain threat of a global catastrophe. For such a threat in the next 100 years, international contingency planning is warranted, especially to determine urgently and conclusively whether a collision will occur.
	CERTAIN COLLISIONS (red)
8	A collision is certain, capable of causing localized destruction for an impact over land or possibly a tsunami if close offshore. Such events occur on average between once per 50 years and once per several 1,000 years.
9	A collision is certain, capable of causing unprecedented regional devastation for a land impact or the threat of a major tsunami for an ocean impact. Such events occur on average between once per 10,000 years and once per 100,000 years.
10	A collision is certain, capable of causing global climatic catastrophe that may threaten the future of life as we know it, whether impacting land or ocean. Such events occur on average once per 100,000 years, or less often.

Source: Courtesy of NASA.

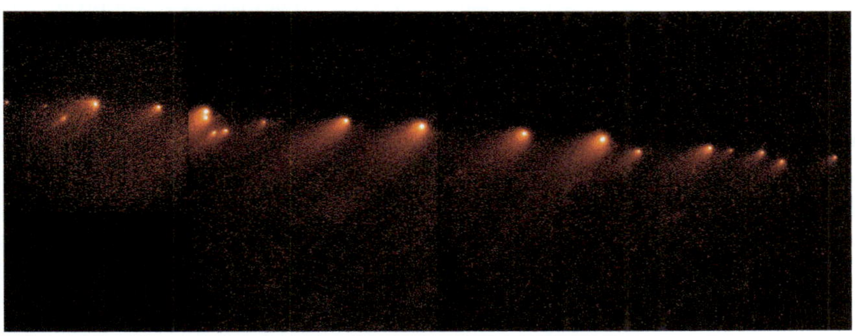

Figure 14.13 Image from the Hubble Space Telescope of 22 pieces of comet Shoemaker-Levy 9. This series of comet fragments passed near Jupiter in 1992, then plunged into the planet's atmosphere 2 years later. Courtesy NASA/STScI; Bond, 2020.

all occurred on the side of Jupiter that was facing away from the Earth and could not be directly observed. However, as rotation of the planet brought the impact sites into view, dark spots marked the location of the collisions. The largest of these dark spots was roughly the diameter of Earth. No crater was generated because Jupiter is a gas-giant and does not have a solid surface; only disturbances in the atmosphere, including plumes and atmospheric waves, resulted from the collision.

In July 2009, a new dark spot, similar to the ones created by Shoemaker-Levy 9, was found on the surface of Jupiter by an amateur astronomer, Australian Anthony Wesley. NASA's Jet Propulsion Laboratory (JPL) was able to confirm that this was indeed an impact event by an undetected object. Based on Hubble images of the dark spot and comparisons with observations in 1994, astronomers believe this impactor to be an asteroid, approximately 0.3 mi, or 500 m, wide. Based on observations of other dark spots, Jupiter likely has 30 to 100 of these impacts every year. In comparison, Earth has impacts only every 6 to 15 years. The reason Jupiter has so many more impacts than the Earth is because of its size and mass. The greater mass of Jupiter translates to a much higher gravitational attraction between Jupiter and asteroids than we experience here on Earth. This is obviously great news for us!

A bit closer to home there are many examples of smaller meteorite impacts that are hazardous on a small scale. There are several examples of meteorites striking vehicles, homes, or even people during the twentieth century. Here are a few examples of meteorites' impacts in the last century.

The Nakhla meteorite fell in Egypt in 1911. This meteorite is a piece of Mars, ejected from the planet during a meteorite impact, which eventually landed on Earth. On 28 June 1911 at 9 a.m. local time, a meteorite impacted the area near El Nakhla El Bahariya village. Stones were collected in this and the surrounding area shortly after the fall. A local farmer claimed to have witnessed one piece of the meteorite strike a dog on impact, vaporizing it. No remains of the animal were found, so the death of the Nakhla dog has never been confirmed.

The Benld meteorite struck Edward McCain's car in Benld, Illinois, on 29 September 1938. The meteorite struck the roof of McCain's garage and imbedded itself in the seat of his Pontiac coupe (Fig. 14.14).

Figure 14.14 The meteorite shown imbedded itself in the cushion of the seat of a Pontiac coupe 1938. From Shsilver at English Wikipedia, CC BY 3.0; https://commons.wikimedia.org/w/index.php?curid=10506633

In 1954 the Sylacauga meteorite struck the home of Ann Hodges on 30 November 1954, at 12:46 p.m. local time. Hodges was sleeping on her couch when the meteorite broke through the roof of the house, struck a radio, bounced off and injured Hodges, leaving Hodge severely bruised. This meteorite created a fireball visible in three states as it descended.

The Peekskill meteorite was recorded by 16 different individuals as it fell on 9 October 1992. Michelle Knapp of Peekskill, New York, had only recently purchased a 1980 Chevrolet Malibu. Knapp heard the impact of the meteorite as it tore through her automobile parked outside her home, narrowly missing the car's gas tank.

The Winchcombe meteorite impacted a driveway in Cloucestershire, England, on 28 February 2021. As the meteor descended, it created a sonic boom around 10 p.m. local time. A fragment with a mass of over 300 g struck the driveway of a family home in Winchcombe. This carbonaceous chondrite was the first of its kind ever recovered in the United Kingdom. The meteorite's descent was recorded by a network of cameras operated by the U.K. Fireball Alliance, and by numerous doorbell cameras.

In each of these cases damage was minimal, but harm to individuals and property occurred. Other large events also occurred in the twentieth century. At 7:17 a.m. on 20 June 1908, several people in the sparsely populated region near the Podkamennaya Tunguska River of Siberia noticed a bright object traveling east-southeast to the west-northwest across the sky. Suddenly, the object exploded with tremendous force between 3 and 6 mi (5–10 km) above the Earth. Scientists estimate the energy released by this explosion was between 10 and 15 Megatons, roughly the same amount of energy as the largest hydrogen bomb ever detonated. The explosion was so bright that the Sun paled in comparison. Buildings shook from shock waves. Windows shattered. Tents of the Evenki people in the area, called chums (pronounced *chooms*), were torn apart. People were thrown as much as 150 ft (40 m) by the blast wave. The landscape was bathed in heat. Tens of millions of trees were uprooted, stripped of their branches and bark, and blown down radially from a central point. The area of felled trees covered approximately 770 mi^2 (2,000 km^2), roughly the size of Tokyo, Japan (Fig. 14.15).

About 70 eyewitnesses to the explosion sent letters to newspapers in 1908 describing their experiences during and immediately after the event. Reports by eyewitnesses in the area during the explosion contain many of the same elements: bright flashes in the sky; loud thunder; a blast wave, fires; ground shaking; fallen trees; burning material on the forest floor, including trees, pine needles, and moss; strong winds; and a smoky haze that burned the eyes.

Figure 14.15 Trees knocked down during the Tunguska Event as they appeared in 1927. (ESA) from Bond, 2020.

Instruments across Europe and Asia recorded the event. Barometric pressure records show atmospheric pressure waves that propagated through the atmosphere. Seismographs in Irkutsk, Tashkent, Tiblisi, and in Jena, Germany, recorded shock waves. The Magnetographic and Meteorological Observatory in Irkutsk, Russia, detected a magnetic storm immediately after the explosion that lasted for 5 hours.

14.4.2 Impact Craters

What happens if something big *does* impact the Earth? Asteroids enter Earth's atmosphere traveling around 27,000 mi/hr (12 km/s)! Comets are much faster, thanks to a boost from the gravitational pull of the Sun. This tremendous velocity corresponds to immense kinetic energy. This energy is released at the point of impact, generating a shock wave that propagates radially from the site much like the detonation of a bomb. In fact, studies of craters formed during nuclear explosion testing were critical in identifying extraterrestrial impact craters on Earth. The shock wave from the impact compresses the rocks it contacts so that they begin to move as a fluid. In some cases, the impactor and the rock beneath it are melted and/or vaporized. High temperatures and pressures deform the internal structure of quartz crystals, creating planes within the structure, such as those in Figure 14.16, which are caused only by these extreme conditions. This is known as shocked quartz. Identification of shocked quartz is a key piece of evidence to validate the impact origin of craters on Earth.

Directly behind the shock wave is an area of low pressure that allows the rocks to be ejected from the rapidly growing crater, while rock layers are often overturned to form the crater rim. The final size and shape of a crater depends on the size, mass, and velocity of the impactor. In general, the crater formed is 8 to 16 times larger than the impactor. Almost all impact craters are circular because of the radial nature of the shock wave. Elongated craters

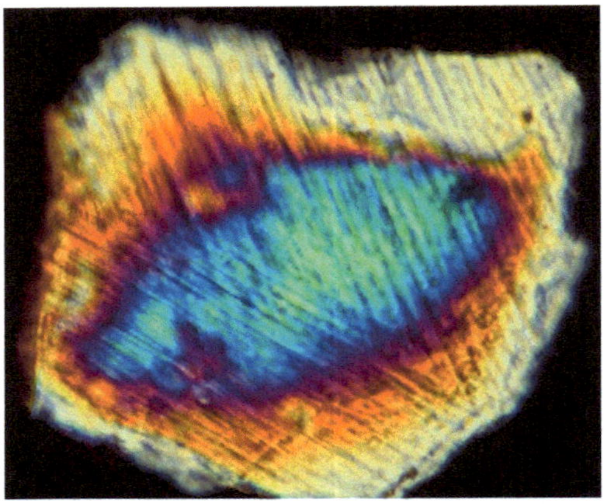

Figure 14.16 Microscopic view of a sand-sized shocked quartz grain from near the edge of the Chesapeake Bay impact crater. Image by Glen A. Izett, Public domain, via Wikimedia Commons. https://commons.wikimedia.org/wiki/File:820qtz.jpg

form only when the impactor enters the atmosphere at a very low angle, very close to parallel to the surface.

Simple craters generated during low energy impacts by small objects, tend to be less than 1.2 mi (< 2 km) in diameter in sedimentary rocks, and less than 2.4 mi (< 4 km) in diameter in crystalline rocks. These are bowl-shaped craters with floors that are lower than the surrounding surface and a rim that is higher than the surface, as shown in Figure 14.17. In larger impacts, the walls of the crater slump back toward the center and are pushed up into a central peak or a peak ring, forming what is known as a complex crater. Complex impact craters are shallower, relative to their diameter, than simple craters. The Clearwater Lakes in Quebec, shown in Figure 14.18, are both complex craters.

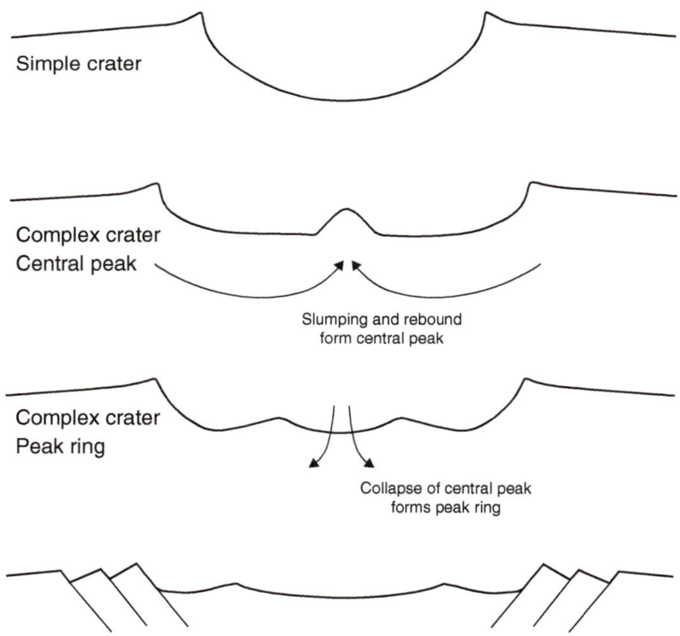

Figure 14.17 Simple craters are bowl-shaped. Complex craters from as the walls of the excavated crater slump inward and push up into a central peak. In large craters, the peak may collapse to form a peak ring. Multi-ring basins are the largest impact structures, with rings formed by normal faults as the walls of the impact crater slump inward. Image by K. Wakefield.

Figure 14.18 Clearwater Lakes in Quebec, Canada, are complex craters. North is to the left in this image. From NASA.

The eastern crater has a central peak that is submerged beneath the lake while the larger western lake has a visible peak ring due to its larger size. The timing of formation of the two craters is not well understood. It has been suggested that they formed due to a binary asteroid that impacted the Earth but the magnetic signatures and chemically determined ages are quite different. It may be a coincidence that these two craters are found next to one another.

The largest impact crater on Earth is the Vredefort crater in South Africa. Vredefort is a multiring basin, a very large crater that has multiple concentric slopes formed due to faulting and slumping along the edges. Originally 300 km in diameter, so much of the crater rim has been eroded that only the central peak (known as Vredefort Dome) is visible as seen in Figure 14.19.

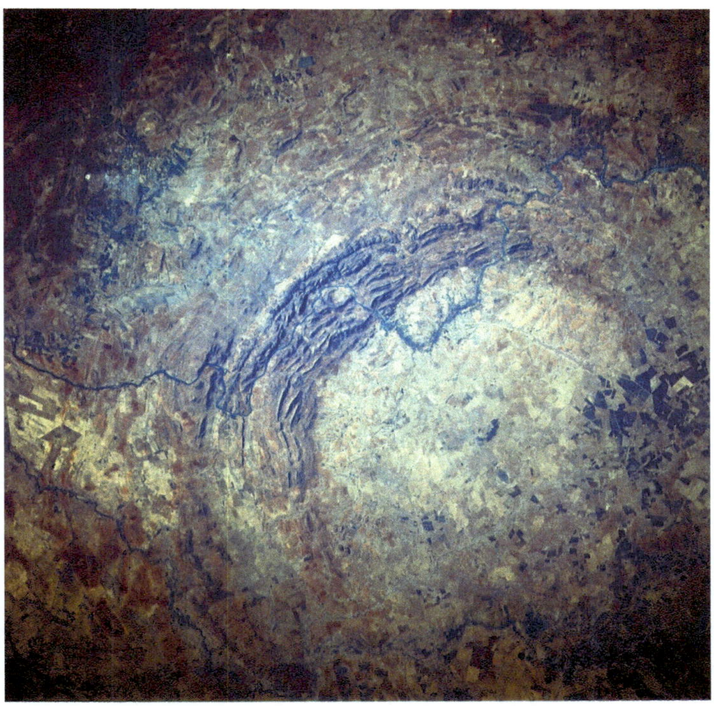

Figure 14.19 Vredefort Dome, which represents the central peak of the crater. From NASA Earth Observatory image by Lauren Dauphin, using Landsat data from the U.S. Geological Survey.

The first crater to be recognized as being generated by an extraterrestrial impact was Barringer Crater, also known as Meteor Crater, in northern Arizona. It is a bowl-shaped, simple crater about 0.75 mi (1.2 km) in diameter. Rather than being round like most craters, it has a rectangular shape when viewed from above as seen in Figure 14.20. This has been attributed to preexisting, local fractures within the rocks. Approximately 50,000 years ago, a 160 ft (50 m) wide iron-rich near Earth asteroid entered the atmosphere above North America. This area of the Colorado Plateau was likely the home to animals such as mammoths, sloths, and bison, which grazed in rolling hills of grasslands, near woodland forests. Moving through the atmosphere in 30 s, the asteroid struck the sandstone and siltstones at the surface. The asteroid would have been largely vaporized upon impact, though fragments, now known as Canyon Diablo meteorites, are found up to 5.6 mi (9 km) from the crater. These iron meteorites are named after a ghost town that shares its name with a nearby canyon.

The impact shock wave ejected the rocks beneath the impactor radially, generating overturned rock layers along the rim. Figure 14.21

Figure 14.21 A layer of sedimentary rock, known as the Moenkopi Formation, was ejected and overturned to form the crater's rim. Large boulders of limestone, some as large as a house, top the crater rim. The floor of the crater is to the right. From Bob Leighty.

Figure 14.20 Barringer Crater, or Meteor Crater, is located in northern Arizona. The shape is more square than round due to preexisting local fracture patterns within the rock. From D. Roddy/USGS, Lunar and Planetary Institute; From Bond, 2020.

shows layers of the dark red Moenkopi siltstone and sandstone along the rim of Meteor Crater. Originally flat-lying, the rocks have been overturned so that they now tilt away from the center of the crater.

Shattered rock inside the crater forms a layer of impact breccia that includes shocked quartz, a phenomena that was first identified in Meteor Crater. An airburst created winds of over 600 mi/hr (1,000 km/hr) that extended 3 mi (5 km) from the site of impact. Any animals within 1.9 mi (3 km) of the impact would have been killed, and severe injuries would have occurred up to 15 mi (24 km) away. Many of these injuries, often to the lungs, would be caused by the higher air pressure created by the impact shock wave. This overpressure would have also been able to pick up debris such as rocks or branches. A pea-sized rock could reach a speed of over 190 mi/hr (300 km/hr), causing injuries as far as 8 mi (13 km) from the impact. A similar event occurring near a populated area today would likely shatter glass and concrete or cinder block walls and cause the collapse of brick walls.

How an impact affects the surrounding environment and organisms depends on the scale of the impactor and the composition of the target. The most common target for impacts on Earth is the ocean simply because it covers 70% of the surface. In most cases, impacts in the ocean strike deep water, where the volume of water causes loss of the impactor's kinetic energy and no crater is formed in the rocks below. However, impacts in shallow water may form a tsunami. The tsunami that is generated may do more damage than a land-based impact site. Shallow water impact sites mean that there is lower loss of kinetic energy and craters on the ocean floor may still form.

Approximately 35 million years ago, an impactor hit the area that is now Chesapeake Bay in the United States, on the submerged continental shelf. It left a crater that is 56 mi (90 km) in diameter but currently buried in sediment. The crater was discovered because the rivers flowing into Chesapeake Bay make sharp turns as they cross the rim. Scientists investigated this phenomenon and found the buried crater. Because the impact occurred in water, it generated a 1,500 ft (about the height of a 150-story building!) tsunami that reached the Blue Ridge Mountains to the west and spread debris along the entire eastern coast of the United States. A map of this can be seen in Figure 14.22.

14.4.3 The Cretaceous Extinction

The dinosaurs were all killed by an asteroid impact, right? Not exactly. The Cretaceous-Paleogene Boundary (K-Pg, formerly referred to as the K-T, or Cretaceous-Tertiary boundary) occurred about 66 million years ago and is marked by a mass extinction that includes non-avian dinosaurs, pterosaurs, mosasaurs, ammonites, and many plant species. But many familiar dinosaurs like *Stegosaur*, *Brachiosaur*, and *Allosaur* had already been extinct for around 80 million years. The K-Pg Extinction event affected the last of the dinosaurs except for the birds, or avian dinosaurs, which survived the extinction event.

Most evidence seems to indicate that a devastating impact occurred at the time of the K-Pg boundary, such as the impact depicted in Figure 14.23. The possibility of a Cretaceous-ending impact event was first suggested in 1980. Walter and Luis Alvarez, Frank Asaro and Helen V. Michel discovered anomalously high levels of the element iridium in ocean sediments deposited during the K-Pg boundary at sites spanning the globe. Iridium is rare in Earth's crust, likely concentrated within the core, but is found in much higher concentrations within asteroids and comets. This suggests an extraterrestrial origin for the spike in concentration at the boundary. The team suggested that the impacting asteroid would have been around 6.2 mi (10 km) in diameter. At the time, there were no known craters corresponding to this size and age.

During further studies, shocked quartz was found in deposits with the iridium anomaly.

(a)

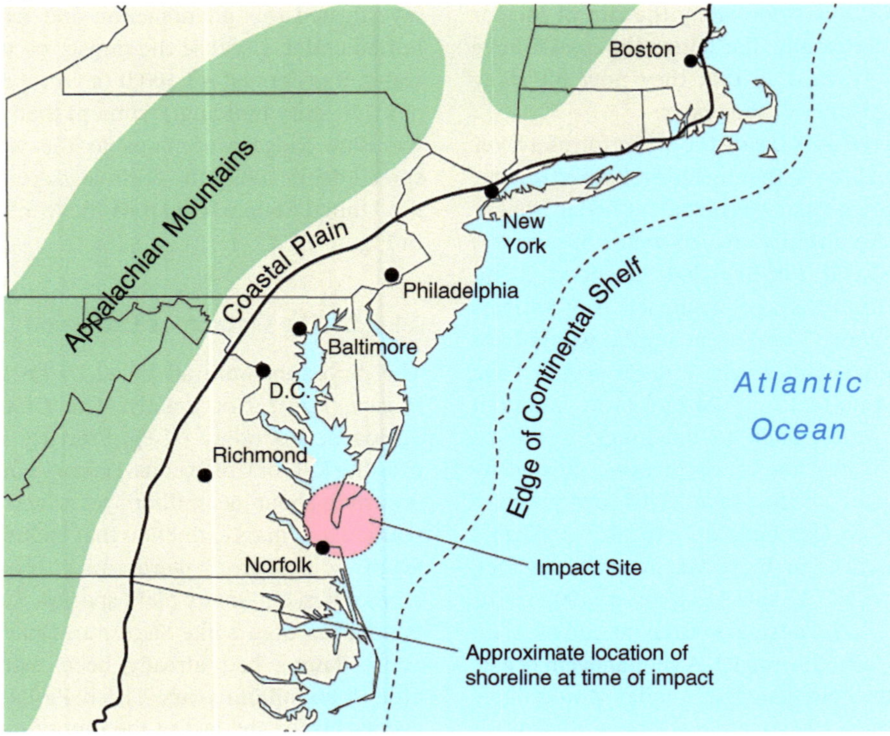

(b)

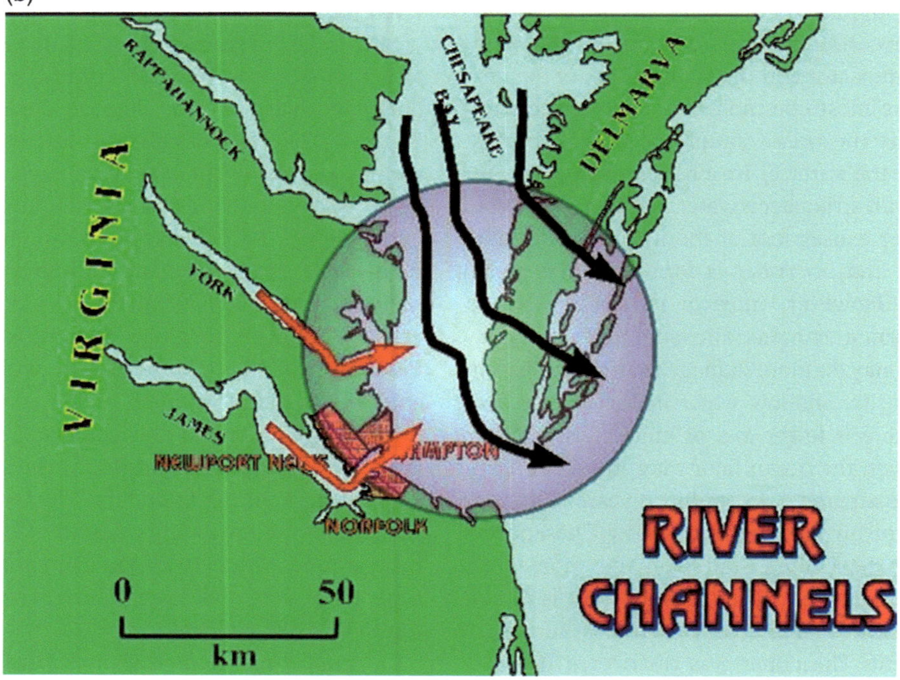

Figure 14.22 (a) The location of the impact site by the Chesapeake. The impact took place on the continental shelf. (b) The sharp turns in the York and James rivers at the edge of the crater.

Figure 14.23 An artist's depiction of the Chicxulub impact on the Yucatan Peninsula of Mexico. From NASA/Don Davis.

This suggested an impact into continental rock rather than oceanic since oceanic crust does not contain much quartz. The shocked quartz was found at higher concentrations in western North America, narrowing down the area to search for the crater. This brought attention to a buried feature that was discovered during geophysical studies on the Yucatan Peninsula of Mexico (Fig. 14.24). Core samples identified shocked quartz, impact breccia, and impact melts within the structure, thereby confirming its impact origin. Radiometric dating of the structure indicated an age of 66 million years, coinciding with the K-Pg boundary. The Chicxulub Crater, named for a town near its center, is a large, multi-ring basin. The rings are evident in Figure 14.24, a map of the gravity anomaly created by the buried crater.

Gravity anomalies like these can and have been used to map out many craters. Using this technique, extremely accurate measurements of the local acceleration due to gravity are made, and the variations of local gravity from a simple theoretical model can describe if there is a surplus or deficit of mass in the area. These measurements can be used to map out craters, even those under water.

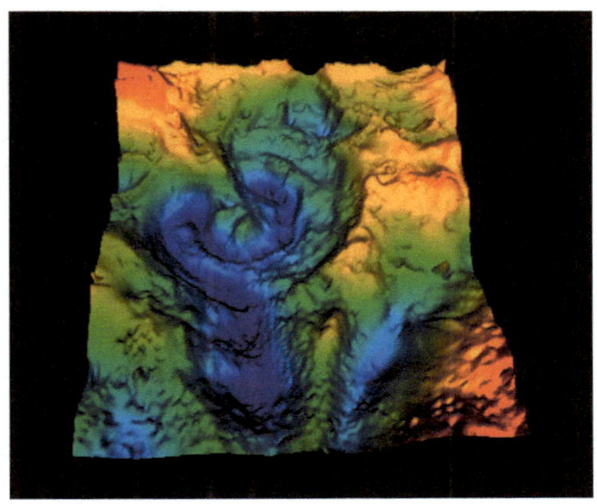

Figure 14.24 A gravity anomaly marks the Chicxulub impact crater buried beneath the Yucatan Peninsula of Mexico. Gravity measurements indicate a multi-ring basin with a fourth, outer ring about 185 miles (300 km) in diameter. Blues in the image indicate areas where gravity is lower than expected, and reds are areas where gravity is higher than expected. From LIP/NASA; Bond, 2020.

When the impact occurred, the northern part of the Yucatan Peninsula was covered by a shallow ocean. Layers of marine rocks were deposited on the granite continental crust. The crater is 93 mi (150 km) in diameter, suggesting an impactor of around 6 mi (10 km) in diameter, and confirming previous prediction of the crater diameter by Alvarez and coauthors. This impact had global effects, leading to the observed mass extinction.

An airburst would have been generated during this impact, similar to, but much stronger than, that of Meteor Crater, and may have reached 1,100 mi (1,800 km) away, with hurricane-force winds affecting the entire southern part of North America. The extreme heat caused by the airburst and the impact increased ground temperatures, sparking wildfires as far as 2,500 mi (4,000 km) away. Wind-blown soot from the fires was deposited globally and the fires would have released a considerable amount of carbon dioxide into the atmosphere. The soot, dust, and sulfur aerosols within the atmosphere likely blocked sunlight for months, causing a drop in global temperature. Once all of this settled, water vapor from the wet sediments and carbon dioxide from the carbonate rocks at the impact site would have led to global warming of between 1.8° and 13.5° F (1° and 7.5° C). Heating of the atmosphere would have created nitric acid rain that lasted for months or years. The sulfur released from the marine deposits of the Yucatan Peninsula would have turned to sulfuric acid rain. Both of these acids led to the death of plants on the continents and increased weathering and erosion of rocks.

The impact onto the continental shelf would have generated tsunami waves with a height of 300 to 1,000 feet (100–300 m) as they reached the shores of the Gulf of Mexico and then traveled as much as 180 mi (300 km) inland, through areas already devastated by the airburst.

Life struggled in the months to years after the impact. The lack of sunlight from dust and aerosols in the atmosphere decreased the number of microscopic phytoplankton in the oceans that relied on photosynthesis, crippling the global food chain. Plants dying on continents, particularly near the impact site, caused barren ground and high rates of soil erosion until ferns began to repopulate. This wiped out food sources for large animals on land. Insects, particularly in North America, appear to have disappeared due to the immediate impact effects or the loss of plant life. In the oceans, plankton that survived were small with more simple body shapes and were found only at high latitudes. The organisms that survived were able to scavenge among the devastation. On the continents, freshwater organisms that fed off of detritus were much more likely to survive than those that lived on land. Only lizards, amphibians, and mammals fared well in the aftermath because they could better forage the dead plant materials. Paleontological evidence points to a rapid extinction of 70% of all plants and animals on land and in the oceans.

14.5 What Can We Do?

In January 2016, NASA created the Planetary Defense Coordination Office (PDCO) and tasked the office with coordinating efforts to track near earth objects (NEO) that could pose a threat to the Earth. Using data from the Jet Propulsion Laboratory (JPL) Center for NEO Studies on locations and orbits of NEOs, the PDCO identifies objects larger than 100 to 164 ft (30–50 m) that pose a threat to the United States. The office is tasked with issuing warnings about potential impacts and coordinating efforts with the U.S. government to mitigate the effects, including deflecting, or disrupting an object when necessary. The hunt for NEOs is an effort that spans multiple government offices, international organizations, and many astronomers.

What happens if an object is on a crash course with the Earth? Early detection is the most important aspect, which is why tracking objects is so important. Hollywood often depicts stopping an impactor by setting off a nuclear bomb. In reality, the possible courses of action depend greatly on a number of factors, including but not limited to, the size of the impactor, the type of impactor, where the impact would occur, and how much time is available.

For small impactors of less than 0.6 mi (1 km), damage would likely be regional, similar to that experienced at Chelyabinsk, if it hit in a sparsely populated area on a continent. During a presentation to the U.S. House of Representatives Committee on Science, Space, and Technology, NASA administrator Charles Bolden, Jr., told the committee that if a similar object was found to be approaching New York City in 3 weeks, the only recourse would be prayer. An impactor targeted at a large metropolitan area with little warning would likely be devastating. The chances of detecting a small object prior to the impact are low. Evacuation of the target site, when possible, may be sufficient to save lives and reduce the consequences. The 2 min between the fireball and the shock wave reaching the surface in Russia could be sufficient to reach more stable parts of a building structure or to move away from glass windows. Obviously, this would greatly depend on the location itself in terms of population, infrastructure, and support. This smaller size of impact occurs much more often than the larger ones, indicating that planning for these should be a priority. Events like this are expected roughly once every 60 years.

For impacts of 0.6 to 1.2 mi (1–2 km), there would be considerable regional damage such as that described for Meteor Crater but would also have worldwide effects. However, these are more likely to be identified prior to impact allowing time to generate a mitigation strategy. For any of these smaller impactors, an ocean impact near a shoreline would likely be a worst-case situation due to the regional effect of the tsunami that would likely be generated. This would affect numerous countries and possibly millions of people along the more densely populated coastlines. Impactors over 1.2 mi (2 km) will likely have effects on a global scale. These kinds of impacts are only expected once every 100,000 to 500,000 years.

Response to an impact that is certain to occur would require a massive international approach. NASA is the U.S. based organization that tracks NEOs and predicts impacts, but there is a high probability that an identified impactor would strike another country (since the United States is a small percentage of Earth's land area) or the ocean. There are many international organizations focused on the possibility of impacts as well. The European Union created the NEOShield program specifically to develop the technologies required for mitigation attempts.

The best-case scenario for a confirmed future impact is early detection. To plan and launch a spacecraft, even for an emergency, requires a minimum of 1 to 2 years, and this assumes that we have the means to actually change the orbit of the incoming object. Any mitigation attempts would require a spacecraft, but designing, building, and launching one takes a considerable amount of time. Ideally, detection would take place 20 years or more prior to impact, though the longer the time we have to prepare for an impact, the greater our chances of avoiding one. The Earth orbits the Sun at 67,000 mi/hr (30 km/s) and NEOs travel at velocities near this, so changing the orbit of the object by even a small fraction would deflect it enough to miss the Earth entirely. There are several ways that have been suggested to change the trajectory of a NEO.

14.5.1 Kinetic Impactor

Changing the orbit of a NEO could be as easy as hitting it. A kinetic impactor is a single spacecraft or set of spacecraft that impacts an object in order to change the trajectory. The benefit of this method is the relatively small size of spacecraft required because it would be able to travel at a higher velocity to quickly reach the object. Success depends on how well the orbit and characteristics of the impactor are understood, as well as what the results of the kinetic impact would be. For example, an object that is a highly fragmented rubble-pile would likely not respond as well as a coherent mass.

A NASA mission, Deep Impact, collided with comet Tempel 1 in 2005. However, the small collision and relatively large size of the comet meant that no change in trajectory was measured, though important observations about the structure of the comet were made. An international collaboration known as the Asteroid Impact and Deflection Assessment (AIDA) launched the first full-scale test of the use of kinetic impact to deflect a NEO. The asteroid Didymos and its moonlet Dimorphos are considered a binary asteroid system. Dimorphos was chosen because the 500 ft (150 m) size is in the range of object size that is most likely to impact Earth. On 24 November 2021, the Double Asteroid Redirection Test (DART) spacecraft was launched on a mission to deliberately crash into the moonlet at 4 mi/s (approximately 6 km/s). The DART spacecraft successfully impacted Dimorphos on 22 September 2022 with the objective of changing the speed and orbital period of Dimorphos by 73 s. The mission was considered a smashing success when the orbital period of the moonlet was changed by 32 min.

14.5.2 Gravity Tractor

A spacecraft could change the orbit of the impactor just by being near it, without having to make contact. A gravity tractor is a spacecraft that hovers near the surface of an asteroid and changes the orbit due to gravitational attraction. This method has the benefit of requiring little information about the asteroid itself other than the mass. This addresses the issue of the rubble-pile asteroids. It can be scaled up, with more massive spacecraft or through the deployment of multiple bodies, for more massive asteroids. A major drawback is that this approach takes more time to be effective. It also requires that the spacecraft hover very near the surface, requiring significant technology, like ion drives, and autonomous guidance systems.

14.5.3 Blast Deflection

Without sufficient time for the above methods, a somewhat inelegant but largely effective method available is blast deflection. This method uses an explosion to change the path of the NEO. Rather than an explosion that fractures the object itself, the detonation occurs near the object. A nuclear blast would cause material to evaporate from the surface and jets of this material would act as a propellant to change the trajectory of the object. A nonnuclear explosion likely would not cause this evaporation but would, however, contain enough energy to nudge the object from its current path. In both cases, the composition of the object must be well understood in order to accomplish deflection. An explosion used on a breccia or rubble pile object may fracture the object and, instead of missing Earth entirely, would result in thousands or millions of pieces of radioactive material falling over a large area. A nuclear blast is likely only under the most extreme circumstances due to the political ramifications of the use of nuclear material.

14.5.4 Other Proposals

There are several other ideas that form new approaches, or even combine those already discussed. For example, one proposed strategy is to use a kinetic impactor to create a crater in a space body, and follow that up with subsurface nuclear detonations to break a rigid body apart. Ion bean thrusters directed at an asteroid, focusing solar energy to create thrust by vaporizing material, using a laser system to vaporize a portion of an asteroid to create thrust, using solar sails, or even "painting" an asteroid white to use solar radiation pressure to alter an asteroid's trajectory have all been proposed. Some proposals have even suggested landing rocket engines on asteroids to direct their motion.

The most important factor in all attempts to deflect an asteroid or comet is time. Identifying potential hazards and tracking their orbital motion for as long as possible is essential to identifying a hazard.

14.6 Solar Events

14.6.1 The Role of the Sun

Rocky or icy objects within the solar system are not the only things that pose a threat to Earth. At the center of the solar system is an immense, churning ball of gas. The Sun is a nuclear reactor, ejecting particles and radiation toward the Earth and other planets at incredible temperatures and speed. The features on the Sun, like those in Figure 14.25, are controlled by magnetic forces. The energy emitted by the Sun, such as heat and visible light, is integral to life on Earth, but it is not always benevolent.

On 1 September 1859, astronomers studying sunspots observed two bright, white regions on the surface of the Sun. This solar flare was followed by a magnetic storm that reached the Earth in just under 18 hr. Aurora sightings were reported in the Northern Hemisphere as far south as Hawaii and as far north as Santiago, Chile, in the Southern Hemisphere. Firsthand accounts describe the aurora being so bright on the eastern coast of the United States that observers could read by the light at night. Currents induced along lengths of telegraph wires set fires as the wires sparked, shocked telegraph operators, and had enough energy that they still transmitted messages even when they were unplugged. Now known as the Carrington Event after one of the astronomers observing sunspots, this was one of the most powerful geomagnetic storms ever recorded.

Figure 14.25 Coronal loops on the Sun seen in ultraviolet light by the TRACE spacecraft. The loops are shaped by magnetic field lines. From NASA; From Bond, 2020.

14.6.2 The Sun's Anatomy

The Sun is a star, no different than the ones that twinkle in the night sky; it just happens to be much closer to Earth. Energy from the Sun lights the Earth, heats the surface, oceans, and atmosphere, and drives wind patterns and weather. This energy is created within the core where the temperature (around 27,000,000°F, or 15,000,000°C!) and density are so great, and the atoms are compressed together tightly enough to overcome the electric forces separating them. The compressed material fuses hydrogen atoms together to form helium atoms in a nuclear reaction, emitting energy. The energy is released in the form of light across the electromagnetic spectrum, from radio waves to visible light to even some gamma radiation, and neutrinos. This energy then travels through the layers of the Sun, shown in Figure 14.26, to the surface. It is theorized that the movement of energy from the core to the surface takes over 170,000 years.

In the radiative zone, photons carry energy to areas of lower temperature, while the convective zone contains heated gases that move toward cooler areas. The visible surface of the Sun is the photosphere. Although this layer is not solid, it is the portion of the Sun that we can directly observe. Above the photosphere is the chromosphere, a thin layer that reaches 18,000°F (10,000°C), and the corona, which is only visible as wispy trails during a total solar eclipse. The temperature of the corona can reach 5.4 million degrees F (3 million degrees C).

The Sun has a magnetic field that varies within 11 year cycles from a solar minimum, to a maximum, and back to minimum within

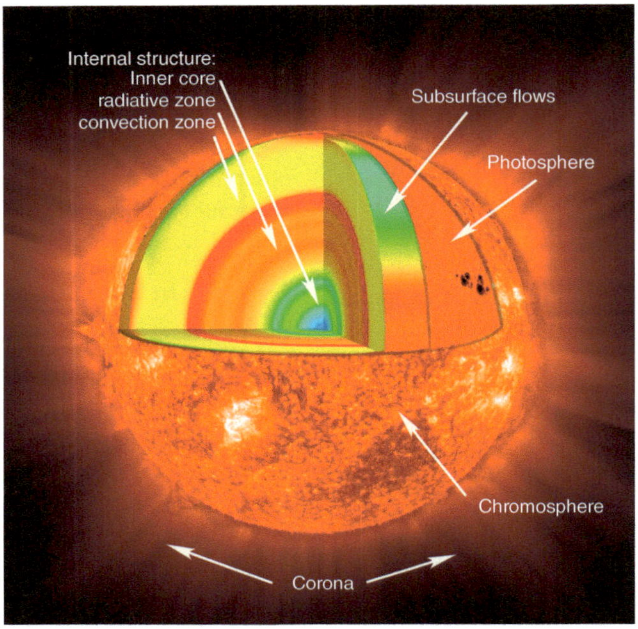

Figure 14.26 The layers of the Sun and its atmosphere. From Bond, 2020.

a single cycle. At the end of the cycle, the polarity flips and the magnetic field is oriented in the opposite direction. Activity in the photosphere and atmosphere of the Sun reacts to variations in the magnetic field. The Sun constantly emits the solar wind, plasma composed mainly of electrons and protons, which flows outward throughout the solar system, and activity peaks at the solar maxima. When the solar wind encounters the Earth's magnetic field, particles are accelerated along magnetic lines and collide with molecules in Earth's atmosphere, transferring part of their energy. This energy is released as light, creating the aurora near the Earth's poles like the one shown in Figure 14.27.

A solar prominence is a large arch of plasma that is aligned with the Sun's magnetic field. It is cooler than the surface and therefore appears as a dark line, or filament, by contrast when viewed against the Sun. Sunspots are dark areas of the Sun's photosphere that are thousands of degrees cooler than the surrounding areas. A cluster of sunspots, like the ones in

Figure 14.27 An aurora formed by the interaction of the solar wind and Earth's magnetic field. From Martincco, CC BY-SA 4.0, via Wikimedia Commons.

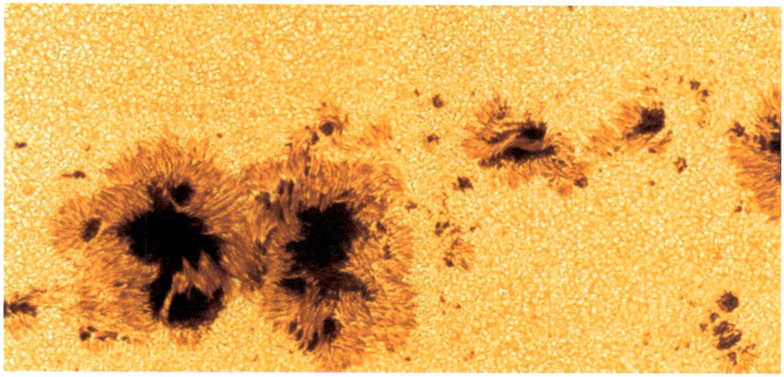

Figure 14.28 Close view of sunspots observed by the Hinode spacecraft in February of 2014. Sunspots have a darker central region (the umbra) surrounded by a lighter outer section (the penumbra) with a structure of linear filaments. From NAOJ/JAXA; From Bond, 2020.

Figure 14.28, can be so large that they can be viewed even without a telescope by looking at the Sun with protective lenses or by using a pinhole projector. At the beginning of a solar cycle, these form close to the Sun's poles, but the location migrates toward the equator as the cycle continues. These are areas of intense magnetism and magnetic field loops between two sunspots, rising into the atmosphere.

Equatorial regions of the Sun rotate faster than the polar regions, which can cause magnetic lines to tangle. Occasionally, sudden changes in the magnetic field in these tangled regions cause a massive release of energy, known as a solar flare. Surrounding material is quickly heated by millions of degrees, emitting an observable flash of light like that seen in Figure 14.29. Solar flares release electromagnetic radiation primarily in the form of visible light, infrared, X-rays, and gamma rays. Additionally, solar flares release streams of electrically charged particles. Solar flares are classified as A (the least energetic), B, C, M, or X (the most energetic). Class A-C solar flares have little impact on the Earth. Electrically charged particles from more energetic flares can produce electromagnetic disturbances, which affect communications and navigation. Class M flares can put astronauts in space at risk and disrupt radio signals near the Earth's poles. Class X flares can have global impacts such as radio blackouts or navigation problems on satellites but generally pose little threat to Earth thanks to the protection of Earth's atmosphere and magnetic field.

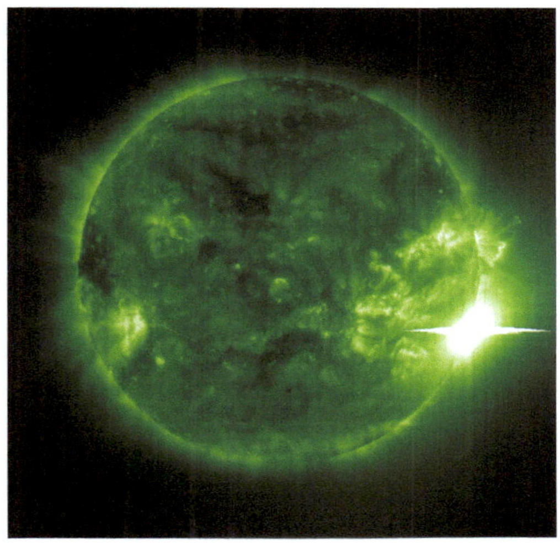

Figure 14.29 A solar flare (bright area) on 4 November 2003 seen in the extreme ultraviolet. This solar flare affected radio communications around the globe and caused power outages in Europe due to fluctuations in Earth's magnetic field. From ESA/NASA; From Bond, 2020.

Coronal Mass Ejections (CMEs) eject plasma, currents, and magnetic fields from the Sun's corona (Fig. 14.30). These ejections can be large, stretching a quarter of the

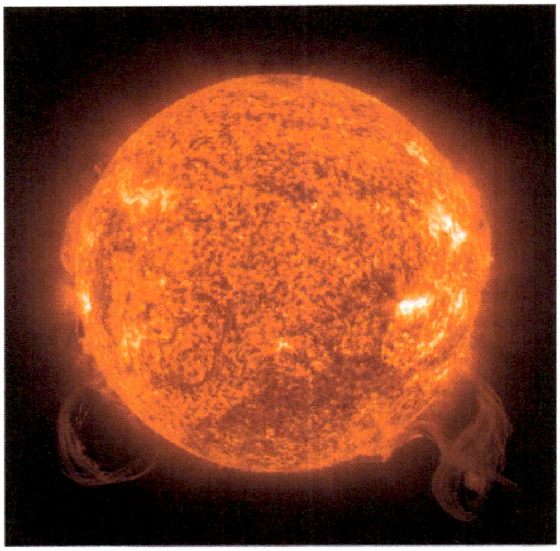

Figure 14.30 Image of the Sun from 18 March 2003. Plasma structures are shaped by magnetic field lines. These loops may have been associated with a flare and coronal mass ejection that happened that day. Image taken in extreme ultraviolet light by the SOHO spacecraft. From EIT Consortium, ESA/NASA; From Bond, 2020.

distance between the Earth and the Sun. They are also fast, reaching Earth in as little as 15 hr. The ejected materials interact with, and cause disturbances in, the Earth's magnetic field, known as geomagnetic storms. CMEs pose a significant threat to communication, satellites, and power generation. They are often associated with sunspots, but each can occur independent of the other. From 18 October to 8 November 2003, intense activity on the Sun during a solar maximum produced 80 CMEs and 143 solar flares.

In 1989, a CME triggered a geomagnetic storm that reached North America and induced currents in the ground because of Earth's wildly fluctuating magnetic field. This tripped a protection system at the Hydro-Quebec transmission grid, causing a blackout that lasted for over 9 hr during the Canadian winter. Disruptions occurred in New England as well, but did not result in blackouts.

The National Oceanic and Atmospheric Administration (NOAA) houses the Space Weather Prediction Center, which forecasts solar storms and releases alerts about events. In order to better communicate risks from solar activity, the Center created the NOAA Space Weather Scale, a scale that describes current space weather and the effects of events. Three scales explain the risk from geomagnetic storms, solar radiation storms, and radio blackouts. The scales range from level 1, a minor event with minimal impacts, to level 5, an extreme event with significant risk to satellites, the electric grid, and astronauts in space and people on high altitude flights near the poles.

Solar flares, coronal mass ejections, and the resulting geomagnetic storms pose a threat to humans mainly because of our reliance on electricity. The Carrington Event had little impact in the fledgling days of telegraph communication. But as we have become more dependent on electrical grids and satellites, we have also become more vulnerable to these events.

As of right now, there is no way to predict when a solar flare or a CME will occur, so preparedness is key to mitigating the effects of a geomagnetic storm. CMEs can reach the Earth in less than a day, allowing very little time for action after detection, but solar winds take almost a week. CMEs are ejected from the Sun in one direction, often not in the direction of the Earth and pose no threat at all, such as the small CME in Figure 14.30. Both the impact of the material in the CME with the atmosphere as well as the UV and X-ray will heat the upper atmosphere. The worst-case scenario is a powerful CME that causes atmospheric drag on low Earth orbit satellites due to this heating of the atmosphere, affecting their navigation and communication systems. This could cripple communication and navigation on the surface. Currents induced in the ground damage power supply infrastructure (particularly transformers), causing blackouts and requiring replacement of expensive equipment that may take a year to manufacture.

Industries such as health care, prisons, nuclear energy generators, communication, and even municipal water supplies that require a constant, reliable energy supply often have backup generators. There is a limit to how long the fuel available for the generators lasts and there may be no way to refuel in the short term. A blackout in a large, metropolitan area like New York City can strand people on the upper floors of buildings that are their homes or places of business. This puts those with limited mobility or emergency medical situations in a particularly vulnerable situation. Without electric pumps, water supplies to buildings and homes would be cut off, limiting a critical resource. Without power for a significant period of time, food spoilage would lead to shortages. Medications requiring refrigeration are lost. Lack of communication and navigation systems would interrupt the ability to effectively move supplies to locations of greatest need. It has been estimated that a severe geomagnetic storm could have economic and societal costs of $1–2 billion in just the first year and may require 4–6 years to recover.

The most important thing we as a society can do to mitigate the effects of a solar storm is to upgrade the aging power grid with technology that can withstand the effects of a solar storm. This is particularly important for locations at high latitudes such as Canada or the northeastern United States where they are more vulnerable to perturbations in the magnetic field because of their proximity to the magnetic pole. The Northeast Power Coordinating Council, Inc. (NPCC), oversees the power grid for the northeastern United States, including particularly vulnerable locations such as Maine. A task force created a set of procedures for geomagnetic storms after the event in 1989 that include carefully managing the loading of transmissions, generators, and interconnections. These actions hinge on detection of solar events and prediction of the ensuing geomagnetic storm. Sufficient mitigation of the effects of geomagnetic storms must include a large supply of temporary electric transformers to restore the grid as quickly as possible.

14.7 Earth's Future

The Earth has experienced many hazards from the other bodies within the solar system and it will continue to do so. The effects of these hazards are celebrated when the impact is small, such as shooting stars and auroras. Large collisions with solid bodies or particles from the Sun could alter the Earth and society as we know it. Only through extensive study of our neighbors in space and cooperation with our neighbors on Earth can we understand the threat, predict an event before it occurs, and successfully mitigate these hazards.t

14.8 Chapter in Review

The Earth is just one body within a solar system that contains the Sun, planets, moons, dwarf planets, asteroids, meteors, comets, and cosmic dust. The solar system was formed by accretion of gas and dust into the bodies we have now. Asteroids, meteors, comets, and cosmic dust still impact larger bodies within the solar system, and therefore constitute a hazard to the Earth.

The Earth, as with all bodies in the solar system, experienced impacts from solid bodies much more frequently in the distant past than it does today. However, impacts still occur and can cause dramatic effects on the Earth's surface. When a body such as an asteroid or meteor enters the Earth's atmosphere, the molecules of the atmosphere experience a great deal of friction with the object. This causes the object to erode, or burn up in the atmosphere. The streak of light caused when this occurs is often referred to as a falling star. Most objects completely burn up in the atmosphere, but a small percentage of objects are large enough to withstand the friction of the atmosphere, and they fall to Earth. At this point they are called meteorites. Objects that impact the Earth are known as impactors or bolides.

Meteorites come in three varieties: stony, iron, and stony-iron. Those that are more than a few meters across have the ability to cause damage to the Earth and human-made structures. A 62 ft (19 m) meteor that fell near Chelyabinsk, Russia, in 2013 resulted in a brilliant streak of light in the sky followed by a strong pressure wave that hit the Earth's surface 2 min later. The blast wave shattered windows and caused other types of damage to buildings. The larger the meteor, the larger the crater it produces, and the more severe and widespread damage will be.

Craters tend to be between 8 and 16 times the diameter of the object that creates them. Craters may be simple, complex, or multiringed based on the size of the impactor. The Earth's surface does not have many visible impact craters compared with other bodies in the solar system. This is because the Earth has an atmosphere. The atmosphere causes many bolides to burn up due to friction with the atmosphere. The atmosphere also produces weather, which causes weathering and erosion of rocks. This erosion can erase craters from the surface of the planet. The Earth also experiences plate tectonics, so much of the planet's surface is young. The Moon and Mercury do not have atmospheres or any detectable plate tectonics, so these bodies have numerous craters on their surfaces.

Approximately 66 million years ago, a large asteroid struck the Yucatan Peninsula of Mexico, resulting in devastating worldwide effects. An estimated 70% of species, including pterosaurs and the last of the nonavian dinosaurs, went extinct because of dust and aerosols in the atmosphere, acid rain, a major tsunami, and intense warming of the planet. The crater is 93 mi (150 km) in diameter, suggesting an impactor of around 6 mi (10 km) in diameter.

Comets are composed of water and carbon dioxide ice, ammonia, dust, and organic compounds. They typically have elongated orbits and are only visible on the Earth when they approach the Sun and grow tails through the process of sublimation. In 1994, a comet broke into pieces and impacted the planet Jupiter, causing a string of dark patches in the atmosphere.

The United States (and other countries around the world) track near Earth objects (NEOs) in space larger than about and calculate their trajectories. The current goal of the NEO- tracking program seeks to identify 90% of objects that are larger than 460 ft (140 m) in diameter, which could potentially cause large-scale, though not worldwide, effects. Approximately 900 NEOs over 0.62 mi (1 km) in diameter and 8,600 over 459 ft (140 m) have been identified. The Torino Scale was introduced as a way to communicate the risk due to impact of each object. No objects large enough to be tracked are expected to impact the Earth. This does not mean there will be no impacts, only that no impacts of large objects capable of producing damage on a regional scale are expected.

Scientists have proposed several ways of preventing the impact of a large object with the Earth. With enough lead time, spacecraft can be designed to intercept an asteroid, allowing the spacecraft to alter the object's trajectory. In one scenario, known as the kinetic impactor method, spacecraft would be used to impact the object to change its trajectory. In a second scenario, a massive spacecraft could be positioned adjacent to the object, using gravity to slightly alter the trajectory of the object. Third, a spacecraft could be used to deliver a large nuclear explosion in close proximity to an object in an effort to shift its trajectory. In all cases, identifying the object well ahead of a potential impact is necessary to achieve a deflection of the incoming object.

The Sun is a source of hazards for the Earth. It not only emits heat and light, but also charged particles that have a strong magnetic effect. Solar flares involve a massive release of energy, including gamma and X-ray radiation and charged particles, which can cause problems with satellites, endanger astronauts in orbit, and disrupt radio signals. Coronal mass ejections can cause geomagnetic storms that disturb satellites, endanger astronauts in orbit, cause radio blackouts, and disrupt power grids. The effects are greater closer to the poles. In 1859, a large geomagnetic storm called the Carrington Event caused magnetic and electrical disturbances. Since the use of wires was limited to the those of telegraphs at that time, the effects were minimal for the average person. However, in today's world where we are all dependent on electricity and satellites, a geomagnetic storm of the magnitude of the Carrington Event would cause major disruptions to daily life as power grids failed, and satellites were affected.

Bibliography

Aagard, B. T., Blair, J. L., Boatwright, J., Garcia, S. H., Harris, R. A., Michael, A. J., Schwartz, D. P., et al. (2014). Earthquake outlook for the San Francisco Bay region, 2014–2043. https://pubs.usgs.gov/fs/2016/3020/fs20163020.pdf

Abbott, P. L. (2017). *Natural disasters, 10th ed.* McGraw Hill Education.

Adamo, N., Al-Ansari, N., & Sissakian, V. (2021). Review of climate change impacts on human environment: Past, present and future projections. *Engineering 13.*

Agle, D. C., & Brown, D. (2014). New map shows frequency of small asteroid impacts, provides clues on larger asteroid population. https://www.jpl.nasa.gov/news/new-map-shows-frequency-of-small-asteroid-impacts-provides-clues-on-larger-asteroid-population

AJMC Staff (2021). A timeline of COVID-19 developments in 2020. https://www.ajmc.com/view/a-timeline-of-covid19-developments-in-2020

Alvarez, L. W., Albarez, W. A., Asaro, F., & Michel, H. V. (1980). Extraterrestrial cause for the Cretaceous-Tertiary extinction. *Science, 208* (4448), 1095–1108.

Arbogast, A. F. (2017). *Discovering physical geography, 4th ed.* Wiley.

Associated Press (2021). Why are earthquakes so devastating in Haiti? https://www.voanews.com/a/americas_why-are-earthquakes-so-devastating-haiti/6209611.html

Asteroid Impact & Deflection Assessment (AIDA). Collaboration. *ESA*, https://www.esa.int/Space_Safety/Hera/Asteroid_Impact_Deflection_Assessment_AIDA_collaboration

Atwater, T. (1970). Implications of plate tectonics for the Cenozoic tectonic evolution of western North America. *Geological Society of America Bulletin, 81*, 3513–3536.

Babcock, L. E. (2009). *Visualizing Earth history.* John Wiley & Sons.

Balch, J., Bradley, B., Abatzoglou, J., Nagy, R., Fusco, E., & Mahood, A. (2017). Human-started wildfires expand the fire niche across the United States. *Proceedings of the National Academy of Sciences, 114*(11), 2946–2951. https://www.pnas.org/doi/epdf/10.1073/pnas.1617394114

Barber, G. (2016). Photos: Big Thompson River flood of 1976. *Denver Post.* https://www.denverpost.com/2016/07/31/big-thompson-river-flood-1976-photos/

Bardan, R. (2022). NASA confirms DART Mission impact changed asteroid's motion in space. https://www.nasa.gov/news-release/nasa-confirms-dart-mission-impact-changed-asteroids-motion-in-space/

Barlow, P. P. M. (2003). Groundwater in freshwater-saltwater environments of the Atlantic coast. *Groundwater in freshwater-saltwater environments of the Atlantic coast.* https://pubs.usgs.gov/circ/2003/circ1262/

BBC News (2003). Iran earthquake kills thousands. http://news.bbc.co.uk/2/hi/middle_east/3348613.stm

Earth's Natural Hazards and Disasters, Advanced Textbook 7, First Edition. Bethany D. Hinga.

BBC News (2017). Haiti to rebuild quake-damaged National Palace. https://www.bbc.com/news/world-latin-america-39649680

BBC News (2018). Indonesia tsunami: Death toll from Anak Krakatau volcano rises. https://www.bbc.com/news/world-asia-46674490

Beaumont, H. (2023). Alaska native community relocates as climate crisis ravages homes. *Al Jazeera News*. https://www.aljazeera.com/news/2022/12/15/alaska-native-community-relocates-as-climate-crisis-ravages-homes

Berke, J. (2018). San Francisco is building skyscrapers like crazy–and there could be a deadly downside. *Business Insider*. http://www.businessinsider.com/san-francisco-bay-area-earthquake-skyscraper-risk-2018-4

Berkeley Seismology Lab (2008). Today in earthquake history: Mexico City 1985. *Seismo Blog*. http://seismo.berkeley.edu/blog/2008/09/19/mexico-city-1985.html

Berkeley Seismology Lab (2009). Today in earthquake history: Loma Prieta 1989. *Seismo Blog*. http://seismo.berkeley.edu/blog/2009/10/16/today-in-earthquake-history-loma-prieta-1989.html

Bhattacharya, S. (2003). European heat wave caused 35,000 deaths. *New Scientist*. https://www.newscientist.com/article/dn4259-european-heatwave-caused-35000-deaths/

Binzel, R. P. (2000). The Torino impact hazard scale. *Planetary and Space Science*, *48*(4), 297–303.

Bisi, M. M., et al. (2010). From the Sun to the Earth: The 13 May 2005 coronal mass ejection. *Solar Physics*, *265*(1–2), 49–127.

Black, J.G. & Black, L.J. (2015). *Microbiology Principles and Explorations*, 9th ed. John Wiley & Sons. 940 p.

Bland, P. A.Artemieva, N. A., (2006). The rate of small impacts on Earth. *Meteoritics & Planetary Science*, *41*, 607–631.

Bliss, L. (2015). 10 years later, There's so much we don't know about where Katrina survivors ended up. https://www.citylab.com/equity/2015/08/10-years-later-theres-still-a-lot-we-dont-know-about-where-katrina-survivors-ended-up/401216/

Board, Space Studies, and National Research Council (2009). *Severe space weather events: Understanding societal and economic impacts: A workshop report*. National Academies Press.

Boehnlein, D. (2020). The Benld meteorite: An ordinary space rock that slammed into a car. *Astronomy.com*, 18 Sept. 2020. https://www.astronomy.com/science/the-benld-meteorite-an-ordinary-space-rock-that-slammed-into-a-car/

Bolt, B. A. (2004). *Earthquakes, 5th ed.* W. H. Freeman and Company.

Bombardelli, C., & Peláez, J. (2011). Ion beam shepherd for asteroid deflection. *Journal of Guidance, Control, and Dynamics*, *34*(4), 1270–1272.

Bond, P. (2020). *Exploring the solar system, 2nd ed.* Wiley/Blackwell.

Borovicka, J., et al. (2013). The trajectory, structure and origin of the Chelyabinsk asteroidal impactor. *Nature*, *503*, 235–237.

Boston.com (2009). The big picture: Red River flooding. http://archive.boston.com/bigpicture/2009/03/red_river_flooding.html

Bottomley, R. J., York, D., & Grieve, R. A. F. (1990). Argon-40-argon-39 dating of impact craters. *Lunar and Planetary Science Conference Proceedings*, *20*.

Brown, P.G. & others (2013). A 500-kiloton airburst over Chelyabinsk and an enhanced hazard from small impactors. *Nature*, *503*, 238–241.

Brueck, H. (2018). The record-breaking year of deadly fires in California was stoked by distinctly devilish winds: Here's how they work. *Business Insider*. https://www.businessinsider.com/santa-ana-winds-why-they-spark-fires-2018-12?r=UK&IR=T

Bryant, E. (2008). *Tsunami: The underrated hazard, 2nd ed.* Springer.

Bryner, J. (2020). 1st known case of coronavirus traced back to November in China. *LiveScience*. https://www.livescience.com/first-case-coronavirus-found.html

Buis, A., Veraverbeke, S., & Bell, B. (2017). Lightning sparking more boreal forest fires. earth research findings. https://climate.nasa.

gov/news/2602/lightning-sparking-more-boreal-forest-fires/

Burnett, J. (2017). The tempest at Galveston: "We knew there was a storm coming, but we had no idea." *National Public Radio*. https://www.npr.org/2017/11/30/566950355/the-tempest-at-galveston-we-knew-there-was-a-storm-coming-but-we-had-no-idea

Cal OES News (2022). Four years later: Paradise residents continue to return home. https://news.caloes.ca.gov/four-years-later-paradise-residents-continue-to-return-home/

Cannon Township (2018). What is a Watershed? https://www.cannontwp.org/department/division.php?structureid=74

Carrington, R. C. (1859). Description of a singular appearance seen in the Sun on September 1, 1859. *Monthly Notices of the Royal Astronomical Society*, *20*, 13–15.

CBC News (2019). Stifling heat wave descends on Australia, killing dozens of horses. https://www.cbc.ca/news/technology/australia-heat-wave-1.4991164

CBS News (2019). Tropical Storm Lane: Hawaii battered by torrential rains. https://www.cbsnews.com/live-news/hawaii-lane-tropical-storm-latest-forecast-path-track-live-updates-today-2018-08-25/

CBS News (2021). U.S. scientist says he's found the real COVID patient zero, and strong evidence pandemic started at animal market. https://www.cbsnews.com/news/origin-covid-19-us-scientist-patient-zero-wuhan-china-evidence-market/

Centers for Disease Control (2022). CDC Museum COVID-19 timeline. https://www.cdc.gov/museum/timeline/covid19.html#:~:text=January%2020%2C%202020%20CDC,18%20in%20Washington%20state

Centers for Disease Control (2022). Understanding mRNA COVID-19 vaccines. https://www.cdc.gov/coronavirus/2019-ncov/vaccines/different-vaccines/mRNA.html

Centers for Disease Control (2022). Understanding viral vector COVID-19 vaccines. https://www.cdc.gov/coronavirus/2019-ncov/vaccines/different-vaccines/viralvector.html

Charis-Carlson, J. (2016). Online maps let Iowans see their risk of flooding. *Des Moines Register*. https://www.desmoinesregister.com/story/news/education/2016/09/23/online-maps-let-iowans-see-their-risk-flooding/90914026/

Cheng, A. F., et al. (2016). Asteroid impact & deflection assessment mission: Kinetic impactor. *Planetary and Space Science*, *121*, 27–35.

Choi, B. H., Hong, S. J., & Pelinovsky, E. (2006). Distribution of runup heights of the December 26, 2004 tsunami in the Indian Ocean. *Geophysical Research Letters*, *33*(L13601).

Chyba, C. F., Thomas, P. J., & Zahnle, K. J. (1993). The 1908 Tunguska explosion: Atmospheric disruption of a stony asteroid. *Nature*, *361* (6407), 40–44.

City of Los Angeles (2016). Los Angeles River revitalization. https://www.lariver.org/blog/about-la-river

Clark, P., Shakun, J., Marcott, S., et al. (2016). Consequences of twenty-first-century policy for multi-millennial climate and sea-level change. *Nature Climate Change*, *6*, 360–369. https://doi-org.oregonstate.idm.oclc.org/10.1038/nclimate2923

Cliver, E. W., & Leif S. (2004). The 1859 solar–terrestrial disturbance and the current limits of extreme space weather activity. *Solar Physics*, *224*(1–2), 407–422.

Comet Shoemaker-Levy 9 Reports (March 3, 1995). *Science*, *267*, 1277–1323.

Confederated Salish & Kootenai Tribes (2021). Fire on the land: Native people and fire in the northern Rockies. http://fwrconline.csktnrd.org/Fire/index.html

Cook, B. Mankin, J., Williams, A., Marvel, K., Smerdon, J., & Liu, H. (2021). Uncertainties, limits, and benefits of climate change mitigation for soil moisture drought in southwestern North America. *Earth's Future*, *9*.

Cook, B. I., Miller, R. I., & Seager, R. (2009). Amplification of the North American Dust

Bowl through human-induced land degradation. *Proceedings of the National Academy of Science*, *106*(13), 4997–5001.

Cook, J. (2017). 7 years after Haiti's earthquake, millions still need aid. *Huffington Post*. https://www.huffingtonpost.com/entry/haiti-earthquake-anniversary_us_5875108de4b02b5f858b3f9c

Cook, J. (2020). Comparing CO_2 emissions to CO_2 levels. *Skeptical Science*. https://skepticalscience.com/CO2-emissions-correlation-with-CO2-concentration.htm

Courtney, C., (2020). Central China flood, *1931*. http://www.disasterhistory.org/central-china-flood-1931

Crawl, D., Block, J., Lin, K., & Altintas, I. (2017). Firemap: A dynamic data-drive predictive wildfire modeling and visualization environment. *Procedia Computer Science 108C* (2017), 2230–2239.

Dangerfield, K (2017). Timeline of the 1997 flood in Grand Forks. *Global News*. https://globalnews.ca/news/3383532/timeline-of-1997-flood-in-grand-forks-north/

Daniels, J. (2018). Global disasters in 2017 caused an estimated $306 billion in economic losses, says Swiss Re. http://www.swissre.com/library/publication-sigma/sigma_1_2018_en.html

Dart, T., & Helmore, E. (2017). Hurricane Harvey claims second victim in Texas as threat of flooding rises. *The Guardian*. https://www.theguardian.com/us-news/2017/aug/26/texas-cities-catastrophic-flooding-hurricane-harvey

David, L. (2013). Russian fireball explosion shows meteor risk greater than thought. https://www.space.com/23423-russian-fireball-meteor-airburst-risk.html

Deedler, W. R. (2018). A great storm is upon Michigan: The great blizzard of 1978! http://www.wilx.com/blogs/weatherblog/82723922.html

Deep Impact Mission (n.d.). http://deepimpact.jpl.nasa.gov/home/index.html. Deep Impact Special Section (2005). *Science*, *310*, 258–283.

Denver Post (2013). 1965 South Platte floods. http://blogs.denverpost.com/library/2013/06/14/1965-south-platte-river-floods/8735/

Dickie, G. (2022). Tsunami models underestimated shockwave from Tonga eruption. *Reuters*. https://www.reuters.com/world/asia-pacific/tsunami-models-underestimated-shockwave-tonga-eruption-2022-01-28/?utm_campaign=trueAnthem%3A+Trending+Content&utm_medium=trueAnthem&utm_source=facebook&fbclid=IwAR3uLi0KxWSadtdNobWElQkWq3rV3_hVBTLW_YDx5o7JctNxYbvESbGD52I

Dietz, R. S. (1961). Vredefort ring structure: meteorite impact scar? *The Journal of Geology*, *69*(5), 499–516.

Doocy, S., Rofi, A., Moodie, C., Spring, E., Bradley, S., Burnham, G., & Robinson, C. (2007). Tsunami mortality in Aceh Province, Indonesia. *Bulletin of the World Health Organization*, *85*(4), 273–278. https://www.ncbi.nlm.nih.gov/pmc/articles/PMC2636329/

Drabek, T. E. (2010). *The human side of disaster*: CRC Press.

Drye, W. (2018). The strongest hurricanes in recorded history: See how Michael stacks up. *National Geographic*. https://www.nationalgeographic.com/environment/2018/10/strongest-hurricanes-in-history/

Dunbar, B. (2013). Russia meteor not linked to asteroid flyby. *NASA*, 6 June 2013.

Durda, D. D. (2013). The Chelyabinsk super-meteor. *Sky & Telescope*, June 2013.

Earth Impact Database. http://www.passc.net/EarthImpactDatabase/

Earthquake Country Alliance. https://www.earthquakecountry.org/step5/

Earthquake Research Institute Outreach and Public Relations Office (2012). 2011 Tohoku Earthquake. http://www.eri.u-tokyo.ac.jp/PREV_HP/outreach/eqvolc/201103_tohoku/eng/#tsunamiphoto

Editors (August 26, 2011). Hayabusa: Dust from Itokawa. *Science Special Issue*, *333*, 1113–1131.

Editors (2022). Waves in the ocean. *Exploring Florida: Teaching resources for science.*

https://fcit.usf.edu/florida/teacher/science/mod2/resources/waves.pdf

Editors of Encyclopaedica Britannica. (2022). Hurricane Katrina. https://www.britannica.com/event/Hurricane-Katrina

Editors of Encycopaedia Brittanica (2022). Huang He floods. https://www.britannica.com/event/Huang-He-floods

Egger, A. E. (2022). Visionlearning: Plates, plate boundaries, and driving forces. https://www.visionlearning.com/en/library/Earth-Science/6/Plates-Plate-Boundaries-and-Driving-Forces/66

Eggler, B. (2011). Hurricane Katrina levee breach sites floated for National Register of Historic Places. https://www.nola.com/katrina/index.ssf/2011/05/hurricane_katrina_levee_breach.html

Eschner, K. (30 Nov. 2016). For the only person ever hit by a meteorite, the real trouble began later. *Smithsonian.com*. Smithsonian Institution. https://www.smithsonianmag.com/smart-news/only-person-ever-hit-meteorite-real-trouble-began-later-180961238/#:~:text=Only%20one%20person%20in%20recorded,hitting%20her%20in%20the%20thigh

Facts and Details (2022). Facts and details: Great tsunami of 2004 in India and the Andaman Islands. http://factsanddetails.com/asian/cat63/sub411/item2542.html#chapter-3

Fastovsky, D. E., & Sheehan, P. M. (2005). The extinction of the dinosaurs in North America. *GSAToday*, *15*(3), 4–10.

Fletcher, C. (2014). *Physical geology: The science of Earth, 2nd ed.* John Wiley & Sons.

Florensky, K. P., Vronsky, B. I., & Emel'yanov Yu, M. (1960). Preliminary results of the Tunguska meteoritic expedition of 1958. *Meteoritika*, *19*, 103–134.

Flynn, C. (2022). Nearly 1,100 homes destroyed in Marshall Fire, valued at over $500 million. *Fox31 Denver*. https://kdvr.com/news/local/more-than-1000-homes-destroyed-in-marshall-fire-valued-at-over-500-million/amp/

Formby, B. (2018). Nearly 10% of Texans displaced by Harvey still haven't gone home, survey says. *The Texas Tribune*. https://www.texastribune.org/series/in-harveys-wake/

Foster, A. (2023). Turkey earthquake: Survivors living in fear on streets. *BBC News*. https://www.bbc.com/news/world-europe-64824516

Foster, G. L., Royer, D. L., Lunt, D. J. (2017). Future climate forcing potentially without precedent in the last 420 million years. *Nature Communications*, *8*(1), 14845.

Fountain, H. (2022). Here's what scientists know about the Tonga volcano eruption. *The New York Times*. https://www.nytimes.com/2022/01/19/climate/scientists-tonga-volcano-eruption-effects.html

Fox 26 News Houston (2017). Final Harvey rainfall totals, peak winds from National Hurricane Center. http://www.fox26houston.com/news/harvey-coverage/final-harvey-rainfall-totals-peak-winds-from-national-hurricane-center

Framingham, J. L. & Teasley, M. L. (Eds.) (2012). *Behavioral response to disasters*. CRC Press.

Francis, E., Parker, C., & Pannett, R. (2022). Underwater Pacific volcano sends tsunami waves to Tonga, cutting off contact with outer islands. *Washington Post*. January 15, 2022. Updated January 16, 2022. https://www.washingtonpost.com/world/2022/01/15/tonga-tsunami-new-zealand-volcano-eruption/

Frazier, I. (2017). The day the great plains burned. *The New Yorker*. https://www.newyorker.com/magazine/2018/11/05/the-day-the-great-plains-burned

Frutos, R., Serra-Sobo, J., Chen, T., & Devaux, C. A. (2020). COVID-19: Time to exonerate the pangolin from the transmission of SARS-CoV-2 to humans. *Infection, Genetics and Evolution*, *84*, 104493.

Gabbert, B. (2014). Tree killed by 1988 Yellowstone fire causes death of visitor. *Wildfire Today*. https://wildfiretoday.com/2014/06/11/tree-killed-by-1988-yellowstone-fire-causes-death-of-visitor/

Gabion1 (2019). https://www.gabion1.com/gabion-erosion-control/

Galloway, D., & Riley, F. S. (2022). San Joaquin Valley, California: Largest human alteration of the Earth's surface. United States Geological

Survey Circular 1182. https://pubs.usgs.gov/circ/circ1182/pdf/06SanJoaquinValley.pdf?fbclid=IwAR0ZkbfDluF8eM4gk9Zw78JdZ6NjrhqUw0rCxQt0KTnCBN7FGhqBTHQDd0w

Gasperini, L, Bonatti, E., & Longo, G. (2008). The Tunguska mystery. *Scientific American, 19*(3), 80–86.

Geology.com (2022). World's largest recorded earthquake. https://geology.com/records/largest-earthquake/

Ghebrezgabher, M., Yang, T., & Yang, X. (2016). Long term trend of climate change and drought assessment in the Horn of Africa. *Advances in Meteorology*, 2016, Article ID: 8057641.

Gibian, J. (2019). Blizzard of '78, what happened in Ohio: A meteorological review. http://www.bceo.org/78blizzardrev.html

Goltrant, O., et al. (1992). Formation mechanisms of planar deformation features in naturally shocked quartz. *Physics of the Earth and Planetary Interiors, 74*(3–4), 219–240.

Goodson, C., & Adams, B.(Eds.) (1998). *Fundamentals of wildland fire fighting, 3rd ed.* Fire Protection Publications, Oklahoma State University.

Goodwin, P. (2014). How to calculate the impact of carbon emissions. World Economic Forum, 3 Dec. https://www.weforum.org/agenda/2014/12/how-to-calculate-the-impact-of-carbon-emissions/

Gopalswamy, N., et al. (2005). Coronal mass ejections and other extreme characteristics of the 2003 October–November solar eruptions. *Journal of Geophysical Research: Space Physics, 110*(A9). https://doi.org/10.1029/2004JA010958

Government of Tonga (2022). Media release: First official update following the volcanic eruption. January 18, 2022. https://twitter.com/ConsulateKoT/status/1483384039826464768/photo/1

Guan, W. J., Ni, Z. Y., Hu, Y., Liang, W. H., Ou, C. Q., et al. (2020). Clinical characteristics of coronavirus disease 2019 in China. *New England Journal of Medicine, 382*, 1708–1720.

Guardian (1999). Thousands killed and injured as earthquake shakes Turkey. https://www.theguardian.com/world/1999/aug/17/marktran

Guardian (2019). Last year's wildfires were the most expensive in California history. https://www.theguardian.com/us-news/2019/may/08/california-2018-wildfires-most-expensive

Guardian (2020). "Devastating": More than 61,000 koalas among 3 billion animals affected by bushfire crisis. https://www.theguardian.com/australia-news/2020/dec/07/devastating-more-than-61000-koalas-among-3-billion-animals-affected-by-bushfire-crisis

Guardian. (2023). Las Vegas residents dry out after heavy rainfall and floods. https://www.theguardian.com/us-news/2023/sep/03/las-vegas-rain-flooding

Gutenberg, B., & Richter, C. F. (1956). Earthquake magnitude, intensity, energy, and acceleration. *Bulletin of the Seismological Society of America, 46*(2), 105–145.

Halliday, A. N. (2000). Terrestrial accretion rates and the origin of the Moon. *Earth and Planetary Science Letters, 176*(1), 17–30.

Hammel, H. B., et al. (1995). HST imaging of atmospheric phenomena created by the impact of comet Shoemaker-Levy 9. *Science, 267* (5202), 1288–1296.

Hanks, T. C., & Kanamori, H. (1979). A moment magnitude scale. *Journal of Geophysical Research, 84*(B5), 2348–2350.

Harnsberger, S. (2020). The great floods of 1931 at Gaoyou. http://aboutgaoyou.com/history/floods/the_floods.aspx

Harris, A. W., et al. (2015). Asteroid impacts and modern civilization: Can we prevent a catastrophe? In P. Michel, F. E. DeMeo, & W. F. Bottke (Eds.), *Asteroids IV* (pp. 835–854). Tucson: University of Arizona Press.

Hartmann, W. K., & Davis, D. R. (1975). Satellite-sized planetesimals and lunar origin. *Icarus, 24*(4), 504–515.

Hartmann, W. K., & Wood, C. A. (1971). Moon: Origin and evolution of multi-ring basins. *The Moon, 3*(1), 3–78.

Hasnawir, H. (2013). Early warning of rainfall-induced landslides and debris flows on Mt. Bawakaraeng, South Sulawesi, Indonesia. *Indonesian Journal of Forestry Research, 10*(1), 1–10.

Hawai'i News Now (2018). 26 years ago, Hurricane Iniki devastated Kaua'i. http://www.hawaiinewsnow.com/story/39072291/26-years-ago-hurricane-iniki-devastated-kauai/

Healy, J. H., Rubey, W. W., Griggs, D. T., & Raleigh, C. B. (1968). The Denver earthquakes. *Science, 161*(3848), 1301–1310.

Heezen, B. C., Tharp, M., & Ewing, M. (1959). The floors of the oceans: I. The North Atlantic (Vol. 65). Geological Society of America.

Hermoso, C. (2017). Masses to be held for "Yolanda" victims and survivors. *Manila Bulletin*. https://news.mb.com.ph/2017/11/05/masses-to-be-held-for-yolanda-victims-and-survivors/

Hess, H. H. (1960). *Evolution of ocean basins.* Princeton University, Department of Geology.

Heward, A. (2017). Lava tubes as hidden sites for future human habitats on the moon and Mars. *Phys.org*. https://phys.org/news/2017-09-lava-tubes-hidden-sites-future.html

History.com Editors (2018). Yangtze River peaks in China. https://www.history.com/this-day-in-history/yangtze-river-peaks-in-china

History.com Editors. (2022). Bangladesh cyclone of 1991. https://www.history.com/topics/natural-disasters-and-environment/bangladesh-cyclone-of-1991

History.com Editors. (2022). Dust Bowl. https://www.history.com/topics/great-depression/dust-bowl

History.com Editors. (2022). First confirmed case of COVID-19 found in U.S. https://www.history.com/this-day-in-history/first-confirmed-case-of-coronavirus-found-in-us-washington-state

History.com Editors (2022). Hurricane Katrina. https://www.history.com/topics/natural-disasters-and-environment/hurricane-katrina

Hogg, H. S. (1962). Out of old books: The Tunguska meteoric event. *Royal Astronomical Society of Canada, 56*(4), 174–179.

Holder, S. (2019). Should these train tracks be on fire? *Bloomberg*. https://www.bloomberg.com/news/articles/2019-01-30/polar-vortex-2019-chicago-s-train-tracks-are-on-fire

Hollandsworth, S. (2017). The day the fire came. *Texas Monthly*. https://features.texasmonthly.com/editorial/the-day-the-fire-came/

Holshue, M. L., et al. (2020). First case of 2019 novel coronavirus in the United States. *New England Journal of Medicine, 382*, 929–936.

Hopewell, J. (2016). Remembering the 2011 Virginia earthquake that rocked the eastern U. S. *The Washington Post*. https://www.washingtonpost.com/news/capital-weather-gang/wp/2016/08/23/remembering-the-2011-virginia-earthquake-that-rocked-the-eastern-u-s/?utm_term=.39b53648617d

hurricanescience.org (2022). 1970: The great Bhola cyclone. http://www.hurricanescience.org/history/storms/1970s/greatbhola/

Iceland 24 Blog (2022). Thingvellir National Park: Guide to National Park Thingvellir. http://www.iceland24blog.com/2015/01/thingvellir-national-park-iceland.html

Institute of Computational Mathematics and Mathematical Geophysics, Tsunami Laboratory (2022). Analysis of the tsunami travel time maps for damaging tsunamis in the world ocean. http://tsun.sscc.ru/TTT_rep.htm

Interesting Engineering (2017). Top 5 earthquake resistant structures around the world. https://interestingengineering.com/top-5-earthquake-resistant-structures-around-world

Intergovernmental Panel on Climate Change (2022). Assessment reports https://www.ipcc.ch/reports/

Ivashkin, V. V., Stikhno, C. A., & Guo, P. (2017). On the structure of the set of probable Earth-impact trajectories for the asteroid Apophis in 2036. *Doklady Physics, 62*(8), 387–391.

Japan Times (2011). 90% of disaster casualties drowned. https://www.japantimes.co.jp/news/2011/04/21/national/90-of-disaster-casualties-drowned/#.W2ihU7gnZPZ

Jeffery, A. (2017). 2017: A year of disaster. https://www.cnbc.com/2017/12/26/2017--a-year-of-disaster.html

Jenniskens, P. et al. (2019). Tunguska eyewitness accounts, injuries, and casualties. *Icarus, 327*, 4–18. https://doi.org/10.1016/j.icarus.2019.01.001

Jervis, R. (2015). Katrina Q&A: New Orleans before and after the historic storm. https://www.usatoday.com/story/news/2015/08/27/katrina-new-orleans-hurricane-louisiana/32431935/

Jibson, R.W. (2006). Photographs, correspondence, and presentations related to the La Conchita, California, Landslide. U.S. Geological Survey Open-File Report 2006-1278. https://pubs.usgs.gov/of/2006/1278/downloads/pdf/of06-1278_508.pdf

Johnston, I. (2017). Ethiopia drought: Millions of people urgently in need of food aid after string of natural disasters. https://www.independent.co.uk/news/world/africa/ethiopia-drought-floods-starvation-malnourished-crisis-catastrophe-east-africa-a7901156.html

Kahler, S. W. (1992). Solar flares and coronal mass ejections. *Annual Review of Astronomy and Astrophysics*, *30*(1), 113–141.

Kanamori, K. (1978). Quantification of earthquakes. *Nature*, *27* (5644), 411–414.

Kandasamy, A., & Tomei, L. (2009). Earthquake readiness tips for 2010. *San Francisco Public Press*. https://sfpublicpress.org/http%3A/%252Fsfpublicpress.org/news/2009-11/much-of-city-unprepared-for-next-big-quake

Kearey, P., Klepeis, K. A., & Vine, F. J. (2009). *Global tectonics, 3rd ed.* Wiley.

Keller, G. (1988). Extinction, survivorship and evolution of plankontic foraminifera across the Cretaceous/Tertiary boundary at El Kef, *Tunisia. Marine Micropaleontology*, *13*(3), 239–263.

Kimmelman, M. (2013). Going with the flow. *The New York Times*. https://www.nytimes.com/slideshow/2013/02/17/arts/design/20130217-WATERWAYS/s/20130217-WATERWAYS-slide-QUI0.html

King, H. M. (2020). Pictures of igneous rocks. https://geology.com/rocks/igneous-rocks.shtml

King, K. (2015). 2011 Va. earthquake identifies previously unknown fault line. *WTOP*. https://wtop.com/virginia/2015/02/2011-va-earthquake-identifies-previously-unknown-fault-line/

Koren, J. R. (2018). California plans takeover of property insurer overwhelmed by Camp fire claims. *Los Angeles Times*. https://www.latimes.com/local/lanow/la-fi-merced-insurance-paradise-20181203-story.html

Kourounis, G. (2012). Avalanche control British Columbia 2012. http://www.stormchaser.ca/Avalanche/BC_2012/BC_2012.html

Kring, D. A. (1997). Air blast produced by the Meteor Crater impact event and a reconstruction of the affected environment. *Meteoritics & Planetary Science*, *32*(4), 517–530.

Kring, D. A. (2007). The Chicxulub impact event and its environmental consequences at the Cretaceous-Tertiary boundary. *Palaeogeography, Palaeoclimatology, Palaeoecology 255*(1–2), 4–21.

Kring, D. A. (2007). *Guidebook to the geology of Barringer Meteorite Crater, Arizona (aka Meteor Crater)*. Houston: Lunar and Planetary Institute.

Kuzmin, A. (2013). Meteorite explodes over Russia, more than 1,000 injured. *Reuters*. https://www.reuters.com/article/us-russia-meteorite/meteorite-explodes-over-russia-more-than-1000-injured-idUSBRE91E05Z20130215

LaGanga, M. L., Newberry, L., St. John, P., & Lin, R-G., II (2018). Many victims of California's worst wildfire were elderly and died in or near their homes, new data show. *Los Angeles Times*. https://www.latimes.com/local/lanow/la-me-ln-paradise-fire-dead-map-20181213-story.html

Laskin, D. (2005). The children's blizzard. HarperPerennial.

Lauer, S. A., Grantz, K. H., Bi, Q., Jones, F.K., Zheng, Q., et al. (2020). The incubation period of coronavirus disease 2019 (COVID-19) from publicly reported confirmed cases: Estimation and application. *Annals of Internal Medicine*, *172*(9), 577–582.

Lee, J. F. (2013). Revisiting the Johnstown flood. *Washington Post*. https://www.washingtonpost.com/lifestyle/travel/revisiting-the-johnstown-flood/2013/05/16/49e6c468-b80b-11e2-b94c-b684dda07add_story.html?utm_term=.0db5a17293d2

Levin, H.L. (2006). *The Earth Through Time*, 8th Edition. John Wiley & Sons.

Lindsay, R. (2022). Climate change: Global sea level. NOAA Climate.gov. http://www.climate.gov/news-features/understanding-climate/climate-change-global-sea-level

Lipman, P. W., & Mullineaux, D. R. (Eds.) (1981). The 1980 eruptions of Mount St. Helens, Washington. Geological Survey Professional Paper 1250: United States Geological Survey.

Longo G. (2007). The Tunguska event. In P.T. Bobrowsky & H. Rickman (Eds.), *Comet/asteroid impacts and human society*. Berlin, Heidelberg; Springer. https://doi.org/10.1007/978-3-540-32711-0_18

Loomis, I. (2018). Lava clues chronicled Kilauea's unusual 2018 eruption. *EOS (99)*. https://doi.org/10.1029/2018EO111315

Lou, M. (2019). High-tide flooding is only going to get worse, NOAA says. *CNN*. https://www.cnn.com/2019/07/10/weather/noaa-high-tide-flooding-increasing-report-trnd/index.html?fbclid=IwAR2JRjv2ugGe1HNH8E-kELXirozY-fECN-i19PpU8H9wjdZWDjtJZCeCOlQ

Luber, G., & McGeehin, M. (2008). Climate change and extreme heat events. *American Journal of Preventive Medicine*, *35*(5), 429–435.

Lundstrom, K. (2018). Study says to fix Harvey-related housing damage, Houston needs $2 billion more in federal funds. *The Texas Tribune*. https://www.texastribune.org/2018/10/11/hurricane-harvey-housing-damage-houston-2-billion/

Maslin Nir, S. (2018). In California, houses burned. so did the toxic chemicals they contained. *New York Times*. https://www.nytimes.com/2018/11/29/us/california-fire-chemicals.html

Mason, B. (1967). The Bununu meteorite, and a discussion of the pyroxene-plagioclase achondrites. *Geochimica et Cosmochimica Acta*, *31*(2), 107–115.

Masson-Delmotte, V., et al. (Eds.) (2021). *Climate change 2021: The physical science basis. Contribution of Working Group I to the Sixth Assessment Report of the Intergovernmental Panel on Climate Change*. Cambridge University Press.

McGladrey, L. (2017). The 5 components of psychological first aid. https://blog.nols.edu/2017/05/22/5-components-psychological-first-aid

McKenzie, D., & Jackson, J. (2012). Tsunami earthquake generation by the release of gravitational potential energy. *Earth and Planetary Science Letters*,45–348, 1–8.

McKirdy, E. (2019). Satellite images show extent of Indonesian volcano's collapse. *CNN*. https://www.cnn.com/2019/01/03/asia/indonesia-anak-krakatau-satellite-images-intl/index.html

McMurtry, G. M., Watts, P., Fryer, G. J., Smith, J. R., & Imamura, F. (2004). Giant landslides, mega-tsunamis, and paleo-sea level in the Hawaiian Islands. *Marine Geology*, *203*, 219–233.

Melosh, H. J. (1989). *Impact cratering: A geologic process*. New York: Oxford University Press

Melosh, H. J. (1993), Solar asteroid diversion. *Nature*, *366* (6450), 21–22.

Melosh, H. J., & Ivanov, B. A. . (1999). Impact crater collapse. *Annual Review of Earth and Planetary Sciences*, *27*(1), 385–415.

Meseret, E. (2017). Death toll in Ethiopia landfill collapse rises to 113. *Chicago Tribune*. http://www.chicagotribune.com/news/nationworld/ct-ethiopia-landfill-collapse-20170315-story.html

Messier, D. (2013). Nuking Dangerous Asteroids Might Be the Best Protection, Expert Says. https://www.space.com/21333-asteroid-nuke-spacecraft-mission.html

Mielke, B., & Hutchinson, B. (2018). "This was a firestorm": Deadly California wildfire leaves entire Paradise town council homeless. *ABC*

News. https://abcnews.go.com/beta-story-container/US/deadly-camp-fire-leaves-entire-paradise-town-council/story?id=59159481

Miller, M. E. (2022). Koalas are getting harder to find. Scientists in Australia are on a quest to uncover a hidden population. *Washington Post*. https://www.washingtonpost.com/world/2022/01/13/australia-koalas-fires-climate-change/

Miranda, S. (2023). What we know about the flooding at Burning Man. *Axios*. https://www.axios.com/2023/09/03/burning-man-festival-flooding-weather

Mitchell, H. (2016). The Rivers of the Mississippi watershed. *NASA Scientific Visualization Studio*. https://svs.gsfc.nasa.gov/4493

Mittlefehldt, D. W., et al. (1998). Non-chondritic meteorites from asteroidal bodies. *Reviews in Mineralogy and Geochemistry*, *36*(1), 4.1–4.195.

Montañez, A., & Fischetti, M. (2017). Humans start most wildfires. *Scientific American*. https://www.scientificamerican.com/article/humans-start-most-wildfires/

Mooney, C. (2018). Hurricane Harvey was year's costliest U.S. disaster at $125 billion in damages. *The Texas Tribune*. https://www.texastribune.org/2018/01/08/hurricane-harvey-was-years-costliest-us-disaster-125-billion-damages/

Moran, J. M. (2012). *Weather studies: An introduction to atmospheric science, 5th ed.* American Meteorological Society.

Morella, C. (2018). Why super typhoon Yolanda was so deadly. *ABS CBS News*. https://news.abs-cbn.com/news/11/05/18/why-super-typhoon-yolanda-was-so-deadly

Morrison, D., Chapman, C. R., Steel, D., & Binzel, R. P. (2004). Impacts and the public: communicating the nature of the impact hazard. In M. J. S. Belton, T. H. Morgan, N. H. Samarasinha, & D. K. Yeomans (Eds), *Mitigation of hazardous comets and asteroids*. Cambridge University Press.

Mukherjee, S., Mishra, A., & Trenberth, K. E. (2018). Climate change and drought: A perspective on drought indices. *Current Climate Change Reports*, *4*, 145–163.

Murchie, S. L., et al. (2015). *Phobos and Deimos. Asteroids IV*, *4*, 451.

Murck, B. W., & Skinner, B. J. (2016). *Visualizing geology, 4th ed.* Wiley.

Murck, B. W., Skinner, B. J., & Mackenzie, D. (2008). *Visualizing geology*. John Wiley & Sons.

Murdock, V. (2023), Stafford Township, New Jersey, helping residents elevate homes to combat rising sea levels. *CBS News New York*. https://www.cbsnews.com/newyork/news/new-jersey-rising-sea-levels-elevating-homes-manahawkin-stafford-township-climate-change/

NASA, 6 Nov. 2013, https://www.nasa.gov/content/nasa-and-international-researchers-collect-clues-to-meteoroid-science/#.U_vfLlfvmlA

National Aeronautics and Space Administration Earth Observatory (2023). Patterns of lightning activity. https://earthobservatory.nasa.gov/images/6679/patterns-of-lightning-activity

National Hurricane Center (2022). Storm surge overview. https://www.nhc.noaa.gov/surge/

National Oceanic and Atmospheric Administration (2022). Avalanche safety. http://www.nws.noaa.gov/om/winter/avalanche.shtml

National Park Service (2022). Hawai'i Volcanoes National Park. https://www.nps.gov/havo/index.htm

National Park Service (2022). Capulin volcano. https://www.nps.gov/cavo/index.htm

National Park Service (2023). Indigenous fire practices shape our land. https://www.nps.gov/subjects/fire/indigenous-fire-practices-shape-our-land.htm

National Precast Concrete Association. Retaining walls image gallery. https://precast.org/precast-product/retaining-walls/

National Solar Observatory: http://www.nso.edu/

National Weather Service (1993). The Great Flood of 1993. https://www.weather.gov/dvn/071993_greatflood

National Weather Service (2017). Major Hurricane Harvey, 25– 29 August 2017. https://www.weather.gov/crp/hurricane_harvey

National Weather Service (2023). How lightning is created. https://www.weather.gov/jetstream/lightning

Near-Earth Object Science Definition Team (2017). Update to determine the feasibility of enhancing the search and characterization of NEOs. https://www.nasa.gov/sites/default/files/atoms/files/2017_neo_sdt_final_e-version.pdf

Near-Earth Object Survey and Definition (2007). Analysis of alternatives. Report to Congress, NASA. http://www.nasa.gov/pdf/171331main_NEO_report_march07.pdf

Needham, K. (2022). Tonga reconnects to world as submarine cable restored after tsunami. *Reuters*. https://www.reuters.com/world/asia-pacific/tonga-reconnects-world-submarine-cable-restored-after-tsunami-2022-02-22/

Neufer, D. (2022). Panorama of Mission Street. http://www.dualmoments.com/pansfromfilms/sfq1906pano/sfq1906lev1/02-mission.htm

Newhall, C. G., & Punongbayan, R. S. (Eds.) (1996). Fire and mud: Eruptions and lahars of Mount Pinatubo, Philippines. Philippine Institute of Volcanology and Seismology and University of Washington Press.

Nield, D., (2022). It's official: Tonga's volcano was the largest explosive eruption of the 21st century. *Science Alert*. https://www.sciencealert.com/tonga-s-volcanic-eruption-was-the-largest-explosion-of-the-21st-century

Nishikiori, N., Abe, T., Costa, D., Dharmaratne, S., Kunii, O., & Moji, K. (2006). Who died as a result of the tsunami? Risk factors of mortality among internally displaced persons in Sri Lanka: a retrospective cohort analysis. *BMC Public Health*. https://www.ncbi.nlm.nih.gov/pmc/articles/PMC1435747/

NOAA Space Environment Center: http://www.sec.noaa.gov/

Norton, R. (1998). *Rocks from space*. Missoula, MT: Mountain Press Publishing Company. 85–87

Oregon State University (2010). Columbia River flood basalts. *Volcano World*. http://volcano.oregonstate.edu/columbia-river-flood-basalts

Osanai, N., & Shimizu, T., & Kuramoto, K., Kojima, S., & Noro, T. (2010). Japanese early-warning for debris flows and slope failures using rainfall indices with Radial Basis Function Network. *Landslides*, *7*, 325–338. 10.1007/s10346-010-0229-5.

Owen, D. (2017). *Where the water goes: Life and death along the Colorado River*. New York: Riverhead Books/Penguin.

Owen, P., Khomami, N., & Russell, G. (2017). Hurricane Harvey: Fatality confirmed as Texas battered by "marathon" storm, rolling report. *The Guardian*. https://www.theguardian.com/us-news/live/2017/aug/26/hurricane-harvey-makes-landfall-in-texas-latest-updates

Pacific Marine Environmental Laboratory (2022). Understand a unique tsunami event caused by the Tonga volcano eruption. https://www.pmel.noaa.gov/news-story/understanding-unique-tsunami-event-caused-tonga-volcano-eruption

Paine, M., & Peiser, B. (2002). The frequency and consequences of cosmic impacts since the demise of the dinosaurs. https://www.semanticscholar.org/paper/The-Frequency-and-Consequences-of-Cosmic-Impacts-of-Paine-Peiser/fb78a56757dc8412a5d1ca006d0084c3e47308c2

Pamies, R. J. (Ed.) (2011). Help and hope for Haiti: Omaha Books.

Paris, R., Ramalho, R. S., Madeira, J., Avila, S., May, S. M., Rixhon, G., Engel, et. al. (2018). Mega-tsunami conglomerates and flank collapses of ocean island volcanoes. *Marine Geology*, *395*, 168–187.

Paul, J. Vo, T., Prentzel, O., & Najmabadi, S. (2021). As many as 600 homes lost, 6 people injured as Marshall fire quickly spreads across Boulder County. *The Colorado Sun*. https://coloradosun.com/2021/12/30/boulder-grass-fire-evacuations/

Pearce, R. (2017). Analysis: Why scientists think 100% of global warming is due to humans. *carbon brief, 13 Dec*. https://www.carbonbrief.org/analysis-why-scientists-think-100-of-global-warming-is-due-to-humans/

Petley, D. (2019). Anak Krakatau: Planet Labs imagery of the aftermath of the landslide. *The Landslide Blog*. American Geophysical Union. https://blogs.agu.org/landslideblog/2019/01/03/anak-krakatau-3/

Phillips, F. M., et al. (1991). Age and geomorphic history of Meteor Crater, Arizona, from cosmogenic ^{36}Cl and ^{14}C in rock varnish. *Geochimica et Cosmochimica Acta*, *55*(9), 2695–2698.

Phillips, R. J., et al. (1992). Impact craters and Venus resurfacing history. *Journal of Geophysical Research: Planets*, *97*(E10), 15,923–15,948.

Pilger, C., et al. (2019). Large meteoroids as global infrasound reference events. In A. Le Pichon, E. Blanc, & A. Hauchecorne (Eds.), *Infrasound monitoring for atmospheric studies*. Springer, Cham. https://doi.org/10.1007/978-3-319-75140-5_12

Poag, C. W. (1998). The Chesapeake Bay bolide: Modern consequences of an ancient cataclysm. https://pubs.er.usgs.gov/publication/7000063

Popova, O. P., et al. (2013). Chelyabinsk airburst, damage assessment, meteorite recovery, and characterization. *Science*, *342*(6162), 1069–1073. https://www.science.org/doi/10.1126/science.1242642

Powars, D. S. (2000). The effects of the Chesapeake Bay impact crater on the geologic framework and the correlation of hydrogeologic units of southeastern Virginia, south of the James River. U.S. Geological Survey Professional Paper 1662.

Prell, A. (2016). Wildfire 101: Dead fuel moisture. *Redzone blog*. https://www.redzone.co/2016/12/19/wildfire-101-dead-fuel-moisture/

Pyne, S. J., Andrews, P. L., & Laven, R. D. (1996). *Introduction to wildland fire, 2nd ed.* John Wiley & Sons.

Reeves, R. V., & Deng, B. (2021). At least 65,000 more men than women have died from COVID-19 in the U.S. Brookings Institute. https://www.brookings.edu/blog/up-front/2021/10/19/at-least-65000-more-men-than-women-have-died-from-covid-19-in-the-us/

Regan, H. (2022). Tonga remains cut off after a massive eruption and tsunami. *Here's what we know* https://www.cnn.com/2022/01/17/asia/tonga-volcanic-eruption-tsunami-explainer-intl-hnk/index.html

Reimold, W. U., Grieve, R. A. F., & Palme, H. (1981). Rb-Sr dating of the impact melt from East Clearwater, *Quebec. Contributions to Mineralogy and Petrology*, *76*(1), 73–76.

Richter, C. F. (1935). An instrumental earthquake magnitude scale. *Bulletin of the Seismological Society of America*, *25* (1), 1–32.

Riemer, M. F., Collins, B. D., Badger, T. C., Toth, C., & Yu, Y. C. (2015). Geotechnical soil characterization of intact Quaternary deposits forming the March 22, 2014 SR-530 (Oso) landslide, Snohomish County, Washington. USGS Open-File Report 2015-1089. https://pubs.usgs.gov/of/2015/1089/pdf/ofr20151089.pdf

Rinard Hinga, B. D. (2015). *Ring of fire: An encyclopedia of the Pacific Rim's earthquakes, tsunamis, and volcanoes*. ABC-CLIO Publishers.

Ripple, W. J., Wolf, C., Gregg, J., Levin, K., Rockström, J., Newsome, T. M. Betts, M. G., et al. (2022). World Scientists' Warning of a Climate Emergency 2022, *BioScience*, *72*(12): 149–1155.

Ripple, W. J., Wolf, C., Lenton, T. M., Gregg, J. W., Natali, S. M., Duffy, P. B., Rockström, et al. (2023). Many risky feedback loops amplify the need for climate action. *One Earth*, *6*(2), 86–91.

Robinson, C., Dilkina, B., & Moreno-Cruz, J. (2020). Modeling migration patterns in the USA under sea level rise. *PLoS ONE*. *15*(1), e0227436. https://doi.org/10.1371/journal.pone.0227436

Rosetta mission. http://sci.esa.int/rosetta/

Roth, S. (2020). Meet the New PG&E. *It looks a lot like the old PG&E*. https://www.latimes.com/environment/story/2020-06-17/pge-bankruptcy-new-pge-looks-like-old-pge

Rothman, H. K. (2007). *Blazing heritage: A history of wildland fire in the national parks*. Oxford University Press.

Rousi, E., Kornhuber, K., Beobide-Arsuaga, G., Luo, E., & Coumou, D. (2022). Accelerated western European heat wave trends linked to more-persistent double jets over Eurasia. *Nature Communications*, *13*, 3851.

Rubstov, V. V. (2008). *The Tunguska mystery*. Springer.

Sample, I. (2021). Rare meteorite chunk traced by scientists to Gloucestershire driveway. *The Guardian*. www.theguardian.com/science/2021/mar/09/rare-meteorite-chunk-traced-by-scientists-to-gloucestershire-driveway

Sastry, N., & Gregory, J. (2014). The location of displaced New Orleans residents in the year after Hurricane Katrina. *Demography*, *51*(3), 753–775. https://www.ncbi.nlm.nih.gov/pmc/articles/PMC4048822/

Satake, K., Shimazaki, K., Tsuji, Y., & Ueda, K. (1996). Time and size of a giant earthquake in Cascadia inferred from Japanese tsunami records of January 1700. *Nature*, *379*, 246–249.

Schleifstein, M. (2012). Hurricane Katrina flooding in lower 9th Ward, St. Bernard Parish heading to court. https://www.nola.com/katrina/index.ssf/2012/09/hurricane_katrina_flooding_in.html

Schulte, P., et al. (2010). The Chicxulub asteroid impact and mass extinction at the Cretaceous-Paleogene boundary. *Science*, *327*(5970), 1214–1218.

Schwartz, M. S. (2020). Iowa derecho this August was most costly thunderstorm event in modern U.S. history. https://www.npr.org/2020/10/18/925154035/iowa-derecho-this-august-was-most-costly-thunderstorm-event-in-modern-u-s-histor

Schwenn, R., Dal Lago, A., Huttunen, E., & Gonzalez, W. D. (2005). The association of coronal mass ejections with their effects near the Earth. *Annales Geophysicae, European Geosciences Union*, *23*(3), 1033–1059.

Science Education Resource Center (SERC) at Carleton College (2022). EarthLabs. https://serc.carleton.edu/eslabs/climate/1a.html

Scott, R. G., Pilkington, M., & Tanczyk, E. I. (1995). Magnetic studies of the Clearwater Lake impact structures. *Lunar and Planetary Science Conference*, 26.

Sharma, M., & Scarr, S. (2022). How big was the Tonga eruption? *Reuters*. https://graphics.reuters.com/TONGA-VOLCANO/lgpdwjyqbvo/

Shepherd, D. (2007). Stellar origins: From the cold depths of space. *Sky & Telescope*, *113*(6).

Shoemaker, E. M. (1959). Impact mechanics at Meteor crater, Arizona. U.S. Geological Survey Open-File Report 59–108.

Shoemaker, E. M. (1987). Meteor crater, Arizona. *Centennial Field Guide 2*, 399–404.

Simms, J. R. Z., Waller, H. L., Brunet, C., et al. (2021). The long goodbye on a disappearing, ancestral island: A just retreat from Isle de Jean Charles. *Journal of Environmental Studies and Sciences*, *11*, 316–328. https://doi.org/10.1007/s13412-021-00682-5

Skinner, B. J., & Murck, B. M. (2011). *The blue planet: An introduction to Earth System science, 3rd ed.* Wiley.

Smithsonian Institution (2020). Paricutin. https://naturalhistory.si.edu/onehundredyears/expeditions/Paricutin.html

SOHO mission. http://soho.nascom.nasa.gov/

Solar Dynamics Observatory: https://sdo.gsfc.nasa.gov/

Solar wind loses power, hits 50-year low, Science@NASA, 23 September 2008. http://science.nasa.gov/headlines/y2008/23sep_solarwind.htm

Space Weather Prediction Center. https://www.swpc.noaa.gov/

Space.com, Future US, Inc., 1 Nov. 2013. https://www.space.com/23423-russian-fireball-meteor-airburst-risk.html

Space.com, Space, 29 May 2013. https://www.space.com/21333-asteroid-nuke-spacecraft-mission.html

St. John, P., Phillips, A. M., Serna, J., Kohli, S., & Newberry, L. (2018). California fire: What started as a tiny brush fire became the state's deadliest wildfire. Here's how. *Los Angeles Times*. https://www.latimes.com/local/california/la-me-camp-fire-tictoc-20181118-story.html

St. John, P., Serna, J., & Lin, R.-G., II (2018). Here's how Paradise ignored warnings and became a deathtrap. *Los Angeles Times*. https://www.latimes.com/local/california/la-me-camp-fire-deathtrap-20181230-story.html

Stamps, D. S., Calais, E., Saria, E., Hartnady, C., Nocquet, J.-M., Ebinger, C. J., & Fernandes, R. M. (2008). A kinematic model for the East African Rift. *Geophysical Research Letters*, *35*(5), L05304.

State of Alaska, Department of Homeland Security and Emergency Management (2019). Earthquake mitigation. https://www.ready.alaska.gov/Plans/Mitigation/Equake

Strahler, A., (2013). *Introducing physical geography*, *6th ed.* Wiley.

Stendel, M., Francis, J., White, R., Williams, P. D., Woollings, T. (2021). The jet stream and climate change, In T. M. Letcher (Ed.), *Climate change: Observed impacts on planet Earth*, *3rd ed.* (pp. 327–357). Elsevier.

Stuiver, M., & Quay, P. D. (1981). Atmospheric ^{14}C changes resulting from fossil fuel CO_2 release and cosmic ray flux variability. *Earth and Planetary Science Letters*, *53*(3), 349–362. https://doi.org/10.1016/0012-821X(81)90040-6

Swiss Re Institute (2018). Natural catastrophes and man-made disasters in 2017: A year of record-breaking losses. http://www.swissre.com/library/publication-sigma/sigma_1_2018_en.html

Swiss Re Institute (2021). Natural catastrophes in 2020: Secondary perils in the spotlight, but don't forget primary-peril risks. https://www.swissre.com/institute/research/sigma-research/sigma-2021-01.html#

Talbert, T. (2018). Five years after the Chelyabinsk meteor: NASA leads efforts in planetary defense. *NASA*. https://www.nasa.gov/feature/five-years-after-the-chelyabinsk-meteor-nasa-leads-efforts-in-planetary-defense

Tan, Q. (2022). Tonga eruption heard in New Zealand, pressure waves picked up in Europe. *New Zealand Herald*. https://www.nzherald.co.nz/nz/tonga-eruption-heard-in-new-zealand-pressure-waves-picked-up-in-europe/WQYYVD3EPXZGEFXOLEHFHU4TO4/

Tauzin, B., et al. (2013). Seismoacoustic coupling induced by the breakup of the 15 February 2013 Chelyabinsk meteor. *Geophysical Research Letters*, *40*(14), 3522–3526.

Taylor, A. (2016). Photos of the 1906 San Francisco earthquake. *The Atlantic*. https://www.theatlantic.com/photo/2016/04/photos-of-the-1906-san-francisco-earthquake/477750/

Taylor, D. B. (2021). A timeline of the coronavirus pandemic. *New York Times*. https://www.nytimes.com/article/coronavirus-timeline.html

Therriault, A. M., Grieve, R. A. F., & Reimold, W. U. (1997). Original size of the Vredefort structure: Implications for the geological evolution of the Witwatersrand Basin. *Meteoritics & Planetary Science*, *32*(1), 71–77.

Thomas, G., & Morgan-Witts, M. (1969). *The day the world ended*: Stein and Day Publishers.

Thomas, P. C., et al. (2005). Differentiation of the asteroid Ceres as revealed by its shape. *Nature*, *437*, 224–226.

Thompson, B. (2016). Though year ahead for great plains wheat farmers. *Kansas Public Radio*. http://kansaspublicradio.org/kpr-news/tough-year-ahead-great-plains-wheat-farmers

Timm, J. C. (2019). Fact check: Trump says Puerto Rico got $92 billion. They've seen only a fraction. *CBS News*. https://www.nbcnews.com/politics/donald-trump/fact-check-trump-says-puerto-rico-got-92-billion-they-n1031276

Trayner, C. (1997). The Tunguska event. *Journal of the British Astronomical Association*, *107*, 117–130.

Tsurutani, B. T., et al. (2003). The extreme magnetic storm of 1–2 September 1859. *Journal of Geophysical Research: Space Physics*, *108* (A7). https://doi.org/10.1029/2002JA009504

Turtle, E. P., et al. (2005). Impact structures: What does crater diameter mean. *Large meteorite impacts III: Geological Society of America Special Paper*, *384*, 1–24.

25143 Itokawa. https://solarsystem.nasa.gov/asteroids-comets-and-meteors/asteroids/25143-itokawa/in-depth/

U.S. Congressional House Committee on Science, Space, and Technology (2013). Threats from space: A review of U.S. government efforts to track and mitigate asteroids and meteors, *March 19, 2013.*

U.S. Department of the Interior Office of Wildland Fire. (2021). Embracing indigenous knowledge to address the wildfire crisis. https://www.doi.gov/wildlandfire/embracing-indigenous-knowledge-address-wildfire-crisis

U.S. Geological Survey (1997). 1997 Floods in the Red River of the North and Missouri River basins in North Dakota and Western Minnesota. https://pubs.usgs.gov/of/1997/0575/report.pdf

U.S. Geological Survey (2001). Natural hazards on alluvial fans: The Venezuela debris flow and flash flood disaster. *U.S. Geological Survey Fact Sheet FS 103 01.* https://pubs.usgs.gov/fs/fs-0103-01/fs-0103-01.pdf

U.S. Geological Survey. (2005). Landslide hazards: A national threat. *U.S. Geological Survey Fact Sheet 2005-3156.* https://pubs.usgs.gov/fs/2005/3156/2005-3156.pdf

U.S. Geological Survey. (2008). The landslide handbook: A guide to understanding landslides. *Circular 1325.* https://pubs.usgs.gov/circ/1325/

U.S. Geological Survey. (2009). This dynamic Earth: The story of plate tectonics. https://pubs.usgs.gov/gip/dynamic/

U.S. Geological Survey (2018). Preliminary summary of Kīlauea volcano's 2018 lower East Rift Zone eruption and summit collapse. https://home.nps.gov/havo/planyourvisit/upload/PrelimSum_LERZ-Summit_2018_508.pdf

U.S. Geological Survey (2019). Snowmelt: The water cycle. https://water.usgs.gov/edu/watercyclesnowmelt.html

U.S. Geological Survey (2022). Crater Lake. https://www.usgs.gov/volcanoes/crater-lake

U.S. Geological Survey (2022). Future eruptions at Mount St. Helens. https://volcanoes.usgs.gov/volcanoes/st_helens/

U.S. Geological Survey (2022). The Pu`u`o`ō eruption lasted 35 years. https://volcanoes.usgs.gov/volcanoes/kilauea/geo_hist_1983.html

U.S. Geological Survey (2022). Yellowstone. https://volcanoes.usgs.gov/volcanoes/yellowstone/

U.S. Geological Survey (2023). M 7.5 Elbistan earthquake, Kahramanmaras earthquake sequence. https://earthquake.usgs.gov/earthquakes/eventpage/us6000jlqa/executive

U.S. Geological Survey (2024). Winter Park Florida sinkhole of 1981. https://www.usgs.gov/media/images/winter-park-florida-sinkhole-1981

U.S. Geological Survey Earthquake Hazards Program. http://earthquake.usgs.gov/

Vera, S. (2023). Turkish undertaker's family move to cemetery as he buries earthquake dead. *Reuters.* https://www.reuters.com/investigates/special-report/turkey-quake-cemetery?fbclid=IwAR2OsJbUYAFjIO07IA1DwMqP-pgkVjAxEYbOIBDseWXuS7nV5Ryi6lRoEjk&utm_medium=Social&utm_source=Facebook

Voiland, A. (2021). Anticipating future sea levels. NASA Earth Observatory, 8 July. https://earthobservatory.nasa.gov/images/148494/anticipating-future-sea-levels

Waikato Regional Council (2022). Tsunami. https://www.waikatoregion.govt.nz/services/regional-services/regional-hazards-and-emergency-management/coastal-hazards/tsunami/

Wartman, J., Montgomery, D. R., Anderson, S. A., Keaton, J. R., Benoit, J., dela Chapelle, J., & Gilbert, R. (2016). The 22 March 2014 Oso landslide, Washington, USA. *Geomorphology, 253*, 275–288. http://dx.doi.org/10.1016/j.geomorph.2015.10.022

Weather.com Meteorologists. (2018). Hurricane Michael recap: Historic category 4 Florida Panhandle landfall; Swath of wind damage and flooding into the Carolinas, Mid-Atlantic. https://weather.com/storms/hurricane/news/2018-10-11-hurricane-michael-recap-gulf-coast-southeast

Wegener, A. (1966). The origin of continents and oceans. Dover Publications.

Wegener, A. (2011). *The origin of continents and oceans (translated from the 1929 fourth edition, originally published 1915)*. New York: Dover Earth Science.

Weston, P. (2017). Massive earthquake could hit Istanbul at any moment with just SECONDS warning, say scientists. *Daily Mail*. http://www.dailymail.co.uk/sciencetech/article-4513880/Istanbul-overdue-potentially-devastating-earthquake.html

Wieczorek, G. F., Larsen, M. C., Eaton, L. S., Morgan, B. A. & Blair, J. L. (2001). Debris-flow and flooding hazards associated with the December 1999 storm in coastal Venezuela and strategies for mitigation. *U.S. Geological Survey Open File Report 01-0144*. https://pubs.usgs.gov/of/2001/ofr-01-0144/

Wikipedia (2023). Effects of climate change. https://en.wikipedia.org/w/index.php?title=Effects_of_climate_change&oldid=1161188589

Wilf, P., & Johnson, K. R. (2004). Land plant extinction at the end of the Cretaceous: A quantitative analysis of the North Dakota megafloral record. *Paleobiology*, *30*(3), 347–368.

Williams, S. (2020). A brief history of human coronaviruses. *The Scientist*. https://www.the-scientist.com/news-opinion/a-brief-history-of-human-coronaviruses-67600

Williams Walsh, M. (2017). A broke, and broken, flood insurance program. *The New York Times*. https://www.nytimes.com/2017/11/04/business/a-broke-and-broken-flood-insurance-program.html

Wolbach, W. S., Gilmour, I., & Anders, E. (1990). Major wildfires at the Cretaceous/Tertiary boundary. *Geological Society of America Special Paper*, *247*, 391–400.

World Health Organization (2020). *COVID-19-China*. https://www.who.int/emergencies/disease-outbreak-news/item/2020-DON229

World Nuclear Association (2018). Fukushima accident. http://www.world-nuclear.org/information-library/safety-and-security/safety-of-plants/fukushima-accident.aspx

World Vision (2022). 1980s Ethiopia famine: Facts, FAQs, and how to help. https://www.worldvision.org/disaster-relief-news-stories/1980s-ethiopia-famine-facts

Worster, D. (2004). *Dust bowl: Southern plains in the 1930s* (25th Anniversary Edition). Oxford University Press, Oxford.

Wünnemann, K., Collins, G. S., & Weiss, R. (2010). Impact of a cosmic body into Earth's ocean and the generation of large tsunami waves: Insight from numerical modeling. *Reviews of Geophysics*, *48*(4).

Yeats, R. S. (2004). *Living with earthquakes in the Pacific northwest*. Oregon State University Press.

Yeomans, D., & Chodas, P. Additional details on the large fireball event over Russia on Feb. 15, 2013. *Additional details on the large fireball event over Russia on Feb. 15, 2013*, NASA/JPL, 1 Mar. 2013, https://web.archive.org/web/20130430164941/http://neo.jpl.nasa.gov/news/fireball_130301.html

Zalzal, K. S. (2016). Kīlauea increases asthma risk. *Earth Magazine*. https://www.earthmagazine.org/article/kilauea-increases-asthma-risk

Zhang, M. (2013). Russia meteor 2013: Damage to top $33 million; Rescue, cleanup team travels to meteorite-hit Urals. https://www.ibtimes.com/russia-meteor-2013-damage-top-33-million-rescue-cleanup-team-heads-meteorite-hit-urals-1090104

Index

Note: Tables are indicated by **bold** page references. Figures are indicated by *italicized* page references.

Earth's Natural Hazards and Disasters, Advanced Textbook 7, First Edition. Bethany D. Hinga.

y